Ultra-Processed Foods, Diet Quality and Human Health

Ultra-Processed Foods, Diet Quality and Human Health

Editors

Monica Dinu
Daniela Martini

MDPI

Basel • Beijing • Wuhan • Barcelona • Belgrade • Novi Sad • Cluj • Manchester

Editors

Monica Dinu
University of Florence
Firenze, Italy

Daniela Martini
University of Milan
Milan, Italy

Editorial Office
MDPI
St. Alban-Anlage 66
4052 Basel, Switzerland

This is a reprint of articles from the Special Issue published online in the open access journal *Nutrients* (ISSN 2072-6643) (available at: https://www.mdpi.com/journal/nutrients/special_issues/Ultraprocessed_Foods).

For citation purposes, cite each article independently as indicated on the article page online and as indicated below:

Lastname, A.A.; Lastname, B.B. Article Title. *Journal Name* **Year**, *Volume Number*, Page Range.

ISBN 978-3-0365-8464-5 (Hbk)
ISBN 978-3-0365-8465-2 (PDF)
doi.org/10.3390/books978-3-0365-8465-2

Contents

About the Editors

Monica Dinu

Dr. Monica Dinu is currently Assistant Professor of Human Nutrition at the Department of Experimental and Clinical Medicine, University of Florence, Italy. She holds a master's degree in Nutrition Sciences and a PhD in Physiological and Nutritional Sciences. Dr. Dinu is coordinator of the youth working group of the Italian Society of Human Nutrition (SINU) and secretary of the regional section. Her research activities include: (a) the design and implementation of clinical trials and observational studies to evaluate the role of food and dietary patterns on health-related biomarkers in primary and secondary prevention; (b) the development of systematic reviews and meta-analyses on the role of diet in the prevention and treatment of overweight, obesity and chronic diseases; and (c) the development and validation of questionnaires to assess food consumption and adherence to dietary patterns. She has participated in many research projects of international and national interest. She is the author of 70 scientific articles published in international peer-reviewed journals, and of 2 book chapters.

Daniela Martini

Prof. Daniela Martini is currently Associate Professor of Human Nutrition at the University of Milan. She is also Secretary of the Italian Society of Human Nutrition (SINU). She completed a master's degree in Human Nutrition at the University of Milan (2007) and a PhD in Food Science and Nutrition (2015) at the University Campus Bio-Medico of Rome. Her expertise includes the development and the optimization of strategies to favor the transition towards healthy and sustainable diets, including aspects related to food regulation (mainly nutrition labelling, nutrition and health claims), and the evaluation of the role of foods and dietary models in the modulation of health markers (e.g., oxidative stress in in vivo models). She is the author of 98 scientific manuscripts published in international peer-reviewed journals, and of 3 book chapters.

Preface

Ultra-processed foods (UPF) are food products that undergo extensive processing and contain multiple ingredients and additives and often have a long shelf life. They are typically manufactured through industrial processes and may bear little resemblance to their original whole food ingredients. In recent years, numerous observational studies have linked the consumption of UPF with poor diet quality and increased risk of chronic diseases, such as obesity, cardiovascular disease, type 2 diabetes, and certain types of cancer. However, the evidence still has many limitations and criticisms, both because of definitional and data collection problems and because general dietary patterns, lifestyle, and other health behaviors may play a role in the observed associations.

The purpose of this Special Issue was to collect new studies that investigated various aspects related to UPF, including the development of more specific questionnaires for data collection, the evaluation of possible differences in consumption in various population groups, the relationship with markers of health status and possible mechanisms underlying associations between UPF consumption and health, and the application of the concept of "ultra-processed food" in front-of-pack labeling. A total of 21 contributions were collected, including 18 original articles, 1 narrative review, 1 systematic review, and 1 meta-analysis. By providing up-to-date assessments of UPF consumption and health implications, and highlighting some criticisms of the overlap of the NOVA classification with other types of information provided to consumers, this Special Issue supports the importance of conducting new research on this topic. To achieve this goal, collaboration among the different parties involved in the production of these products, research, and consumers is encouraged.

The Guest Editors hope that this volume will be a useful resource for scientists in academia and industry, the general public, and policy makers. They would also like to thank all of the authors, the reviewers who contributed to the success of this Special Issue, and the *Nutrients* team for their valuable and continued support.

Monica Dinu and Daniela Martini
Editors

nutrients

Editorial

Ultra-Processed Foods, Diet Quality and Human Health

Monica Dinu [1,*] and Daniela Martini [2,*]

1 Department of Experimental and Clinical Medicine, University of Florence, 50134 Florence, Italy
2 Department of Food, Environmental and Nutritional Sciences (DeFENS), University of Milan, 20122 Milan, Italy
* Correspondence: monica.dinu@unifi.it (M.D.); daniela.martini@unimi.it (D.M.)

The increase in the volume of industrially processed products in the global food supply has coincided with an increasing prevalence of obesity and non-communicable diseases in many countries, suggesting that ultra-processed foods (UPF) consumption may be detrimental to human health. However, studies are still limited and underline the need to better understand the main determinants of their consumption and the mechanisms that may explain the associations between these products and human health. The Special Issue "Ultra-Processed Foods, Diet Quality and Human Health" aimed to collect new studies investigating the relationship between the consumption of UPF, diet quality and human health, including those aiming to: (i) develop new tools to better determine the rate of consumption of UPF in the population; (ii) investigate the rate of consumption of UPF in different subgroups of the population, including subjects following different dietary patterns; (iii) analyze the relationship between the consumption of UPF and markers of health status; and (iv) explore possible mechanisms behind associations between the consumption of processed foods and health.

This Special Issue provides a series of 21 contributions, with 18 original articles, 1 narrative review, 1 systematic review, and 1 meta-analysis. Some of the articles were devoted to the analysis of the amount of UPF consumed in different populations and over the years. Romero Ferreiro and colleagues estimated UPF consumption in Spain from 1991 to 2008, finding a 10.8% increase in UPF consumption between 1991 and 2008 [1]. Bertoni Maluf et al. described UPF consumption in adults living in Switzerland, finding a median UPF energy intake of 587 kcal/day (range 364–885), corresponding to 28.7% (range 19.9–38.9) of total energy intake [2]. Both studies found higher UPF intake in young participants and large differences between different geographical areas of countries.

It has been hypothesized that the harmful effects of UPF on human health may be related to the worse diet quality of subjects with higher UPF intake. In this regard, Dinu et al. studied UPF consumption in a group of Italian adults, observing a significant inverse association between adherence to the Mediterranean diet (as assessed by the Medi-Lite score) and the percentage of UPF in the diet [3]. Similar results were found by Tristan Asensi et al. who observed an inverse trend between UPF consumption and adherence to the Mediterranean diet in adults with celiac disease [4]. The association between UPF consumption and diet was also considered by Nansel and colleagues, who found that UPF intake during pregnancy and postpartum was inversely related to 8 of 13 component scores of the 2015 Healthy Eating Index [5]. This suggests that the higher the UPF intake, the lower the diet quality. Finally, analyzing data on 8688 Italians from the Italian Nutrition & Health Survey (INHES), Bonaccio et al. observed that late eaters had higher UPF intake and lower adherence to the Mediterranean diet than early eaters [6]. Overall, these data seem to suggest that a reduction in UPF intake can also be achieved by promoting the Mediterranean diet, adherence to which was correlated with a higher quality of life during the COVID-19 lockdown among Brazilian and Spanish youth aged 3–17 years [7].

Other contributions included in the Special Issue were devoted to exploring the association between UPF consumption and markers of human health or disease risk. In

Citation: Dinu, M.; Martini, D. Ultra-Processed Foods, Diet Quality and Human Health. *Nutrients* **2023**, *15*, 2890. https://doi.org/10.3390/nu15132890

Received: 24 May 2023
Accepted: 16 June 2023
Published: 26 June 2023

particular, the systematic review of Mambrini et al. [8] evaluated the association between UPF consumption and the incidence of obesity and cardiometabolic risk factors. The analysis of 17 studies showed substantial agreement in defining UPF consumption as being associated with the incident risk of general and abdominal obesity. More limited was the evidence on cardiometabolic risk. The other narrative review [9] summarized the available evidence on the possible relationship between excessive consumption of UPF and low-grade inflammation, focusing on nutritional and non-nutritional components that may explain this relationship. Among different markers of inflammation, Lane and colleagues focused on high-sensitivity C-reactive protein (hsCRP) by analyzing data from the Melbourne Collaborative Cohort Study (MCCS), finding that each 100 g increase in UPF intake was associated with a 4.0% increase in hsCRP concentration [10].

Two studies conducted in China analyzed data from the China Health and Nutrition Survey 1997–2015. In the first study, the authors observed that the higher the UPF consumption, the higher the incident rates of hypertension, with hazard ratios (HRs) for UPF intake of 1–49, 50–99, and ≥100 g/day of 1.00 (95% CI: 0.90–1.12), 1.17 (95% CI: 1.04–1.33), and 1.20 (95% CI: 1.06–1.35), respectively, compared with non-consumers [11]. In the second study, the same authors used the data to assess the association between UPF consumption and diabetes in Chinese adults. The odds ratios (ORs) of diabetes for people with a mean UPF consumption of 1–19, 20–49, and ≥50 g/day were 1.21 (95% CI: 0.98, 1.48), 1.49 (95% CI: 1.19, 1.86), and 1.40 (95% CI: 1.08, 1.80), respectively, compared with non-consumers [12]. Diabetes was also considered in a systematic review with meta-analysis that investigated maternal consumption of UPF and perinatal outcomes [13]. Authors observed that maternal consumption of UPF-rich diets was associated with an increased risk of gestational diabetes mellitus and preeclampsia, highlighting the need to reduce UPF consumption during the gestational period to prevent adverse perinatal outcomes.

Other associations between UPF and detrimental health effects were explored by Koniecza and coworkers who observed that each 10% daily increment in UPF consumption in 1 year was associated with higher levels of biomarkers related to non-alcoholic fatty liver diseases (i.e., non-invasive fatty liver index and hepatic steatosis index) in a cohort of 5867 older participants with overweight/obesity and metabolic syndrome from the PREDIMED-Plus trial, following for 1 year [14].

High UPF consumption was also found to be associated with an increased risk of depression and depressive symptoms. In a sample of 596 young Italian adults, Godos and colleagues showed that individuals in the highest quartile of UPF consumption were more likely to have depressive symptoms [15]. As this association became stronger when adjusted for other confounding factors, including adherence to the Mediterranean diet as a proxy for diet quality, the authors suggest that dietary components other than nutritional quality may play a role in the reported association. Similar results were observed in the Korean population of The Korea National Health and Nutrition Examination Survey, where 9463 subjects were analyzed [16]. However, in a sex-specific stratification, only women showed a significant association between higher UPF consumption and depression.

The other studies published in this Special Issue focused on different aspects, such as the overlap between the NOVA classification and other systems. In this regard, Angelino and colleagues compared the level of processing (assessed by NOVA) and nutritional quality (assessed by nutritional values, the Nutri-Score and the NutrInform battery) of breakfast cereals available on the Italian market and found a partial overlap between the NOVA classification and the systems based on the nutritional quality of foods [17]. Similarly, Grech et al. used data from the 2011–2012 National Nutrition and Physical Activity Survey, a large cross-sectional study representative of the Australian population, to compare the NOVA classification system with the Australian Dietary Guidelines (ADG) in classifying foods as healthy and unhealthy through their effectiveness in predicting energy overconsumption and body mass index (BMI) [18]. The analysis demonstrated considerable overlap between the NOVA and ADG classification systems, but some discrepancy emerged between the system that best identifies foods to avoid in Australia, with many culinary ingredients classified as unhealthy in ADG and most international dietary guidelines, but not in NOVA.

Finally, Krois and colleagues focused on dietitians' knowledge of the NOVA food classification system and their attitudes toward the classification of products containing whole grains [19], while Camargo et al. conducted a descriptive and exploratory analysis of the healthiness of 823 culinary recipes shared during a 6-month period on popular Brazilian YouTube® cooking channels, considering the degree of ingredient processing [20]. Only one study was conducted in animal models, analyzing the effects of Zn supplementation on the gut microbiota, intestinal barrier, and blood–brain barrier in Wistar rats fed a cafeteria diet (CAF) rich in UPF [21]. The results showed that chronic consumption of CAF causes dysbiosis, morphological changes and decreased levels of SCFA in the colon, as well as increased saturated fatty acids.

The Guest Editors would like to thank all the authors, the reviewers who contributed to the success of this Special Issue, and the *Nutrients* team for their valuable and constant support. By providing up-to-date assessments of UPF consumption and health implications, but at the same time highlighting some criticisms about the overlap of NOVA classification with other types of information provided to the consumers, these reports support the importance of providing new research on this topic. In particular, a better elucidation of the mechanisms through these products may exert a detrimental effect on human health and evidence from intervention studies seem to be crucial to better understand if future public health nutrition policies are needed.

Author Contributions: Conceptualization, M.D. and D.M.; methodology, M.D. and D.M.; writing—original draft preparation, M.D. and D.M.; writing—review and editing, M.D. and D.M. All authors have read and agreed to the published version of the manuscript.

Conflicts of Interest: The authors declare no conflict of interest.

References

1. Romero Ferreiro, C.; Cancelas Navia, P.; Lora Pablos, D.; Gómez de la Cámara, A. Geographical and Temporal Variability of Ultra-Processed Food Consumption in the Spanish Population: Findings from the DRECE Study. *Nutrients* **2022**, *14*, 3223. [CrossRef]
2. Bertoni Maluf, V.A.; Bucher Della Torre, S.; Jotterand Chaparro, C.; Belle, F.N.; Khalatbari-Soltani, S.; Kruseman, M.; Marques-Vidal, P.; Chatelan, A. Description of Ultra-Processed Food Intake in a Swiss Population-Based Sample of Adults Aged 18 to 75 Years. *Nutrients* **2022**, *14*, 4486. [CrossRef]
3. Dinu, M.; Tristan Asensi, M.; Pagliai, G.; Lotti, S.; Martini, D.; Colombini, B.; Sofi, F. Consumption of Ultra-Processed Foods Is Inversely Associated with Adherence to the Mediterranean Diet: A Cross-Sectional Study. *Nutrients* **2022**, *14*, 2073. [CrossRef]
4. Tristan Asensi, M.; Pagliai, G.; Lotti, S.; Corrao, A.; Colombini, B.; Giangrandi, I.; Sofi, F.; Dinu, M. Adherence to the Mediterranean Diet and Ultra-Processed Foods Consumption in a Group of Italian Patients with Celiac Disease. *Nutrients* **2023**, *15*, 938. [CrossRef]
5. Nansel, T.R.; Cummings, J.R.; Burger, K.; Siega-Riz, A.M.; Lipsky, L.M. Greater Ultra-Processed Food Intake during Pregnancy and Postpartum Is Associated with Multiple Aspects of Lower Diet Quality. *Nutrients* **2022**, *14*, 3933. [CrossRef] [PubMed]
6. Bonaccio, M.; Ruggiero, E.; Di Castelnuovo, A.; Martínez, C.F.; Esposito, S.; Costanzo, S.; Cerletti, C.; Donati, M.B.; de Gaetano, G.; Iacoviello, L., on behalf of the INHES Study Investigators. Association between Late-Eating Pattern and Higher Consumption of Ultra-Processed Food among Italian Adults: Findings from the INHES Study. *Nutrients* **2023**, *15*, 1497. [CrossRef]
7. Victoria-Montesinos, D.; Tárraga-Marcos, A.; Brazo-Sayavera, J.; Jiménez-López, E.; Gutiérrez-Espinoza, H.; Panisello Royo, J.M.; Tárraga-López, P.J.; López-Gil, J.F. Adherence to the Mediterranean Diet and Health Related Quality of Life during the COVID-19 Lockdown: A Cross-Sectional Study including Preschoolers, Children, and Adolescents from Brazil and Spain. *Nutrients* **2023**, *15*, 677. [CrossRef]
8. Mambrini, S.P.; Menichetti, F.; Ravella, S.; Pellizzari, M.; De Amicis, R.; Foppiani, A.; Battezzati, A.; Bertoli, S.; Leone, A. Ultra-Processed Food Consumption and Incidence of Obesity and Cardiometabolic Risk Factors in Adults: A Systematic Review of Prospective Studies. *Nutrients* **2023**, *15*, 2583. [CrossRef] [PubMed]
9. Tristan Asensi, M.; Napoletano, A.; Sofi, F.; Dinu, M. Low-Grade Inflammation and Ultra-Processed Foods Consumption: A Review. *Nutrients* **2023**, *15*, 1546. [CrossRef] [PubMed]
10. Lane, M.M.; Lotfaliany, M.; Forbes, M.; Loughman, A.; Rocks, T.; O'Neil, A.; Machado, P.; Jacka, F.N.; Hodge, A.; Marx, W. Higher Ultra-Processed Food Consumption Is Associated with Greater High-Sensitivity C-Reactive Protein Concentration in Adults: Cross-Sectional Results from the Melbourne Collaborative Cohort Study. *Nutrients* **2022**, *14*, 3309. [CrossRef] [PubMed]
11. Li, M.; Shi, Z. Ultra-Processed Food Consumption Associated with Incident Hypertension among Chinese Adults—Results from China Health and Nutrition Survey 1997–2015. *Nutrients* **2022**, *14*, 4783. [CrossRef]

12. Li, M.; Shi, Z. Association between Ultra-Processed Food Consumption and Diabetes in Chinese Adults—Results from the China Health and Nutrition Survey. *Nutrients* **2022**, *14*, 4241. [CrossRef] [PubMed]
13. Paula, W.O.; Patriota, E.S.O.; Gonçalves, V.S.S.; Pizato, N. Maternal Consumption of Ultra-Processed Foods-Rich Diet and Perinatal Outcomes: A Systematic Review and Meta-Analysis. *Nutrients* **2022**, *14*, 3242. [CrossRef] [PubMed]
14. Konieczna, J.; Fiol, M.; Colom, A.; Martínez-González, M.Á.; Salas-Salvadó, J.; Corella, D.; Soria-Florido, M.T.; Martínez, J.A.; Alonso-Gómez, Á.M.; Wärnberg, J.; et al. Does Consumption of Ultra-Processed Foods Matter for Liver Health? Prospective Analysis among Older Adults with Metabolic Syndrome. *Nutrients* **2022**, *14*, 4142. [CrossRef]
15. Godos, J.; Bonaccio, M.; Al-Qahtani, W.H.; Marx, W.; Lane, M.M.; Leggio, G.M.; Grosso, G. Ultra-Processed Food Consumption and Depressive Symptoms in a Mediterranean Cohort. *Nutrients* **2023**, *15*, 504. [CrossRef]
16. Lee, S.; Choi, M. Ultra-Processed Food Intakes Are Associated with Depression in the General Population: The Korea National Health and Nutrition Examination Survey. *Nutrients* **2023**, *15*, 2169. [CrossRef]
17. Angelino, D.; Dinu, M.; Gandossi, B.; Pellegrini, N.; Martini, D., on behalf of the SINU Young Working Group. Processing and Nutritional Quality of Breakfast Cereals Sold in Italy: Results from the Food Labelling of Italian Products (FLIP) Study. *Nutrients* **2023**, *15*, 2013. [CrossRef] [PubMed]
18. Grech, A.; Rangan, A.; Allman-Farinelli, M.; Simpson, S.J.; Gill, T.; Raubenheimer, D. A Comparison of the Australian Dietary Guidelines to the NOVA Classification System in Classifying Foods to Predict Energy Intakes and Body Mass Index. *Nutrients* **2022**, *14*, 3942. [CrossRef] [PubMed]
19. Krois, N.; Hughes, J.; Grafenauer, S. Dietitians' Attitudes and Understanding of the Promotion of Grains, Whole Grains, and Ultra-Processed Foods. *Nutrients* **2022**, *14*, 3026. [CrossRef] [PubMed]
20. Camargo, A.M.d.; Botelho, A.M.; Camargo, Á.M.D.; Dean, M.; Fiates, G.M.R. Ultra-Processed Foods as Ingredients of Culinary Recipes Shared on Popular Brazilian YouTube Cooking Channels. *Nutrients* **2022**, *14*, 3689. [CrossRef] [PubMed]
21. Squizani, S.; Jantsch, J.; Rodrigues, F.d.S.; Braga, M.F.; Eller, S.; de Oliveira, T.F.; Silveira, A.K.; Moreira, J.C.F.; Giovenardi, M.; Porawski, M.; et al. Zinc Supplementation Partially Decreases the Harmful Effects of a Cafeteria Diet in Rats but Does Not Prevent Intestinal Dysbiosis. *Nutrients* **2022**, *14*, 3921. [CrossRef] [PubMed]

nutrients MDPI

Article

Consumption of Ultra-Processed Foods Is Inversely Associated with Adherence to the Mediterranean Diet: A Cross-Sectional Study

Monica Dinu [1,*], Marta Tristan Asensi [1], Giuditta Pagliai [1], Sofia Lotti [1], Daniela Martini [2], Barbara Colombini [1] and Francesco Sofi [1,3]

[1] Department of Experimental and Clinical Medicine, University of Florence, 50134 Florence, Italy; mtristanasensi@gmail.com (M.T.A.); giuditta.pagliai@unifi.it (G.P.); sofia.lotti@unifi.it (S.L.); barbara.colombini@unifi.it (B.C.); francesco.sofi@unifi.it (F.S.)

[2] Department of Food, Environmental and Nutritional Sciences (DeFENS), University of Milan, 20122 Milan, Italy; daniela.martini@unimi.it

[3] Unit of Clinical Nutrition, Careggi University Hospital, 50134 Florence, Italy

* Correspondence: monica.dinu@unifi.it; Tel.: +39-0557949420

Abstract: Information on the consumption of ultra-processed foods (UPF) in relation to an adherence to the Mediterranean diet (MD) is limited. Our aim was to assess UPF consumption in a group of Italian adults and to evaluate the relationship with the MD adherence. A total of 670 participants (median age: 30 years) were included in the analysis. The consumption of UPF was assessed through the NOVA Food Frequency Questionnaire (NFFQ). Adherence to the MD was assessed through the Medi-Lite score. The percentage of UPF in the diet was 16.4% corresponding to 299 g of UPF per day. These amounts were significantly ($p < 0.05$) higher in men than in women and came mainly from ready-to-eat meals or pre-packaged bread, bread alternatives, pizza, frozen potato chips (24.5% of total UPF intake), pre-packaged biscuits and sweets (20.7%), soft drinks (15.8%), and dairy products such as flavored yogurt (12%). As to the MD adherence, a significant inverse association between the Medi-Lite score and the percentage of UPF in the diet ($R = -0.35$; $p < 0.001$) was observed. Participants with a low adherence to the MD had a significantly higher contribution of UPF in the diet (22.2%) compared to those with a moderate (16.2%) and high (12.6%) adherence. In terms of individual UPF, the largest difference between low and high MD adherents was observed for pre-packaged biscuits and sweets, soft and energy drinks, sausages and other reconstituted meat products, and pre-packaged bread and bread alternatives. These results suggest that public health strategies are needed to implement more effective actions to promote healthy eating habits in the population.

Keywords: ultra-processed foods; NOVA classification; NFFQ; Mediterranean diet; Medi-Lite

Citation: Dinu, M.; Tristan Asensi, M.; Pagliai, G.; Lotti, S.; Martini, D.; Colombini, B.; Sofi, F. Consumption of Ultra-Processed Foods Is Inversely Associated with Adherence to the Mediterranean Diet: A Cross-Sectional Study. *Nutrients* **2022**, *14*, 2073. https://doi.org/10.3390/nu14102073

Academic Editor: Licia Iacoviello

Received: 22 April 2022
Accepted: 12 May 2022
Published: 15 May 2022

1. Introduction

Ultra-processed foods (UPF) are defined within the NOVA classification system, which groups foods according to the extent and purpose of industrial processing [1]. They are "formulations of ingredients, mostly for industrial use only, derived from a series of industrial processes" [2]. Examples of UPF are pre-packaged frozen meals, ready-to-eat meals, fast food, mass-produced bread, savory snacks, breakfast cereals with added ingredients, reconstituted meat products, instant soups and noodles, and soft and sweetened beverages.

Available literature suggests that higher UPF consumption is associated with a lower diet quality, leading to diets high in calories, free sugars, fat, and salt, and low in dietary fiber [3]. This, in turn, may be associated with a worse cardio—metabolic risk profile and an increased risk of cardiovascular disease, cerebrovascular disease, depression, and all-cause mortality, as suggested by a recent meta-analysis by our group [4]. It is also known that increased UPF intake can replace unprocessed foods and freshly prepared meals and dishes [2], which are the basis of traditional dietary patterns recognized to promote long and healthy lives [5–7].

The Mediterranean diet (MD) is one of the most studied dietary patterns in the scientific community. Despite its health benefits, widely documented in epidemiological and clinical studies [8,9], many Mediterranean countries are experiencing a progressive shift away from this dietary pattern [10]. In this regard, some recent studies have suggested an association between MD adherence and UPF consumption, but evidence is still limited [11–15]. Furthermore, available studies have not used specifically validated tools to estimate UPF consumption and this could potentially lead to a misinterpretation of the associations found [16]. The aim of this study was to evaluate UPF consumption in a group of Italian adults and to assess the possible relationship between MD adherence and UPF consumption using validated tools specifically designed for these purposes.

2. Materials and Methods

2.1. Study Design and Data Collection

A web survey was conducted between January and October 2021. Data were collected anonymously from an online questionnaire prepared on SurveyMonkey (www.surveymonkey.com, accessed on 11 May 2022) [17], which is a free tool with an easy-to-use web interface. To enroll as many participants as possible, the questionnaire was disseminated and shared with a link among personal and non-personal contacts using advertisements in local media, social media, and on websites. Before starting the questionnaire, participants were asked to read the project information sheet, which contained an explanation of the study's objectives. Then, participants were asked to complete a brief questionnaire focused on sociodemographic characteristics (age, sex, marital status, education level, weight, and height), and two validated questionnaires suitable for collecting information on UPF consumption and MD adherence. Marital status was categorized as unmarried/single and married/partner, and education level was categorized as primary/secondary school, high school, and university. Body mass index (BMI) was calculated as weight in kilograms divided by the square of height in meters for each participant. Participants were classified as overweight if their BMI ranged between 25–30 kg/m^2, and obese if their BMI was 30 kg/m^2 or more. This study was conducted according to the guidelines laid down in the Declaration of Helsinki and all procedures involving human subjects were approved by the Ethics Committee of the University of Florence, Florence, Italy [n 199/2022, protocol 11917]. Informed consent was obtained from all participants.

2.2. NOVA Food Frequency Questionnaire (NFFQ)

The NFFQ is a validated questionnaire specifically designed to estimate food intake according to NOVA groups in the Italian adult population [18]. The questionnaire includes 94 items divided into nine categories: (1) fruits and nuts; (2) vegetables and legumes; (3) cereals and tubers; (4) meat and fish; (5) milk, dairy products, and eggs; (6) oils, fats, and seasonings; (7) sweets and sweeteners; (8) beverages; and (9) other. For each item, the participant should indicate the frequency of consumption and usual portion size considering the diet in a typical month over the past 12 months. Answers correspond to one out of ten options: (1) "never or less than once a month"; (2) "one to three times a month"'; (3) "once a week"; (4) "twice a week"; (5) "three times a week"; (6) "four times a week"; (7) "five times a week"; (8) "six times a week"; (9) "every day"; and (10) "if every day, how many times a day?". Because all 94 items included in the FFQ were pre-categorized according to the NOVA classification, each food fell into one of the following categories: unprocessed and minimally processed food or beverage (MPF); processed culinary ingredients (PCI); processed food or beverage (PF); and UPF.

When processing the NFFQ data, the amount of food consumed is calculated in grams per week and in grams per day for each food item and category. Then, the proportion of MPF, PCI, PF, and UPF in the diet is determined by calculating the weight ratio. As previously reported, the weight ratio is considered rather than the energy ratio because it allows for a better accounting for processed foods that do not provide energy (e.g., artificially sweetened beverages) and non-nutritional factors related to food processing (e.g., newly formed contaminants, food additives and alterations in the structure of raw foods) [18]. Finally, because PCI products are not intended to be consumed alone as food, but are supposed to be used to prepare and season other foods, they are grouped with PF.

2.3. The Medi-Lite Adherence Score

Adherence to MD was assessed through the validated Medi-Lite questionnaire, developed by Sofi et al. [19]. The questionnaire includes nine domains, based on daily and/or weekly consumption of fruits, vegetables, cereals, legumes, fish, meat and meat products, dairy products, alcohol, and olive oil. For fruits, vegetables, cereals, legumes, and fish (typical foods of MD), two points are given for the highest consumption category, one point for the medium category and zero points for the lowest category. For olive oil, two points are given for regular use, one for frequent use, and zero for occasional use. On the other hand, for meat and meat products and dairy products (foods not typical of MD), two points are given for the lowest category of consumption, one point for the medium category, and zero for the highest category. Finally, for alcohol consumption, two points are given for the medium category, one for the lowest category, and zero for the highest category. The questionnaire score ranges from 0 to 18, where the highest value corresponds to the highest adherence to the MD.

2.4. Statistical Analysis

Descriptive statistics were used to analyze and report the data. Results were reported as mean ± standard deviation (SD), median and range, or geometric mean with 95% confidence intervals (CIs), as appropriate. Categorical variables were presented in terms of frequencies and percentages. The Mann—Whitney test was used for comparisons between women and men, while the Chi—Square test was used to test for proportions. The correlation between the percentage of UPF in the diet and the Medi-Lite score was estimated using the Spearman (R) test.

The possible relationship between UPF consumption and MD adherence was analyzed by grouping the participants according to UPF contribution in the diet (1st tertile $\leq$ 10%; 2nd tertile = 10–19%; 3rd tertile $\geq$ 19%), and by MD adherence into low (1st tertile = Medi-Lite score < 9), moderate (2nd tertile = Medi-Lite score between 9 and 11), and high (3rd tertile = Medi-Lite score > 11) adherence. Then, a general linear model adjusted for age, sex, BMI, education level, marital status, and daily food intake was conducted to compare dietary habits according to the percentage of UPF in the diet, and daily UPF intake according to the MD adherence. Since these tests assume normal data distribution, non-distributed data were transformed into logs and further analyses were performed with the processed data. However, to facilitate interpretation, the log data were again converted to the original scale (antilog) and presented as geometric means with 95% CIs.

Finally, to evaluate the influence of individual UPF on Medi-Lite score, a linear regression model was performed, and results were expressed as regression coefficient (b) ± SE. The b coefficient estimated in the linear regression analysis indicates the expected mean change in Medi-Lite score associated with 1-unit change in the independent variables. p-values < 0.05 were considered statistically significant. The statistical package IBM SPSS Statistics for Macintosh, version 28.0 (IBM Corp., Armonk, NY, USA) was used.

3. Results

A total of 670 participants (70.4% women) with a mean age of 35.8 years were included in the analysis. Table 1 reports their socio-demographic characteristics, according to sex. Overall, the study population was highly educated and more than half of the participants were single. Compared to men, women had significantly ($p < 0.001$) lower BMI and lower prevalence of overweight and obesity (21.1% vs. 40.9%). No significant differences were observed for education level or marital status.

Table 1. Characteristics of the study population.

	All (n = 670)	Women (n = 472)	Men (n = 198)	p-Value
Age, year	35.8 ± 13.4	35.3 ± 13	36.9 ± 14.1	0.160
Body weight, kg	66.5 ± 14.1	61.5 ± 11.7	78.6 ± 11.9	<0.001
BMI, kg/m^2	23.2 ± 4	22.6 ± 4.1	24.7 ± 3.3	<0.001
BMI ≥ 25 kg/m^2	176 (26.3)	95 (20.1)	81 (40.9)	<0.001
Education level				
Secondary school	23 (3.4)	15 (3.2)	8 (4)	0.744
High school	289 (43.1)	198 (41.9)	91 (46)	0.384
University	358 (53.4)	259 (54.9)	99 (50)	0.285
Marital status				
Single	372 (55.5)	253 (53.6)	119 (60.1)	0.144
Married/partner	256 (38.2)	189 (40)	67 (33.8)	0.155
Divorced/widowed	42 (6.3)	30 (6.4)	12 (6.1)	0.886

Legend: BMI = Body Mass Index. Data are reported as mean ± standard deviation or number and percentage (%), as appropriate

3.1. Nova Food Frequency Questionnaire (NFFQ)

The percentage of UPF in the diet was 16.4 ± 10.4%, with significantly ($p < 0.05$) lower values in women (15.6 ± 10%) than in men (18.2 ± 11.1%). No significant differences were observed according to BMI, education level, and marital status. In terms of daily UPF intake, the percentage of UPF in the diet corresponded to 299.0 ± 233.5 g/day in the total sample, 285.5 ± 231.0 g/day in women, and 331.2 ± 236.7 g/day in men. The food categories that contributed most to this amount were ready-to-heat meals or pre-packaged bread, bread alternatives, pizza, and frozen potato chips (24.5% of total UPF intake), pre-packaged sweets and biscuits (20.7%), soft drinks (15.8%), flavored yogurt (12.0%), followed by ready-to-eat vegetables and legumes (10.6%), dairy products and meat substitutes (8%), sausages and other reconstituted meat and fish products (5.8%), and fats and seasonings (2.5%). The contribution of individual UPF, according to sex, is reported in Supplementary Table S1. Significant differences between sex were observed for ready-to-heat pasta/gnocchi, sausages and other reconstituted meat and fish products, and for alcoholic beverages (e.g., liquor), with higher percentages in men than in women.

Food consumption (g/day) according to the percentage of UPF in the diet is reported in Table 2. After adjustment for possible confounding factors such as age, sex, and daily food intake, participants in the highest UPF tertile showed a lower consumption of fruits, nuts, and vegetables than those in the first tertile, and a higher consumption of cereals and tubers, fats and seasonings, sweets and sweeteners, beverages, and dairy substitutes. For meat and fish, the total consumption did not differ significantly between groups, but subjects in the highest UPF tertile consumed more meat and poultry and less fish and seafood. A similar trend was observed for beverages, with subjects in the highest UPF tertile reporting a higher consumption of soft and energy drinks, and subjects in the lowest UPF tertile reporting a higher consumption of tea and coffee. No significant differences were observed for milk and dairy products, except for cheese.

3.2. Medi-Lite and UPF Consumption

Regarding MD adherence, the mean Medi-Lite score was 10.3 ± 2.5, with no significant differences (p = 0.151) between women (10.4 ± 2.4) and men (10 ± 2.6). Correlation analyses showed a significant inverse association between the Medi-Lite score and percentage of UPF in the diet (R = −0.35; $p < 0.001$) (Figure 1), which was confirmed when women (R = −0.35; $p < 0.001$) and men (R = −0.36; $p < 0.001$) were analyzed separately.

Table 2. Food consumption (g/day) according to the percentage of UPF in the diet.

	%UPF in the Diet			
	<10% $n = 222$	10–19% $n = 227$	>19% $n = 221$	p Trend †
Fruits and nuts	275.9 (251.1–302.8)	173.1 (157.9–190)	117.1 (106.6–128.8) *	<0.001
Fruits	258.3 (233.9–285.4)	159.0 (144.2–175.4)	104.5 (94.5–115.5) *	<0.001
Dried and syrup fruits	3.3 (2.6–4.2)	3.4 (2.7–4.3)	4.2 (3.3–5.3)	0.304
Nuts	9.8 (8.3–11.4)	7.6 (6.5–8.9)	6.5 (5.5–7.7) *	0.002
Vegetables and legumes	387.2 (353.9–423.3)	298.9 (273.7–326.4)	231.6 (211.7–253.4) *	<0.001
Vegetables	360.0 (326.7–396.6)	276.4 (251.4–304.0)	211.0 (191.3–232.8) *	<0.001
Legumes	20.4 (17.9–23.3)	20.0 (17.4–23.1)	22.8 (19.6–26.6)	0.432
Cereals and tubers	218.8 (206.2–232.1)	268.0 (252.9–284)	299.2 (281.7–317.3) *	<0.001
Grains (e.g., rice, spelt, barley, wheat)	19.0 (16.9–21.4)	18.3(16.3–20.6)	14.9 (13.1–16.8) *	0.010
Pasta, bread, and pizza	146.5 (136.2–157.4)	183.1 (170.4–196.6)	220.3 (204.8–237.0) *	<0.001
Potatoes and tubers	37.6 (33.9–41.8)	49.3 (44.5–54.4)	50.1 (45.2–55.6) *	<0.001
Breakfast cereals	9.9 (8.2–12.0)	8.0 (6.7–9.5)	9.0 (7.5–10.7)	0.246
Meat and fish	105.6 (97.9–114)	117.8 (109.3–127)	119.2 (110.4–128.8)	0.054
Meat and poultry	54.6 (49.8–59.8)	69.5 (63.5–76.0)	78.9 (72.1–86.5) *	<0.001
Fish and seafood	45.0 (40.8–49.7)	40.7 (36.9–44.8)	33.6 (30.4–37.2) *	<0.001
Milk, dairy products, and eggs	130.7 (115.5–148.0)	138.2 (122.4–156.3)	122.2 (107.8–138.7)	0.389
Milk and milk beverages (e.g., probiotic milk)	74.6 (62.1–89.7)	62.5 (52.2–74.7)	60.7 (50.4–73.0)	0.244
Yogurt	38.2 (32.7–44.7)	42.4 (36.7–49.1)	41.3 (35.3–48.3)	0.620
Cheese	26.3 (23.0–30.0)	32.3 (28.3–36.8)	32.5(28.5–37.1) *	0.040
Eggs	14.7 (13.6–16.0)	14.1 (13.1–15.3)	12.9 (11.8–14.0) *	0.067
Oil, fats, and seasonings	38.7 (36.6–40.9)	43.8 (41.4–46.2)	51.5 (48.7–54.4) *	<0.001
Olive oil and vegetable oils	25.7 (24.7–26.8)	25.8 (24.8–26.8)	25.2 (24.2–26.2)	0.692
Other fats (e.g., butter, margarines)	1.3 (1.1–1.5)	1.7 (1.5–1.9)	1.9 (1.7–2.2) *	<0.001
Sauces	11.3 (10.0–12.7)	14.1 (12.5–15.8)	21.4 (19.0–24.1) *	<0.001
Sweets and sweeteners	39.7 (35.6–44.2)	66.6 (60.0–74.2)	73.6 (66.0–82.1) *	<0.001
Biscuits, cakes, snacks, and ice-cream	29.4 (26.3–32.9)	50.5 (45.2–56.3)	59.8 (53.4–67) *	<0.001
Chocolate, spreads, and candies	6.9 (5.9–8.1)	9.4 (8.1–11.0)	9.4 (8.0–11.0) *	0.006
Sugar	2.8 (2.1–3.7)	2.4 (1.8–3.1)	3.0 (2.3–3.9)	0.499
Beverages	246.7 (225.0–270.2)	273.7 (250.1–299.5)	308.0 (280.6–337.6) *	0.004
Tea and coffee	139.9 (124.6–157.3)	120.7 (107.4–135.4)	97.7 (86.7–110.1) *	<0.001
Fruit and vegetable juice	42.1 (32.7–54.1)	57.1 (47.0–69.3)	61.1 (51.0–73.3) *	0.057
Soft and energy drinks	27.1 (22.0–33.4)	41.5 (35.7–48.2)	87.1 (76.3–99.4) *	<0.001
Alcoholic beverages	54.7 (45.6–65.6)	67.3 (55.8–81.0)	49.1 (40.8–59.1)	0.060
Other	19.1 (14.3–25.6)	31.4 (24.5–40.4)	45.0 (35.2–57.5) *	<0.001
Plant-based dairy substitutes	22.0 (15.4–31.5)	42.3 (31.1–57.5)	49.4 (37.3–65.4) *	0.003
Plant-based meat substitutes	11.6 (9.2–14.6)	11.9 (10.0–14.2)	10.5 (8.9–12.5)	0.596

Legend: UPF = ultra-processed foods. Data are reported as geometric mean and 95% confidence interval (CI). † Adjusted for age, sex, BMI, education level, marital status, and total food consumed (g/day). * $p < 0.05$ for differences between the 1st and the 3rd tertile adjusted for age, sex, BMI, education level, marital status, and total dietary intake (g food/day)

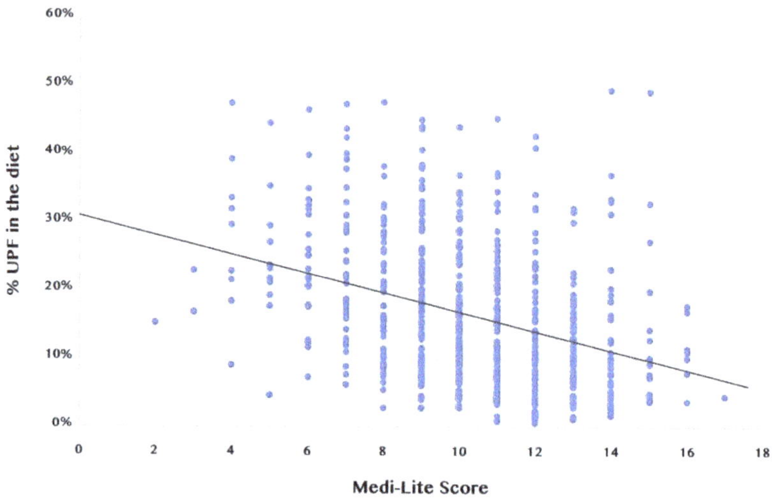

Figure 1. Correlation between Medi-Lite score and the percentage of UPF in the diet. Legend: UPF: ultra-processed foods.

To better explore the possible relationship between MD adherence and UPF consumption, three adherence groups to MD (low, moderate, and high) were considered, and the mean percentage of total UPF in these groups was calculated. As showed in Figure 2, subjects with low MD adherence had a significantly higher contribution of UPF in the diet (22.2 ± 10.3%) than those with moderate (16.2 ± 9.8%) and high (12.6 ± 9.5%) adherence. This result was confirmed when women and men were analyzed separately.

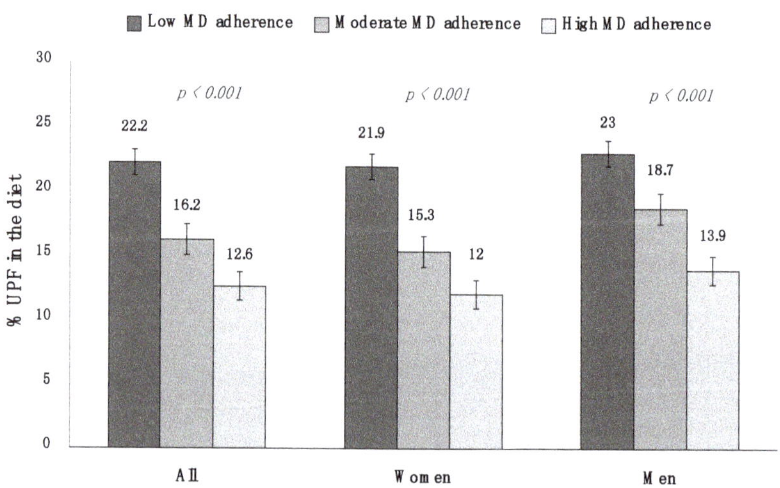

Figure 2. Percentage of UPF in the diet according to MD adherence. Legend: MD: Mediterranean diet; UPF: ultra-processed foods.

Regarding individual UPF, their intake in terms of daily amounts was significantly higher in subjects with a low MD adherence for most of the items considered, with the largest differences observed for pre-packaged biscuits and other sweets, soft and energy drinks, sausages and other reconstituted meat products, and pre-packaged bread and bread alternatives (Table 3).

Table 3. UPF intake (g/day) according to the MD adherence.

	Adherence to the MD			
	Low $n = 143$	Moderate $n = 325$	High $n = 203$	p Trend †
Vegetables and legumes UPF	38.2 (30.8–47.6)	43.8 (37.8–50.7)	34.8 (28.5–42.4)	0.167
Ready-to-heat vegetables and legumes (with added ingredients)	38.2 (30.8–47.6)	43.8 (37.8–50.7)	34.8 (28.5–42.4)	0.167
Cereals and tubers UPF	70.5 (59.1–84.3)	43.2 (38.7–48.7)	33.5 (28.8–39) *	<0.001
Ready-to-heat pasta/gnocchi dishes	17.9 (14.9–21.6)	17.5 (15.3–19.9)	15.1 (12.5–18.1)	0.363
Pre-packaged breads, buns, and bread alternatives	18.2 (15.1–22.0)	13.8 (12.2–15.6)	11.9 (10.1–13.9) *	0.004
Pre-packaged pizza, focaccia, sandwich, and savory pies	33.5 (27.4–41.0)	24.5 (21.2–28.3)	25.6 (20.6–31.8)	0.043
Pre-packaged instant rice, soups, noodles	13.4 (10.5–17.1)	12.3 (10.3–14.7)	11.8 (9.2–15.0)	0.754
Breakfast cereals and energy bars (with added sugar)	7.7 (5.9–9.9)	6.5 (5.5–7.6)	6.8 (5.4–8.5)	0.547
Pre-packaged potatoes (e.g., frozen potato chips)	17.9 (15.8–20.3)	14.9 (13.6–16.4)	13.9 (12.2–15.9) *	0.022
Meat and fish UPF	21.5 (18.8–24.5)	15.7 (14.3–17.3)	14.9 (13.0–17) *	<0.001
Nuggets, sticks, sausages, burgers, and other reconstituted meat products	18.9 (16.6–21.5)	13.9 (12.6–15.2)	13.0 (11.4–14.9) *	<0.001
Fish nuggets, fish sticks, and other reconstituted fish products	7.9 (6.9–9)	8.4 (7.6–9.3)	7.9 (7.0–9.1)	0.626
Milk and dairy products UPF	20.7 (16.3–26.3)	23.4 (19.9–27.5)	18.9 (14.9–26.9)	0.302
Milk beverages (e.g., probiotic milk with added sugar)	15.8 (11.5–21.9)	22.6 (17.7–28.9)	18.3 (13.4–25.1)	0.205
Fruit or flavored yogurts (e.g., vanilla flavored)	25.0 (20.0–31.1)	27.1 (23.5–31.2)	24.3 (19.8–29.8)	0.632
Melted cheese (also used to stuff sandwich)	3.4 (2.9–4.0)	2.9 (2.5–3.3)	3.2 (2.6–3.8)	0.240
Fats and seasonings UPF	6.7 (5.5–8.1)	5.6 (4.9–6.5)	5.5 (4.6–6.6)	0.309
Margarines and other spreads	0.8 (0.4–1.6)	0.9 (0.5–1.7)	0.9 (0.4–1.6)	0.940
Pre-packaged or instant sauces (e.g., mayonnaise, ketchup, meat sauce)	7.0 (5.8–8.6)	5.7 (4.9–6.5)	5.5 (4.6–6.5)	0.138
Sweets and sweeteners UPF	54.6 (46.8–63.7)	35.8 (32.5–39.6)	29.8 (63.7–33.8) *	<0.001
Pre-packaged biscuits, cakes, snacks, and ice-cream	39.4 (33.1–46.9)	25.0 (22.3–28)	21.9 (18.9–25.5) *	<0.001
Chocolate, spreads (e.g., nut spread), and candies	10.2 (8.4–12.4)	8.8 (7.7–10.0)	7.2 (6.1–8.5) *	0.033
Beverages UPF	57.9 (44.6–75.3)	35.3 (29.5–42.1)	25.0 (19.8–31.5) *	<0.001
Soft and energy drinks (e.g., iced tea, coke)	72.2 (59.0–89.0)	51.4 (44.4–59.4)	45.2 (36.9–55.4) *	0.005
Alcoholic beverages (e.g., rum, gin, spirits)	5.7 (4.3–7.4)	4.9 (4.1–5.7)	4.8 (4.0–5.8)	0.589
Other UPF	29.3 (19.1–44.8)	27.4 (22.2–33.9)	33.7 (25.3–44.8)	0.527
Plant-based dairy substitutes (e.g., soy yogurt, tofu)	35.2 (21.3–58.4)	33.7 (26.4–42.9)	49.3 (35.3–68.9)	0.195
Plant-based meat substitutes (e.g., veggie burger)	11.8 (8.2–16.9)	10.9 (9.2–12.8)	10.1 (8.1–12.8)	0.783

Legend: MD: Mediterranean diet; UPF: ultra-processed foods. Data are reported as geometric mean and 95% confidence interval (CI). † Adjusted for age, sex, BMI, education level, marital status, and total food consumed (g/day). * $p < 0.05$ for differences between low and high adherence to the MD adjusted for age, sex, BMI, education level, marital status, and total dietary intake (g food/day).

Finally, to evaluate the possible influence of individual UPF on Medi-Lite score, a linear regression analysis with the Medi-Lite score as a dependent variable was performed. As shown in Table 4, sausages and other reconstituted meat products, pre-packaged pizza, potatoes, biscuits and sweets, and soft and energy drinks were revealed to influence the Medi-Lite score significantly and negatively after an adjustment for age, sex, BMI, education level, marital status, and the total food consumed.

Table 4. Linear regression analysis relating UPF intake (g/day) and Medi-Lite score.

	Medi-Lite Score			
	Model 1		Model 2 [a]	
	ß (SE)	p-Value	ß (SE)	p-Value
Pre-packaged breads, buns, and bread alternatives	−0.035 (0.003)	0.369	-	-
Pre-packaged pizza, focaccia, sandwich, and savory pies	−0.116 (0.003)	0.003	−0.156 (0.003)	<0.001
Pre-packaged potatoes (e.g., frozen potato chips)	−0.143 (0.005)	<0.001	−0.201 (0.005)	<0.001
Nuggets, sticks, sausages, burgers, and other reconstituted meat products	−0.204 (0.004)	<0.001	−0.243 (0.004)	<0.001
Pre-packaged biscuits, cakes, snacks, and ice-cream	−0.089 (0.002)	0.021	−0.149 (0.002)	<0.001
Chocolate, spreads (e.g., nut spread), and candies	−0.017 (0.006)	0.669	-	-
Soft and energy drinks (e.g., iced tea, coke)	−0.098 (0.001)	0.011	−0.150 (0.001)	<0.001

[a] Model 2 includes age, sex, BMI, education level, marital status, and total food consumed (g/day) as covariates.

4. Discussion

In this study we used a validated questionnaire to assess UPF consumption in relation to an adherence to the MD in a group of adult subjects living in the Mediterranean area. In our study group, we observed that subjects reporting a low adherence to MD were also those showing a higher consumption of UPF in their diet. In fact, an increased intake of UPF was associated with a significantly lower intake of some typical products of the MD such as fruits, vegetables, nuts and fish, and a higher intake of meat, fats, seasonings, and sugary products. Furthermore, we observed that some UPF, such as sausages and other reconstituted meat products, pre-packaged pizza, frozen potato chips, industrial biscuits and sweets, and soft and energy drinks, were shown to influence the adherence score to MD.

Since Monteiro and colleagues proposed the NOVA classification to identify the level of food processing [20], several studies have investigated UPF consumption and its effects on health. A meta-analysis by our group recently reported a significantly increased risk of the occurrence of major chronic diseases in association with a higher UPF intake [4]. In terms of daily UPF consumption, a significant increase has been reported worldwide in recent decades. Data from most of the industrialized countries show a range of percentages from 20–50% of UPF present in the diet of the general population. To date, countries bordering the Mediterranean Sea have shown a lower UPF consumption in contrast to Western countries, but the intake of UPF is increasing rapidly, leading to a gradual displacement of long-established diets. In our sample population, the percentage of UPF in the diet was 16% of the total energy intake, which corresponded to a daily consumption of almost 300 g of UPF. These values are slightly higher than those reported also in Italy but in the Southern part by Bonaccio et al., (10% of total energy intake, corresponding to 182 g/day) [13] but substantially lower compared with other countries such as Spain (24%) [21], France (36%) [22], the United Kingdom (57%) [23], or the United States (58%) [24]. Such a difference among countries, in particular between Mediterranean and non-Mediterranean countries, let us to hypothesize that the traditional dietary patterns of the Mediterranean basin could have influenced the findings.

Currently, a higher adherence to MD is known to reduce the risk of all-cause mortality, cardiovascular diseases, coronary heart disease, myocardial infarction, overall cancer incidence, neurodegenerative diseases, and diabetes [8]. In line with the literature [11–15], the results obtained from this study confirm a significant inverse association between UPF consumption and adherence to MD, highlighting that higher UPF consumption is associated with a lower adherence to MD. This could be explained as a nutritional transition from fresh meals and dishes that are part of the traditional cuisine towards a higher intake of ready-to consume and hyper-palatable food and beverages products. Indeed, an impact on the intake of some of the foods known to be the basis of MD was reported due to the high UPF consumption. Specifically, participants with greater UPF intake reported a lower intake of fruits, vegetables, nuts, and fish, and a higher intake of cereals, fats, meat, seasonings, and sweets in their diets, in contrast to participants with a lower UPF consumption. Similar findings were found in other studies of Mediterranean populations, such as Spain [11] or France [22]. Furthermore, UPF consumption has not only impacted eating habits, but also beverage quality, with a higher consumption of soft and energy beverages and a lower consumption of water, tea, and coffee in participants with a greater UPF consumption [14,15,22].

To better understand the association between UPF intake and MD adherence in our study population, we also investigated the possible influence of individual UPF on the MD adherence score, observing that the consumption of sugary products, processed meats, soft and energy drinks, and pre-packaged potatoes and pizza negatively influenced the MD adherence score. Interestingly, most of these foods were major contributors to UPF consumption in our population and are present in all countries of the world, supporting the hypothesis of a nutritional transition from traditional diets such as MD to Westernized dietary patterns due to an increased UPF consumption. There is an urgent need to raise awareness on the negative health effects of excessive UPF consumption and new public health strategies to prevent the progressive loss of traditional diets.

The present study has several limitations that deserve discussion. First, the cross-sectional design does not allow us to establish any cause-effect relationship. Prospective cohort studies are needed to confirm these preliminary findings. Second, the use of self-administered online questionnaires may have led to recall bias and misclassification. The emergency response due to the COVID-19 pandemic did not allow us to conduct another type of study with a stronger statistical power and less biases. On the other hand, several strengths are present since a validated questionnaire specifically designed to estimate food intake according to NOVA classification was used, avoiding the misclassification of foods into UPF categories. Furthermore, the variables used in the main analyses included the proportion, by weight, of UPF in the diet. This approach is more appropriate than the use of energy proportion, as it takes a better account of non-nutritional factors pertaining to food processing (e.g., neo-formed contaminants, additives, alterations to the structure of raw foods).

5. Conclusions

In conclusion, in our study population of middle-aged Italian adults we were able to observe an inverse relation between adherence to MD and UPF consumption. In considering the over-consumption of UPF as an important risk factor for non-communicable diseases, overweightness and obesity, our results reinforce the importance of public health strategies to improve the population's health by promoting MD and limiting the intake of UPF, which is also proposed by the World Health Organization (WHO). Examples of such strategies could be taxation, the regulation of marketing, or the control of food labeling standards, some of which have already shown good results [25,26]. Moreover, particular attention should be paid to the consumption of specific UPF (sugary products, processed meats, soft and energy beverages, pre-packaged potatoes, and pizza) due to their association with a low adherence to MD.

Supplementary Materials: The following supporting information can be downloaded at: https://www.mdpi.com/article/10.3390/nu14102073/s1, Table S1: Contribution (%) of individual foods to the total intake of UPF.

Author Contributions: F.S. and M.D. designed the research; M.D. and M.T.A. performed the statistical analysis; M.D., M.T.A., G.P. and S.L. wrote the paper; D.M., B.C. and F.S. critically reviewed the manuscript; M.D. had primary responsibility for final content. All authors have read and agreed to the published version of the manuscript.

Funding: This research received no specific grant from any funding agency, commercial or not-for-profit sectors.

Institutional Review Board Statement: All procedures involving human subjects were approved by the Ethics Committee of the University of Florence, Florence, Italy [n 199/2022, protocol 11917].

Informed Consent Statement: Informed consent was obtained from all subjects involved in the study.

Data Availability Statement: Additional data are available from the corresponding author on reasonable request.

Conflicts of Interest: The authors declare no conflict of interest.

References

1. Monteiro, C.A.; Cannon, G.; Levy, R.B.; Moubarac, J.C.; Louzada, M.L.; Rauber, F.; Khandpur, N.; Cediel, G.; Neri, D.; Martinez-Steele, E.; et al. Ultra-processed foods: What they are and how to identify them. *Public Health Nutr.* **2019**, *22*, 936–941. [CrossRef] [PubMed]
2. Monteiro, C.A.; Cannon, G.; Moubarac, J.C. The UN Decade of Nutrition, the NOVA food classification and the trouble with ultra-processing. *Public Health Nutr.* **2018**, *21*, 5–17. [CrossRef] [PubMed]
3. Martini, D.; Godos, J.; Bonaccio, M.; Vitaglione, P.; Grosso, G. Ultra-Processed Foods and Nutritional Dietary Profile: A Meta-Analysis of Nationally Representative Samples. *Nutrients* **2021**, *13*, 3390. [CrossRef] [PubMed]
4. Pagliai, G.; Dinu, M.; Madarena, M.P.; Bonaccio, M.; Iacoviello, L.; Sofi, F. Consumption of ultra-processed foods and health status: A systematic review and meta-analysis. *Br. J. Nutr.* **2021**, *125*, 308–318. [CrossRef]
5. Sho, H. History and characteristics of Okinawan longevity food. *Asia Pac. J. Clin. Nutr.* **2001**, *10*, 159–164. [CrossRef]
6. Sofi, F.; Abbate, R.; Gensini, G.F.; Casini, A. Accruing evidence on benefits of adherence to the Mediterranean diet on health: An updated systematic review and meta-analysis. *Am. J. Clin. Nutr.* **2010**, *92*, 1189–1196. [CrossRef]
7. Jung, S.J.; Park, S.H.; Choi, E.K.; Cha, Y.S.; Cho, B.H.; Kim, Y.G.; Kim, M.G.; Song, W.O.; Park, T.S.; Ko, J.K.; et al. Beneficial effects of Korean traditional diets in hypertensive and type 2 diabetic patients. *J. Med. Food* **2014**, *17*, 161–171. [CrossRef]
8. Dinu, M.; Pagliai, G.; Casini, A.; Sofi, F. Mediterranean diet and multiple health outcomes: An umbrella review of meta-analyses of observational studies and randomised trials. *Eur. J. Clin. Nutr.* **2018**, *72*, 30–43. [CrossRef]
9. Dinu, M.; Pagliai, G.; Angelino, D.; Rosi, A.; Dall'Asta, M.; Bresciani, L.; Ferraris, C.; Guglielmetti, M.; Godos, J.; Del Bo', C.; et al. Effects of Popular Diets on Anthropometric and Cardiometabolic Parameters: An Umbrella Review of Meta-Analyses of Randomized Controlled Trials. *Adv. Nutr.* **2020**, *11*, 815–833. [CrossRef]
10. Pereira-da-Silva, L.; Rêgo, C.; Pietrobelli, A. The Diet of Preschool Children in the Mediterranean Countries of the European Union: A Systematic Review. *Int. J. Environ. Res. Public Health* **2016**, *13*, 572. [CrossRef]
11. Da Rocha, B.R.S.; Rico-Campà, A.; Romanos-Nanclares, A. Adherence to Mediterranean diet is inversely associated with the consumption of ultra-processed foods among Spanish children: The SENDO project. *Public Health Nutr.* **2021**, *24*, 3294–3303. [CrossRef] [PubMed]
12. Sandoval-Insausti, H.; Jiménez-Onsurbe, M.; Donat-Vargas, C.; Rey-García, J.; Banegas, J.R.; Rodríguez-Artalejo, F.; Guallar-Castillón, P. Ultra-Processed Food Consumption Is Associated with Abdominal Obesity: A Prospective Cohort Study in Older Adults. *Nutrients* **2020**, *12*, 2368. [CrossRef] [PubMed]
13. Bonaccio, M.; Di Castelnuovo, A.; Costanzo, S.; De Curtis, A.; Persichillo, M.; Sofi, F.; Cerletti, C.; Donati, M.B.; de Gaetano, G.; Iacoviello, L. Ultra-processed food consumption is associated with increased risk of all-cause and cardiovascular mortality in the Moli-sani Study. *Am. J. Clin. Nutr.* **2021**, *113*, 446–455. [CrossRef]
14. Cordova, R.; Kliemann, N.; Huybrechts, I.; Rauber, F.; Vamos, E.P.; Levy, R.B.; Wagner, K.H.; Viallon, V.; Casagrande, C.; Nicolas, G.; et al. Consumption of ultra-processed foods associated with weight gain and obesity in adults: A multi-national cohort study. *Clin. Nutr.* **2021**, *40*, 5079–5088. [CrossRef]

15. Martinez-Perez, C.; San-Cristobal, R.; Guallar-Castillon, P.; Martínez-González, M.Á.; Salas-Salvadó, J.; Corella, D.; Castañer, O.; Martinez, J.A.; Alonso-Gómez, Á.M.; Wärnberg, J.; et al. Use of Different Food Classification Systems to Assess the Association between Ultra-Processed Food Consumption and Cardiometabolic Health in an Elderly Population with Metabolic Syndrome (PREDIMED-Plus Cohort). *Nutrients* **2021**, *13*, 2471. [CrossRef] [PubMed]
16. Marino, M.; Puppo, F.; Del Bo', C.; Vinelli, V.; Riso, P.; Porrini, M.; Martini, D. A Systematic Review of Worldwide Consumption of Ultra-Processed Foods: Findings and Criticisms. *Nutrients* **2021**, *13*, 2778. [CrossRef] [PubMed]
17. Available online: https://www.surveymonkey.com/ (accessed on 31 October 2021).
18. Dinu, M.; Bonaccio, M.; Martini, D.; Madarena, M.P.; Vitale, M.; Pagliai, G.; Esposito, S.; Ferraris, C.; Guglielmetti, M.; Rosi, A.; et al. Reproducibility and validity of a food-frequency questionnaire (NFFQ) to assess food consumption based on the NOVA classification in adults. *Int. J. Food Sci. Nutr.* **2021**, *72*, 861–869. [CrossRef] [PubMed]
19. Sofi, F.; Macchi, C.; Abbate, R.; Gensini, G.F.; Casini, A. Mediterranean diet and health status: An updated meta-analysis and a proposal for a literature-based adherence score. *Public Health Nutr.* **2014**, *17*, 2769–2782. [CrossRef]
20. Monteiro, C.A.; Levy, R.B.; Claro, R.M.; Castro, I.R.; Cannon, G. A new classification of foods based on the extent and purpose of their processing. *Cad. Saude Publica* **2010**, *26*, 2039–2049. [CrossRef]
21. Blanco-Rojo, R.; Sandoval-Insausti, H.; López-Garcia, E.; Graciani, A.; Ordovás, J.M.; Banegas, J.R.; Rodríguez-Artalejo, F.; Guallar-Castillón, P. Consumption of Ultra-Processed Foods and Mortality: A National Prospective Cohort in Spain. *Mayo Clin. Proc.* **2019**, *94*, 2178–2188. [CrossRef]
22. Julia, C.; Martinez, L.; Allès, B.; Touvier, M.; Hercberg, S.; Méjean, C.; Kesse-Guyot, E. Contribution of ultra-processed foods in the diet of adults from the French NutriNet-Santé study. *Public Health Nutr.* **2018**, *21*, 27–37. [CrossRef] [PubMed]
23. Rauber, F.; da Costa Louzada, M.L.; Steele, E.M.; Millett, C.; Monteiro, C.A.; Levy, R.B. Ultra-Processed Food Consumption and Chronic Non-Communicable Diseases-Related Dietary Nutrient Profile in the UK (2008–2014). *Nutrients* **2018**, *10*, 587. [CrossRef] [PubMed]
24. Martínez Steele, E.; Baraldi, L.G.; Louzada, M.L.; Moubarac, J.C.; Mozaffarian, D.; Monteiro, C.A. Ultra-processed foods and added sugars in the US diet: Evidence from a nationally representative cross-sectional study. *BMJ Open* **2016**, *6*, e009892. [CrossRef] [PubMed]
25. Popkin, B.M.; Barquera, S.; Corvalan, C.; Hofman, K.J.; Monteiro, C.; Ng, S.W.; Swart, E.C.; Taillie, L.S. Towards unified and impactful policies to reduce ultra-processed food consumption and promote healthier eating. *Lancet Diabetes Endocrinol.* **2021**, *9*, 462–470. [CrossRef]
26. Taillie, L.S.; Hall, M.G.; Popkin, B.M.; Ng, S.W.; Murukutla, N. Experimental Studies of Front-of-Package Nutrient Warning Labels on Sugar-Sweetened Beverages and Ultra-Processed Foods: A Scoping Review. *Nutrients* **2020**, *12*, 569. [CrossRef] [PubMed]

nutrients

MDPI

Article

Dietitians' Attitudes and Understanding of the Promotion of Grains, Whole Grains, and Ultra-Processed Foods

Natasha Krois [1], Jaimee Hughes [2] and Sara Grafenauer [3,*]

1 School of Medical, Indigenous and Health Sciences, University of Wollongong, Northfields Avenue, Wollongong 2522, Australia; natasha.krois1@gmail.com
2 Grains & Legumes Nutrition Council, 1 Rivett Rd, North Ryde 2113, Australia; jaimee.hughes2@gmail.com
3 School of Health Science, Faculty of Medicine and Health, University of New South Wales, Randwick 2052, Australia
* Correspondence: s.grafenauer@unsw.edu.au; Tel.: +61-2-9348-0864

Abstract: NOVA is a food-classification system based on four levels of processing, from minimally processed to ultra-processed foods (UPFs). Whole-grain-containing commercial breads and ready-to-eat breakfast cereals are considered ultra-processed within NOVA, despite being considered core foods in the Australian Dietary Guidelines. These food categories contribute the greatest quantities of whole grain in the Australian diet, although consumption is less than half of the 48 g/day daily target intake. Dietitians are key to disseminating messages about nutrition and health; therefore, an accurate understanding of whole grains and the effects of processing is critical to avoid the unnecessary exclusion of nutritionally beneficial foods. The aim was to utilise an online structured questionnaire to investigate dietitians' attitudes to the promotion of grains and whole grains and understand their level of knowledge about and attitudes towards NOVA and the classification of specific whole-grain foods. Whole-grain foods were perceived positively and are regularly promoted in dietetic practice (n = 150). The dietitians tended not to consider whole-grain breads and ready-to-eat breakfast cereals as excessively processed, although most generally agreed with the classification system based on the extent of processing. If dietitians intend to incorporate NOVA and concepts of UPFs in their counselling advice, the anomalies regarding the categorisation of whole-grain choices and optimum intakes should be addressed.

Keywords: whole grain; grains; ultra-processed; dietitian; education; NOVA

Citation: Krois, N.; Hughes, J.; Grafenauer, S. Dietitians' Attitudes and Understanding of the Promotion of Grains, Whole Grains, and Ultra-Processed Foods. *Nutrients* **2022**, *14*, 3026. https://doi.org/10.3390/nu14153026

Academic Editors: Monica Dinu and Daniela Martini

Received: 14 June 2022
Accepted: 21 July 2022
Published: 23 July 2022

1. Introduction

Consumers are faced with an ever-increasing array of products available to purchase, and there is considerable marketing of packaged and convenience foods. Most countries have developed food-based dietary guidelines to make it easier for consumers to choose healthier options. However, while it may be easier for consumers to judge the degree of healthfulness of single-ingredient food items, foods with multiple ingredients may pose a challenge and necessitate further interpretation [1]. There may be multiple mechanisms through which individuals make food choices; however, the dietitian's role in educating and supporting individuals, groups, and populations to make healthful food choices has become increasingly important [1]. The responsibility of a dietitian to translate nutrition science into dietary patterns extends across a spectrum of domains, from individualised advice and public health settings through to working alongside food industries. Understanding the determinants that influence dietitians' perceptions of foods and the impact this has on practice may be helpful in considering how anomalies in messaging around grain foods and processing might be managed, such as the classification of whole-grain breads and cereals as ultra-processed according to the NOVA food-classification system.

An increasingly widespread food-classification system termed NOVA was developed in Brazil by Monteiro et al. (2016) [2]. The system has been integrated into the Brazilian

Dietary Guidelines, and directs consumer choice by grouping food products into four main categories based on their degree of processing [3]. According to recent research, a higher consumption of ultra-processed foods is associated with a higher risk of developing chronic diseases, such as cardiovascular disease, type 2 diabetes, and obesity [4]. Monteiro et al. also argued that the classification of foods by their nutritional composition and origin—as in the Australian Dietary Guidelines (ADG) [5]—is a lesser indicator of the relationship between food and health than classification by the extent of food processing [6]. The four categories are as follows: (1) unprocessed and minimally processed foods, (2) processed culinary ingredients, (3) processed foods, and (4) ultra-processed foods (UPFs). NOVA researchers describe UPFs as, typically, energy-dense foods of poor nutritional quality that are low in dietary fibre and contain excessive amounts of sodium, simple sugars, and saturated and trans fats [7], and recommended their avoidance [3]. When NOVA and the ADGs are compared, many of their overall dietary messages are similar; however, discrepancies between the two systems exist [8]. An analysis of the 2011–12 National Nutrition and Physical Activity Survey (NNPAS) indicated that 23.5% of core foods (foods that should form the basis of a healthy diet [9]) were classified as ultra-processed, and 31.2% of discretionary foods were classified as not ultra-processed [8]. Notably, NOVA categorises core foods, such as mass-produced packaged breads and buns, as well as ready-to-eat breakfast cereals, as UPFs; however, discretionary choices, such as butter, cream, sugar, honey, homemade cakes, and biscuits were not classified within this category [8,10]. This is important, as breads, bread rolls, and ready-to-eat breakfast cereals are the largest contributors to Australian's whole-grain intake [11]. Therefore, to the advice to avoid these foods, as they are classified as UPFs, is likely to decrease whole-grain consumption. As a study by Monteiro et al. (2019) states: "Processes and ingredients used to manufacture ultra-processed foods are designed to create highly profitable (low-cost ingredients, long shelf-life, emphatic branding), convenient (ready-to-consume), hyper-palatable products liable to displace all other NOVA food groups, notably unprocessed or minimally processed foods" [12]. However, this description does not consider the nutrient content of the foods, or the related health-outcome research, and implies that convenient and packaged food products are automatically classified as ultra-processed without further examination through translational research.

Evidence suggests that consumers' ability to accurately interpret nutrition information is poor, particularly in relation to bread choices [13]. This accentuates the responsibility of dietitians, particularly in public-health and food-industry roles, to advocate for products to be clearly labelled, and for nutrition information provided to be unambiguous. Dietitians also have a responsibility to distinguish misconceptions from evidence-based nutrition and assist consumers in obtaining reliable information to inform their dietary decisions. Those working in the broader fields of public health or industry also have a responsibility to advocate for the distribution of health messages and resources that provide accurate and reliable information. If dietitians are to be viewed as informed and credible sources, then they are required to have a sound understanding of nutrients, foods, and a range of dietary concepts, which includes an understanding of food processing, its effects on health, and the established links with disease for individuals, sub-groups, and populations. Dietitians can also provide insight into the barriers that consumers may face in implementing dietary advice; thus, research on dietitians' understanding is a valuable method to inform health-promotion strategies and educate health professionals to provide support accordingly.

The Theory of Planned Behaviour (TPB) is a behavioural model that can be applied to understand and predict an individual's actions and intentions. In relation to this study, the TPB suggests that the intent of a dietitian to promote and/or exclude grains, whole grains, and the NOVA food classification system is influenced by personal attitudes, subjective normative beliefs, and perceived behavioural control [14]. Subjective normative beliefs are a key constituent of the TPB, implying that a dietitian's intent to promote whole grains may be influenced by the perception of whole-grain promotion by other dietitians and behaviours that they take to be normative [15]. A study conducted by Chase et al. (2003), prior to the release of NOVA, identified that subjective normative beliefs were the greatest predictors of dietitians' intent to promote whole-grain foods, since they were 11.9 times more important than attitudes and 2.3 times more important than perceived behavioural control [14]. Therefore, the aim of this study was to utilise an online structured questionnaire to investigate dietitians' attitudes to the promotion of grains and whole grains and understand the extent of dieticians' knowledge of and attitudes towards NOVA and the classification of specific whole-grain foods. A further aim was to examine dietitians' familiarity with the NOVA food-classification system and their knowledge and attitudes towards the classification of specific foods within this system, particularly those that contain whole grains.

2. Materials and Methods

2.1. Survey Design

The Qualtrics XM Platform ™ (Provo, UT, USA) was utilised to distribute an anonymous online structured questionnaire [16] targeting dietitians from a number of countries, including Australia, New Zealand (NZ), Canada, the United States (US), United Kingdom (UK), and South Africa between April 2021 and July 2021. Survey questions were pilot-tested in consultation with stakeholders, including dietitians, to test construct and content validity. The final 10 min survey consisted of 41 questions and utilised an open and closed questionnaire design with free text and multiple-choice responses, Likert scales, matrix, and rank-order questions (Supplementary Materials File S1). The survey was divided into four parts: (1) demographic questions, including age, gender, level of education, self-reported dietetic credentials, country where dietetics education program was completed, and main area of dietetic practice (2, 3) advice that survey respondents might provide in one-on-one consultations, in group sessions, or via media about grain foods and whole-grain foods (4), and questions related to the NOVA food-classification system (Supplementary Materials File S2). The TPB was incorporated into the survey design and was utilised to propose questions aimed at drawing out the key factors that may influence dietitians' perceptions and the integration of grains, whole grains and/or ultra-processing in their normal practice advice (Supplementary Materials File S2). Questions were assigned to one or more of the three constituents of the TPB—attitudes, subjective normative beliefs, and perceived behavioural control—to systematically investigate dietitians' understanding and perceptions of and attitudes to grains, processing, and the NOVA food-classification system, how these were influenced, and how this translated into practice (Supplementary Materials File S2). The attitudes were investigated in questions regarding the promotion of grains, including sources, which were classified as ultra processed, specifically aiming to explore whether any particular grain foods were considered excessively processed, as well as underlying reasons for limiting grain consumption. Subjective normative beliefs were investigated in questions exploring participants' perceptions of the understanding and prioritisation of whole grains by other dietitian colleagues. Perceived behavioural control was investigated in questions exploring participant confidence and the frequency with which advice about whole grains and NOVA in practice was provided. Ethical approval for this study was obtained from the University of Wollongong Human Research Ethics Committee (HREC), approval number 2021/038.

2.2. Participants

This study employed voluntary-response sampling, including a combination of convenience, snowballing, and purposive sampling, as it is a relatively fast and inexpensive means of response collection. Dietitians within several countries—Australia, NZ, Canada, US, UK, and South Africa—were recruited to participate in the study. The countries were chosen based on similar dietetic education systems and, to some extent, similar dietary patterns. Participant recruitment was achieved in several ways, as follows: online advertisement on Grains & Legumes Nutrition Council (GLNC); social media platforms, including LinkedIn, Facebook, Twitter, and Instagram; GLNC and Dietitians Australia (DA); and e-newsletters, including those published by and e-newsletters international dietetic networks, such as the organisations, Oldways (US), Dietitian Connection (Australia), and Education in Nutrition (Australia). To encourage participation, eligible participants were offered the opportunity to enter their email in a separate survey to be placed into a draw to win one of six Portion Perfection gift cards at a value of AUD 100 for professional materials. Inclusion criteria required participants to be over the age of eighteen, a registered dietitian (RD), or an accredited practising dietitian (APD) with access to the internet and online technologies, such as a computer or smart phone, to undertake the questionnaire.

2.3. Response Analysis

Participant response data were exported from Qualtrics (Provo, UT, USA) to a Microsoft Excel™ spreadsheet (Version 16.53, Washington, DC, USA) for data analysis. Using Microsoft Excel™ (Version 16.53, Washington, DC, USA), descriptive statistics were applied for analysis of quantitative data. Content analysis was conducted for qualitative data from free-text responses, in which responses were assigned to reoccurring themes by the researcher (NK). A summary report of free-text responses was also generated by Qualtrics (Provo, UT, USA) to assist with the qualitative data analysis.

3. Results

A total of 199 respondents attempted the survey, of which 123 completed the survey in full and 76 provided partial responses. However, of the 199 respondents, three participants did not meet the inclusion criteria, as they did not hold appropriate dietetics credentials, and 46 participants did not respond to the questions beyond those in the demographic characteristics, which left a total 150 responses included in the analysis (122 full and 28 partial responses).

3.1. Demographics

The number of participants who had completed their dietetics education in Australia was far greater than the number of participants who completed education elsewhere, accounting for 68% of the total number of participants (Table 1). Therefore, we compared Australia to other countries. However, as the initial analysis showed minimal variation in the responses, no further analysis by country was undertaken. Similarly, the analysis of the responses by age category indicated minimal variation between the groups, and in limited instances of dissimilarity, large disparities in the number of participants between age categories were evident, such as 25–34 years (n = 51) versus 65+ years (n = 8), limiting the value of the comparisons. The number of participants in each age category was compared to the age profiles of the current Dietitians Australia members, identifying the greatest number of dietitians in both instances to be in the 25–34-years category.

Table 1. Demographic characteristics of participants.

Demographic Characteristics	Count (%)
Gender	
Female	142 (94.7)
Male	6 (4.0)
Prefer not to answer	2 (1.3)
Prefer to self-describe	0 (0)
Age	
18–24	19 (12.7)
25–34	51 (34.0)
35–44	31 (20.7)
45–54	25 (16.7)
55–64	16 (10.7)
65+	8 (5.3)

Table 1. *Cont.*

Demographic Characteristics	Count (%)
Level of Education	
Certificate/diploma	3 (2.0)
Bachelor degree	64 (42.7)
Masters degree	77 (51.3)
Ph.D.	6 (4.0)
Dietetic credential	
Accredited practicing dietitian	110 (73.3)
Registered dietitian	35 (23.3)
Qualified dietitian but not registered	3 (2.0)
Other	2 (1.3)
Country dietetics education was completed	
Australia	102 (68.0)
New Zealand	5 (3.3)
Canada	3 (2.0)
United States	33 (22.0)
United Kingdom	1 (0.7)
South Africa	4 (2.7)
Other	2 (1.3)
Country currently practising as a dietitian	
Australia	110 (73.3)
New Zealand	2 (1.3)
Canada	0 (0.0)
United States	30 (20.0)
United Kingdom	2 (1.3)
South Africa	2 (1.3)
Other	4 (2.7)
Years practised as a dietitian	
≤5 years	50 (33.3)
6–10 years	38 (25.3)
11–20 years	24 (16.0)
>20 years	38 (25.3)
Main area of work	
Community/public health	31 (20.7)
Food service	3 (2.0)
Academia/education	8 (5.3)
Research	4 (2.7)
Clinical (hospital)	32 (21.3)
Clinical (primary care)	13 (8.7)
Private practice	38 (25.3)
Corporate nutrition	4 (2.7)
Food industry	3 (2.0)
Retail	3 (2.0)
Other	11 (7.3)

3.2. *Perceived Value, Attitudes, and Recommendations of Grains, including Whole Grains*

Grain foods, specifically whole-grain varieties, were perceived positively by the dietitians, and are regularly promoted in advice (Table 2). The participants frequently recommended whole grains (134/150) and high-fibre grains (114/150). This is in line with the fact that most of the participants (102/150) encouraged the consumption of grain foods based on national dietary guidelines, such as the Australian Dietary Guidelines (Table 2).

Table 2. Dietitian recommendations, sources and frequency of advice related to grains and whole grains.

Question	Response	Count (%)
Do you recommend or discuss grain foods in consultation, groups sessions or via media messages?	Yes No	149 (99.3) 1 (0.7)
Are grain foods prioritised in your advice for general healthy eating?	Yes No Other	116 (77.3) 19 (12.7) 15 (10.0)
Do you promote amounts of grain foods based on National Dietary Guidelines?	Yes No	121 (80.7) 29 (19.3)
Do you recommend whole grain foods?	Yes No	148 (99.3) 1 (0.7)
Considering your advice on general healthy eating, how often do you recommend whole grain foods in dietetic practice?	Always Most of the time About half of the time Sometimes Never	74 (49.7) 63 (42.3) 6 (4.0) 6 (4.0) 0 (0.0)
What sources of information do you most often use for your advice relating to whole grain food intake? *	National Dietary Guidelines Government Resources Resources from professional organisations Resources from non-government organisations Other	102 (68.5) 26 (17.5) 82 (55.0) 44 (29.5) 11 (7.4)

* Question allowed respondents to select more than one answer; consequently, values presented are the proportion of respondents selecting each point.

On the other hand, some dietitians specifically stated they did not recommend refined or non-whole -grain foods (57/150). When recommending whole-grain foods in practice, the dietitians most commonly used a specific suggestion (e.g., swap refined ready-to-eat cereal for oats; swap white for wholemeal/whole-grain bread), promoting the substitution of refined grains with whole grain varieties. Notably, the specific grain-based foods most frequently recommended by the dietitians were bread (132/150) and breakfast cereal (115/150).

Generally, the dietitians were aware of the benefits of whole-grain consumption; however, some of these benefits were poorly recognised (Table 3). Almost all the dietitians were familiar with the benefits associated with the high fibre content of whole grains (144/149), as well as their benefits for blood-glucose control (133/149). However, some of the benefits of whole grains for cardiovascular health were more poorly recognised in comparison (Table 3). The prioritisation of whole grains for weight control (81/149) and their potential for reducing inflammation (62/149) were also less well known. The participants recognised suitable contraindications of whole grains (e.g., gluten containing grains in coeliac disease) in some instances (87/149).

When the participants were asked to identify whole-grain sources by responding 'yes', 'no', or 'unsure' to a list of whole-grain- and non-whole-grain-containing foods, the majority of the responses showed a good knowledge of whole grains (Figure 1). However, the categorization of some foods showed a poor understanding; for example, some dietitians (39/124) presumed that quick cook-oats would not be whole-grain and that multi-grain bread and wheat-bran cereal were whole-grain (95/124 and 98/124, respectively). The dietitians generally recognised refined foods as non-whole-grain, such as white bread (116/124).

Table 3. Dietitians' understanding of the benefits of whole grains and situations in which they should be prioritised.

Question	Response	Count (%)
In your opinion, what are the nutrition and health benefits of whole grain foods? *	High fibre	144 (96.6)
	Low GI	120 (80.5)
	Improves weight control	104 (69.8)
	Improves blood glucose control	133 (89.3)
	Reduces insulin resistance	78 (52.4)
	Increases HDL-cholesterol	40 (26.9)
	Decreases LDL-cholesterol	95 (63.8)
	Lowers blood pressure	51 (34.2)
	Reduces inflammation	62 (41.6)
	Reduces risk of heart disease	102 (68.5)
	Management and reduced risk of type 2 diabetes	112 (75.2)
	Protective factor against colorectal cancer	119 (79.9)
	Other	18 (12.1)
Typically, in what situations might you prioritise whole grain foods in dietetic practice? *	I do not prioritise whole-grain foods in practice	1 (0.7)
	In general dietary advice	112 (75.2)
	To increase dietary fibre intake	130 (87.3)
	For weight control	81 (54.4)
	For diabetes management	106 (71.1)
	For blood-glucose control	105 (70.5)
	For cholesterol management	99 (66.4)
	For blood-pressure management	43 (29.9)
	Other	8 (5.4)
Are there any reasons why you would not recommend whole grain foods to a patient/client/group?	Contraindicated	87 (56.9)
	No reasons	46 (30.1)
	Individual taste preferences	10 (6.5)
	Low-carbohydrate diet	5 (3.3)
	Other client priorities	2 (1.3)
	Weight loss	2 (1.3)

* Question allowed respondents to select more than one answer; consequently values presented are the proportion of respondents selecting each point.

The majority of the dietitians were confident in providing whole-grain education (with 137/149 scoring 4 or 5, with 5 being the most confident) and agreed (130/142 claimed to agree or strongly agree) that dietitians are well educated about the importance and benefits associated with whole-grain-food consumption. In a question assessing the perception of whole-grain promotion by other dietitians, most of the dietitians (90/142) believed that other dietitians regularly promote and prioritise the intake of whole-grain foods, and 39/142 believe this occurs 'somewhat'. The dietitians suggested that whole-grain education for dietitians could be improved by 'better resources for clients' (102/142), 'CPD/online learning' (90/142), 'better resources from dietary guidelines/national policy' (70/142), and 'marketing campaigns' (41/142).

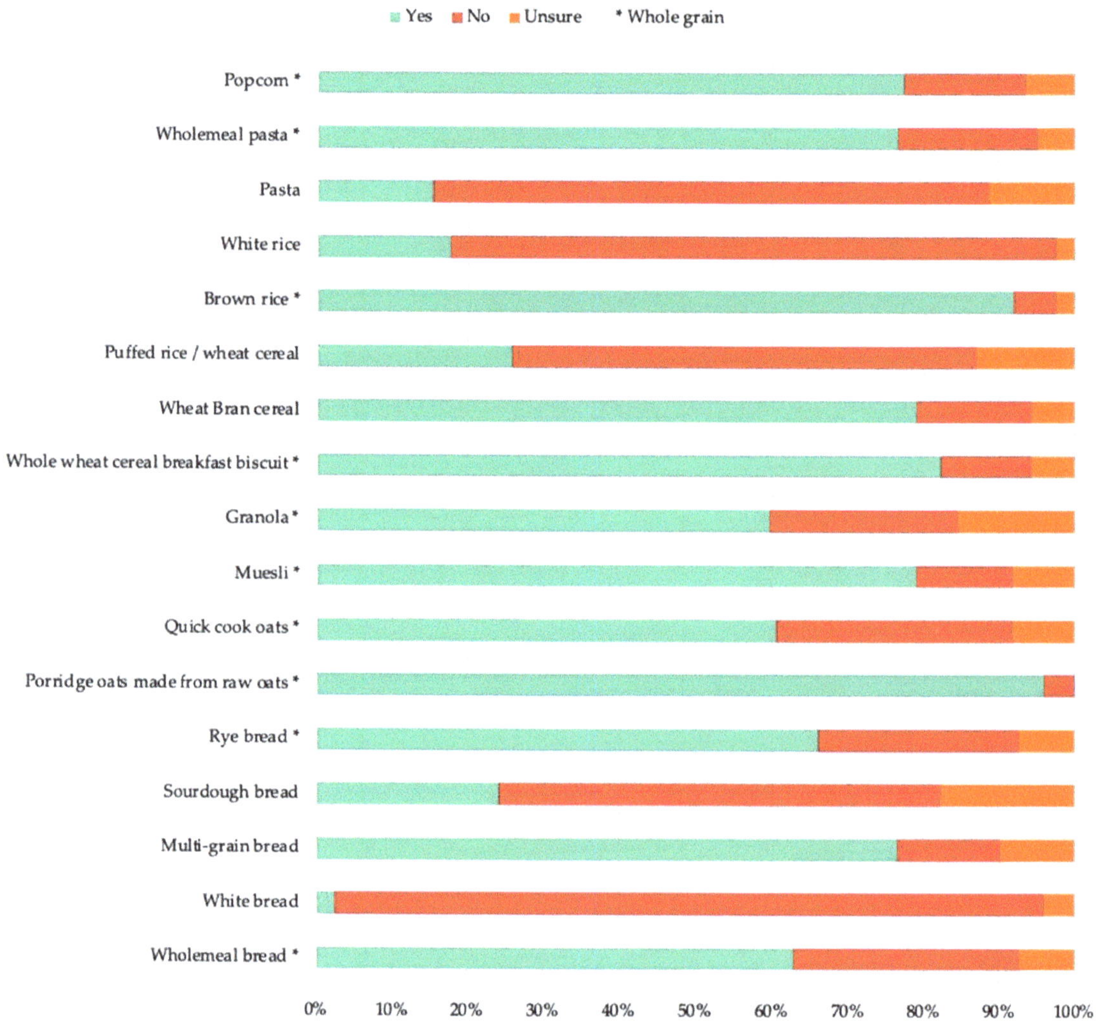

Figure 1. Identification of whole-grain- and non-whole-grain-containing foods by dietitians. *: indicates that the food is whole-grain.

3.3. *Perceived Barriers to Whole-Grain Consumption*

The dietitians perceived there to be several key barriers to whole-grain intake. Alongside this, the majority were not confident that the public is well educated on the importance and benefits associated with whole-grain-food consumption (86/142). Some were ambivalent (43/142), and only thirteen thought the public was well educated. The barriers to whole-grain consumption cited by the dietitians frequently related to 'taste' (83/116) and 'concerns about carbohydrate intake' (66/116) (Table 4). When asked which strategies could help to overcome the barriers to whole-grain consumption, several themes were identified across the responses, which were most commonly related to education (86/116) (Table 4). The participants were asked to rank the strategies they had previously used to improve whole-grain intake (on a scale of 1–6, with 1 being most effective and 6 being the least effective) (Figure 2). Promotion via the media and the promotion of the health benefits were perceived as the two most effective strategies. Improved front-of-pack scoring systems and changes to dietary guidelines were ranked as the least effective.

Table 4. Perceived barriers to whole-grain intake, and suggested strategies to overcome them.

Question	Response	Count (%)
In your opinion, what are the barriers to whole grain food consumption? *	Taste	83 (71.6)
	Concerns about carbohydrate intake	66 (56.9)
	Culinary skills (e.g., easy recipes)	65 (56.0)
	Time taken to prepare	52 (44.8)
	Price	46 (39.7)
	Other (please specify)	37 (31.9)
	Availability	26 (22.4)
	There are no barriers	2 (1.7)
What strategies could help overcome the barriers to whole grain consumption?	Education	86 (51.5)
	Public-health-promotion messages	31 (18.6)
	Food-industry action	24 (14.4)
	Improving individual acceptability	22 (13.2)
	Using evidence-based practice	2 (1.2)
	Not sure	2 (1.2)

* Question allowed respondents to select more than one answer; consequently, values presented are the proportion of respondents selecting each point.

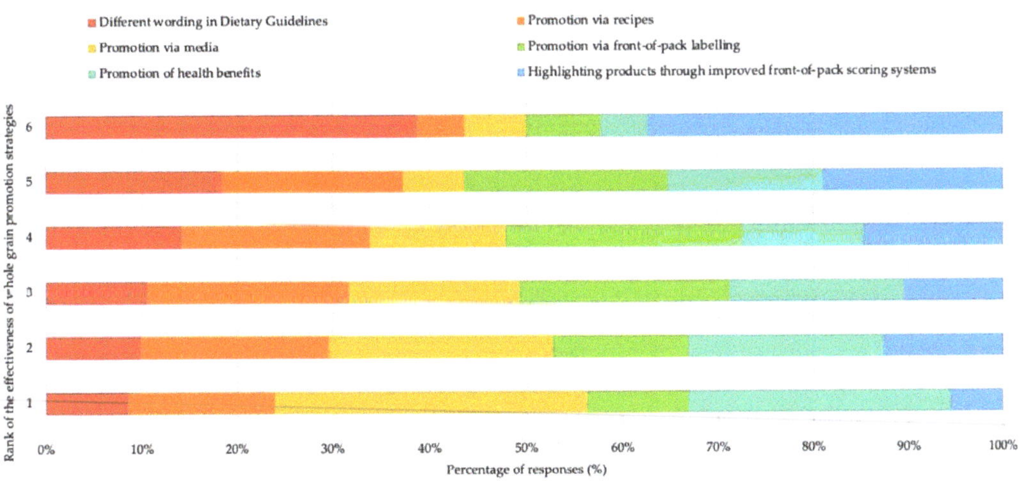

Figure 2. Dietitians' responses ranking the effectiveness of previously used whole-grain-promotion strategies on a scale of 1–6 (1 being most effective, 6 being least effective).

3.4. Knowledge, Attitudes and Use of the NOVA Classification System

Many of the dietitians (75/124) were unfamiliar with the NOVA classification system, although the majority (116/124) were familiar with the advice to limit the intake of highly processed/ultra-processed foods. When asked if they incorporated and/or referred to NOVA or the processing of food in practice, the results were mixed, with 18/124 dietitians reporting that they ''always' referred to NOVA, 29/124 reporting that they referred to NOVA 'most of the time', 9/124 claiming to do so 'about half of the time', 18/124 claiming to do so 'sometimes', and 50/124 claiming to never refer to NOVA in practice. Furthermore, many of the dietitians (75/124) agreed (agree or strongly agree) with NOVA's classification of foods, while 45/124 were ambivalent and 4/124 disagreed. The participants were asked to elaborate on why they responded in the manner that they did, and a content analysis was undertaken to categorise the responses. Many of the reported reasons related to unfamiliarity with NOVA (45/124), but some also related to support for a food-classification system based on processing (32/124). Moreover, when dietitians were asked to identify which grain foods are classified as ultra-processed, most dietitians considered whole grain foods to not fit into this classification, whereas their refined counterparts were generally considered to be ultra-processed (Figure 3). For example, wholemeal bread and white bread were considered as UPFs by 10/124 and 73/124, respectively.

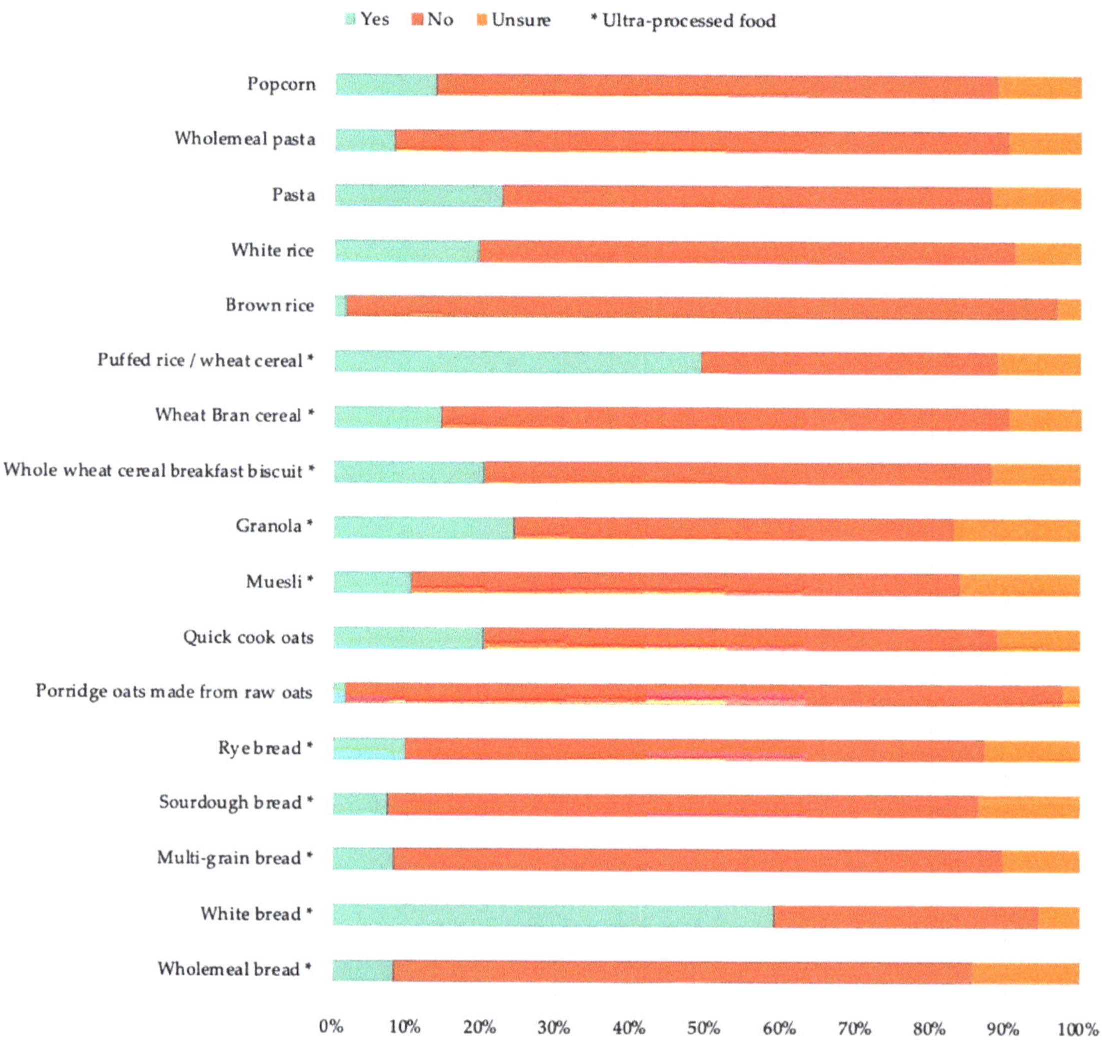

Figure 3. Identification of ultra-processed (NOVA classifications) and non-ultra-processed grain foods by dietitians. A star indicates that a food is classified as ultra-processed.

The dietitians' attitudes to and perceptions of whole grains, NOVA, and UPFs were further explored in a matrix-style question prompting the participants to indicate the extent to which they agreed or disagreed with the listed statements (Table 5). The majority (92/123) of the dietitians agreed (agreeing or strongly agreeing) that UPFs should generally be avoided; however, the whole-grain breads and cereals should not be included in this classification. Furthermore, 47/123 of the dietitians agreed (agreeing or strongly agreeing) that they were less inclined to recommend the avoidance of UPFs, knowing that this classification may include whole-grain breads and cereals. On the other hand, some dietitians (30/123) agreed (agreeing or strongly agreeing) that knowledge of the classification of these grain foods as ultra-processed had negatively affected their perception of these sources of whole grains, and 16/123 agreed (agreeing or strongly agreeing) that they were less inclined to recommend these foods in practice.

Table 5. Dietitians' attitudes to and perceptions of whole grains, NOVA, and UPFs.

Statement	Count (%)			
	Disagree	Neither Agree-Nor Disagree	Agree	Not Sure
I agree with the classification in NOVA for breads as "ultra-processed foods" if they are packaged and fortified.	75 (61.0)	16 (13.0)	29 (23.6)	3 (2.4)
I agree with the classification in NOVA for ready-to-eat breakfast cereals as "ultra-processed foods" even if they are fortified.	62 (50.4)	23 (18.7)	34 (27.6)	4 (3.3)
Knowing that some whole grain breads and ready-to-eat cereals are classified as "ultra-processed foods" has negatively impacted my perception of these sources of whole grains.	63 (51.2)	28 (22.8)	30 (24.4)	2 (1.6)
I am less inclined to recommend whole grain breads and ready-to-eat cereals in dietetic practice knowing that they are classified as "ultra-processed foods".	83 (67.5)	21 (17.1)	16 (13.0)	3 (2.4)
I am less inclined to recommend avoidance of ultra-processed foods knowing that they may include some whole grain foods (such as some ready-to-eat cereals and breads).	43 (35.0)	29 (23.6)	47 (38.2)	4 (3.3)
I generally agree to avoid ultra-processed foods but do not agree that whole grain breads and cereals should be included in this classification.	10 (8.1)	16 (13.0)	92 (74.8)	5 (4.1)

4. Discussion

Grains, especially whole grains, were perceived positively by the dietitians and were regularly promoted in their advice. Encouragingly, almost all the dietitians recommended grain foods in consultation, group sessions and/or through media messages, and, more specifically, recommended whole-grain varieties over refined-grain options. One quarter of the dietitians did not prioritise grain foods in advice for general healthy eating, suggesting that while the majority of the dietitians were aware of the role of grains in a healthy dietary pattern, others may have prioritised other foods or food groups in their practice. A study by Chase et al. (2003) reported that dietitians perceive the prioritisation of other dietary changes, a lack of time, and a lack of resources to use with clients to be barriers to whole-grain promotion [14]. Furthermore, the majority of the dietitians participating were confident in their provision of whole-grain education (137/149) and agreed that dietitians were well educated regarding the importance and benefits associated with whole-grain-food consumption. This was positive, as the TPB implied that an individual's perceived behavioural control, including feelings of self-efficacy, influence the dietitian's intent to promote whole grains [17]. As confidence increases, whole-grain promotion is more likely to occur. The dietitians advised that whole-grain education could be improved for dietitians through better resources for clients, CPD/online learning, and better resources from dietary guidelines/national policy, indicating that the dietitians may have perceived the current quality of these resources to be insufficient for optimal learning. Future education strategies for dietitians may benefit from targeting these areas.

The dietitians tended to be aware of and refer to national dietary guidelines (NDG) to inform whole-grain advice. NDGs are important educational tools, translating the complex matrix of nutrition science into simple messages that enable consumers to make positive food choices and form healthy dietary patterns [18]. Country-specific guidelines exist worldwide, incorporating positive messages about grains, especially whole grains. For example, the US Dietary Guidelines recommend that 'at least half of total grains should be whole grains', while the UK Dietary Guidelines recommend choosing 'whole grain or higher fibre versions with less added fat, salt and sugar' and, similarly, the ADG recommend choosing 'mostly whole grain and/or high cereal fibre varieties' [5,19,20]. These results indicated that NDGs are meaningful resources that influence dietitians' understanding and recommendations in practice. Interestingly, the participants did not report that any other wording in the dietary guidelines would specifically further encourage consumer intake of whole grains and appeared satisfied with the current guidelines.

Generally, the dietitians displayed a good level of knowledge when identifying whole grains; however, some items were less well understood. For example, many of the dietitians were able to correctly identify cereals such as raw oats and muesli as whole grains, although some incorrectly perceived that quick cooking may be a factor that differentiates whole-grain content. This suggests that dietitians may perceive whole grains that have undergone any processing, including the addition of other ingredients, to have inferior properties. It is true that processing may mean a product is not whole grain, but when these components (bran, germ, and endosperm) are included in the same proportion as the original grain, this constitutes a whole grain, and includes wholemeal. In this study, the dietitians appeared to be less informed of the nuanced information relating to what defines a whole grain or whole-grain food. Correspondingly, a study by Botelho et al. (2018) concluded that the nutritional quality of a product is associated with the product formulation, rather than the degree of processing [21]. Furthermore, many dietitians incorrectly presumed that wheat-bran cereal and multi-grain bread were whole grains. However, it is also notable that bran cereals are high in cereal fibre, one of the elements of whole grains that confer the most health effects [22]. Comparatively, a study by Chase et al. (2003) reported that dietitians' ability to identify whole-grain products were low, with only 60% of dietitians correctly identifying whole-grain products according to a corresponding sample food label [14]. It is clear that the dietitians in this survey appeared to have a greater knowledge of what constitutes a whole-grain food, indicating that whole-grain education for dietitians may have improved over time.

Although more than half of the dietitians in this survey were unfamiliar with NOVA, the majority agreed with NOVA's classification of foods based on the brief statement that the foods in this system are classified based on the extent of their processing. A study by Sadler et al. (2022) highlighted the ambiguities associated with concepts of processing, (ultra) processed food, and healthy food [23]. For example, an exploration of health professionals' views of processed foods identified tensions between the concept of 'food processing', which was recognised as necessary for food sustainability and security, and that of 'processed foods' which was perceived as less healthy or natural [23]. Whilst the participants agreed that these are broad concepts that require differentiation, it was highlighted that consumers tend to view foods through a dichotomous classification of 'good' or 'bad', and that when they refer to so-called 'processed foods', they do not refer to the degree to which a food is processed but, rather, to their perception of its healthfulness [23]. Unsurprisingly, the dietitians generally agreed that the more a food is processed, the less favourable it may be for consumption. In line with this, a recent study demonstrated that a higher percentage of energy from UPFs was inversely associated with diet quality when applied to the ADG features of a healthy diet (for example, enjoying a variety of foods from each of the core food groups) [24]. Consequently, it is likely that as UPF consumption increases and whole-grain consumption decreases, reinforcing the divergence from concepts of healthy eating, as defined by the ADG.

Although the dietitians were familiar with the advice to limit the intake of highly processed/UPFs, it appeared that they demonstrated a knowledge gap in the identification of whole-grain foods that are classified as ultra-processed. For example, the dietitians in the survey were mostly unaware that commercial breads, including whole-grain varieties and ready-to-eat breakfast cereals, were included in the classification of ultra-processed. This may have been the result of a predilection towards the NOVA food-classification system, potentially overlooking the anomalies between the benefits of whole-grain consumption and the processing characteristics defined by NOVA. Generally, the dietitians associated UPFs with discretionary choices, such as those containing nutrients unfavourable to health, including added sugars, salt, or saturated fat Where dietitians perceived grain foods may apply to this category, these tended to be refined varieties, for example white bread, highlighting that this may be a differentiating factor in the participant's identification of UPFs. Whole-grain foods were often considered to not be ultra-processed; for example, wholemeal bread and muesli were incorrectly identified by the majority of the dietitians as not ultra-processed, suggesting that dietitians do not associate whole grains with UPFs. Research has demonstrated that replacing refined grains with whole grains reduces the risk of cardiometabolic disease [25]. Research has also found nutrient-dense UPFs, including whole-grain breads and ready-to-eat breakfast cereals, to contribute to greater dietary quality [26,27], and that the message that they are to be 'avoided' may not be appropriate.

While it was encouraging that the majority of the dietitians remained positive towards sources of whole grains, worryingly, a small number (16/123) became dubious about advising their consumption, when alerted to the potential for these to be classified as ultra-processed. Further education may be necessary to guide dietitians in promoting the avoidance of UPFs, to ensure that they are well informed on the whole-grain-food sources included in this classification and the consequences of their avoidance. A study assessing the factors that influence Australian dietitians' perceptions of packaged foods (n = 117) reported that alongside ingredients and nutrition composition, the shelf-life and storage of packaged food items were indicators of product healthfulness, taking into consideration the potential of a food item to cause food-borne illness [1]. This is notable given that the addition of additives, which enhance shelf stability, results in the classification of ultra-processed according to the NOVA food classification system, despite the benefits of additives for food safety. It is essential that dietitians recognise the factors that influence individual and population eating practices, such as accessibility, time, income, and skills [28]. For example, NOVA classifies commercial breads as ultra-processed, but not homemade breads. However, it is questionable whether it is feasible for individuals to make their own bread, and whether doing so provides worthwhile nutritional and health benefits compared with purchasing a commercial loaf. The findings of this study provide the basis for the reclassification of whole-grain breads and ready-to-eat breakfast cereals within NOVA, utilising evidence-based health-outcome data to reevaluate the classification rules.

Limitations

The findings of this survey should be considered in light of its limitations. This study used voluntary response sampling rather than randomised sampling; therefore, it is unlikely that the results are representative of the dietetic profession as a whole. It is also possible that the distribution of the survey link via the GLNC social media channels attracted nutrition professionals that have an interest in or prior knowledge of grains, whole grains, and the anomalies of the NOVA classification system in relation to whole-grain foods, as previously mentioned. It is known that some individuals may be inherently more likely to participate than others; for example, those who respond may have stronger opinions or be more invested in the subject matter than those who do not, further limiting the generalisability of the results to the profession as a whole. The small number of responses from the participants from selected countries outside of Australia limited the ability to conduct a statistical analysis to compare differences between countries.

5. Conclusions

Although the findings are not representative of the entire dietetic profession, whole-grain foods were perceived positively by the dietitians who were included in the survey, who reported regularly promoting them in their practice. The dietitians appeared confident in their ability to provide whole-grain education and believed they were well educated about whole grains. This is important, as dietitians are key health professionals in the dissemination of messages about nutrition and health and the provision of specific food advice. The dietitians generally had a good ability to identify whole grains and their health benefits, although there were some gaps in this knowledge and some errors. Further education may be required in these areas. The dietitians acknowledged that the public may not be well educated as to the benefits of whole grains, and they perceived several barriers to whole-grain consumption. The dietitians believed that improving knowledge of the health benefits of whole grains and their promotion via the media have been the most effective methods for their promotion, indicating approaches for future promotional strategies.

The dietitians tended to use NDG to inform their whole-grain advice. However, if NOVA were to be more widely promoted, given that it classifies whole-grain breads and ready-to-eat breakfast cereals as ultra-processed, these valuable sources of whole grain may be discouraged. However, the majority of the dietitians did not consider these sources as ultra-processed and remained positive towards recommending these foods in dietary advice. Rather, the dietitians tended to associate UPFs with discretionary foods and foods containing nutrients that negatively affect health, although they generally considered a system based on the degree of processing as an effective way to categorise foods. If dietitians are to refer to and incorporate NOVA and concepts of UPFs in advice, the anomalies between messages promoting the avoidance of UPFs and messages promoting whole-grain intake must be addressed.

Supplementary Materials: The following supporting information can be downloaded at: https://www.mdpi.com/xxx/s1, Supplementary Materials File S1: Survey Questions; Supplementary Materials File S2: Application of survey questions relative to the constituents of the Theory of Planned Behaviour.

Author Contributions: Conceptualization, S.G.; formal analysis, N.K.; writing—original draft preparation, N.K.; writing—review and editing, J.H. and S.G. All authors have read and agreed to the published version of the manuscript.

Funding: This research received no external funding but the publication costs were supported by the Grains & Legumes Nutrition Council, not-for-profit charity.

Institutional Review Board Statement: The study was conducted in accordance with the Declaration of Helsinki, and approved by the Ethics Committee of University of Wollongong (protocol code 2021/038).

Informed Consent Statement: Informed consent was obtained from all subjects involved in the study.

Data Availability Statement: Not applicable.

Acknowledgments: The authors acknowledge the Honours research supervision of NK by Eleanor Beck, including the original data analysis and review of this work as part of an Honours thesis.

Conflicts of Interest: J.H. and S.G. were employed by the Grains & Legumes Nutrition Council, a not-for-profit charity, during the period of time when this research was planned and conducted; N.K. declares no conflict of interest.

References

1. Thurecht, R.L.; Pelly, F.E.; Cooper, S.L. The influence of current food and nutrition trends on dietitians' perceptions of the healthiness of packaged food. *Public Health Nutr.* **2020**, *23*, 2124–2131. [CrossRef] [PubMed]
2. Monteiro, C.A.; Cannon, G.; Levy, R.; Moubarac, C.; Jaime, P.; Martins, A.P.; Canella, D.; Louzada, M.; Parra, D. NOVA. The star shines bright. *World Nutr.* **2016**, *7*, 28–38.
3. Dietary Guidelines for the Brazilian Population Ministry of Health of Brazil 2015. Available online: http://bvsms.saude.gov.br/bvs/publicacoes/dietary_guidelines_brazilian_population.pdf (accessed on 11 February 2021).
4. Elizabeth, L.; Machado, P.; Zinöcker, M.; Baker, P.; Lawrence, M. Ultra-processed foods and health outcomes: A narrative review. *Nutrients* **2020**, *12*, 1955. [CrossRef] [PubMed]
5. National Health and Medical Research Council. *Australian Dietary Guidelines*; National Health and Medical Research Council: Canberra, Australia, 2013.
6. Monteiro, C.A.; Cannon, G.; Moubarac, J.-C.; Levy, R.B.; Louzada, M.L.C.; Jaime, P.C. The UN Decade of Nutrition, the NOVA food classification and the trouble with ultra-processing. *Public Health Nutr.* **2018**, *21*, 5–17. [CrossRef] [PubMed]
7. Monteiro, C.A. Nutrition and health. The issue is not food, nor nutrients, so much as processing. *Public Health Nutr.* **2009**, *12*, 729–731. [CrossRef] [PubMed]
8. Mackerras, D. Mis-match between the healthy food and the ultra-processed food classifications in Australia (OR14-03-19). *Curr. Dev. Nutr.* **2019**, *3* (Suppl. 1), nzz038.OR14-03-19. [CrossRef]
9. National Health and Medical Research Council. *Eat For Health Educator Guide*; National Health and Medical Research Council: Canberra, Australia, 2013.
10. Monteiro, C.A.; Cannon, G.; Lawrence, M.; Costa Louzada, M.L.; Pereira Machado, P. *Ultra-Processed Foods, Diet Quality, and Health Using the NOVA Classification System*; FAO: Rome, Italy, 2019.
11. Galea, L.M.; Beck, E.J.; Probst, Y.C.; Cashman, C.J. Whole grain intake of Australians estimated from a cross-sectional analysis of dietary intake data from the 2011–13 Australian Health Survey. *Public Health Nutr.* **2017**, *20*, 2166–2172. [CrossRef]
12. Monteiro, C.A.; Cannon, G.; Levy, R.B.; Moubarac, J.C.; Louzada, M.L.; Rauber, F.; Khandpur, N.; Cediel, G.; Neri, D.; Martinez-Steele, E.; et al. Ultra-processed foods: What they are and how to identify them. *Public Health Nutr.* **2019**, *22*, 936–941. [CrossRef]
13. Sandvik, P.; Nydahl, M.; Kihlberg, I.; Marklinder, I. Consumers' health-related perceptions of bread—Implications for labeling and health communication. *Appetite* **2018**, *121*, 285–293. [CrossRef]
14. Chase, K.; Reicks, M.; Jones, J.M. Applying the theory of planned behavior to promotion of whole-grain foods by dietitians. *J. Am. Diet. Assoc.* **2003**, *103*, 1639–1642. [CrossRef]
15. Manning, M. The effects of subjective norms on behaviour in the theory of planned behaviour: A meta-analysis. *Br. J. Soc. Psychol.* **2009**, *48*, 649–705. [CrossRef] [PubMed]
16. Qualtrics. Provo, Utah, USA. 2021. Available online: https://www.qualtrics.com (accessed on 30 February 2022).
17. Ajzen, I. Perceived behavioral control, self-efficacy, locus of control, and the theory of planned behavior. *J. Appl. Soc. Psychol.* **2002**, *32*, 665–683. [CrossRef]
18. Phillips, F.; Ruxton, C.H.S. Scientific evidence and daily food for a better life: Milan, 19 June 2015. *Public Health* **2016**, *140*, 73–79. [CrossRef]
19. The Eatwell Guide: Public Health England. 2016. Available online: https://www.gov.uk/government/publications/the-eatwell-guide (accessed on 20 July 2021).
20. Dietary Guidelines for Americans 2020–2025: U.S. Department of Agriculture and U.S. Department of Health and Human Services. 2020. Available online: https://www.dietaryguidelines.gov/ (accessed on 20 July 2022).
21. Botelho, R.; Araújo, W.; Pineli, L. Food formulation and not processing level: Conceptual divergences between public health and food science and technology sectors. *Crit. Rev. Food Sci. Nutr.* **2018**, *58*, 639–650. [CrossRef] [PubMed]
22. Barrett, E.M.; Batterham, M.J.; Ray, S.; Beck, E.J. Whole grain, bran and cereal fibre consumption and CVD: A systematic review. *Br. J. Nutr.* **2019**, *121*, 914–937. [CrossRef] [PubMed]
23. Sadler, C.R.; Grassby, T.; Hart, K.; Raats, M.M.; Sokolović, M.; Timotijevic, L. "Even We Are Confused": A thematic analysis of professionals' perceptions of processed foods and challenges for communication. *Front. Nutr.* **2022**, *9*, 826162. [CrossRef] [PubMed]
24. Marchese, L.; Livingstone, K.M.; Woods, J.L.; Wingrove, K.; Machado, P. Ultra-processed food consumption, socio-demographics and diet quality in Australian adults. *Public Health Nutr.* **2021**, *25*, 94–104. [CrossRef]
25. Sawicki, C.M.; Jacques, P.F.; Lichtenstein, A.H.; Rogers, G.T.; Ma, J.; Saltzman, E.; McKeown, N.M. Whole- and refined-grain consumption and longitudinal changes in cardiometabolic risk factors in the Framingham Offspring Cohort. *J. Nutr.* **2021**, *151*, 2790–2799. [CrossRef]
26. Blumfield, M.; Starck, C.; Keighley, T.; Petocz, P.; Roesler, A.; Abbott, K.; Cassettari, T.; Marshall, S.; Fayet-Moore, S. Diet and economic modelling to improve the quality and affordability of the Australian diet for low and medium socioeconomic households. *Int. J. Environ. Res. Public Health* **2021**, *18*, 5771. [CrossRef]
27. Cole, M.B.; Augustin, M.A.; Robertson, M.J.; Manners, J.M. The science of food security. *NPJ Sci. Food* **2018**, *2*, 1–8. [CrossRef]
28. Bisogni, C.A.; Jastran, M.; Seligson, M.; Thompson, A. How people interpret healthy eating: Contributions of qualitative research. *J. Nutr. Educ. Behav.* **2012**, *44*, 282–301. [CrossRef] [PubMed]

nutrients

MDPI

Article

Geographical and Temporal Variability of Ultra-Processed Food Consumption in the Spanish Population: Findings from the DRECE Study

Carmen Romero Ferreiro [1,2,3,*], Pilar Cancelas Navia [1,2], David Lora Pablos [1,2,4,5,†] and Agustín Gómez de la Cámara [1,2,4,†]

1 Scientific Support Unit, Instituto de Investigación Sanitaria Hospital Universitario 12 de Octubre (imas12), 28041 Madrid, Spain
2 Spanish Clinical Research Network (SCReN), 28040 Madrid, Spain
3 Faculty of Health Sciences, Universidad Francisco de Vitoria, Pozuelo de Alarcón, 28223 Madrid, Spain
4 Consorcio de Investigación Biomédica en Red de epidemiología y salud pública (CIBEResp), 28029 Madrid, Spain
5 Faculty of Statistical Studies, Universidad Complutense de Madrid (UCM), 28040 Madrid, Spain
* Correspondence: mcromero.imas12@h12o.es
† These authors contributed equally to this work.

Abstract: The consumption of ultra-processed foods (UPFs) has increased in recent decades, worldwide. Evidence on the negative impacts of food processing on health outcomes has also been steadily increasing. The aim of this study is to describe changes in consumption patterns of ultra-processed foods in the Spanish population over time and their geographical variability. Data from four representative cohorts of the Spanish population were used (1991–1996–2004–2008). Dietary information was collected using a validated frequency questionnaire and categorized using the NOVA classification. A total increase of 10.8% in UPF consumption between 1991 and 2008 was found in Spain (p-value < 0.001). The products contributing most to UPF consumption were sugar-sweetened beverages, processed meats, dairy products, and sweets. Those who consumed more ultra-processed foods were younger (p-value < 0.001) and female (p-value = 0.01). Significant differences between the different geographical areas of Spain were found. The eastern part of Spain was the area with the lowest UPF consumption, whereas the north-western part was the area with the highest increase in UPF consumption. Given the negative effect that the consumption of ultra-processed foods has on health, it is necessary to implement public health policies to curb this increase in UPF consumption.

Keywords: ultra-processed foods; NOVA classification; geographic variability; dietary patterns

Citation: Romero Ferreiro, C.; Cancelas Navia, P.; Lora Pablos, D.; Gómez de la Cámara, A. Geographical and Temporal Variability of Ultra-Processed Food Consumption in the Spanish Population: Findings from the DRECE Study. *Nutrients* **2022**, *14*, 3223. https://doi.org/10.3390/nu14153223

Academic Editors: Monica Dinu and Daniela Martini

Received: 13 July 2022
Accepted: 3 August 2022
Published: 6 August 2022

1. Introduction

Non-communicable diseases (NCDs) are the leading causes of disability and death worldwide and currently account for more than half of the global burden of disease [1,2]. One of the main public health objectives is to prevent and combat the development of the most prevalent non-communicable chronic diseases (cardiovascular disease, diabetes, obesity, high blood pressure, chronic respiratory disease, and some types of cancer), which are largely the result of excessive or unbalanced consumption of certain foods and/or nutrients [3,4], among other factors. Conventional teaching and practice on nutrition and health usually focuses on nutrients, or else on specific foods and drinks [5]. However, the issue of food processing is largely ignored or minimized in food and nutrition, and also in public health policies. It is now acknowledged that some of these chronic diseases have as one of their major causes increased consumption of ultra-processed foods [6–8].

Ultra-processed foods (UPF) are industrial formulations performed from substances derived from food or synthesized in laboratories (dyes, flavorings, and other additives). These foods generally contain little or no natural foods, have also high amounts of fat, salt, or sugar, and low fiber, protein and micronutrients content [9,10]. They are distinguished as food products of low nutritional quality [11–15]. In this group, a large variety of industrially processed food products, such as some pastries, savory snacks, reconstituted meat products, pre-prepared frozen dishes, and soft drinks, among other food items, are included.

Evidence on the relationships between food processing and health outcomes has been increasing steadily in the last years. UPFs are prevalent in diets worldwide, contributing from 20% to more than 60% of total energy intake, depending on the country and age range [16–18]. UPFs account for more than 50% of total daily energy consumption in some high-income countries, such as the United States [19], the United Kingdom [20], Australia [21], and Canada [22]. The consumption of UPF has been associated with unhealthy dietary patterns [11–13,15,23–28] and with overweight and obesity in studies conducted in the United States [29], Canada [30], France [31], Brazil [32,33], and in most Latin American [34,35] and European [36] countries. Other recent cohort studies from Spain and France found relationships between UPF and hypertension [37,38] and cancer [39], respectively. In addition, some studies reported results on the negative effect of ultra-processed food consumption on all-cause mortality [40–44].

Globally, between 1990 and 2010, the consumption of unhealthy food items worsened, with heterogeneity across regions and countries [45]. Among unhealthy foods, consumption of ultra-processed foods is on the rise [8,34,46] around the world. In Spain, the percentage of ultra-processed foods of all food purchases almost tripled between 1990 and 2010 (from 11.0% to 31.7%) [47]. In addition, the burden of chronic non-communicable diseases also increased by approximately 4% between 1990 and 2010 in Spain [48,49], and is estimated to increase further in the forthcoming years. Several studies report that consumption of ultra-processed foods in Spain accounts for approximately 24.4% of total energy intake [43,44], but these studies calculate consumption at a given point in time. There are no previous reports on the evolution of ultra-processed consumption over time (just about purchases) and its geographical distribution in Spain. In this context of the growing trends in chronic diseases, it is important to know the pattern of consumption of these products over time in order to understand the connection between diet and public health. In addition, factors such as cultural differences, education, personal tastes and traditions, geographic location, access to technology, and health and health attitudes are known to influence food availability and food preferences [50], so it is of particular interest to study the geographical distribution of food consumption.

The aim of the study was to describe changes in the consumption pattern of ultra-processed foods in the Spanish population over time (1991–1996–2004–2008), according to eight geographical regions.

2. Materials and Methods

2.1. Design and Participants

The multicentre population-based study Diet and Risk of Cardiovascular Disease in Spain (DRECE) was used as a substrate for analysis. DRECE [51] was designed in 1991 to determine the real situation of the Spanish population with regard to the risk of cardiovascular disease (CVD), based on the prevalences of risk factors and their relationships with dietary habits. DRECE I (1991) was a representative sample of the Spanish population stratified by age, sex, and geographical areas. After 5 and 12 years, DRECE II (1996) and DRECE III (2004), two subgroups of the original DRECE cohort, were undertaken. Nearly 20 years after the start of DRECE, the capacity to locate and re-screen cohort participants for follow-up was reduced and biased to scientifically unprofitable extremes. For this reason, in 2008 the DRECE Institute for Biomedical Studies formulated a new breakthrough strategy and undertook the DRECE IV study. To this end, a new cohort was recruited, with respect to the initial distribution in eight geographical regions and the same conditions of DRECE I to make it a representative sample of the current Spanish population and an extension of the DRECE project. This study will compare the above mentioned DRECE cohorts. DRECE I (1991) consists of 4787 persons, DRECE II (1996) consists of 1079 persons, DRECE

III (2004) consists of 2009 persons, and DRECE IV (2008) consists of 5038 subjects with the same geographical and population strata design as the initial population. All cohorts have answered a food frequency questionnaire, designed and validated for epidemiological studies in the Spanish population [52,53].

2.2. Geographical Areas

The geographical distribution was structured according to the area scheme of the food consumption panel of the Ministry of Agriculture, Fisheries, and Food (MAPA, acronym in Spanish) [54], previously described in Gómez Jerique et al. [51], and included the Canary Islands, north-east, Levante (East), Andalucía (South), central-south, Castilla y León (west), north-west, and north areas (Figure 1).

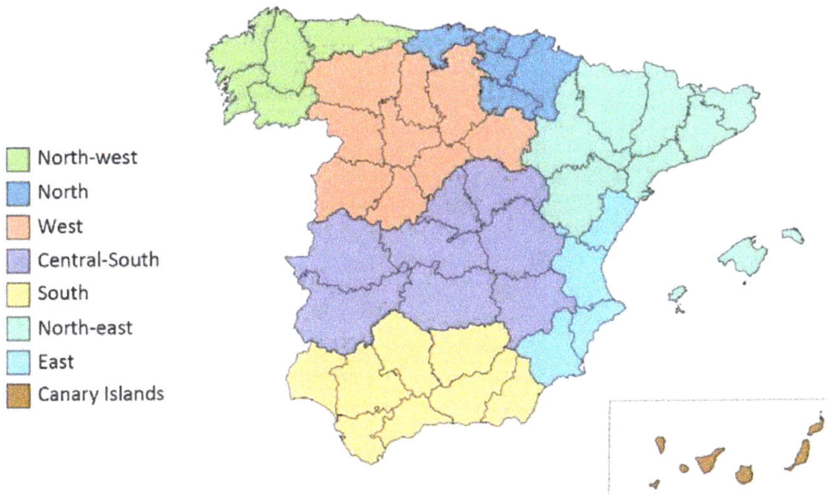

Figure 1. Geographical distribution of Spain in eight areas according to the Ministry of Agriculture, Fisheries, and Food (MAPA).

2.3. Dietary Assessment

The estimation of ultra-processed food consumption was carried out through the data collected in the dietary questionnaires. The first step in modelling dietary changes was to classify all foods according to the NOVA classification, developed in Brazil and used internationally in research [10,55]. The NOVA classification divides foods into four groups according to their degrees of processing: Group 1, unprocessed/minimally processed foods; Group 2, processed culinary ingredients; Group 3, processed products; Group 4, all ultra-processed foods. The full list of the recorded foods in the food frequency questionnaire and their NOVA classification is shown in supplemental Table S1. The kcal/day consumed from ultra-processed foods and their percentages of total kcal were then determined. Respondents with extreme total energy intakes (<200 kcal and > 5000 kcal) were excluded from the analysis [15]. Those with an extremely low BMIs (BMI < 13) were also excluded.

2.4. Statistical Analysis

All statistical analyses were performed using SAS© software (SAS Institute Inc., Cary, NC, USA), version 9.4 of the SAS System for Windows. Descriptive data are presented as mean and standard deviation (SD) for continuous variables, and categorical variables are expressed as absolute or relative frequencies. Food consumption according to the NOVA classification in the different cohorts globally and by geographical area was described by simple correspondence analysis. A ternary diagram represents this relationship [56,57]. A ternary diagram is a triangular graph that visualizes in a two-dimensional way the relationships between cohorts (represented by dots in the diagram) and the percentage of food consumption according to the NOVA classification (represented on each of the three axes). The study of the change in UPF consumption over time (between the four different cohorts) was carried out using a multivariate mixed model adjusted for age, sex, body mass index (BMI), and total energy intake. An unstructured covariance matrix was used. The intercept was considered a random effect, and the rest of the variables were used as fixed effects [58]. Comparisons between geographical areas were estimated using the chi-square test or Fisher's exact test for categorical variables, and for continuous variables were estimated using ANOVAs. In each cohort, the consumption of ultra-processed foods is represented by density maps according to the eight geographical areas. p-values < 0.05 were considered statistically significant.

3. Results

The final sample size included 4679 individuals in DRECE I, 928 individuals in DRECE II, 1065 individuals in DRECE III, and 4835 individuals in DRECE IV. The demographic characteristics of the four cohorts are shown in supplemental Table S2. Between 1991 and 2008, there was a general increase in total energy intake (kcal/day) in the Spanish population (Table 1). Average consumption of ultra-processed foods (NOVA group 4) was found to be 24.44% of the total energy intake in 1991 (DRECE I), 25.61% in 1996 (DRECE II), 27.48% in 2004 (DRECE III), and 31.09% in 2008 (DRECE IV) (Table 1). UPF consumption changed over time also in both sexes, from 24.48% in males and 24.39% in females in 1991, to 31.03% and 31.39%, respectively, in 2008. In addition, the same evolution was observed according to age group and BMI (Table 1).

Table 1. Food intake according to the NOVA classification over time (DRECE cohorts) and distribution of ultra-processed food consumption (NOVA 4) by sex, age, and BMI class.

	DRECE I 1991	DRECE II 1996	DRECE III 2004	DRECE IV 2008
Total energy intake (kcal/day)	2024.80 (727.09)	2362.49 (1197)	2373.91 (1068)	2441.01 (948.75)
NOVA classification (% of energy)				
NOVA 1	45.91 (13.33)	47.96 (15.58)	51.47 (14.01)	55.21 (12.13)
NOVA 3+2	29.65 (13.24)	26.43 (15.96)	21.05 (16.48)	13.70 (15.37)
NOVA 4	24.44 (13.95)	25.61 (16.29)	27.48 (19.17)	31.09 (19.24)
UPF consumption (NOVA 4) (% of energy)				
By sex				
Male	24.48 (13.89)	23.71 (16.76)	26.14 (19.72)	31.03 (17.57)
Female	24.39 (14.01)	27.83 (15.16)	29.01 (15.18)	31.39 (18.47)
By age group				
5–24	32.79 (12.83)	31.69 (14.84)	33.72 (14.04)	34.12 (11.48)
25–49	19.81 (11.62)	24.70 (16.63)	26.70 (17.26)	27.91 (21.01)
50–75	16.13 (11.41)	19.39 (19.01)	22.01 (12.29)	25.14 (19.83)
By BMI class				
Normal weight	16.96 (11.41)	23.67 (17.30)	22.76 (18.74)	27.93 (19.22)
Overweight	19.99 (12.45)	26.04 (16.15)	25.88 (19.86)	31.77 (18.90)
Obese	28.27 (13.85)	26.11 (15.06)	28.67 (18.67)	33.31 (21.15)

Data is shown as mean (SD).

The mixed model shows a significant upward trend (all adjusted p-values <0.001) in the consumption of ultra-processed products over the 17 years of the study, and a 10.79% ± 0.39 increase in the consumption of this type of product in Spain between 1991 and 2008 (Table 2). This increase over time can be seen in the ternary diagram (Figure 2). In the ternary diagram, for better representation, the NOVA 2 and NOVA 3 groups are shown together, as NOVA 2 represents a very low percentage of consumption, and it was decided to unify processed culinary ingredients (NOVA 2) and processed foods (NOVA 3) into one category. The axes of the diagram correspond to the percentages of foods belonging to NOVA 1, NOVA 3+2, and NOVA 4 (these percentages are also shown in Table 1). The points represented in the diagram correspond to the four cohorts (1991, 1996, 2004, and 2008) according to the amounts of products they included from each of the different NOVA groups. As an example of an interpretation, using the 2008 cohort (DRECE IV), represented with dashed lines in Figure 2, 31.09% of the food consumed corresponded to ultra-processed foods (NOVA 4), 13.70% to processed foods (NOVA 3+2), and 55.21% to unprocessed or minimally processed foods (NOVA 1). This interpretation can be made in the same way for the rest of the points in the diagram.

As a result of the mixed model, it was also found that participants who consumed the most UPF had significantly higher intakes of total energy ($\beta = 1.86$, p-value < 0.001) and were mostly female ($\beta = 1.06$, p-value = 0.01) (Table 2). In addition, individuals who consumed more ultra-processed foods were younger ($\beta = -0.15$, p-value < 0.001). UPF consumption in young people remained above 30% at all time points (Table 1). No association was found between UPF consumption and BMI ($\beta = -0.05$, p-value = 0.19) (Table 2).

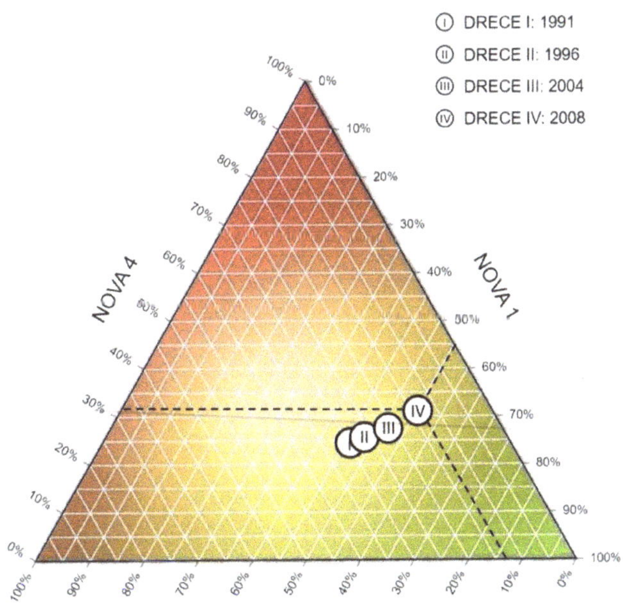

Figure 2. Ternary diagram of the average percentage of energy intake from the NOVA classification by the Spanish population over time.

Table 2. Mixed model coefficients for UPF consumption over time adjusted for age, sex, BMI, and total energy intake.

	Estimate	Standard Error	*p*
Intercept	24.49	1.21	<0.001
Time (Cohorts)			
DRECE I 1991	Ref.		
DRECE II 1996	5.31	0.62	<0.001
DRECE III 2004	9.63	0.66	<0.001
DRECE IV 2008	10.79	0.39	<0.001
Age (years)	−0.15	0.01	<0.001
Sex			
Male	Ref.		
Female	1.06	0.33	0.01
BMI (kg/m^2)	−0.05	0.04	0.19
Total energy intake (kcal/day)	1.86	0.19	<0.001

AIC:3156 8154 subjects included.

The main food groups contributing to ultra-processed food intake (>10% energy contribution) were sugar-sweetened beverages (i.e., soft drinks) (18.41%), milkshakes and juice boxes (17.53%), meat and meat products (16.38%), and dairy products (13.50%) in 1991; dairy products (i.e., yogurts, ice cream, or Petit Suisse) (17.51%), meat and meat products (15.06%), and sweets and cookies (11.79%) in 1996; meat and meat products (17.92%), dairy products (14.01%), and sugar sweetened beverages (13.64%) in 2004; and industrial cakes and pastries (19.69%), dairy products (17.41%), and sugar sweetened beverages (11.73%) in 2008.

The geographical study shows that in all cohorts the sample was homogeneous in terms of age and sex across the eight geographical areas (all p-values > 0.05) (Table 3). Significant differences in BMI, total energy intake, and ultra-processed food consumption were found between geographical areas at all time points (Table 3). When studying the consumption of ultra-processed foods by geographical area, the same trend was observed in all of them as in Spain as a whole: an increase over time in the consumption of this type of product (Figure 3). During the 17 years of the study, there was an overall increase in the consumption of ultra-processed foods of 11% in the north-west and north regions, 10.10% in the north-east, 9.41% in the west, 8.38% in the east, 6.70% in the Canary Islands, 6.13% in the south, and 5.20% in the central-south region.

Table 3. Ultra-processed food (NOVA 4) intake and demographic characteristic by geographical area.

	Geographical Areas								p
DRECE I 1991	**North-West**	**North**	**North-East**	**West**	**Central-South**	**East**	**South**	**Canary Islands**	
n	514	422	683	341	913	553	1075	178	
Age (years)	30.04 (15.61)	30.38 (15.59)	31.74 (15.59)	30.22 (15.50)	30.54 (15.73)	31.57 (15.58)	29.89 (15.50)	29.48 (16.04)	0.175
Sex (male)	249 (48.44%)	206 (48.82%)	331 (48.46%)	171 (50.15%)	440 (48.19%)	263 (47.56%)	539 (50.14%)	89 (50.00%)	0.977
BMI (kg/m^2)	24.43 (4.61)	23.53 (4.48)	24.43 (4.83)	23.65 (4.47)	24.05 (5.10)	24.14 (4.73)	24.75 (5.68)	24.09 (5.63)	0.002
Total energy intake (kcal/day)	1996.47 (641.84)	1942.73 (593.48)	2037.55 (707.23)	2152.04 (686.01)	2023.84 (693.92)	1964.44 (802.26)	2040.11 (796.11)	2109.13 (846.20)	<0.001
NOVA classification GROUP 4 (% of energy)	24.47 (14.45)	25.03 (12.95)	22.65 (13.75)	24.60 (12.94)	24.97 (14.00)	22.64 (14.02)	25.15 (14.05)	28.10 (15.16)	<0.001
DRECE II 1996	**North-West**	**North**	**North-East**	**West**	**Central-South**	**East**	**South**	**Canary Islands**	
n	78	124	83	83	162	123	223	47	
Age (years)	48.29 (13.75)	46.03 (14.45)	48.95 (13.64)	46.05 (15.11)	47.92 (14.95)	45.65 (15.35)	45.53 (14.93)	44.26 (15.88)	0.248
Sex (male)	51 (65.38%)	73 (58.87%)	54 (61.36%)	51 (61.45%)	97 (59.88%)	84 (68.29%)	130 (58.30%)	26 (55.32%)	0.415
BMI (kg/m^2)	27.79 (3.76)	26.69 (4.03)	28.61 (3.70)	26.08 (4.06)	26.86 (4.68)	27.21 (4.14)	28.36 (5.49)	27.75 (6.02)	<0.001
Total energy intake (kcal/day)	2474.43 (1061)	2212.04 (751.12)	2576.62 (2016)	2359.88 (769.19)	2178.05 (862.04)	2625.60 (1149)	2441.92 (1459)	1919.39 (770.33)	<0.001
NOVA classification GROUP 4 (% of energy)	25.91 (15.78)	28.87 (15.90)	25.53 (16.11)	25.06 (17.82)	26.95 (17.41)	21.85 (15.71)	23.13 (15.28)	29.33 (15.12)	0.010
DRECE III 2004	**North-West**	**North**	**North-East**	**West**	**Central-South**	**East**	**South**	**Canary Islands**	
n	89	135	192	104	178	45	257	65	
Age (years)	44.18 (15.66)	44.68 (15.13)	47.80 (16.14)	44.96 (16.01)	45.29 (17.18)	44.93 (14.66)	44.17 (14.05)	51.17 (15.40)	0.061
Sex (male)	42 (47.19%)	64 (47.41%)	86 (44.79%)	46 (44.23%)	72 (40.45%)	21 (46.67%)	116 (45.14%)	28 (43.08%)	0.954
BMI (kg/m^2)	28.52 (5.02)	26.59 (4.72)	27.80 (5.36)	26.70 (4.74)	26.34 (4.59)	28.23 (5.11)	28.38 (5.85)	28.40 (4.91)	<0.001
Total energy intake (kcal/day)	2286.15 (1080)	2580.85 (1353)	2485.81 (1109)	2408.26 (969.91)	2518.23 (1254)	2368.07 (771.36)	2114.92 (756.77)	2311.56 (906.83)	0.002
NOVA classification GROUP 4 (% of energy)	34.13 (18.02)	35.34 (15.59)	32.48 (12.55)	34.67 (14.78)	29.99 (19.68)	25.71 (11.57)	25.85 (13.51)	34.42 (12.85)	<0.001

Table 3. *Cont.*

	Geographical Areas								*p*
DRECE IV 2008	**North-West**	**North**	**North-East**	**West**	**Central-South**	**East**	**South**	**Canary Islands**	
n	562	370	833	373	1037	548	922	190	
Age (years)	44.06 (14.91)	45.58 (15.13)	43.51 (14.32)	43.28 (14.45)	43.81 (14.70)	44.58 (15.32)	42.88 (14.21)	42.14 (14.25)	0.067
Sex (male)	251 (44.66%)	176 (47.57%)	404 (48.50%)	180 (48.26%)	469 (45.23%)	271 (49.45%)	411 (44.58%)	95 (50.00%)	0.372
BMI (kg/m^2)	26.26 (4.13)	24.67 (3.82)	25.30 (4.23)	26.88 (3.91)	25.17 (4.13)	24.84 (3.96)	26.60 (4.46)	25.87 (4.54)	0.007
Total energy intake (kcal/day)	2400.25 (910.08)	2382.53 (959.17)	2386.65 (985.88)	2490.63 (855.33)	2453.17 (860.99)	2432.02 (1021)	2491.67 (1003)	2530.22 (1001)	0.005
NOVA classification GROUP 4 (% of energy)	35.47 (16.94)	36.02 (18.33)	32.75 (19.76)	34.01 (17.48)	30.17 (17.14)	31.03 (17.60)	31.28 (18.33)	34.80 (16.12)	<0.001

Data is shown as mean (SD) or *n* (%).

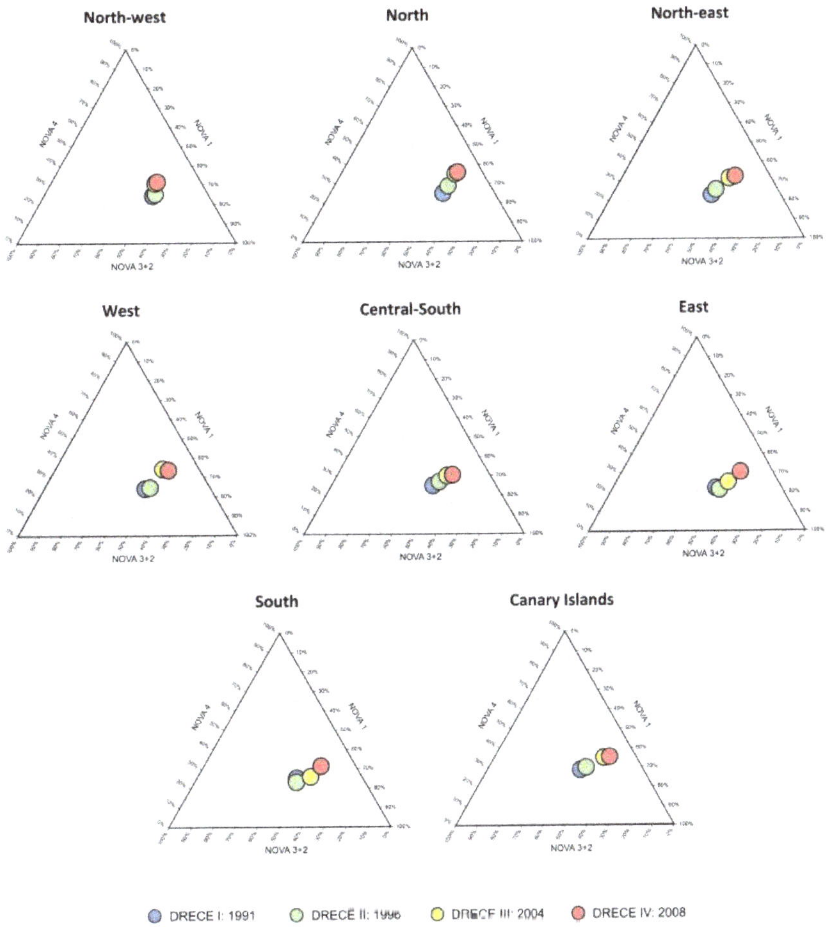

Figure 3. Ternary diagram of the average percentage of energy intake in the NOVA classification over time by geographical area of Spain.

In 1991, the region with the highest consumption of ultra-processed foods was the Canary Islands, and in 2008 it was the northern region. As can be seen in Figure 4, the region with the lowest consumption of ultra-processed foods was the east, which was the region with the lowest consumption in 1991 (22.64%), 1996 (21.85%), and 2004 (25.75%), and had the second lowest in 2008 (31.03%). The Canary Islands was the region with the highest consumption of ultra-processed foods in 1991 (28.10%) and 1996 (29.33%), and then the northern region was the region with the highest consumption of ultra-processed foods in 2004 (35.34%) and 2008 (36.03%). The central-south region went from having intermediate consumption in the early years to becoming the region with the lowest consumption of ultra-processed foods in 2008, at 30.17%. The southern region started as one of the regions with the highest consumption of ultra-processed foods in 1991, and ended up as one of the regions with lower consumption compared to the rest. The western and north-western regions started with intermediate consumption but were among the regions with the highest consumption in 2004 and 2008, respectively. The north-east region retained intermediate consumption values compared to the rest of the regions consistently (Figure 4).

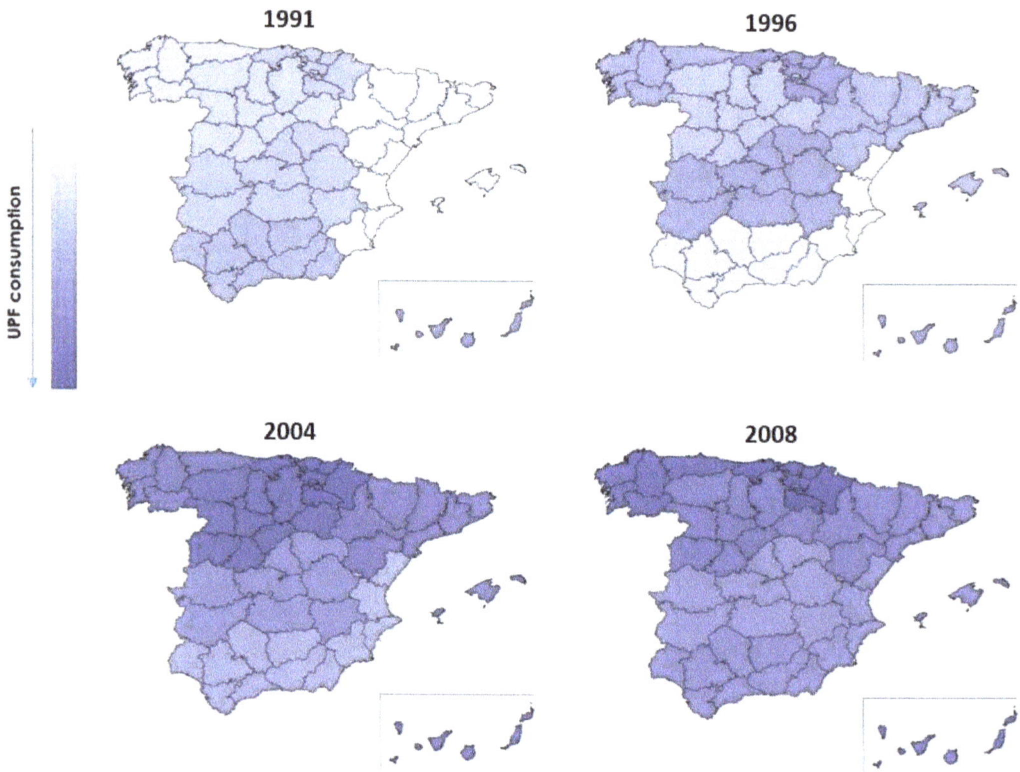

Figure 4. Geographical distribution of ultra-processed food consumption in Spain over time.

4. Discussion

About one third of daily energy intake was found to be provided by ultra-processed foods (UPF) in the Spanish population. Estimates of UPF purchases calculated from national household budget surveys (conducted in Europe between 1991 and 2008) showed that the average household availability of UPF ranged from 10% of total purchased dietary energy in Portugal to 50% in the UK [36]. In Spain, UPFs were found to contribute about 24–31% to total dietary energy (between 1991 and 2008), which is slightly higher than the average usual proportion of daily energy intake from UPFs (26.4%) found in this study. However, food consumption surveys often provide more details on the foods consumed compared to household budget surveys, which are based on purchases. When looking at published consumption data rather than household budget survey data, Spain is shown to be a country with a low consumption of ultra-processed foods compared to other countries, such as Canada (48%) [11], the United States (57.9%) [19], the United Kingdom (56.8%) [20], Belgium (about 33%) [28], and France (35.9%) [31]. These differences may be due to the fact that the data published in other countries correspond to different periods of time. They also could be due to the Mediterranean diet, which is characterized by high consumption of plant-based foods and fresh fruits, low consumption of red meat and other processed foods, the use of olive oil as the main source of fat, and a moderate intake of wine during meals [59]. In addition, other Mediterranean countries, such as Italy, also have lower UPF consumption (18%) [60].

On the other hand, a negative shift in the pattern of food consumption was found. UPF consumption has increased over time across the country. An increase of 10.79% in UPF consumption was found between 1991 and 2008 in Spain, from 1 in 4 foods being ultra-processed in 1991 to 1 in 3 in 2008, which is in line with the previously reported increase in UPF purchases between 1990 and 2010 in Spanish households [47]. As the nutrition literature increasingly recognizes ultra-processed foods (UPF) to be unhealthy, the diet in Spain can be considered increasingly unhealthy. This supports the evidence that between 1990 and 2010, diets based on unhealthy items worsened worldwide [45]. This trend has also been shown in other countries, such as Belgium [28], Sweden [61], the United Kingdom [20], and the United States [62]. This increase also parallels the growing burden in Spain and worldwide of non-communicable diseases [48,49], of which excessive consumption of ultra-processed foods is known to be one of the main causes [8,63]. The exact reasons for this increase in UPF consumption are not known, but may include the increased availability and accessibility of such products, as they are highly palatable and inexpensive, increased consumption of prepared foods outside the home over the past few decades, and aggressive and unregulated advertising of convenience foods, which may promote overconsumption [46,64]. The main groups of UPFs consumed in Spain were sugar-sweetened beverages; processed meats; dairy products; and sweets, biscuits, and cakes. These data are in line with those provided by the European household budget surveys (conducted between 1991 and 2008), where the most purchased UPFs were packaged breads, cakes, sweets and cookies, meat products, and sugar-sweetened beverages [36]. This also agrees with the most consumed UPFs in the United Kingdom, Belgium, Canada, and the United States [20,28,65]. It is worth noting that the consumption of processed meats decreased between 1991 and 2008 in Spain, from 16.38% to less than 10%, and the consumption of sugar-sweetened beverages from 18.41% to 11.73%. On the other hand, consumption of processed dairy products increased from 13.50% to 17.41%, and consumption of sweets from less than 10% to 19.69%. Similar results were found in young people in the United States between 1999 and 2018, where there was also a decrease in the consumption of sugar-sweetened beverages and an increase in the consumption of sweets [62]; and also in Sweden where there was a slight decrease in consumption of sugar-sweetened beverages between 2002 and 2010 [61]. This highlights the types of ultra-processed products for which there is most need to reduce consumption in the population and to implement policies to reduce their sales. Some countries, such as Uruguay [66] and Brazil [67], already include the concept of UPFs in food guidelines; and other countries, such as Mexico [68] and Hungary [69], have taken actions to limit the marketing of UPFs through taxation. Such policies do not exist in Spain and should start to be implemented in view of the evidence of the growing consumption of UPFs.

Young people consume the highest proportion of ultra-processed foods in their diets in the Spanish population, consistently—above 30%. Other studies, such as those from Belgium [28], the United States [70], Canada [11], Colombia [71], and Chile [26], have also found that children consume the highest amounts of UPF compared to other age groups. Given that young people are the highest consumers of UPFs, it could be beneficial to implement health policies targeting this population stratum in order to raise awareness of healthy food consumption. Higher UPF consumption was associated with higher BMI in other studies [29,30,32,36,61,72,73], but no such association was detected in Spain. Females consumed more UPF than males; this may be influenced by gender differences in food choices. Females appear to exhibit more stress-related eating behaviors [74], which may lead to higher UPF consumption.

Consumption of ultra-processed foods is high in all regions of Spain (21–36%). It is notorious that factors such as palatability and the high commercialization of these foods contribute to their presence in the eating habits of all families [75]. In addition, all regions saw a progressive increase in the consumption of this type of food (5.2–11%) during the 17 years of the study, similar to the overall increase in Spain. The Canary Islands is one of the regions with higher relative consumption of ultra-processed foods, which is in agreement with the dietary pattern found in other studies on this region, in which it has been characterized by high intakes of fats and carbohydrates (present at high levels in UPFs) with respect to other regions of Spain [76]. The north, north-west, and west regions showed worsening in their dietary patterns, being the regions with the highest increases in UPF consumption over time, and reaching the highest percentages of intake in 2008 (36%, 35.5%, and 34% of total intake, respectively) together with the Canary Islands. This may be due to the high carbohydrate and high fat consumption patterns of these regions, whose citizens have also been reported to have high HDL lipid profiles [76]. The eastern region remained over time one of the regions with the lowest consumption of UPFs, probably because it is geographically located on the Mediterranean coast and may be more deeply linked to the culture and traditions of a quality Mediterranean diet [77]. This has been evidenced by recent studies finding an inverse association between UPF consumption and adherence to the Mediterranean diet [78]. The north-east region retained average consumption over time, probably also due to its adherence to the Mediterranean diet because of its geographical position. Particularly, the southern and north-central regions are characterized by improved consumption patterns compared to the rest of the regions, being the regions with the lowest increases in UPF consumption over time. This is reflected in the micronutrient patterns of these regions, where low carbohydrate and protein intake and a low HDL lipid profile are reported [76]. The geographical variability found in UPF consumption in Spain has some consistency with the economic data provided by the National Statistics Institute (INE) [79]. The regions with the highest consumption of UPF in 2008 were those with the lowest growth in per capita household income in the 2000s. Along the same lines, the southern and central areas had the highest growth in per capita household income and the lowest growth in UPF consumption.

All these results reinforce the increase in the consumption of ultra-processed foods over the last few decades and the need for health policies that take into account the degree of food processing to address the increasing intake of UPFs.

There are several strengths to this study. The use of a large, nationally representative sample of the Spanish population maximizes generalizability. The testing of the same hypothesis both cross-sectionally and over time lends credibility to our results. Self-reported dietary intake data are less biased than purchasing data, as all meals consumed are included, including those consumed away from home, which are more likely to be ultra-processed. However, the study also has some limitations. Although the NOVA classification has been questioned sometimes, it is simple and clear to apply; no better alternative has yet been proposed. The food frequency questionnaire was not designed to collect data on consumption of UPFs according to the NOVA classification. Each food item was classified into its most likely NOVA group, but we cannot rule out misclassification of some foods. Finally, to minimize information bias, validated procedures were used, and subjects with inconsistent intake data were excluded. Finally, future studies in this field of research could consider including more qualitative data.

5. Conclusions

There has been an increase in UPF consumption over time in Spain, namely, of approximately 10.8% between 1991 and 2008. About 21–36% of the average daily energy intake is provided by UPFs, with differences depending on the geographical area. The products contributing most to UPF consumption are sugar-sweetened beverages, processed meats, dairy products, and sweets. Young people and females have the highest intakes of ultra-processed foods. No correlation was found between UPF consumption and BMI. The eastern part of Spain is the area with the lowest UPF consumption, and the north-western part of Spain is the area with the highest increase in UPF consumption. Given the robust scientific evidence associating UPF consumption with various adverse health outcomes, realistic public health policies are needed to limit the availability, affordability, and marketing of UPFs. In addition, raising awareness through educational programs that promote healthier food environments to individuals of all socio-demographic and socio-economic categories, but especially to the youngest, would be useful to prevent further increases in UPF consumption in Spain.

Supplementary Materials: The following supporting information can be downloaded at: https://www.mdpi.com/article/10.3390/nu14153223/s1. Table S1: Classification of items of the food frequency questionnaires according to degree of processing (NOVA classification). Table S2: Demographic characteristics of the DRECE cohorts.

Author Contributions: Conceptualization, D.L.P. and C.R.F.; methodology, D.L.P. and C.R.F.; software, C.R.F.; validation, D.L.P. and C.R.F.; formal analysis, C.R.F.; investigation, C.R.F. and P.C.N.; resources, D.L.P.; data curation, C.R.F.; writing—original draft preparation, C.R.F.; writing—review and editing, C.R.F., P.C.N., D.L.P. and A.G.d.l.C.; visualization, A.G.d.l.C.; supervision, D.L.P. and A.G.d.l.C.; project administration, P.C.N. and A.G.d.l.C.; funding acquisition, A.G.d.l.C. All authors have read and agreed to the published version of the manuscript.

Funding: This research was funded by Instituto de Salud Carlos III, grant numbers FIS 03/0014 and FIS 08/90643; and by Fundación MMA de Investigación biomédica, grant number P-MMA2004/19 and P-MMA2008/88.

Institutional Review Board Statement: The study was conducted in accordance with the Declaration of Helsinki, and approved by the Ethics Committee of the Hospital Universitario 12 de Octubre on 5 November 2010 (ref. 10/292).

Informed Consent Statement: Informed consent was obtained from all subjects involved in the study.

Data Availability Statement: Additional data are available from the corresponding author on reasonable request.

Acknowledgments: The authors thank all collaborators for their involvement in the DRECE study. The authors report no conflict of interest.

Conflicts of Interest: The authors declare no conflict of interest. The funders had no role in the design of the study; in the collection, analyses, or interpretation of data; in the writing of the manuscript; or in the decision to publish the results.

References

1. Mendis, S.; Davis, S.; Norrving, B. Organizational Update: The World Health Organization Global Status Report on Noncommunicable Diseases 2014; One More Landmark Step in the Combat against Stroke and Vascular Disease. *Stroke* **2015**, *46*, e121–e122. [CrossRef] [PubMed]
2. United Nations Political Declaration of the High-Level Meeting of the General Assembly on the Prevention and Control of Non-Communicable Diseases 2011. Available online: https://digitallibrary.un.org/record/710899/ (accessed on 28 May 2022).
3. Cena, H.; Calder, P.C. Defining a Healthy Diet: Evidence for the Role of Contemporary Dietary Patterns in Health and Disease. *Nutrients* **2020**, *12*, 334. [CrossRef] [PubMed]
4. Hawkes, C. Food Policies for Healthy Populations and Healthy Economies. *BMJ* **2012**, *344*, e2801. [CrossRef]
5. Monteiro, C.A. Nutrition and Health. The Issue Is Not Food, nor Nutrients, so Much as Processing. *Public Health Nutr.* **2009**, *12*, 729–731. [CrossRef]
6. World Health Organization. *Diet, Nutrition, and the Prevention of Chronic Diseases: Report of a WHO-FAO Expert Consultation*; [Joint WHO-FAO Expert Consultation on Diet, Nutrition, and the Prevention of Chronic Diseases, 2002, Geneva, Switzerland]; WHO Technical Report Series; World Health Organization: Geneva, Switzerland, 2003; ISBN 978-92-4-120916-8.
7. Wiseman, M.; Cannon, G.; Butrum, R.; Martin, G.; Higginbotham, S.; Jones, C.; Fletcher, M. *Policy and Action for Cancer Prevention. Food, Nutrition and Physical Activity: A Global Perspective*; World Cancer Research Fund/American Institute for Cancer Research: London, UK, 2009.
8. Stuckler, D.; McKee, M.; Ebrahim, S.; Basu, S. Manufacturing Epidemics: The Role of Global Producers in Increased Consumption of Unhealthy Commodities Including Processed Foods, Alcohol, and Tobacco. *PLoS Med.* **2012**, *9*, e1001235. [CrossRef] [PubMed]
9. Monteiro, C.A.; Cannon, G.; Levy, R.; Moubarac, J.-C.; Jaime, P.; Martins, A.P.; Canella, D.; Louzada, M.; Parra, D. NOVA. The Star Shines Bright. *World Nutr.* **2016**, *7*, 28–38.
10. Monteiro, C.A.; Cannon, G.; Levy, R.B.; Moubarac, J.-C.; Louzada, M.L.; Rauber, F.; Khandpur, N.; Cediel, G.; Neri, D.; Martinez-Steele, E.; et al. Ultra-Processed Foods: What They Are and How to Identify Them. *Public Health Nutr.* **2019**, *22*, 936–941. [CrossRef] [PubMed]
11. Moubarac, J.-C.; Batal, M.; Louzada, M.L.; Martinez Steele, E.; Monteiro, C.A. Consumption of Ultra-Processed Foods Predicts Diet Quality in Canada. *Appetite* **2017**, *108*, 512–520. [CrossRef]
12. Martínez Steele, E.; Popkin, B.M.; Swinburn, B.; Monteiro, C.A. The Share of Ultra-Processed Foods and the Overall Nutritional Quality of Diets in the US: Evidence from a Nationally Representative Cross-Sectional Study. *Popul. Health Metr.* **2017**, *15*, 6. [CrossRef]

13. da Costa Louzada, M.L.; Ricardo, C.Z.; Steele, E.M.; Levy, R.B.; Cannon, G.; Monteiro, C.A. The Share of Ultra-Processed Foods Determines the Overall Nutritional Quality of Diets in Brazil. *Public Health Nutr.* **2018**, *21*, 94–102. [CrossRef]
14. Marrón-Ponce, J.A.; Flores, M.; Cediel, G.; Monteiro, C.A.; Batis, C. Associations between Consumption of Ultra-Processed Foods and Intake of Nutrients Related to Chronic Non-Communicable Diseases in Mexico. *J. Acad. Nutr. Diet.* **2019**, *119*, 1852–1865. [CrossRef] [PubMed]
15. Parra, D.C.; da Costa-Louzada, M.L.; Moubarac, J.-C.; Bertazzi-Levy, R.; Khandpur, N.; Cediel, G.; Monteiro, C.A. Association between Ultra-Processed Food Consumption and the Nutrient Profile of the Colombian Diet in 2005. *Salud Publica Mex.* **2019**, *61*, 147–154. [CrossRef] [PubMed]
16. Elizabeth, L.; Machado, P.; Zinöcker, M.; Baker, P.; Lawrence, M. Ultra-Processed Foods and Health Outcomes: A Narrative Review. *Nutrients* **2020**, *12*, 1955. [CrossRef] [PubMed]
17. Kelly, B.; Jacoby, E. Public Health Nutrition Special Issue on Ultra-Processed Foods. *Public Health Nutr.* **2018**, *21*, 1–4. [CrossRef] [PubMed]
18. Casas, R. Moving towards a Healthier Dietary Pattern Free of Ultra-Processed Foods. *Nutrients* **2022**, *14*, 118. [CrossRef]
19. Martínez Steele, E.; Baraldi, L.G.; da Costa Louzada, M.L.; Moubarac, J.-C.; Mozaffarian, D.; Monteiro, C.A. Ultra-Processed Foods and Added Sugars in the US Diet: Evidence from a Nationally Representative Cross-Sectional Study. *BMJ Open* **2016**, *6*, e009892. [CrossRef]
20. Rauber, F.; da Costa Louzada, M.L.; Steele, E.M.; Millett, C.; Monteiro, C.A.; Levy, R.B. Ultra-Processed Food Consumption and Chronic Non-Communicable Diseases-Related Dietary Nutrient Profile in the UK (2008–2014). *Nutrients* **2018**, *10*, 587. [CrossRef]
21. Moodie, R.; Stuckler, D.; Monteiro, C.; Sheron, N.; Neal, B.; Thamarangsi, T.; Lincoln, P.; Casswell, S.; Lancet NCD Action Group. Profits and Pandemics: Prevention of Harmful Effects of Tobacco, Alcohol, and Ultra-Processed Food and Drink Industries. *Lancet* **2013**, *381*, 670–679. [CrossRef]
22. Moubarac, J.-C.; Martins, A.P.B.; Claro, R.M.; Levy, R.B.; Cannon, G.; Monteiro, C.A. Consumption of Ultra-Processed Foods and Likely Impact on Human Health. Evidence from Canada. *Public Health Nutr.* **2013**, *16*, 2240–2248. [CrossRef]
23. Adams, J.; White, M. Characterisation of UK Diets According to Degree of Food Processing and Associations with Socio-Demographics and Obesity: Cross-Sectional Analysis of UK National Diet and Nutrition Survey (2008–12). *Int. J. Behav. Nutr. Phys. Act.* **2015**, *12*, 160. [CrossRef]
24. Batal, M.; Johnson-Down, L.; Moubarac, J.-C.; Ing, A.; Fediuk, K.; Sadik, T.; Tikhonov, C.; Chan, L.; Willows, N. Quantifying Associations of the Dietary Share of Ultra-Processed Foods with Overall Diet Quality in First Nations Peoples in the Canadian Provinces of British Columbia, Alberta, Manitoba and Ontario. *Public Health Nutr.* **2018**, *21*, 103–113. [CrossRef] [PubMed]
25. Bielemann, R.M.; Motta, J.V.S.; Minten, G.C.; Horta, B.L.; Gigante, D.P. Consumption of Ultra-Processed Foods and Their Impact on the Diet of Young Adults. *Rev. Saude Publica* **2015**, *49*, 28. [CrossRef]
26. Cediel, G.; Reyes, M.; da Costa Louzada, M.L.; Martinez Steele, E.; Monteiro, C.A.; Corvalán, C.; Uauy, R. Ultra-Processed Foods and Added Sugars in the Chilean Diet (2010). *Public Health Nutr.* **2018**, *21*, 125–133. [CrossRef] [PubMed]
27. Cornwell, B.; Villamor, E.; Mora-Plazas, M.; Marin, C.; Monteiro, C.A.; Baylin, A. Processed and Ultra-Processed Foods Are Associated with Lower-Quality Nutrient Profiles in Children from Colombia. *Public Health Nutr.* **2018**, *21*, 142–147. [CrossRef] [PubMed]
28. Vandevijvere, S.; De Ridder, K.; Fiolet, T.; Bel, S.; Tafforeau, J. Consumption of Ultra-Processed Food Products and Diet Quality among Children, Adolescents and Adults in Belgium. *Eur. J. Nutr.* **2019**, *58*, 3267–3278. [CrossRef] [PubMed]
29. Juul, F.; Martinez-Steele, E.; Parekh, N.; Monteiro, C.A.; Chang, V.W. Ultra-Processed Food Consumption and Excess Weight among US Adults. *Br. J. Nutr.* **2018**, *120*, 90–100. [CrossRef]
30. Nardocci, M.; Leclerc, B.-S.; Louzada, M.-L.; Monteiro, C.A.; Batal, M.; Moubarac, J.-C. Consumption of Ultra-Processed Foods and Obesity in Canada. *Can. J. Public Health* **2019**, *110*, 4–14. [CrossRef]
31. Julia, C.; Martinez, L.; Allès, B.; Touvier, M.; Hercberg, S.; Méjean, C.; Kesse-Guyot, E. Contribution of Ultra-Processed Foods in the Diet of Adults from the French NutriNet-Santé Study. *Public Health Nutr.* **2018**, *21*, 27–37. [CrossRef]
32. da Costa Louzada, M.L.; Baraldi, L.G.; Steele, E.M.; Martins, A.P.B.; Canella, D.S.; Moubarac, J.-C.; Levy, R.B.; Cannon, G.; Afshin, A.; Imamura, F.; et al. Consumption of Ultra-Processed Foods and Obesity in Brazilian Adolescents and Adults. *Prev. Med.* **2015**, *81*, 9–15. [CrossRef]
33. Canella, D.S.; Levy, R.B.; Martins, A.P.B.; Claro, R.M.; Moubarac, J.-C.; Baraldi, L.G.; Cannon, G.; Monteiro, C.A. Ultra-Processed Food Products and Obesity in Brazilian Households (2008–2009). *PLoS ONE* **2014**, *9*, e92752. [CrossRef]
34. Pan American Health Organization. Ultra-Processed Food and Drink Products in Latin America: Trends, Impact on Obesity, Policy Implications. Available online: https://iris.paho.org/handle/10665.2/7699 (accessed on 19 November 2020).
35. Neri, D.; Steele, E.M.; Khandpur, N.; Cediel, G.; Zapata, M.E.; Rauber, F.; Marrón-Ponce, J.A.; Machado, P.; da Costa Louzada, M.L.; Andrade, G.C.; et al. Ultraprocessed Food Consumption and Dietary Nutrient Profiles Associated with Obesity: A Multicountry Study of Children and Adolescents. *Obes. Rev.* **2022**, *23* (Suppl. 1), e13387. [CrossRef] [PubMed]
36. Monteiro, C.A.; Moubarac, J.C.; Levy, R.B.; Canella, D.S.; da Costa Louzada, M.L.; Cannon, G. Household Availability of Ultra-Processed Foods and Obesity in Nineteen European Countries. *Public Health Nutr.* **2018**, *21*, 18–26. [CrossRef] [PubMed]
37. de Deus Mendonça, R.; Lopes, A.C.S.; Pimenta, A.M.; Gea, A.; Martinez-Gonzalez, M.A.; Bes-Rastrollo, M. Ultra-Processed Food Consumption and the Incidence of Hypertension in a Mediterranean Cohort: The Seguimiento Universidad de Navarra Project. *Am. J. Hypertens.* **2017**, *30*, 358–366. [CrossRef] [PubMed]

38. Martinez-Perez, C.; San-Cristobal, R.; Guallar-Castillon, P.; Martínez-González, M.Á.; Salas-Salvadó, J.; Corella, D.; Castañer, O.; Martinez, J.A.; Alonso-Gómez, Á.M.; Wärnberg, J.; et al. Use of Different Food Classification Systems to Assess the Association between Ultra-Processed Food Consumption and Cardiometabolic Health in an Elderly Population with Metabolic Syndrome (PREDIMED-Plus Cohort). *Nutrients* **2021**, *13*, 2471. [CrossRef] [PubMed]
39. Fiolet, T.; Srour, B.; Sellem, L.; Kesse-Guyot, E.; Allès, B.; Méjean, C.; Deschasaux, M.; Fassier, P.; Latino-Martel, P.; Beslay, M.; et al. Consumption of Ultra-Processed Foods and Cancer Risk: Results from NutriNet-Santé Prospective Cohort. *BMJ* **2018**, *360*, k322. [CrossRef]
40. Kim, H.; Hu, E.A.; Rebholz, C.M. Ultra-Processed Food Intake and Mortality in the USA: Results from the Third National Health and Nutrition Examination Survey (NHANES III, 1988–1994). *Public Health Nutr.* **2019**, *22*, 1777–1785. [CrossRef]
41. Bonaccio, M.; Di Castelnuovo, A.; Costanzo, S.; De Curtis, A.; Persichillo, M.; Sofi, F.; Cerletti, C.; Donati, M.B.; de Gaetano, G.; Iacoviello, L. Ultra-Processed Food Consumption Is Associated with Increased Risk of All-Cause and Cardiovascular Mortality in the Moli-Sani Study. *Am. J. Clin. Nutr.* **2021**, *113*, 446–455. [CrossRef]
42. Rico-Campà, A.; Martínez-González, M.A.; Alvarez-Alvarez, I.; de Deus Mendonça, R.; de la Fuente-Arrillaga, C.; Gómez-Donoso, C.; Bes-Rastrollo, M. Association between Consumption of Ultra-Processed Foods and All Cause Mortality: SUN Prospective Cohort Study. *BMJ* **2019**, *365*, l1949. [CrossRef]
43. Blanco-Rojo, R.; Sandoval-Insausti, H.; López-Garcia, E.; Graciani, A.; Ordovás, J.M.; Banegas, J.R.; Rodríguez-Artalejo, F.; Guallar-Castillón, P. Consumption of Ultra-Processed Foods and Mortality: A National Prospective Cohort in Spain. *Mayo Clin. Proc.* **2019**, *94*, 2178–2188. [CrossRef]
44. Romero Ferreiro, C.; Martín-Arriscado Arroba, C.; Cancelas Navia, P.; Lora Pablos, D.; Gómez de la Cámara, A. Ultra-Processed Food Intake and All-Cause Mortality: DRECE Cohort Study. *Public Health Nutr.* **2021**, *25*, 1854–1863. [CrossRef]
45. Imamura, F.; Micha, R.; Khatibzadeh, S.; Fahimi, S.; Shi, P.; Powles, J.; Mozaffarian, D. Dietary Quality among Men and Women in 187 Countries in 1990 and 2010: A Systematic Assessment. *Lancet Glob. Health* **2015**, *3*, e132–e142. [CrossRef]
46. Monteiro, C.A.; Moubarac, J.-C.; Cannon, G.; Ng, S.W.; Popkin, B. Ultra-Processed Products Are Becoming Dominant in the Global Food System. *Obes Rev.* **2013**, *14* (Suppl. 2), 21–28. [CrossRef] [PubMed]
47. Latasa, P.; da Coasta Louzada, M.L.; Martinez Steele, E.; Monteiro, C.A. Added Sugars and Ultra-Processed Foods in Spanish Households (1990–2010). *Eur. J. Clin. Nutr.* **2018**, *72*, 1404–1412. [CrossRef] [PubMed]
48. Haro, J.M.; Tyrovolas, S.; Garin, N.; Diaz-Torne, C.; Carmona, L.; Sanchez-Riera, L.; Perez-Ruiz, F.; Murray, C.J. The Burden of Disease in Spain: Results from the Global Burden of Disease Study 2010. *BMC Med.* **2014**, *12*, 236. [CrossRef]
49. Shojaei, E.; Rexachs, D.; Wong, A.; Epelde, F.; Luque, E. A Method for Projections of the Emergency Department Behaviour by Non-Communicable Diseases From 2019 to 2039. *IEEE J. Biomed. Health Inform.* **2020**, *24*, 2490–2498. [CrossRef]
50. Trichopoulou, A.; Naska, A.; Costacou, T.; DAFNE III Group. Disparities in Food Habits across Europe. *Proc. Nutr. Soc.* **2002**, *61*, 553–558. [CrossRef]
51. Gómez-Jerique, J.A.; Rubio Herrera, M.A.; Gómez De La Cámara, A.; Gutiérrez Fuentes, J.A. Diet and Cardiovascular Risk in Spain Study (DRECE) Capítulo 2. El proyecto DRECE. *Med. Clin.* **2011**, *12*, 3–5.
52. Martin-Moreno, J.M.; Boyle, P.; Gorgojo, L.; Maisonneuve, P.; Fernandez-Rodriguez, J.C.; Salvini, S.; Willett, W.C. Development and Validation of a Food Frequency Questionnaire in Spain. *Int. J. Epidemiol.* **1993**, *22*, 512–519. [CrossRef]
53. Rodríguez, I.T.; Ballart, J.F.; Pastor, G.C.; Jordà, E.B.; Val, V.A. Validation of a short questionnaire on frequency of dietary intake: Reproducibility and validity. *Nutr. Hosp.* **2008**, *23*, 242–252.
54. Panel de Consumo Alimentario. Available online: https://www.mapa.gob.es/es/alimentacion/temas/consumo-tendencias/panel-de-consumo-alimentario/ (accessed on 17 November 2021).
55. Monteiro, C.A.; Cannon, G.; Moubarac, J.-C.; Levy, R.B.; Louzada, M.L.C.; Jaime, P.C. The UN Decade of Nutrition, the NOVA Food Classification and the Trouble with Ultra-Processing. *Public Health Nutr.* **2018**, *21*, 5–17. [CrossRef]
56. Greenacre, M.J. *La Práctica del Análisis de Correspondencias*; Fundación BBVA: Barcelona, Spain, 2008; ISBN 978-84-96515-71-0.
57. Romero Ferreiro, C.; Lora Pablos, D.; Gómez de la Cámara, A. Two Dimensions of Nutritional Value: Nutri-Score and NOVA. *Nutrients* **2021**, *13*, 2783. [CrossRef] [PubMed]
58. Hedeker, D.; Gibbons, R.D. *Longitudinal Data Analysis*; Wiley-Interscience: Hoboken, NJ, USA, 2006; p. 337, ISBN 978-0-471-42027-9.
59. Willett, W.C.; Sacks, F.; Trichopoulou, A.; Drescher, G.; Ferro-Luzzi, A.; Helsing, E.; Trichopoulos, D. Mediterranean Diet Pyramid: A Cultural Model for Healthy Eating. *Am. J. Clin. Nutr.* **1995**, *61*, 1402S–1406S. [CrossRef] [PubMed]
60. Bonaccio, M.; Costanzo, S.; Di Castelnuovo, A.; Persichillo, M.; Magnacca, S.; De Curtis, A.; Cerletti, C.; Donati, M.B.; de Gaetano, G.; Iacoviello, L.; et al. Ultra-Processed Food Intake and All-Cause and Cause-Specific Mortality in Individuals with Cardiovascular Disease: The Moli-Sani Study. *Eur. Heart J.* **2022**, *43*, 213–224. [CrossRef] [PubMed]
61. Juul, F.; Hemmingsson, E. Trends in Consumption of Ultra-Processed Foods and Obesity in Sweden between 1960 and 2010. *Public Health Nutr.* **2015**, *18*, 3096–3107. [CrossRef] [PubMed]
62. Wang, L.; Martínez Steele, E.; Du, M.; Pomeranz, J.L.; O'Connor, L.E.; Herrick, K.A.; Luo, H.; Zhang, X.; Mozaffarian, D.; Zhang, F.F. Trends in Consumption of Ultraprocessed Foods Among US Youths Aged 2–19 Years, 1999–2018. *JAMA* **2021**, *326*, 519–530. [CrossRef]
63. de Araújo, T.P.; de Moraes, M.M.; Magalhães, V.; Afonso, C.; Santos, C.; Rodrigues, S.S.P. Ultra-Processed Food Availability and Noncommunicable Diseases: A Systematic Review. *Int. J. Environ. Res. Public Health* **2021**, *18*, 7382. [CrossRef]

64. Boyland, E.J.; Nolan, S.; Kelly, B.; Tudur-Smith, C.; Jones, A.; Halford, J.C.; Robinson, E. Advertising as a Cue to Consume: A Systematic Review and Meta-Analysis of the Effects of Acute Exposure to Unhealthy Food and Nonalcoholic Beverage Advertising on Intake in Children and Adults. *Am. J. Clin. Nutr* **2016**, *103*, 519–533. [CrossRef]
65. Marti, A. Ultra-Processed Foods Are Not "Real Food" but Really Affect Your Health. *Nutrients* **2019**, *11*, 1902. [CrossRef]
66. Oxandabarat, A. OPS/OMS Uruguay—Guía Alimentaria para la Población Uruguaya | OPS/OMS. Available online: https://www3.paho.org/uru/index.php?option=com_content&view=article&id=1375:guia-alimentaria-para-la-poblacion-uruguaya&Itemid=451 (accessed on 30 June 2022).
67. Monteiro, C.A.; Cannon, G.; Moubarac, J.-C.; Martins, A.P.B.; Martins, C.A.; Garzillo, J.; Canella, D.S.; Baraldi, L.G.; Barciotte, M.; Louzada, M.L. da C.; et al. Dietary Guidelines to Nourish Humanity and the Planet in the Twenty-First Century. A Blueprint from Brazil. *Public Health Nutr.* **2015**, *18*, 2311–2322. [CrossRef]
68. Mauricio, H.-F.; Batis, C.; Rivera, J.A.; Colchero, M.A. Reduction in Purchases of Energy-Dense Nutrient-Poor Foods in Mexico Associated with the Introduction of a Tax in 2014. *Prev. Med.* **2019**, *118*, 16–22. [CrossRef]
69. Bíró, A. Did the Junk Food Tax Make the Hungarians Eat Healthier? *Food Policy* **2015**, *54*, 107–115. [CrossRef]
70. Baraldi, L.G.; Martinez Steele, E.; Canella, D.S.; Monteiro, C.A. Consumption of Ultra-Processed Foods and Associated Sociodemographic Factors in the USA between 2007 and 2012: Evidence from a Nationally Representative Cross-Sectional Study. *BMJ Open* **2018**, *8*, e020574. [CrossRef] [PubMed]
71. Khandpur, N.; Cediel, G.; Obando, D.A.; Jaime, P.C.; Parra, D.C. Factores sociodemográficos asociados al consumo de alimentos ultraprocesados en Colombia. *Rev. Saúde Pública* **2020**, *54*. [CrossRef]
72. de Deus Mendonça, R.; Pimenta, A.M.; Gea, A.; de la Fuente-Arrillaga, C.; Martinez-Gonzalez, M.A.; Lopes, A.C.S.; Bes-Rastrollo, M. Ultraprocessed Food Consumption and Risk of Overweight and Obesity: The University of Navarra Follow-Up (SUN) Cohort Study. *Am. J. Clin. Nutr.* **2016**, *104*, 1433–1440. [CrossRef] [PubMed]
73. Beslay, M.; Srour, B.; Méjean, C.; Allès, B.; Fiolet, T.; Debras, C.; Chazelas, E.; Deschasaux, M.; Wendeu-Foyet, M.G.; Hercberg, S.; et al. Ultra-Processed Food Intake in Association with BMI Change and Risk of Overweight and Obesity: A Prospective Analysis of the French NutriNet-Santé Cohort. *PLoS Med.* **2020**, *17*, e1003256. [CrossRef] [PubMed]
74. Pestoni, G.; Habib, L.; Reber, E.; Rohrmann, S.; Staub, K.; Stanga, Z.; Faeh, D. Ultraprocessed Food Consumption Is Strongly and Dose-Dependently Associated with Excess Body Weight in Swiss Women. *Obesity* **2021**, *29*, 601–609. [CrossRef] [PubMed]
75. Chandon, P.; Wansink, B. Does Food Marketing Need to Make Us Fat? A Review and Solutions. *Nutr. Rev.* **2012**, *70*, 571–593. [CrossRef]
76. Gómez de la Cámara, A.; De Andrés Esteban, E.; Urrútia Cuchí, G.; Calderón Sandubete, E.; Rubio Herrera, M.Á.; Menéndez Orenga, M.; Lora Pablos, D. Variability of Nutrients Intake, Lipid Profile and Cardiovascular Mortality among Geographical Areas in Spain: The DRECE Study. *Geospat. Health* **2017**, *12*, 524. [CrossRef]
77. Moreiras-Varela, O. The Mediterranean Diet in Spain. *Eur. J. Clin. Nutr.* **1989**, *43* (Suppl. 2), 83–87.
78. Dinu, M.; Tristan Asensi, M.; Pagliai, G.; Lotti, S.; Martini, D.; Colombini, B.; Sofi, F. Consumption of Ultra-Processed Foods Is Inversely Associated with Adherence to the Mediterranean Diet: A Cross-Sectional Study. *Nutrients* **2022**, *14*, 2073. [CrossRef]
79. Instituto Nacional de Estadística. (National Statistics Institute). Available online: https://www.ine.es/dynt3/inebase/index.htm?padre=1928&capsel=1928 (accessed on 30 June 2022).

nutrients

MDPI

Systematic Review

Maternal Consumption of Ultra-Processed Foods-Rich Diet and Perinatal Outcomes: A Systematic Review and Meta-Analysis

Walkyria O. Paula [1], Erika S. O. Patriota [1], Vivian S. S. Gonçalves [2] and Nathalia Pizato [1,*]

[1] Graduate Program in Human Nutrition, Faculty of Health Sciences, University of Brasilia, Brasilia 70910-900, Brazil

[2] Graduate Program in Public Health, Faculty of Health Sciences, University of Brasilia, Brasilia 70910-900, Brazil

* Correspondence: pizatonat@unb.br

Abstract: The consumption of ultra-processed food (UPF)-rich diets represents a potential threat to human health. Considering maternal diet adequacy during pregnancy is a major determinant for perinatal health outcomes, this study aimed to systematically review and meta-analyze studies investigating the association between maternal consumption of a UPF-rich diet and perinatal outcomes. Conducted according to the Preferred Reporting Items for Systematic Reviews and Meta-Analysis (PRISMA) guidelines, five electronic databases and gray literature using Google Scholar and ProQuest Dissertations and Theses Global were searched up to 31 May 2022. No restrictions were applied on language and publication date. Two reviewers independently conducted the study selection and data extraction process. Meta-analysis was conducted according to the random-effects model. In total, 61 studies were included in the systematic review and the overall population comprised 698,803 women from all gestational trimesters. Meta-analysis of cohort studies showed that maternal consumption of UPF-rich diets was associated with an increased risk of gestational diabetes mellitus (odds ratio (OR): 1.48; 95% confidence interval (CI): 1.17, 1.87) and preeclampsia (OR: 1.28; 95% CI: 1.15, 1.42). Neonatal outcomes showed no association. The overall GRADE quality of the evidence for the associations was very low. The findings highlight the need to monitor and reduce UPF consumption, specifically during the gestational period, as a strategy to prevent adverse perinatal outcomes.

Keywords: maternal diet; NOVA classification; perinatal outcomes

Citation: Paula, W.O.; Patriota, E.S.O.; Gonçalves, V.S.S.; Pizato, N. Maternal Consumption of Ultra-Processed Foods-Rich Diet and Perinatal Outcomes: A Systematic Review and Meta-Analysis. *Nutrients* **2022**, *14*, 3242. https://doi.org/10.3390/nu14153242

Academic Editors: Bruce W. Hollis, Daniela Martini and Monica Dinu

Received: 2 July 2022
Accepted: 1 August 2022
Published: 8 August 2022

1. Introduction

Significant metabolic and physiological changes occur during pregnancy, to support fetal growth and development [1]. Maternal diet quality is a major determinant for perinatal outcomes including hypertensive disorders, gestational diabetes, low birth weight, large gestational age, and preterm birth [2]. Furthermore, inadequate diet quality during pregnancy is associated with chronic diseases in later life such as type 2 diabetes mellitus, obesity, hypertension, and cardiovascular disorders [3].

Additionally to the evidence of the relationship between maternal diet quality and perinatal outcomes, several studies have reported high consumption of unhealthy and ultra-processed foods (UPFs) by pregnant women indicating a generally worse quality of diet [4–7].

The NOVA food classification system has been applied worldwide to evaluate the impact of modern industrial food systems on human diet and health according to the nature, extent, and purpose of food processing [8]. NOVA categorizes foods according to the degree of processing: in natura or minimally processed, processed culinary ingredients, processed food, and UPFs. UPFs are defined as industrial formulations manufactured from processed substances extracted or refined from whole foods. They are typically energy-dense products, with high amounts of sugar, fat, and salt, and low in dietary fiber, protein, vitamins, and minerals. UPFs also include industrial ingredients, such as hydrogenated fat, protein isolates, and additives such as colors, flavors, artificial sweeteners, and emulsifiers [9]. Some examples include products such as fast foods, cereal bars, cakes, ice cream, pizza, sausages, and soft drinks [10].

UPF intake is considered a hallmark of the Western diet and other unhealthy eating patterns such as the Prudent diet, characterized by a high intake of energy-dense and processed food, and rich in industrialized food-like products that are typically made with low-quality ingredients and deliver little nutritional value [11]. UPFs have become increasingly prevalent in the food supply system globally since they are designed to be attractive, palatable, cheap, and convenient products [12]. They account for more than 50% of the energy intake in developed countries such as the USA [13] and the UK [14] and are widely prominent in the diets of populations in lower-middle-income countries [15,16]. A recent meta-analysis of nationally representative samples showed an inverse linear relation between UPFs and less-processed foods when considered in relation to other food groups. The study also indicated that the increase in UPF intake was correlated with an increase in nutrients such as free sugars, total fats, and saturated fats, as well as a decrease in fiber, protein, potassium, zinc, and magnesium, and vitamins A, C, D, E, B3 and B12 [17]. Considering that during pregnancy women need a higher amount of the majority of nutrients to achieve optimal fetal growth and birth weight, varied diets and increased nutrient intake are needed to cope with the extra demand. Associations between maternal UPF consumption and perinatal outcomes have been investigated during the past years, however the findings are limited and inconsistent. Some studies have reported a significant association between consumption of UPF-rich diets during pregnancy and excessive gestational weight gain (GWG) [4,18], higher gestational diabetes mellitus (GDM) risk [19], hypertensive disorders of pregnancy (HDP) such as preeclampsia [20], low birth weight (LBW) [21] and preterm birth [22], while others have shown no association [7,23].

Previous systematic reviews have explored the association between maternal dietary patterns and maternal or infant outcomes [24–26]. However, these studies did not consider the degree of food processing, which has become an important aspect of diet quality [10].

A recent systematic review [27] reported that the highest UPF consumption negatively impacts nutrition and disease development indicators in pregnant, lactating women and children. However, a meta-analysis of the results was not conducted, and no other dietary patterns characterized by high UPF consumption were explored during the pregnancy period.

Since the pregnancy period is considered a window of opportunity to improve dietary intake which is considered a modifiable risk factor [28], a better understanding of maternal UPF consumption effects on perinatal outcomes is crucial to promoting mother and infant health. Thus, this study aimed to determine the association between UPF-rich diet consumption by pregnant women and perinatal (maternal and neonatal) outcomes through a comprehensive systematic review with meta-analysis. The hypothesis was that a higher intake of UPF-rich diet during pregnancy is associated with adverse perinatal outcomes.

2. Materials and Methods

This systematic review was conducted in accordance with the Preferred Reporting Items for Systematic Reviews and Meta-Analyses (PRISMA) statement for reporting systematic reviews [29] and its protocol was registered on the International Prospective Register of Systematic Reviews (PROSPERO) under registry number CRD42021257210. The PECOS acronym (Population, Exposure, Comparison, Outcome, and Study design) was used to elaborate the guiding research question as follows: "Is consumption of a UPF-rich diet during pregnancy associated with adverse perinatal outcomes?" (Supplementary Materials Table S1).

2.1. Eligibility Criteria

This review included observational studies (cross-sectional, longitudinal, case-control) that reported a measure of association (relative risk, odds ratio, or β-coefficients with confidence interval) between UPF-rich diet consumption and perinatal outcomes. For this review, we considered it UPF-rich diet consumption when the evaluated food, diet, or dietary pattern included at least one food from the UPF group defined by the NOVA Food Classification System [9], such as fast foods, junk foods, processed meats, soft drinks, confectionaries, pizzas, hamburgers, candies and sweets, sweetened beverages and cookies. Diet patterns described as unhealthy dietary patterns compared to healthy patterns, and Western and Prudent diet patterns which are characterized by a higher intake of red and processed meats, beverages sweetened with sugar, sweets, desserts, industrialized food-like products, and refined grains with a high intake of energy-dense and processed foods, were also considered as a proxy for high UPF intake. No date of publication or language restriction was applied.

Studies including pregnant women with pre-existing diseases, animal studies, letters to editors, reviews, personal opinions, reviews, book chapters, editorials, congress abstracts, or any publication without primary data were excluded. Studies that evaluated individual nutrient or diet scores and studies without the required data being available even after at least two attempts to contact the authors by e-mail were also excluded.

2.2. Information Sources and Search Strategy

A systematic literature search was performed on 10 June 2021, and updated on 31 May 2022, using the following databases: Medline, Embase, Scopus, Web of Science, and Lilacs (BVS). Furthermore, a gray literature search was also performed using ProQuest Dissertations and Theses Global and Google Scholar (limited to the first 200 most relevant results). The reference lists of selected articles were hand-searched to identify additional relevant publications.

The search strategy was comprised of free text words and identified terms in Medical Subject Headings and Health Sciences Descriptors for participants, exposure, and outcomes. The following terms and words combinations were searched: (pregnancy OR pregnancies OR gestation OR "pregnant women" OR "pregnant woman" OR maternal OR antenatal) AND (ultraprocessed food OR "ultra-processed food" OR "industrialized food" OR "processed food" OR "ready-to-eat meal" OR "ready-to-eat food" OR "ready-prepared food" OR "salty food" OR "high-fat diet" OR "highly processed foods" OR "refined food" OR "fast food" OR "junk food" OR "sugar-sweetened beverages" OR "soft drink" OR "unhealthy eating" OR "unhealthy diet" OR "poor diet" OR "processed meat") AND ("perinatal outcome" OR "pregnancy outcome" OR "pregnancy complications" OR "gestational weight gain" OR "pregnancy weight gain" OR "birth outcomes" OR "birth weight" OR "neonatal weight" OR "newborn weight" OR "birth size" OR "pregnancy-induced hypertension" OR "hypertensive disorders" OR "gestational diabetes" OR "glycemic outcomes" OR "premature birth" OR "preterm birth" OR "fetal growth"). The search strategy quality was assessed by an investigator with experience in systematic reviews and expertise in the subject in accordance with the Peer Review of Electronic Search Strategies (PRESS) checklist [30]. The full search strategy for each database is available in Supplementary Materials Table S2.

2.3. Study Selection

The selection process for the review was independently conducted by two reviewers (WOP and ESOP) in two steps. First, the titles and abstracts of all retrieved articles were screened, according to the eligibility criteria. Then, the selected potentially eligible studies were submitted for full-text analysis. Articles that met the eligibility criteria were included in the review. Disagreements were resolved by consensus. Duplicates were identified and removed using the reference management tool Mendeley Desktop (version 1.19.8). The Rayyan QCRI software (Qatar Computing Research Institute®, Doha, Qatar) was used for the screening of articles.

2.4. Data Extraction

Data extraction was carried out by one author and cross-checking of all information was performed by a second author using a standardized spreadsheet. The following data were extracted from the original selected articles: authors and year of publication, data collection year, follow-up time, year of publication, study design, the country in which the study was conducted, sample size, age of participants, gestational age, denomination and composition of dietary components, dietary assessment methods, main outcomes, outcome measures, measures of effect size with confidence interval (CI), details of adjustment for confounding factors, and study funding/support information. When multiple estimates were reported, the results with adjustment for the highest number of confounders were used. When necessary, the respective study authors were contacted to retrieve additional information. At least two attempts were made to request missing or additional information.

2.5. Appraisal of Methodological Quality

Two investigators (W.O.P and E.S.O.P.) independently assessed the methodological quality of each included study using the Joanna Briggs Institute Critical Appraisal tools according to each study design (cohort, cross-sectional, and case-control) [31]. The tool consists of questions answered as "yes", "no", "unclear", or "not applicable". In this study, the risk of bias was considered low when all items were answered "yes" or "not applicable"; If the response to any item was "no" or "unclear", a high risk of bias was expected. Disagreements were resolved by consensus. The analysis of the relative frequency of each investigated domain was presented and no scores were assigned.

2.6. Summary Measures and Data Analysis

The primary outcomes were the associations between UPF-rich diet consumption and maternal (GWG, GDM, or HDP) and neonatal (LBW, large for gestational age (LGA), or preterm birth) outcomes along with the respective 95% confidence intervals (CI).

Meta-analysis was conducted when at least three studies provided data for a given outcome. In order to minimize heterogeneity, the meta-analysis included only prospective cohort studies, since it is the most adequate approach to assess associations. The overall associations were analyzed using the DerSimonian and Laird random-effects models. Based on data availability, the odds ratio (OR) and 95% CI were measured for maternal (GWG, GDM, or HDP) and neonatal (LBW, large for gestational age (LGA), or preterm birth) outcomes. If studies reported a measure of relative risk (RR), it was converted to OR using the proposed methods of Zhang and Yu [32]. Studies that report the coefficient (β) of the regression were analyzed separately. Statistical heterogeneity between studies was measured using the I-Square (I^2). Heterogeneity was considered important if I^2 values were higher than 40% [33]. Data analysis was performed using Stata software (StataCorp. 2019. Stata Statistical Software: Release 16.1. College Station, TX, USA: StataCorp LLC). When eligible studies did not report data in a form that could be included in the meta-analysis, they were included in the systematic review and qualitatively analyzed. Cross-sectional and case-control studies were also narratively summarized. Publication bias analyses were performed when at least ten studies were available for an outcome measure using Egger's test with a 5% significance level and funnel plot visual inspection [33].

2.7. Quality of Meta-Evidence

The Grading of Recommendations Assessment, Development, and Evaluation (GRADE) system was used to evaluate the certainty of the evidence for each exposure–outcome association based on the major domains of study limitations. The quality of evidence was downgraded based on five criteria: risk of bias, inconsistency of results, indirectness of evidence, imprecision, and publication bias when it was assessed [34].

3. Results

3.1. Selection of Studies

The flow chart of the study selection process is presented in Figure 1. The database search retrieved 11,089 articles. After the removal of duplicates, 4.918 article titles and abstracts were screened. Of these, 151 full-text articles were further assessed for eligibility and, finally, 61 studies [4,18–22,35–89] met the inclusion criteria and were included in this systematic review. The complete list of reasons for the exclusion of articles is presented in Supplementary Materials Table S3.

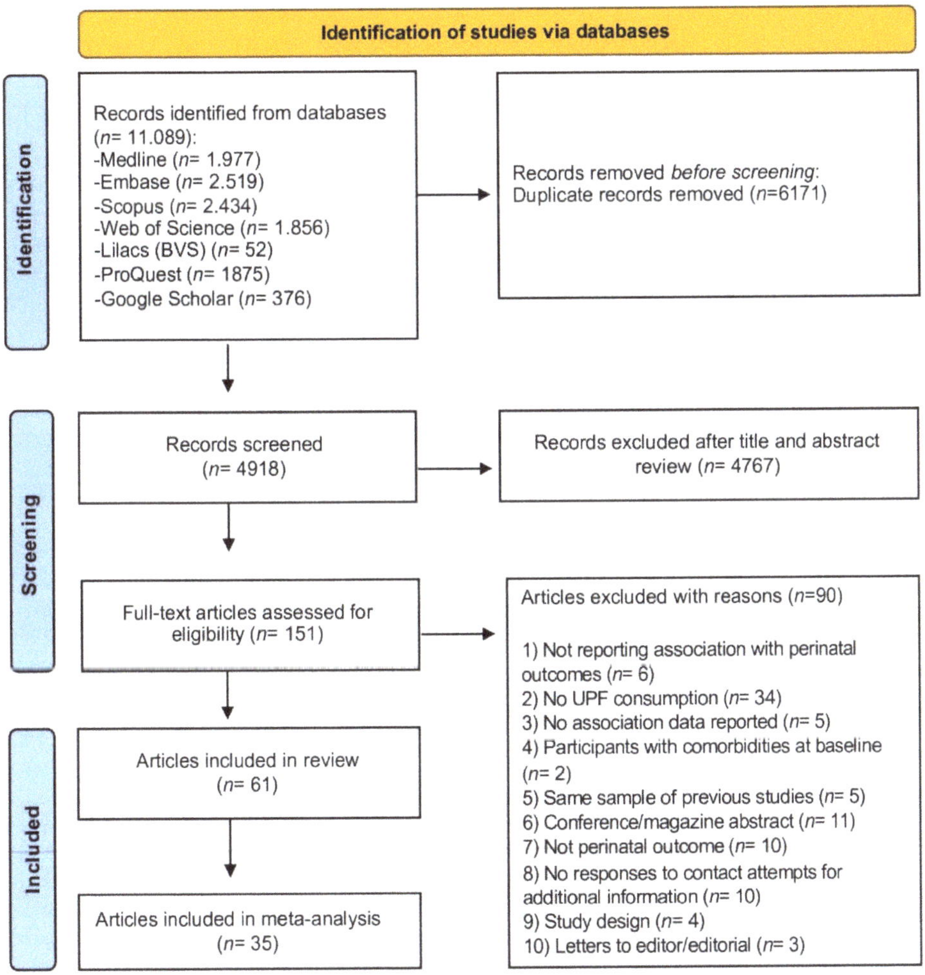

Figure 1. Flowchart of the study selection process. Adapted from PRISMA.

3.2. Study Characteristics

The articles were published between 2006 [57] and 2022 [89]. The sample ranged from 45 [4] to 94.062 [48] with 698.803 pregnant women evaluated in total. The included studies were conducted in Africa [50,51], Asia [19,35–49], America [4,18,52–65,89], Europe [20,21,66–86] and Oceania [22,87,88]. Forty-seven of the studies had a cohort design [4,18–20,36,40,42,44,46,48–60,62,63,66–79,81–88], nine were cross-sectional [22,43,45, 47,61,64,65,80,89] and five case-control [21,35,37,38,41]. Maternal mean age ranged from 24 ± 8 [37] a 37 ± 4 years old [67] and gestational week from ≤6 [19] to 37 [64] in baseline.

Regarding the exposure to UPF-rich diet consumption, seventeen articles assessed Western Diet Pattern (characterized by the presence of unhealthy foods such as savory and sweet snacks, cakes, cookies, desserts, refined grains, processed meats, fast foods, confectionaries and soft drinks) [20,35–41,51,57,62,67,68,71,80,83,85]; the intake of sweetened beverages was explored in twelve articles [46,49,52,56,64,70,72,73,75,78,79,82]; and specific manufactured food groups including UPF were analyzed in twelve articles [4,18,22,43,44,55,59,60,76,81,89]. In addition, studies also reported maternal consumption of junk foods [50,87], processed meats [65,69], snacks [61,84], industrial sweets [21,58,65], fast foods [19,42,50,54,66,74,77], "unhealthy food pattern" [45,86,88], "high salt pattern" [35], and ready-to-eat food [48].

Regarding to maternal outcomes, GWG was investigated in thirteen articles [4,18,36,42,51,58,64,67,77,81,84,89,90]; fifteen explored the association between maternal consumption and GDM [19,38,41,42,49,56,57,61,62,64,69,71,72,74,78]; and eight reported HDP, including maternal hypertension [20,35,39,52] and preeclampsia [20,37,39,45,75,76]. Two articles explored depressive symptoms during pregnancy [46,88]. Neonatal outcomes included LBW, investigated in eleven articles [21,40,43,44,47,48,53,65,73,80,86]; LGA, investigated in eight articles [47,50,54,66,68,73,82,87]; birth length, explored in four articles [48,54,60,86]; one publication reporting body mass index (BMI)/age at birth [59]; five reporting preterm birth [22,48,55,83,85]; and offspring congenital heart defects, examined in two publications [70,79].

3.3. *Results of Individual Studies*

A summary of the characteristics and main results of each study is presented in Table 1.

Regarding the cohort studies evaluating GWG, higher odds ratios of excessive GWG were associated with snack dietary pattern (OR: 1.01; 95% CI:1.004, 1.032) [84], UPF dietary patterns such as margarine, sugar, and chips (OR: 1.45; 95% CI: 1.06, 1.99) [81], and Western dietary pattern (OR: 4.04; 95% CI: 1.07, 15.24) [36]. Gomes et al. [18] showed that each 1% increase in energy intake from UPF was associated with a mean increase of 4.17 g in weekly gestational weight (95% CI: 0.5, 7.79). Other studies also presented an increase in GWG rate associated with a UPF-rich diet consumption. Rohatgi et al. found that each one percent increase in energy intake from UPF was associated with 1.33 kg increase in total GWG (CI: 0.3, 2.4) [4]. Similarly, Maugeri et al. showed that a Western diet consumption was associated with an increase of 1217 kg in total GWG ($p = 0.013$) [67]. A UPF rich-diet was also associated with a slight increase of 0,029 kg (β: 0.029; 95% CI: 0.012, 0,049) [42] and 0,01 kg (β: 0.010; SE: 0.003; $p = 0.004$) in weekly GWG [77]. Conversely, Hirko et al. [58] observed that intake of added sugar (including soft drinks, sugary fruit-flavored drinks, candies and cookies, cakes, pies, or brownies) was associated with a slight reduction in the likelihood of excessive GWG (OR: 0.91; 95% CI: 0.84, 0.99).

Lamyian et al. [19] observed greater chances of developing GDM among pregnant women with higher consumption of fast foods (OR: 2.12; 95% CI: 1.12, 5.43). Six cohort studies also identified an association between the consumption of UPF and a higher risk of GDM [56,57,69,71,74,78]. Three studies [49,62,71] found no significant association.

A Brazilian cohort [52] identified an association between soft drink consumption and hypertension during pregnancy (RR: 1.45; 95% CI: 1.16, 1.82). Ikem et al. [20] showed that higher consumption of the Western dietary pattern increased the odds of gestational hypertension by 18% (OR: 1.18; 95% CI: 1.05, 1.33). On the other hand, Hajianfar et al. [39] observed that consumption of the Western pattern was associated with lower chances of systolic (OR: 0.13, 95% CI: 0.04, 0.42) and diastolic (OR: 0.08; 95% CI: 0.01, 0.67) hypertension. Our results present a positive association between UPF consumption and preeclampsia observed in four cohort studies [20,39,75,76].

Table 1. Summary of included studies characteristics.

Author, Year Country	Study Design	Age (Years)	GW (Range or Mean)	Sample n =	Exposure	Outcome	Main Results
Abbasi et al., 2019 Iran [37]	Case-control	case: 24 ± 8 control: 26 ± 6	>20 weeks	case: 170 control: 340	WDP (red and processed meat, fried potatoes, pickles, sweets, pizza)	Risk of preeclampsia	The Western dietary pattern associated with preeclampsia: (OR: 5.99; 95% CI: 3.414, 10.53; $p < 0.001$)
Alves-Santos et al., 2019 Brazil [54]	Prospective Cohort	26.7 ± 5.5	5–13 weeks	193	Fast foods and candies (fast food and snacks; cakes, cookies, or crackers; and candies or desserts)	LGA Birth Length (BL)	Fast food and candies dietary pattern associated with LGA newborn: OR: 4.38; 95% CI: 1.32, 14.48 Fast food and candies dietary pattern associated with the newborn with BL > 90th percentile: OR: 4.81; 95% CI: 1.77, 13.07
Amezcua-Prieto et al., 2019 Spain [21]	Case-control	NR	NR	518	Industrial sweets	SGA	Intake of industrial sweets associated with odds of having an SGA newborn (OR: 2.70; 95% CI: 1.42, 5.13).
Ancira-Moreno et al., 2020 Mexico [53]	Prospective Cohort	25.08 ± 5.8	2nd and 3rd trimester	660	Mixed dietary patterns (sugary drinks, juices and sodas, red and processed meat, cereals)	LBW	The mixed dietary pattern associated risk LBW infant: (OR: 1.58; 95% CI: 0.63, 3.44)
Angali, Shahri, Borazjani, 2020, Iran [42]	Prospective Cohort	≥18 years	<13 weeks	488	"High fat - fast food" pattern (refined cereal, processed meat and high-fat dairy and juices)	GWG and hyperglycemia	High fat-fast food patterns associated with higher GWG (β: 0.029; 95% CI: 0.012, 0.049).
Asadi et al., 2019 Iran [38]	Case-control	case: 29 ± 5.17 control: 27.5 ± 4.92	24–28 weeks	case: 130 control: 148	WDP (SSB, refined grain products, fast foods, salty snacks, sweets and biscuit, mayonnaise)	GDM	The prudent dietary pattern associated with GDM risk: (OR: 0.88; 95% CI: 0.44, 0.99)
Barbosa et al., 2021 Brazil [52]	Prospective Cohort	>14	22–25 weeks	2750	Soft drinks	Gestational Hypertension (GH)	Soft drink consumption > 7 times per week associated with GH: (RR: 1.45; 95% CI: 1.16, 1.82; $p = 0.001$)
Bärebring et al., 2016 Sweden [84]	Prospective Cohort	32.1 (IQR: 30.8–35.3)	35.9 weeks (IQR: 35.1–36.4)	95	Snacks pattern (sweets, cakes, biscuits, potato chips, popcorn)	GWG	Snacks pattern associated with excessive GWG (OR: 1.018; 95% CI: 1.004, 1.032).
Baskin et al., 2015 Australia [88]	Prospective Cohort	30.55 ± 4.24	16 weeks	167	Unhealthy dietary patterns (sweets and desserts, refined grains, high- energy drinks, fast foods, hot chips, high-fat dairy, fruit juice and red meats)	Depressive symptoms	An unhealthy diet at T2 is associated with depressive symptoms: β: 0.19; 95% CI = 0.04, 0.34; $p < 0.05$
Borgen et al., 2012 Norway [75]	Prospective Cohort	>18 years	15 weeks	32,933	SSB	Preeclampsia	Sugar-sweetened beverages associated with increased risk of preeclampsia: OR: 1.27; 95% CI: 1.05, 1.54
Brantsæter et al., 2009 Norway [76]	Prospective Cohort	>18	20.7 weeks (SD ± 3.7)	23,423	Dietary patterns (Processed meat products, white bread, French fries, salty snacks, and sugar-sweetened drinks)	Risk of preeclampsia	Processed food patterns are associated with increased risk of developing preeclampsia (OR: 1.21; 95% CI: 1.03, 1.42).
Chen et al., 2009 USA [56]	Prospective Cohort	24–44	NR	13,475	SSB	Risk of gestational diabetes mellitus (GDM)	Intake of sugar-sweetened cola associated with risk of GDM (RR: 1.22; 95% CI: 1.01, 1.47).

Table 1. *Cont.*

Author, Year Country	Study Design	Age (Years)	GW (Range or Mean)	Sample n =	Exposure	Outcome	Main Results
Chen et al., 2020 China [35]	Case-control	case: 28 ± 1.3 control: 28 ± 1.5	>22 weeks	case: 1290 control: 1290	High-salt pattern (pickled vegetables, processed and cooked meat, fish and shrimp, bacon and salted fish, bean sauce)	Hypertensive disorder during pregnancy	High-salt pattern diets associated with higher systolic blood pressure: (r: 0.110; $p < 0.05$)
Coelho et al., 2015 Brazil [63]	Prospective Cohort	24.7 ± 6.1	≥22 weeks	1298	Snack dietary patterns (sandwich cookies, salty snacks, chocolate, and chocolate drink)	Birth weight	Snack dietary patterns positively associated with birth weight: (β: 56.64; $p = 0.04$) in pregnant adolescents.
Dale et al., 2019 Norway [79]	Prospective Cohort	≥18	16–18 weeks	88,514	SSB	CHD	25–70 mL/day sucrose-sweetened soft beverages associated with non-severe CHD (RR: 1.30; 95% CI: 1.07, 1.58) and (RR: 1.27; 95% CI: 1.06, 1.52) for ≥70 mL/day.
Dominguez et al., 2014 Spain [74]	Prospective Cohort	>18	NR	3048	Fast food	GDM	Fast food consumption associated with GDM risk: (OR: 1.86; 95% CI: 1.13, 3.06)
Donazar-Ezcurra et al., 2017 Spain [71]	Prospective Cohort	>18	NR	3455	WDP (red meat, high-fat processed meats, potatoes, commercial bakery products, whole dairy products, fast foods, sauces, pre-cooked foods, eggs, soft drinks and sweets, chocolates)	GDM	The Western dietary pattern associated with GDM incidence: (OR: 1.56; 95% CI: 1.00, 2.43; $p = 0.05$)
Donazar-Ezcurra et al., 2017 Spain [72]	Prospective Cohort	>18	NR	3396	Soft drinks	GDM	Sugar-sweetened soft drinks (SSSD) associated with GDM: (OR: 2.06; 95% CI: 1.28, 3.34; p: 0.004)
Englund-Ögge et al., 2014 Norway [85]	Prospective Cohort	<20 to ≥40	15 weeks	66,000	WDP (salty snacks, chocolates and sweets, French fries, cakes, white bread, ketchup, dairy desserts, SSB, mayonnaise, processed meat, waffles, pancakes, cookies)	Preterm delivery	Western diet pattern associated with risk of preterm delivery (Hazard Ratio: 1.02; 95% CI: 0.92, 1.13).
Englund-Ögge et al., 2019 Norway [68]	Prospective Cohort	>18 years	15 weeks	65,904	WDP (salty snacks, chocolate and sweets, cakes, French fries, white bread, ketchup, SSB, processed meat products, and pasta)	LGA	The prudent pattern associated with decreased LGA risk: (OR: 0.84; 95% CI: 0.75, 0.94) The traditional group associated with increased LGA risk: (OR: 1.12; 95% CI: 1.02, 1.24)
Ferreira et al., 2022 Brazil [89]	Cross-sectional	28 (IQR 19–45)	NR	260	Dietary patterns (sweets, snacks and cookies)	GWG	Women with greater adherence to "Pattern 2" (sweets, snacks, and cookies) during pregnancy were less likely to have inadequate GWG (OR: 0.14; 95% CI = 0.03, 0.60)

Table 1. *Cont.*

Author, Year Country	Study Design	Age (Years)	GW (Range or Mean)	Sample n =	Exposure	Outcome	Main Results
Garay et al., 2019 United Kingdom [80]	Cross-sectional	18–45 years	NR	303	WDP (cakes/biscuits/ice cream, chips/crisps, processed meat, takeout, chocolate, soft drinks)	CBWC	Health-conscious dietary pattern associated with increased CBWC (OR: 4.75; 95% CI: 1.17, 8.33; $p = 0.010$) "Western Diet" associated with increased CBWC (β: −2.64; 95% CI: −5.87, 0.59; $p = 0.109$)
Gomes et al., 2020 Brazil [18]	Prospective Cohort	≥18 years	All trimesters	259	UPF energy (cookies, sweets, SSB, reconstituted meats, crackers, packaged chips, frozen dinners, ultra-processed breads)	GWG	Energy percentage derived from UPF associated with average weekly GWG (β: 4.17; 95% CI 0.55, 7.79).
Grieger, et al., 2014 Australia [22]	Cross-sectional	>18	13 weeks	309	Dietary patterns (high-fat/sugar/takeaway: takeaway foods, potato chips, refined grains, and added sugar)	Preterm delivery	High-fat/sugar/takeaway pattern associated with preterm birth: (OR: 1.54; 95% CI: 1.10, 2.15; $p = 0.011$)
Grundt et al., 2016 Norway [73]	Prospective Cohort	>18	15 weeks	50,280	SSC	BW	Each 100 mL intake of SSC associated with: 7.8 g decrease in BW (95% CI: −10.3, 5.3); decreased risk of BW > 4.5 kg (OR: 0.94; 95% CI: 0.90, 0.97) and increased risk of BW < 2.5 kg (OR: 1.05; 95% CI: 0.99, 1.10).
Günther et al., 2019 Germany [66]	Prospective Cohort	30.3 ± 4.4	<12 weeks	1995	Fast foods	LGA	Fast food consumption associated with LGA: (OR 3.14; 95% CI: 1.26, 7.84; $p = 0.014$)
Hajianfar et al., 2018 Iran [39]	Prospective Cohort	20–40	8–16 weeks	812	WDP (processed meats, fruits juice, citrus, nuts, desserts and sweets, potato, legumes, coffee, egg, pizza, high fat dairy, and soft drinks)	Preeclampsia Hypertension	The Western dietary pattern is associated with: Preeclampsia: (OR: 2.08; 95% CI: 1, 4.36, $p = 0.02$) High systolic blood pressure: (OR: 0.13; 95% CI: 0.04, 0.42; $p = 0.002$)
Hajianfar et al., 2018 Iran [40]	Prospective Cohort	29.4 ± 4.85	8–16 weeks	812	WDP (processed meats, fruits juice, citrus, nuts, desserts and sweets, potato, legumes, coffee, egg, pizza, high fat dairy, and soft drinks)	LBW	Western dietary pattern (top quartile) associated with LBW infant: (OR: 5.51; 95% CI: 1.82, 16.66; $p = 0.001$)
Hirko et al., 2020 USA [58]	Prospective Cohort	mean: 27	mean: 13.4 weeks	327	Dietary patterns (added sugar: soda, fruit-flavored drinks with sugar, pastries—donuts, sweet rolls, Danish, and cookies, cake, pie, or brownies)	GWG	Higher added sugar intake associated with excessive GWG (OR: 0.91; 95% CI: 0.84, 0.99)
Ikem et al., 2019 Denmark [20]	Prospective Cohort	25–30	12 weeks	55,139	WDP (potatoes, French fries, bread white, pork, beef veal, meat mixed, meat cold and dressing sauce)	Gestational hypertension Preeclampsia	Western diet associated with GH: (OR: 1.18; 95% CI: 1.05, 1.33) Preeclampsia: (OR: 1.40; 95% CI: 1.11, 1.76)
Itani et al., 2020 United Arab Emirates [36]	Prospective Cohort	19–40	27–42 weeks	242	WDP (sweets, sweetened beverages, added sugars, fast food, eggs, and offal)	GWG	The Western pattern is associated with excessive gestational weight gain (OR: 4.04; 95% CI: 1.07, 15.24) The western pattern is associated with gestational weight gain rate (OR: 4.38; 95% CI: 1.28, 15.03)

Table 1. *Cont.*

Author, Year Country	Study Design	Age (Years)	GW (Range or Mean)	Sample n =	Exposure	Outcome	Main Results
Ker et al., 2021 Taiwan [46]	Prospective Cohort	33.9 ± 4.6	All trimesters	196	SSB	Postpartum depression	SSB intake associated with increased EPDS scores: (β: 0.25; 95% CI: 0.04, 0.45) during the first and second trimesters
Lamyian et al., 2017 Iran [19]	Prospective Cohort	18–45 years	≤6 weeks	1026	Fast food	GDM	Fast food consumption (≥175 g/week) associated with GDM risk: (OR: 2.12; 95% CI: 1.12, 5.43; p-trend: 0.03)
Liu et al., 2021 China [47]	Cross-sectional	26.88 ± 4.62	All trimesters	7934	Dietary patterns (snacks pattern: beverages, sweetmeat, fast-food, dairy and eggs)	Macrossomia SGA	Snacks pattern associated with: risk of macrosomia: (OR: 1.265; 95% CI: 1.000, 1.602) SGA: (OR: 1.260; 95% CI: 1.056, 1.505).
Loy, Marhazlina; Jan 2013 Malaysia [43]	Cross-sectional	29.7 ± 4.8	33.66 ± 3.95 weeks	108	Dietary patterns (confectioneries: cake, cookies, chocolate, candy, sweetened condensed milk)	LBW	Confectioneries food intake associated with lower birth weight: (β: −1.999; p = 0.013)
Marí-Sanchiz et al., 2017 Spain [69]	Prospective Cohort	>18	NR	3298	UPF (Processed meat)	GDM	Processed meat consumption associated with GDM: (OR: 2.01; 95% CI: 1.26, 3.21; p-trend 0.003)
Marquez, 2012 USA [64]	Cross-sectional	18–49	≥37 weeks	290	SSB	GWG	A high intake of regular soda is associated with an increased risk of Excessive GWG (OR: 1.41; 95% CI: 0.60, 3.31).
Martin et al., 2016 Sweden [59]	Prospective Cohort	16–47	39 ± 2 weeks	389	Dietary patterns (latent class 3: white bread, red and processed meats, fried chicken, French fries, and vitamin C–rich drinks)	BMI-for-age at birth	Association between the latent class 3 diet (processed food) and BMI-for-age z-score at birth:(β: −0.41; 95% CI: −0.79, −0.03).
Martin et al., 2015 USA [55]	Prospective Cohort	NR	24–29 weeks	3941	Dietary patterns (hamburgers or cheeseburgers, white potatoes, fried chicken, beans, corn, spaghetti dishes, cheese dishes, processed meats, biscuits, and ice cream)	Preterm birth	Diet characterized by ultra-processed food associated with preterm birth: (OR: 1.53; 95% CI: 1.02, 2.30)
Maugeri et al., 2019 Italy [67]	Prospective Cohort	15–50 (Mean: 37)	4–20 weeks (Mean: 16)	232	WDP (high intake of red meat, fries, dipping sauces, salty snacks and alcoholic drinks)	GWG	Western dietary patterns associated with GWG: (β: 1.217; Standard Error: 0.487; p = 0.013)
Mikeš et al., 2021 Czech Republic [86]	Prospective Cohort	25 ± 5	32 weeks	4320	Unhealthy Dietary pattern: (pizza, fish products, processed meat, sausages, smoked meat, hamburgers, and confectionary foods, sugary drinks, cakes, chocolate and sweets).	Birth Weight Birth Length	A 1-unit increase in the unhealthy pattern score was associated with a mean birth weight reduction of −23.8 g (95% CI: −44.4, −3.3; p = 0.023); a mean birth length reduction of −0.10 cm (95% CI: −0.19, −0.01; p = 0.040).

Table 1. *Cont.*

Author, Year Country	Study Design	Age (Years)	GW (Range or Mean)	Sample n =	Exposure	Outcome	Main Results
Mitku et al., 2020 South Africa [50]	Prospective Cohort	<25 to >30	1st and 2nd trimesters	687	Junk food (sweets, muffins, chips, mixed salad, fruit juice, fizzy soft drinks, vetkoek, coffee creamer, cooking oil, hamburgers, cooked vegetables, cereals rice, margarine)	Birth Weight	Junk food intake is associated with an increase in birth weight ($p < 0.001$).
Nascimento et al., 2016 Brazil [62]	Prospective Cohort	26.2 ± 5.8	26.4 weeks (SD ± 0.8)	841	WDP (white bread, savory, sweet, chocolate, cookies, soft drinks, pasta, fried food, pizza, chicken, canned food)	GDM	Association between GDM incidence and dietary patterns (RR: 0.78; 95% CI: 0.43, 1.43)
Nicolì et al., 2021 Italy [78]	Prospective Cohort	35.75 ± 5.53	NR	376	Soft drink	GDM	Non-nutritive-sweetened soft drink consumption associated with GDM (OR: 1.766; 95% CI: 1.089, 2.863; $p = 0.021$)
Okubo et al., 2012 Japan [44]	Prospective Cohort	≥18	All trimesters	803	Dietary patterns (wheat products pattern: bread, confectioneries, fruit and vegetable juice, and soft drinks)	SGA birth	Wheat products pattern associated with SGA infant: (OR: 5.2; 95% CI: 1.1, 24.4)
Rasmussen et al., 2014 Denmark [83]	Prospective Cohort	21–39	2nd trimester	69,305	WDP (French fries, white bread, meat mixed, margarine, dressing sauce, chocolate milk, soft drink, cakes, chocolate, candy, sweet spread, dessert dairy)	Preterm Birth	Western diet associated with preterm delivery (OR: 1.30; 95% CI: 1.13, 1.49)
Rodrigues, Azeredo, Silva, 2020, Brazil [65]	Cross-sectional	24.9 ± 6.5	39.4 weeks (SD ± 1.2)	99	Processed meat	LBW	Maternal consumption of sausages associated with LBW: (OR: 1.46; 95% CI: 1.02, 2.10)
Rohatgi et al., 2017 USA [4]	Prospective Cohort	27.2 ± 5.1	32–37 weeks	45	UPF energy intake	GWG	Each 1% increase in UPF energy intake associated with increase in GWG: (β: 1.33; 95% CI: 0.3, 2.4; $p = 0.016$)
Schmidt et al., 2020 Denmark [70]	Prospective Cohort	NR	12 weeks	66,387	Soft drinks	CHD	High intake of sugar-sweetened carbonated beverages (≥4 servings) associated with CHD: (OR: 2.41; 95% CI: 1.26, 4.64; p-trend = 0.03)
Sedaghat et al., 2017 Iran [41]	Case-control	case: 29.64 ± 4.52 control: 29.76 ± 4.26	case: 29.39 ± 4.74 weeks control: 31.19 ± 3.53 weeks	case: 122 control: 266	WDP (sweet snacks, mayonnaise, SSB, salty snacks, solid fats, high-fat dairy, red and processed meat, and tea and coffee)	GDM	Western dietary patterns associated with GDM risk: (OR: 1.68; 95% CI: 1.04, 2.27)
Tamada et al., 2021 Japan [48]	Prospective Cohort	30.7 years (SD ± 5.1)	14.4 weeks (SD ± 5.6)	94,062	Ready-made meals (pre-packed foods, instant noodles, soup)	Stillbirth Preterm Birth LBW	Ready-made meals associated with stillbirth: (OR: 2.632; 95% CI: 1.507, 4.597; q = 0.007); Preterm birth: (OR: 0.993; 95% CI: 0.887, 1.125) LBW: (OR: 0.961; 95% CI: 0.875, 10.56)
Teixeira et al., 2020 Brazil [60]	Prospective Cohort	mean: 25.9	10–11 weeks	299	Dietary patterns (processed meats, sandwiches and snacks, sandwich sauces, desserts and sweets, soft drinks)	SGA	Dietary pattern with snacks, sandwiches, sweets, and soft drinks associated with the risk to deliver SGA babies: (RR: 1.92; 95% CI: 1.08, 3.39)

Table 1. *Cont.*

Author, Year Country	Study Design	Age (Years)	GW (Range or Mean)	Sample n =	Exposure	Outcome	Main Results
Tielemans et al., 2015 Netherlands [81]	Prospective Cohort	31.6 (IQR ± 4.3)	13.4 weeks (IQR: 12.2–15.5)	3374	Dietary patterns (margarine—solid and liquid, sugar and confectionary, cakes, chocolate, candy, snacks)	GWG	Margarine, sugar, and snacks pattern are associated with a higher prevalence of excessive GWG: (OR: 1.45; 95% CI: 1.06, 1.99)
Uusitalo et al., 2009 Finland [77]	Prospective Cohort	29.2 ± 5.2	10 weeks	3360	Dietary patterns (fast food: sweets, fast food, snacks, chocolate, fried potatoes, soft drinks, high-fat pastry, cream, fruit juices, white bread, processed meat, sausage)	GWG	Fast food patterns associated with weight gain rate: (β: 0.010; SE: 0.003; $p = 0.004$)
Wen et al., 2013 Australia [87]	Prospective Cohort	>16	24–34 weeks	368	Junk food diet (soft drinks, processed meat, meals, chips or French fries)	LGA	Junk food diet versus without a junk food diet associated with a newborn LGA: (OR: 0.36; 95% CI: 0.14, 0.91; $p = 0.03$)
Wrottesley, Pisa & Norris, 2017; South Africa [51]	Prospective Cohort	≥18	All trimesters	538	WDP (white bread, cheese and cottage cheese, red meat, processed meat, roast potatoes and chips, sweets, chocolate, soft drinks, miscellaneous)	GWG	Western dietary pattern associated with excessive GWG (OR: 1.07; 95% CI: 0.78, 1.45; $p = 0.682$)
Yong et al., 2021 Malaysia [49]	Prospective Cohort	30.01 ± 4.48	1st trimester	452	Beverages (carbonated and juices)	GDM	Higher fruit juice intake associated with GDM (OR: 0.92; 95% CI: 0.89, 0.98).
Zareei et al., 2019 Iran [45]	Cross-sectional	28.96 ± 5.85	NR	82	Dietary patterns (unhealthy dietary patterns: mayonnaise, fries, red meat, soft drinks, pizza, snacks, sweets and dessert, refined cereal, hydrogenated oils, high-fat dairy products, sugar, processed meat, broth)	Preeclampsia	The unhealthy dietary pattern associated with preeclampsia (OR: 1.381; 95% CI: 0.462, 4.126, $p = 0.564$)
Zhang et al., 2006 USA [57]	Prospective Cohort	>18	NR	13,110	WDP (red and processed meat, refined grain products, sweets, French fries and pizza)	GDM	Western pattern score associated with GDM risk (RR: 1.63; 95% CI: 1.20, 2.21; $p = 0.001$); Red meat associated with GDM risk: (RR: 1.61; 95% CI: 1.25, 2.07) Processed meat associated with GDM risk: (RR: 1.64; 95% CI: 1.13, 2.38)
Zhu et al., 2017 Denmark [82]	Prospective Cohort	>18	25 weeks	918	Soft drinks	Birth weight	Daily soft drinks consumption associated with offspring risk of LGA: (RR: 1.57; 95% CI: 1.05, 2.35)
Zuccolotto et al., 2019 Brazil [61]	Cross-sectional	27.6 ± 5.4	24–39 weeks	785	Snack dietary patterns (breads; butter and margarine; Processed meat, sweets, chocolate milk and cappuccino)	GDM	Dietary patterns associated with GDM risk: (OR: 1.01; 95% CI: 0.63, 1.63)

BMI: body mass index; BW: birth weight; CBWC: customized birthweight centiles; CI: confidence interval; CHD: congenital heart defects; EPDS: Edinburgh Postpartum Depression Scores; GDM: gestational diabetes mellitus; GWG: gestational weight gain; IQR: interquartile range; LBW: low birth weight; LGA: large for gestational age; NR: not reported; OR: odds ratio; RR: relative risk; SD: standard deviation; SGA: small for gestational age; SSB: sugar-sweetened Beverage; SSC: sugar-sweetened carbonated beverages; UPF: ultra-processed food; WDP: Western dietary pattern.

Depressive symptoms during pregnancy were also investigated in two cohort studies. Ker et al. [46] reported that increased consumption of sugar-sweetened beverages was associated with higher depression scores (β = 0.25; 95% CI: 0.04, 0.45). Likewise, Baskin et al. [88] found a positive association between an "unhealthy" diet (characterized by the intake of UPF and unhealthy foods such as condiments, sweets and desserts, refined grains, high-energy drinks, fast foods, hot chips, high-fat dairy, fruit juice, and red meats) and increased depressive symptoms during gestation (β = 0.19; 95% CI: 0.04, 0.34).

Regarding neonatal outcomes, Hajianfar et al. [40] and Okubo et al. [44] reported that pregnant women with the highest consumption of UPF were 5.51 (95% CI: 1.82,16.66) and 5.24 (95% CI 1.1, 24.4) times more likely to have children with LBW (<2.5 kg), respectively.

A positive association between maternal UPF consumption and higher birth weight was observed in one cohort [21] whereas no association was observed in four studies [48,53,63,73]. Maternal fast food [54,66] and soft drink [82] intake were associated with LGA birth. Moreover, Grundt et al. [73] observed an inverse association between soft drink consumption and LGA risk.

Two cohorts reported higher odds of preterm birth. Martin et al. [55] and Rasmussen et al. [83] reported that UPF consumption during pregnancy increased preterm birth odds by 53% (OR: 1.53; 95% CI: 1.02, 2.30) and 30% (OR: 1.30; 95% CI: 1.13, 1.49), respectively. In opposition to these results, two cohort studies found no significant association [48,85].

Alves-Santos et al. [54] found that fast food consumption was associated with higher odds of birth length > 90th percentile (OR: 4.81; 95% CI: 1.77, 13.07). Teixeira et al. [60] observed that women who consumed more "snacks, sandwiches, sweets and soft drinks" were significantly more likely to deliver SGA (birth weight and birth length <10th percentile) babies (RR: 1.92; 95% CI: 1.08, 3.39). Mikes et al. [86] showed that higher consumption of unhealthy foods (confectionary, fried, and processed meats) was associated with lower birth length: (β = −0.10 cm; 95% CI: −0.19, −0.01). One study explored BMI-for-age z score at birth and reported a decrease of 20.41 standard deviations (SD) (95% CI: 20.79, 20.03) associated with a diet characterized by a high intake of white bread, red and processed meat, French fries, fried chicken, and vitamin C–rich drinks [59]. Finally, two studies reported a positive association between maternal soft drink intake during pregnancy and higher odds of CHD [70,79].

Selected cross-sectional studies (n = 9) examined the association between maternal UPF consumption and perinatal outcomes. No significant association was observed for excessive GWG [64,89], GDM risk [61,64], preeclampsia [45] and LGA [47]. Three studies [43,47,65] reported a positive association between the consumption of UPF and LBW, while one study [80] (n = 303) showed no significant association. A positive association was also observed for preterm birth (OR: 1.54; 95% CI: 1.10, 2.15) [22].

Of the five included case-control studies, one study observed that higher maternal adherence to Western diet patterns during pregnancy was associated with higher odds of GDM risk (OR: 1.68; 95% CI: 1.04, 2.72) [41]. On the other hand, Asadi et al. did not find such an association [38]. A positive association was observed between higher consumption of UPF and higher systolic blood pressure (r = 0.110, $p < 0.05$) [35], preeclampsia (OR: 5.99; 95% CI: 3.41, 10.53) [37] and LBW (OR: 2.7; 95% CI: 1.42, 5.13) [21].

3.4. Risk of Bias within Individual Studies

The frequency of the items assessed as an indicator of the risk of bias in studies is illustrated according to the study design in Figure 2. Of 47 cohort studies, 24 (51%) were considered at low risk of bias [18–20,36,39,40,44,49–51,54,60,66,67,69–75,79,82,83]. Two indicators were accomplished in all studies: "confounding factors identified" and "strategies to deal with confounding factors stated". Most studies were at high risk of bias due to not presenting the strategies to address incomplete follow-up, which is considered a potential source of bias [4,42,52,53,56,59,63,68,76,78,85–87]. Most of cross-sectional studies (77.7%) were at low risk of bias [22,43,45,61,64,65,80]. Two studies presented a high risk of bias. One article [89] did not use a reliable method to measure the assessed outcome; the other one [47] did not accomplish two of the evaluated parameters: "criteria for inclusion in the sample clearly defined" and "outcomes measured validly and reliably". Three case-control studies (60%) were classified as having a low risk of bias [37,38,41] and two studies presented a high risk of bias due to not reporting the exposure period [21] and statistical analysis [35] clearly. The complete appraisal of the methodological quality of each article is described in Supplementary Materials (Tables S4–S6).

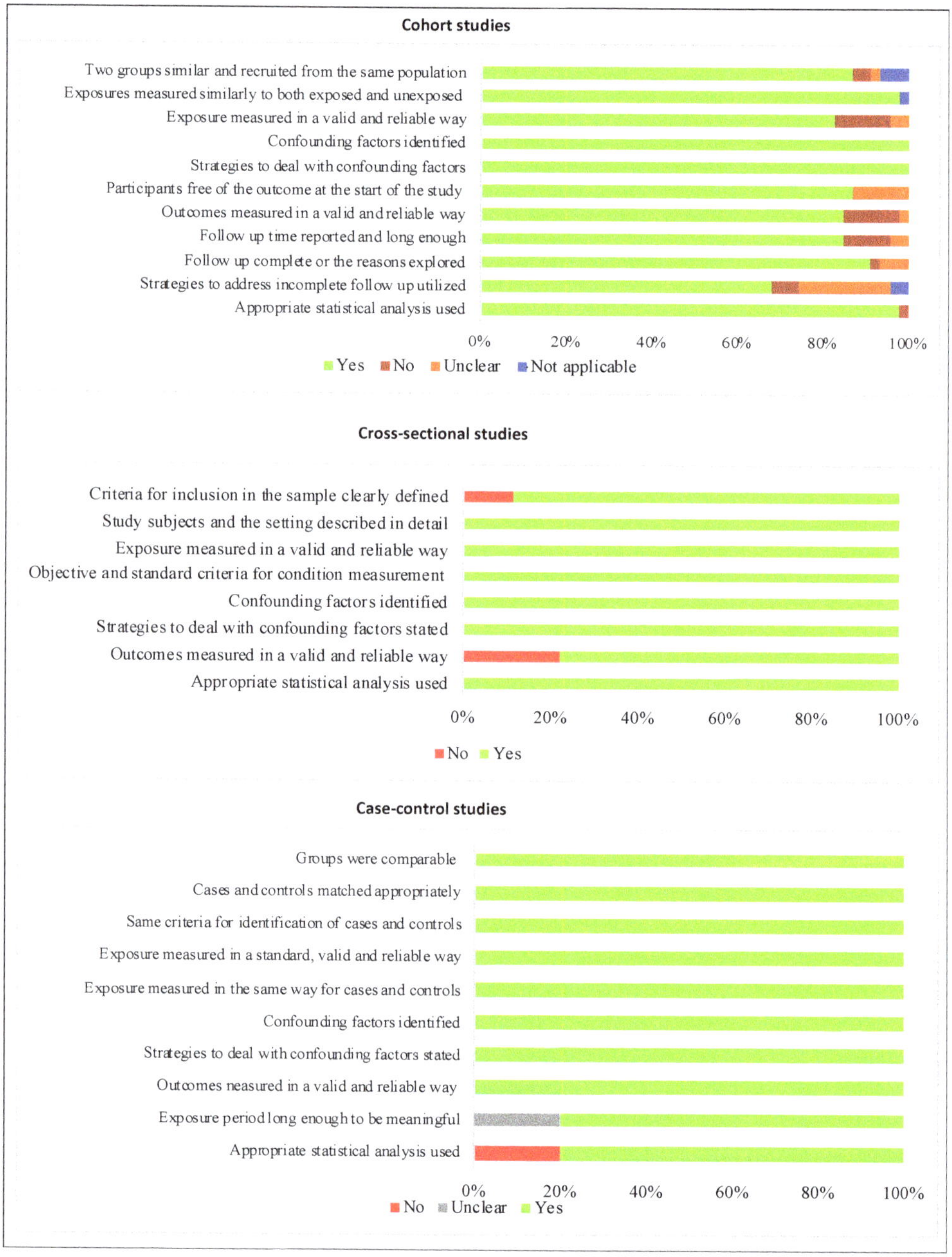

Figure 2. Risk of bias of the included articles according to study design.

3.5. Meta-Analysis of Maternal UPF-Rich Diet Consumption and Maternal Outcomes

3.5.1. Gestational Weight Gain

Five articles were pooled in the meta-analysis, including 4.576 subjects, but no association was found between maternal UPF-rich diet consumption and excessive GWG [(OR: 1.04; 95% CI: 0.92, 1.17) $I^2 = 75.22\%$] [36,51,58,81,84]. This association was also explored using β coefficient in five articles, including 4.384 pregnant women [4,18,42,67,77], but no significant association between UPF-rich diet consumption and GWG was found [(β = 0.02; 95% CI: −0.02, 0,06) $I^2 = 80.63\%$].

3.5.2. Gestational Diabetes Mellitus

Ten cohort studies assessed the association between maternal UPF-rich diet consumption and GDM including 42.477 pregnant women [19,49,56,57,62,69,71,72,74,78]. The meta-analysis showed that higher consumption of diets rich in UPF significantly increased odds of GDM by 48% [(OR: 1.48; 95% CI: 1.17, 1.87) $I^2 = 82.70\%$] (Figure 3). Publication bias analysis by the funnel plot inspection (Supplementary Figure S1) showed asymmetry among the studies, which was confirmed by Egger test ($p = 0.001$).

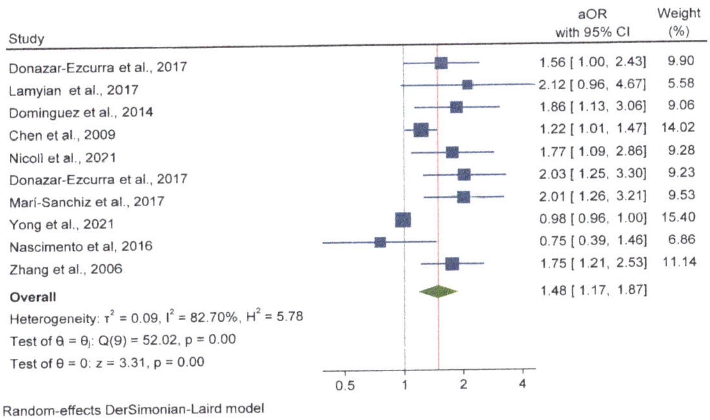

Figure 3. Meta-analysis of ultra-processed food rich diet *vs* gestational diabetes mellitus.

3.5.3. Hypertensive Disorders of Pregnancy

No significant associations were observed between UPF-rich diet consumption and the odds of hypertension during pregnancy of three cohort studies, with 58.701 subjects [20,39,52] [(OR: 0.94; 95% CI: 0.52, 1.70) $I^2 = 88.80\%$].

On the other hand, the consumption of UPF-rich diets was found to be associated with 28% higher odds of preeclampsia in four cohort studies [20,39,75,76] involving 112.307 subjects [(OR: 1.28; 95% CI: 1.15, 1.42) $I^2 = 0.00\%$] (Figure 4).

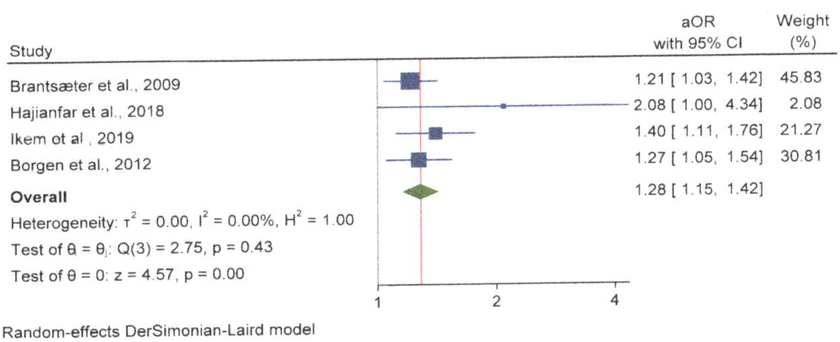

Figure 4. Meta-analysis of ultra-processed food rich diet vs. preeclampsia.

3.6. Meta-Analysis of Maternal UPF-Rich Diet Consumption and Neonatal Outcomes

3.6.1. Low Birth Weight

Five eligible cohort studies that provided an estimate of the association between maternal UPF-rich diet consumption and LBW were included in the meta-analysis [40,44,48,53,73], involving 146.617 subjects. However, no significant association was presented [(OR: 1.08; 95% CI: 0.90, 1.30) I^2 = 74.59%].

3.6.2. Large for Gestational Age

Three eligible cohort studies (n = 52.468) investigated the association between maternal UPF-rich diet consumption and LGA. [54,66,73]. Meta-analysis results revealed no significant association between UPF-rich diet consumption and odds of LGA [(OR: 2.10; 95% CI: 0.71, 6,25) I^2 = 84.61%].

3.6.3. Preterm Birth

The meta-analysis showed no significant association [(OR: 1.13; 95% CI: 0.97, 1.32) I^2 = 76.25%] regarding the association between four cohort studies (n= 233.308) which evaluated the UPF-rich diet consumption and the odds of preterm birth. E [48,55,83,85].

3.7. Certainty of Evidence

The GRADE assessment was moderate for maternal UPF-rich diet consumption and preeclampsia (⊕⊕⊕○) and very low (⊕○○○) for GWG, GDM, LBW, LGA, and preterm birth (Table 2).

Table 2. GRADE evidence profile for maternal UPF consumption and perinatal outcomes.

Outcomes	Studies (n, References)	Risk of Bias	Inconsistency [a]	Indirectness [b]	Imprecision [c]	Publication Bias	Certainty
Maternal Outcomes							
Excessive Gestational Weight Gain	5 [36,51,58,81,84]	Not serious	Serious	Not serious	Not serious	Not assessed [d]	⊕○○○ Very low
Gestational Weight Gain	5 [4,18,42,67,77]	Not serious	Serious	Not serious	Not serious	Not assessed [d]	⊕○○○ Very low
Gestational Diabetes Mellitus	10 [19,49,56,57,62, 69,71,72,74,78]	Not serious	Serious	Not serious	Not serious	strongly suspected [e]	⊕○○○ Very low
Gestational Hypertension	3 [20,39,52]	Not serious	Serious	Not serious	Not serious	Not assessed [d]	⊕○○○ Very low
Preeclampsia	4 [20,39,75,76]	Not serious	Not serious	Not serious	Not serious	Not assessed [d]	⊕⊕⊕○ Moderate
Neonatal Outcomes							
Low Birth Weight	5 [40,44,48,53,73]	Not serious	Serious	Not serious	Not serious	Not assessed [d]	⊕○○○ Very low
Large for Gestational Age	3 [54,66,73]	Not serious	Serious	Not serious	Not serious	Not assessed [d]	⊕○○○ Very low
Preterm Birth	4 [48,55,83,85]	Not serious	Serious	Not serious	Not serious	Not assessed [d]	⊕○○○ Very low

[a] Downgrade 1 level if I^2 was 50% to 75%, and 2 levels if I^2 was 75% to 100%. [b] No downgrade for indirectness because all studies directly measure the outcomes. [c] No downgrade for imprecision because of >2000 participants for each outcome. [d] No downgrade for publication bias, as publication bias could not be assessed due to lack of power for assessing funnel plot asymmetry and small study effects (<10 cohorts included in meta-analysis). [e] Downgrade 1 level for publication bias ($p < 0.05$).

4. Discussion

The present systematic review highlights the role of the maternal diet, including the consequences of UPF-rich diet consumption on perinatal adverse outcomes.

There is growing evidence that high consumption of UPFs is indicative of low diet quality and associated with a higher risk of coronary heart disease, cancer, cerebrovascular and metabolic diseases, hypertension, worse cardiometabolic risk profile, and a higher risk of all-cause mortality in adult and older populations [91–93]. Regarding the pregnancy period, a recent systematic review [27] indicated that high UPF consumption in pregnancy, lactation, and infancy had negative repercussions on health in general but no meta-analysis was performed. To our knowledge, this is the first study with meta-analysis to assess the effect of UPF-rich diet consumption, through unhealthy dietary patterns, Western foods and UPF intake, by pregnant women and perinatal outcomes, and is the most up-to-date and comprehensive systematic review on this topic.

The significant association found between higher maternal consumption of UPF-rich diets and higher risk of GDM is corroborated by previous studies. A meta-analysis of cohort studies showed that the Western dietary pattern, determined by high intakes of red and processed meat, fried foods, and refined grains, could increase the risk of GDM [94]. Quan et al. also showed that consumption of fast food had a positive association with higher GDM risk [95]. Furthermore, diets presenting high amount of UPFs are frequently rich in sugars and refined grains products, recognized risk factors for GDM [15], endorsing the results of this meta-analysis. In contrast to our results, Kibret et al. [96] found no association between the Western diet pattern and GDM, which may be due to the inclusion of studies assessing UPF-rich dietary patterns as well as soft drinks intake and processed meats alone in the present GDM meta-analysis.

Another interesting finding was a significant association between UPF-rich diets consumption and preeclampsia. A previous recent study with meta-analysis investigated the effects of maternal dietary patterns on pregnancy and reported that maternal adherence to an unhealthy diet was associated with 23% higher odds of HDP, including preeclampsia [97]. Another study also found a significant association between higher adherence to a Western dietary pattern, an unhealthy diet pattern characterized by a high amount of UPF such as processed meat, soft drinks, and refined foods, and increased risk of preeclampsia [98], corroborating our results.

Although the causes of preeclampsia are multifactorial, some risk factors are associated with the development of HDP, such as women experiencing their first pregnancy, twin pregnancy, chronic hypertension, GDM, maternal obesity, and maternal age over 35 years. In addition, healthy lifestyle habits before and during pregnancy can influence the severity of the outcomes [99]. UPFs are rich in sodium, free or added sugars, saturated and trans fats, high energy density, and low in fiber, potassium, and micronutrients [15]. In this context, maternal diet quality has clinical significance given the established association of preeclampsia with maternal and fetal complications such as maternal mortality, perinatal deaths, preterm birth, and intrauterine growth restriction. Moreover, pregnant women affected by HDP have a higher risk of cardiovascular disease in later life, regardless of other risk factors [100,101].

Despite the lack of significant association between UPF-rich diets consumption and excessive GWG, evidence indicates that GWG is significantly correlated with maternal energy intake [102–104]. A recent systematic review reported that dietary patterns with ultra-processed components rich in fat and sugars presented an association with higher GWG [89]. Sartorelli et al. [23] also showed that women classified into the highest tertile of UPFs intake had a three times higher chance of obesity when compared to women with the lowest intake of these foods. Thus, monitoring this trend in pregnant women should be an important healthcare concern objective since excessive GWG is associated with greater chances of hypertensive disorders, cesarean delivery, and LGA newborns [105–107], and a strong predictor of postpartum weight retention, contributing to obesity in later life [108,109].

The development of GDM and preeclampsia could be related to the low nutritional quality of the UFP-rich diet. The low quality of carbohydrates found in UPFs may impair glycemic control [110], especially from the second trimester when anti-insulin hormones, such as estrogens, progesterone, and chorionic somatomammotropin, act by decreasing the power of insulin action, making more glucose available in the bloodstream [111]. The risk of pregnancy complications such as preeclampsia has been linked with maternal oxidative stress in the middle of pregnancy [112]. The findings of a multicenter study showed that oxidative stress could be reduced by sufficient intakes of fruit, vegetables, and vitamin C [113], and Pistollato et al. (2015) reported a lower likelihood of pregnancy-induced hypertension or preeclampsia when the diet pattern comprised intake of plant-derived foodstuffs and vegetables [114]. Thus, higher UPFs intake may impact and reduce consumption of antioxidants and foment oxidative stress status during pregnancy.

Regarding neonatal outcomes, the present meta-analysis showed no association between maternal UPF-rich diet consumption and neonatal birth outcomes such as birth weight and preterm birth. Endorsing our results, a study with a meta-analysis conducted by Abdollahi et al. [97] showed no association between an unhealthy pattern and birth weight. Kibret et al. [96] also found that a dietary pattern rich in UPF, a Western dietary pattern, did not increase the odds of preterm birth, corroborating our findings.

Nonetheless, the importance of maternal diet in early pregnancy for neonatal health is well documented. Birth weight is an important parameter for assessing newborn health conditions and development, and also is used as one of the basic indicators in the global reference list of the World Health Organization (WHO) [115]. In a meta-analysis conducted with observational studies, Chia et al. [26] reported that unhealthy dietary patterns, characterized by high intakes of refined grains, processed meat, and foods high in saturated fat or sugar, were associated with lower birth weight and a trend towards a higher risk of preterm birth. The study of Rohatgi et al. [4] reported that higher maternal UPF consumption was associated with increased adiposity in the neonate. Taken together, the evidence suggests that maternal diet quality, including UPF consumption, might affect neonatal health.

The etiology of preterm birth is still not well understood, and most cases do not have clear determinants. Some studies reported greater chances of preterm birth observed in pregnant women with high consumption of highly processed foods high in fat and sugar, while the consumption of a healthy diet, rich in fruits, vegetables, and whole grains, appeared to significantly reduce the risk [22,55,83]. Moreover, a meta-analysis of nine cohort studies indicated that higher adherence to a healthy dietary pattern significantly decreased the odds of preterm birth [96].

The results of the present study indicate important public health implications, since higher UPF consumption may worsen perinatal health outcomes. The positive association between UPF-rich diet consumption and GDM and preeclampsia suggests that the consumption of diets rich in UPFs, such as those with high factor loadings for fast foods, junk foods, processed meats, soft drinks, pizzas, hamburgers, candies and sweets, should be discouraged during pregnancy whereas increasing the proportion of in natura and minimally processed food in the diet should be reinforced. Furthermore, prioritizing a healthy lifestyle, which considers adequate food intake, regular physical exercise, regular sleep, and adequate gestational weight gain is mandatory for this population group. This study provides insights to guide policies on pregnancy healthcare as well as nutritional interventions in prenatal services. Further studies with robust methodological quality, such as larger samples and using a more accurate dietary assessment instrument, are needed to clarify the findings on this topic.

The NOVA food categorization classifies foods and beverages "according to the extent and purpose of industrial processing" and defines UPF as "formulations of ingredients, most of exclusive industrial use, that result from a series of industrial processes" (hence "ultra-processed") [10]. Considering that unhealthy dietary patterns, such as Western and Prudent diets, are characterized by a high consumption of UPF, we speculate that our results provide an effort to measure the UPF consumption association with perinatal outcomes, since diet is a modifiable risk factor. This study has several strengths. To date, this is the first study conducted with a meta-analysis on the topic. A comprehensive search strategy was carried out using a robust and appropriate methodology according to Cochrane Handbook and PRISMA guidelines. Moreover, many subjects were included for each pooled outcome, increasing the generalizability of the results. In addition, the methodological quality of the included studies was assessed independently, and the GRADE system was used to assess the certainty of the evidence of each exposure–outcome association. Despite the few studies in the pregnancy group specifically evaluating UPFs intake, out of the 61 studies included in the review, 83% found a significant association between UPF-rich diets consumption and adverse health outcomes. These data demonstrate the important impact on public health in the maternal and child group and may support future nutritional recommendations for these populations.

Some limitations are also noteworthy. First, the study did not exclusively evaluate UPF consumption, but we speculate that unhealthy and Western dietary patterns may be considered as a proxy for UPF intake. Second, applied dietary assessments of the included studies were not specifically designed for the NOVA classification system. Third, high heterogeneity between studies was observed in many analyses considering the nature of the observational nutritional studies. This is expected because of the diverse characteristics of subjects, the different dietary approaches, and the variance between outcome assessment methods. Fourth, the lack of significant results in perinatal outcomes may be due to the small number of included articles for each outcome, thus it was not possible to perform subgroups analysis to seek the source of heterogeneity. Lastly, publication bias was observed, so, studies that had negative results might not have been submitted for publication and were not included.

Finally, maternal nutrition for successful pregnancy outcomes cannot be addressed during pregnancy alone. A varied diet rich in protein sources, fruit, and vegetables should be consumed by women who intend to become pregnant and during pregnancy as a component of prenatal care. The results presented here suggest that nutritional recommendations should focus not only on foods and nutrients amounts but also on the degree of food processing.

5. Conclusions

This study indicates a positive association between maternal UPF-rich diet consumption during pregnancy and increased risk of developing gestational diabetes mellitus and preeclampsia. These findings corroborate the adverse effects of consumption of diets rich in UPF during pregnancy and highlight the need to monitor and reduce UPF-rich diet consumption specifically during the gestational period, as a strategy to prevent adverse perinatal outcomes.

Supplementary Materials: The following supporting information can be downloaded at: https://www.mdpi.com/article/10.3390/nu14153242/s1, Table S1: PECOS acronym used in the design of the study; Table S2: Database search strategies; Table S3: reasons for exclusion of articles; Table S4: risk of bias of cohort studies; Table S5: risk of bias of cross-sectional studies; Table S6: risk of bias of case-control studies; Figure S1: Publication bias funnel graph for UPF consumption and Gestational Diabetes Mellitus risk.

Author Contributions: Conceptualization, W.O.P., E.S.O.P., V.S.S.G. and N.P.; methodology, W.O.P., E.S.O.P. and V.S.S.G.; conducting the systematic literature search W.O.P. and E.S.O.P.; performed the data extraction and quality assessment, W.O.P. and E.S.O.P.; formal analysis, V.S.S.G.; writing—original draft preparation, W.O.P.; writing—review and editing, E.S.O.P., V.S.S.G. and N.P.; supervision, N.P. All authors have read and agreed to the published version of the manuscript.

Funding: DPG/DPI/University of Brasilia and PPGNH/UnB.

Institutional Review Board Statement: Not applicable.

Informed Consent Statement: Not applicable.

Data Availability Statement: Not applicable.

Conflicts of Interest: The authors declare no conflict of interest.

References

1. Soma-Pillay, P.; Nelson-Piercy, C.; Tolppanen, H.; Mebazaa, A. Physiological Changes in Pregnancy. *Cardiovasc. J. Afr.* **2016**, *27*, 89–94. [CrossRef]
2. WHO. *Good Maternal Nutrition: The Best Start in Life*; WHO Regional Office for Europe: Copenhagen, Denmark, 2016; ISBN 9789289051545.
3. Mendonça, E.L.S.S.; de Lima Macêna, M.; Bueno, N.B.; de Oliveira, A.C.M.; Mello, C.S. Premature Birth, Low Birth Weight, Small for Gestational Age and Chronic Non-Communicable Diseases in Adult Life: A Systematic Review with Meta-Analysis. *Early Hum. Dev.* **2020**, *149*, 105154. [CrossRef] [PubMed]
4. Rohatgi, K.W.; Tinius, R.A.; Cade, W.T.; Steele, E.M.; Cahill, A.G.; Parra, D.C. Relationships between Consumption of Ultra-Processed Foods, Gestational Weight Gain and Neonatal Outcomes in a Sample of US Pregnant Women. *PeerJ* **2017**, *2017*, e4091. [CrossRef] [PubMed]
5. Deierlein, A.L.; Ghassabian, A.; Kahn, L.G.; Afanasyeva, Y.; Mehta-Lee, S.S.; Brubaker, S.G.; Trasande, L. Dietary Quality and Sociodemographic and Health Behavior Characteristics Among Pregnant Women Participating in the New York University Children's Health and Environment Study. *Front. Nutr.* **2021**, *8*, 639425. [CrossRef] [PubMed]
6. Rojhani, A.; Ouyang, P.; Gullon-Rivera, A.; Dale, T.M. Dietary Quality of Pregnant Women Participating in the Special Supplemental Nutrition Program for Women, Infants, and Children. *Int. J. Environ. Res. Public Health* **2021**, *18*, 8370. [CrossRef] [PubMed]
7. Leone, A.; Martínez-González, M.Á.; Craig, W.; Fresán, U.; Gómez-Donoso, C.; Bes-Rastrollo, M. Pre-Gestational Consumption of Ultra-Processed Foods and Risk of Gestational Diabetes in a Mediterranean Cohort. The SUN Project. *Nutrients* **2021**, *13*, 2202. [CrossRef] [PubMed]
8. Monteiro, C.A.; Cannon, G.; Moubarac, J.-C.; Levy, R.B.; Louzada, M.L.C.; Jaime, P.C. The UN Decade of Nutrition, the NOVA Food Classification and the Trouble with Ultra-Processing. *Public Health Nutr.* **2018**, *21*, 5–17. [CrossRef] [PubMed]
9. Monteiro, C.A.; Levy, R.B.; Claro, R.M.; Castro, I.R.R.D.; Cannon, G. A New Classification of Foods Based on the Extent and Purpose of Their Processing. *Cad. Saúde Pública* **2010**, *26*, 2039–2049. [CrossRef] [PubMed]
10. Monteiro, C.A.; Cannon, G.; Levy, R.B.; Moubarac, J.C.; Louzada, M.L.C.; Rauber, F.; Khandpur, N.; Cediel, G.; Neri, D.; Martinez-Steele, E.; et al. Ultra-Processed Foods: What They Are and How to Identify Them. *Public Health Nutr.* **2019**, *22*, 936–941. [CrossRef]
11. Cordain, L.; Eaton, S.B.; Sebastian, A.; Mann, N.; Lindeberg, S.; Watkins, B.A.; O'Keefe, J.H.; Brand-Miller, J. Origins and Evolution of the Western Diet: Health Implications for the 21st Century. *Am. J. Clin. Nutr.* **2005**, *81*, 341–354. [CrossRef]
12. Monteiro, C.A.; Moubarac, J.C.; Cannon, G.; Ng, S.W.; Popkin, B. Ultra-Processed Products Are Becoming Dominant in the Global Food System. *Obes. Rev.* **2013**, *14*, 21–28. [CrossRef] [PubMed]
13. Juul, F.; Parekh, N.; Martinez-Steele, E.; Monteiro, C.A.; Chang, V.W. Ultra-Processed Food Consumption among US Adults from 2001 to 2018. *Am. J. Clin. Nutr.* **2022**, *115*, 211–221. [CrossRef] [PubMed]
14. Rauber, F.; Louzada, M.L.D.C.; Martinez Steele, E.; De Rezende, L.F.M.; Millett, C.; Monteiro, C.A.; Levy, R.B. Ultra-Processed Foods and Excessive Free Sugar Intake in the UK: A Nationally Representative Cross-Sectional Study. *BMJ Open* **2019**, *9*, e027546. [CrossRef] [PubMed]
15. Monteiro, C.A.; Cannon, G.; Lawrence, M.; Costa Louzada, M.L.; Pereira Machado, P. *Ultra-Processed Foods, Diet Quality, and Health Using the NOVA Classification System*; FAO: Rome, Italy, 2019; ISBN 978-92-5-131701-3.
16. Baker, P.; Friel, S. Food Systems Transformations, Ultra-Processed Food Markets and the Nutrition Transition in Asia. *Glob. Health* **2016**, *12*, 80. [CrossRef] [PubMed]
17. Martini, D.; Godos, J.; Bonaccio, M.; Vitaglione, P.; Grosso, G. Ultra-Processed Foods and Nutritional Dietary Profile: A Meta-Analysis of Nationally Representative Samples. *Nutrients* **2021**, *13*, 3390. [CrossRef] [PubMed]
18. Gomes, C.D.B.; Malta, M.B.; Benício, M.H.D.A.; Carvalhaes, M.A.D.B.L. Consumption of Ultra-Processed Foods in the Third Gestational Trimester and Increased Weight Gain: A Brazilian Cohort Study. *Public Health Nutr.* **2020**, *24*, 3304–3312. [CrossRef]
19. Lamyian, M.; Hosseinpour-Niazi, S.; Mirmiran, P.; Banaem, L.M.; Goshtasebi, A.; Azizi, F. Pre-Pregnancy Fast Food Consumption Is Associated with Gestational Diabetes Mellitus among Tehranian Women. *Nutrients* **2017**, *9*, 216. [CrossRef]
20. Ikem, E.; Halldorsson, T.I.; Birgisdóttir, B.E.; Rasmussen, M.A.; Olsen, S.F.; Maslova, E. Dietary Patterns and the Risk of Pregnancy-Associated Hypertension in the Danish National Birth Cohort: A Prospective Longitudinal Study. *BJOG Int. J. Obstet. Gynaecol.* **2019**, *126*, 663–673. [CrossRef]
21. Amezcua-Prieto, C.; Martínez-Galiano, J.M.; Cano-Ibáñez, N.; Olmedo-Requena, R.; Bueno-Cavanillas, A.; Delgado-Rodríguez, M. Types of Carbohydrates Intake during Pregnancy and Frequency of a Small for Gestational Age Newborn: A Case-Control Study. *Nutrients* **2019**, *11*, 523. [CrossRef]
22. Grieger, J.A.; Grzeskowiak, L.E.; Clifton, V.L. Preconception Dietary Patterns in Human Pregnancies Are Associated with Preterm Delivery. *J. Nutr.* **2014**, *144*, 1075–1080. [CrossRef] [PubMed]
23. Sartorelli, D.S.; Crivellenti, L.C.; Zuccolotto, D.C.C.; Franco, L.J. Relationship between Minimally and Ultra-Processed Food Intake during Pregnancy with Obesity and Gestational Diabetes Mellitus. *Cad. Saúde Publica* **2019**, *35*, e00049318. [CrossRef] [PubMed]

24. Miranda, C.; Souza, R.C.V.e.; Santos, L.C. Dos Influência Do Consumo de Alimentos Ultraprocessados Durante a Gestação Nas Medidas Antropométricas Do Bebê, Do Nascimento Ao Primeiro Ano de Vida: Uma Revisão Sistemática (The Influence of the Consumption of Ultra-Processed Foods During Pregnancy on Anthropometric Measurements of the Baby, From Birth to the First Year of Life: A Systematic Review). *Rev. Bras. Saúde Matern. Infant.* **2021**, *21*, 9–26.
25. Kinshella, M.L.W.; Omar, S.; Scherbinsky, K.; Vidler, M.; Magee, L.A.; Von Dadelszen, P.; Moore, S.E.; Elango, R. Maternal Dietary Patterns and Pregnancy Hypertension in Low—A Nd Middle-Income Countries: A Systematic Review and Meta-Analysis. *Adv. Nutr.* **2021**, *12*, 2387–2400. [CrossRef]
26. Chia, A.R.; Chen, L.W.; Lai, J.S.; Wong, C.H.; Neelakantan, N.; Van Dam, R.M.; Chong, M.F.F. Maternal Dietary Patterns and Birth Outcomes: A Systematic Review and Meta-Analysis. *Adv. Nutr.* **2019**, *10*, 685–695. [CrossRef] [PubMed]
27. De Oliveira, P.G.; De Sousa, J.M.; Assunção, D.G.F.; de Araujo, E.K.S.; Bezerra, D.S.; dos Dametto, J.F.S.; da Ribeiro, K.D.S. Impacts of Consumption of Ultra-Processed Foods on the Maternal-Child Health: A Systematic Review. *Front. Nutr.* **2022**, *9*, 821657. [CrossRef] [PubMed]
28. Arabin, B.; Baschat, A.A. Pregnancy: An Underutilized Window of Opportunity to Improve Long-Term Maternal and Infant Health—An Appeal for Continuous Family Care and Interdisciplinary Communication. *Front. Pediatr.* **2017**, *5*, 69. [CrossRef] [PubMed]
29. Page, M.J.; McKenzie, J.E.; Bossuyt, P.M.; Boutron, I.; Hoffmann, T.C.; Mulrow, C.D.; Shamseer, L.; Tetzlaff, J.M.; Akl, E.A.; Brennan, S.E.; et al. The PRISMA 2020 Statement: An Updated Guideline for Reporting Systematic Reviews. *PLoS Med.* **2021**, *18*, e1003583. [CrossRef]
30. McGowan, J.; Sampson, M.; Salzwedel, D.M.; Cogo, E.; Foerster, V.; Lefebvre, C. PRESS Peer Review of Electronic Search Strategies: 2015 Guideline Statement. *J. Clin. Epidemiol.* **2016**, *75*, 40–46. [CrossRef] [PubMed]
31. Moola, S.; Munn, Z.; Tufanaru, C.; Aromataris, E.; Sears, K.; Sfetcu, R.; Currie, M.; Qureshi, R.; Mattis, P.; Lisy, K.; et al. Systematic Reviews of Etiology and Risk. In *JBI Manual for Evidence Synthesis*; Aromataris, E., Munn, Z., Eds.; Available online: https://synthesismanual.jbi.global (accessed on 13 May 2022).
32. Zhang, J.; Yu, K.F. Special Communication What's the Relative Risk? A Method of Correcting the Odds Ratio in Cohort Studies of Common Outcomes. *JAMA* **1998**, *280*, 1690–1691. [CrossRef]
33. Deeks, J.J.; Higgins, J.P.T.; Altman, D.G. Analysing Data and Undertaking Meta-Analyses. In *Cochrane Handbook for Systematic Reviews of Interventions*; Higgins, J.P.T., Thomas, J., Chandler, J., Cumpston, M., Li, T., Page, M.J., Welch, V.A., Eds.; Wiley: Hoboken, NJ, USA, 2022.
34. Schünemann, H.; Brożek, J.; Guyatt, G.; Oxman, A. Quality of Evidence. In *GRADE Handbook*; Schünemann, H., Brożek, J., Guyatt, G., Oxman, A., Eds.; GRADE Working Group: Barcelona, Spain, 2013.
35. Chen, X.; Ding, Y.; Shi, L.; Wu, D.; Wang, L.; Chen, F.; Mo, Y. Dietary Patterns and Gestational Hypertension in Nulliparous Pregnant Chinese Women: A CONSORT Report. *Medicine* **2020**, *99*, e20186. [CrossRef]
36. Itani, L.; Radwan, H.; Hashim, M.; Hasan, H.; Obaid, R.S.; Al Ghazal, H.; Al Hilali, M.; Rayess, R.; Mohamed, H.J.J.; Hamadeh, R.; et al. Dietary Patterns and Their Associations with Gestational Weight Gain in the United Arab Emirates: Results from the MISC Cohort. *Nutr. J.* **2020**, *19*, 36. [CrossRef] [PubMed]
37. Abbasi, R.; Bakhshimoghaddam, F.; Alizadeh, M. Major Dietary Patterns in Relation to Preeclampsia among Iranian Pregnant Women: A Case–Control Study. *J. Matern. Neonatal Med.* **2019**, *34*, 3529–3536. [CrossRef] [PubMed]
38. Asadi, M.; Shahzeidi, M.; Nadjarzadeh, A.; Hashemi Yusefabad, H.; Mansoori, A. The Relationship between Pre-Pregnancy Dietary Patterns Adherence and Risk of Gestational Diabetes Mellitus in Iran: A Case–Control Study. *Nutr. Diet.* **2019**, *76*, 597–603. [CrossRef]
39. Hajianfar, H.; Esmaillzadeh, A.; Feizi, A.; Shahshahan, Z.; Azadbakht, L. The Association between Major Dietary Patterns and Pregnancy-Related Complications. *Arch. Iran. Med.* **2018**, *21*, 443–451.
40. Hajianfar, H.; Esmaillzadeh, A.; Feizi, A.; Shahshahan, Z.; Azadbakht, L. Major Maternal Dietary Patterns during Early Pregnancy and Their Association with Neonatal Anthropometric Measurement. *Biomed Res. Int.* **2018**, *2018*, 4692193. [CrossRef]
41. Sedaghat, F.; Akhoondan, M.; Ehteshami, M.; Aghamohammadi, V.; Ghanei, N.; Mirmiran, P.; Rashidkhani, B. Maternal Dietary Patterns and Gestational Diabetes Risk: A Case-Control Study. *J. Diabetes Res.* **2017**, *2017*, 5173926. [CrossRef] [PubMed]
42. Angali, K.A.; Shahri, P.; Borazjani, F. Maternal Dietary Pattern in Early Pregnancy Is Associated with Gestational Weight Gain and Hyperglycemia: A Cohort Study in South West of Iran. *Diabetes Metab. Syndr. Clin. Res. Rev.* **2020**, *14*, 1711–1717. [CrossRef]
43. Loy, S.L.; Marhazlina, M.; Jan, J.M.H. Association between Maternal Food Group Intake and Birth Size. *Sains Malays.* **2013**, *42*, 1633–1640.
44. Okubo, H.; Miyake, Y.; Sasaki, S.; Tanaka, K.; Murakami, K.; Hirota, Y.; Osaka, M.; Kanzaki, H.; Kitada, M.; Horikoshi, Y.; et al. Maternal Dietary Patterns in Pregnancy and Fetal Growth in Japan: The Osaka Maternal and Child Health Study. *Br. J. Nutr.* **2012**, *107*, 1526–1533. [CrossRef]
45. Zareei, S.; Homayounfar, R.; Naghizadeh, M.M.; Ehrampoush, E.; Amiri, Z.; Rahimi, M.; Tahamtani, L. Dietary Pattern in Patients with Preeclampsia in Fasa, Iran. *Shiraz E-Medical J.* **2019**, *20*, e86959. [CrossRef]
46. Ker, C.R.; Wu, C.H.; Lee, C.H.; Wang, S.H.; Chan, T.F. Increased Sugar-Sweetened Beverage Use Tendency in Pregnancy Positively Associates with Peripartum Edinburgh Postpartum Depression Scores. *Sci. Rep.* **2021**, *11*, 15324. [CrossRef] [PubMed]

47. Liu, Y.; Zhang, H.; Zhao, Y.; Chen, F.; Mi, B.; Zhou, J.; Chen, Y.; Wang, D.; Pei, L. Geographical Variations in Maternal Dietary Patterns during Pregnancy Associated with Birth Weight in Shaanxi Province, Northwestern China. *PLoS ONE* **2021**, *16*, e0254891. [CrossRef]
48. Tamada, H.; Ebara, T.; Matsuki, T.; Kato, S.; Sato, H.; Ito, Y.; Saitoh, S.; Kamijima, M.; Sugiura-Ogasawara, M.; Group, J.E. and C.S. Ready-Meal Consumption During Pregnancy Is a Risk Factor for Stillbirth: The Japan Environment and Children's Study (JECS). *SSRN Electron. J.* **2021**, *14*, 895. [CrossRef]
49. Yong, H.Y.; Shariff, Z.M.; Yusof, B.N.M.; Rejali, Z.; Tee, Y.Y.S.; Bindels, J.; van der Beek, E.M. Beverage Intake and the Risk of Gestational Diabetes Mellitus: The SECOST. *Nutrients* **2021**, *13*, 2208. [CrossRef]
50. Mitku, A.A.; Zewotir, T.; North, D.; Jeena, P.; Naidoo, R.N. The Differential Effect of Maternal Dietary Patterns on Quantiles of Birthweight. *BMC Public Health* **2020**, *20*, 976. [CrossRef] [PubMed]
51. Wrottesley, S.V.; Pisa, P.T.; Norris, S.A. The Influence of Maternal Dietary Patterns on Body Mass Index and Gestational Weight Gain in Urban Black South African Women. *Nutrients* **2017**, *9*, 732. [CrossRef] [PubMed]
52. Barbosa, J.M.A.; da Silva, A.A.M.; Kac, G.; Simões, V.M.F.; Bettiol, H.; Cavalli, R.C.; Barbieri, M.A.; Ribeiro, C.C.C. Is Soft Drink Consumption Associated with Gestational Hypertension? Results from the Brisa Cohort. *Braz. J. Med. Biol. Res.* **2021**, *54*, e10162. [CrossRef] [PubMed]
53. Ancira-Moreno, M.; O'Neill, M.S.; Rivera-Dommarco, J.Á.; Batis, C.; Rodríguez Ramírez, S.; Sánchez, B.N.; Castillo-Castrejón, M.; Vadillo-Ortega, F. Dietary Patterns and Diet Quality during Pregnancy and Low Birthweight: The PRINCESA Cohort. *Matern. Child Nutr.* **2020**, *16*, e12972. [CrossRef] [PubMed]
54. Alves-Santos, N.H.; Cocate, P.G.; Benaim, C.; Farias, D.R.; Emmett, P.M.; Kac, G. Prepregnancy Dietary Patterns and Their Association with Perinatal Outcomes: A Prospective Cohort Study. *J. Acad. Nutr. Diet.* **2019**, *119*, 1439–1451. [CrossRef]
55. Martin, C.L.; Sotres-Alvarez, D.; Siega-Riz, A.M. Maternal Dietary Patterns during the Second Trimester Are Associated with Preterm Birth. *J. Nutr.* **2015**, *145*, 1857–1864. [CrossRef] [PubMed]
56. Chen, L.; Hu, F.B.; Yeung, E.; Willett, W.; Zhang, C. Prospective Study of Pre-Gravid Sugar-Sweetened Beverage Consumption and the Risk of Gestational Diabetes Mellitus. *Diabetes Care* **2009**, *32*, 2236–2241. [CrossRef]
57. Zhang, C.; Schulze, M.B.; Solomon, C.G.; Hu, F.B. A Prospective Study of Dietary Patterns, Meat Intake and the Risk of Gestational Diabetes Mellitus. *Diabetologia* **2006**, *49*, 2604–2613. [CrossRef]
58. Hirko, K.A.; Comstock, S.S.; Strakovsky, R.S.; Kerver, J.M. Diet during Pregnancy and Gestational Weight Gain in a Michigan Pregnancy Cohort. *Curr. Dev. Nutr.* **2020**, *4*, nzaa121. [CrossRef] [PubMed]
59. Martin, C.L.; Siega-Riz, A.M.; Sotres-Alvarez, D.; Robinson, W.R.; Daniels, J.L.; Perrin, E.M.; Stuebe, A.M. Maternal Dietary Patterns during Pregnancy Are Associated with Child Growth in the First 3 Years of Life. *J. Nutr.* **2016**, *146*, 2281–2288. [CrossRef] [PubMed]
60. Teixeira, J.A.; Hoffman, D.J.; Castro, T.G.; Saldiva, S.R.D.M.; Francisco, R.P.V.; Vieira, S.E.; Marchioni, D.M. Pre-Pregnancy Dietary Pattern Is Associated with Newborn Size: Results from ProcriAr Study. *Br. J. Nutr.* **2021**, *126*, 903–912. [CrossRef]
61. Zuccolotto, D.C.C.; Crivellenti, L.C.; Franco, L.J.; Sarotelli, D.S. Padrões Alimentares de Gestantes, Excesso de Peso Materno e Diabetes Gestacional (Dietary Patterns of Pregnant Women, Excessive Maternal Weight and Gestational Diabetes). *Rev. Saúde Publica* **2019**, *53*, 52. [CrossRef]
62. Nascimento, G.R.; Alves, L.V.; Fonseca, C.L.; Figueiroa, J.N.; Alves, J.G. Dietary Patterns and Gestational Diabetes Mellitus in a Low Income Pregnant Women Population in Brazil—A Cohort Study. *Arch. Latinoam. Nutr.* **2016**, *66*, 301–308.
63. De Coelho, N.L.P.; Cunha, D.B.; Esteves, A.P.P.; de Lacerda, E.M.A.; Filha, M.M.T. Dietary Patterns in Pregnancy and Birth Weight. *Rev. Saúde Publica* **2015**, *49*, 1–10. [CrossRef] [PubMed]
64. Marquez, B.V.Y. *Association between Sugar-Sweetened Beverages and Both Gestational Weight Gain and Gestational Diabetes*; The University of Texas School of Public Health: Dallas, TX, USA, 2012.
65. Rodrigues, B.; Azeredo, V.; Silva, A. Relationship between Food Consumption of Pregnant Women and Birth Weight of Newborns. *Rev. Chil. Nutr.* **2020**, *47*, 80–88. [CrossRef]
66. Günther, J.; Hoffmann, J.; Spies, M.; Meyer, D.; Kunath, J.; Stecher, L.; Rosenfeld, E.; Kick, L.; Rauh, K.; Hauner, H. Associations between the Prenatal Diet and Neonatal Outcomes—A Secondary Analysis of the Cluster-Randomised Gelis Trial. *Nutrients* **2019**, *11*, 1889. [CrossRef] [PubMed]
67. Maugeri, A.; Barchitta, M.; Favara, G.; La Rosa, M.C.; La Mastra, C.; Magnano San Lio, R.; Agodi, A. Maternal Dietary Patterns Are Associated with Pre-Pregnancy Body Mass Index and Gestational Weight Gain: Results from the "Mamma & Bambino" Cohort. *Nutrients* **2019**, *11*, 1308. [CrossRef]
68. Englund-Ögge, L.; Brantsæter, A.L.; Juodakis, J.; Haugen, M.; Meltzer, H.M.; Jacobsson, B.; Sengpiel, V. Associations between Maternal Dietary Patterns and Infant Birth Weight, Small and Large for Gestational Age in the Norwegian Mother and Child Cohort Study. *Eur. J. Clin. Nutr.* **2019**, *73*, 1270–1282. [CrossRef] [PubMed]
69. Marí-Sanchis, A.; Díaz-Jurado, G.; Basterra-Gortari, F.J.; de la Fuente-Arrillaga, C.; Martínez-González, M.A.; Bes-Rastrollo, M. Association between Pre-Pregnancy Consumption of Meat, Iron Intake, and the Risk of Gestational Diabetes: The SUN Project. *Eur. J. Nutr.* **2018**, *57*, 939–949. [CrossRef] [PubMed]
70. Schmidt, A.B.; Lund, M.; Corn, G.; Halldorsson, T.I.; Øyen, N.; Wohlfahrt, J.; Olsen, S.F.; Melbye, M. Dietary Glycemic Index and Glycemic Load during Pregnancy and Offspring Risk of Congenital Heart Defects: A Prospective Cohort Study. *Am. J. Clin. Nutr.* **2020**, *111*, 526–535. [CrossRef] [PubMed]

71. Donazar-Ezcurra, M.; Lopez-Del Burgo, C.; Martinez-Gonzalez, M.A.; Basterra-Gortari, F.J.; De Irala, J.; Bes-Rastrollo, M. Pre-Pregnancy Adherences to Empirically Derived Dietary Patterns and Gestational Diabetes Risk in a Mediterranean Cohort: The Seguimiento Universidad de Navarra (SUN) Project. *Br. J. Nutr.* **2017**, *118*, 715–721. [CrossRef]
72. Donazar-Ezcurra, M.; Lopez-del Burgo, C.; Martinez-Gonzalez, M.A.; Basterra-Gortari, F.J.; de Irala, J.; Bes-Rastrollo, M. Soft Drink Consumption and Gestational Diabetes Risk in the SUN Project. *Clin. Nutr.* **2017**, *37*, 638–645. [CrossRef] [PubMed]
73. Grundt, J.H.; Eide, G.E.; Brantsæter, A.L.; Haugen, M.; Markestad, T. Is Consumption of Sugar-Sweetened Soft Drinks during Pregnancy Associated with Birth Weight? *Matern. Child Nutr.* **2016**, *13*, e12405. [CrossRef]
74. Dominguez, L.J.; Martínez-González, M.A.; Basterra-Gortari, F.J.; Gea, A.; Barbagallo, M.; Bes-Rastrollo, M. Fast Food Consumption and Gestational Diabetes Incidence in the SUN Project. *PLoS ONE* **2014**, *9*, e106627. [CrossRef]
75. Borgen, I.; Aamodt, G.; Harsem, N.; Haugen, M.; Meltzer, H.M.; Brantsæter, A.L. Maternal Sugar Consumption and Risk of Preeclampsia in Nulliparous Norwegian Women. *Eur. J. Clin. Nutr.* **2012**, *66*, 920–925. [CrossRef] [PubMed]
76. Brantsæter, A.L.; Haugen, M.; Samuelsen, S.O.; Torjusen, H.; Trogstad, L.; Alexander, J.; Magnus, P.; Meltzer, H.M. A Dietary Pattern Characterized by High Intake of Vegetables, Fruits, and Vegetable Oils Is Associated with Reduced Risk of Preeclampsia in Nulliparous Pregnant Norwegian Women. *J. Nutr.* **2009**, *139*, 1162–1168. [CrossRef] [PubMed]
77. Uusitalo, U.; Arkkola, T.; Ovaskainen, M.L.; Kronberg-Kippilä, C.; Kenward, M.G.; Veijola, R.; Simell, O.; Knip, M.; Virtanen, S.M. Unhealthy Dietary Patterns Are Associated with Weight Gain during Pregnancy among Finnish Women. *Public Health Nutr.* **2009**, *12*, 2392–2399. [CrossRef] [PubMed]
78. Nicolì, F.; Prete, A.; Citro, F.; Bertolotto, A.; Aragona, M.; de Gennaro, G.; Del Prato, S.; Bianchi, C. Use of Non-Nutritive-Sweetened Soft Drink and Risk of Gestational Diabetes. *Diabetes Res. Clin. Pract.* **2021**, *178*, 108943. [CrossRef] [PubMed]
79. Dale, M.T.G.; Magnus, P.; Leirgul, E.; Holmstrøm, H.; Gjessing, H.K.; Brodwall, K.; Haugen, M.; Stoltenberg, C.; Øyen, N. Intake of Sucrose-Sweetened Soft Beverages during Pregnancy and Risk of Congenital Heart Defects (CHD) in Offspring: A Norwegian Pregnancy Cohort Study. *Eur. J. Epidemiol.* **2019**, *34*, 383–396. [CrossRef] [PubMed]
80. Garay, S.M.; Savory, K.A.; Sumption, L.; Penketh, R.; Janssen, A.B.; John, R.M. The Grown in Wales Study: Examining Dietary Patterns, Custom Birthweight Centiles and the Risk of Delivering a Small-for-Gestational Age (SGA) Infant. *PLoS ONE* **2019**, *14*, e0213412. [CrossRef] [PubMed]
81. Tielemans, M.J.; Erler, N.S.; Leermakers, E.T.M.; van den Broek, M.; Jaddoe, V.W.V.; Steegers, E.A.P.; Kiefte-de Jong, J.C.; Franco, O.H. A Priori and a Posteriori Dietary Patterns during Pregnancy and Gestational Weight Gain: The Generation R Study. *Nutrients* **2015**, *7*, 9383–9399. [CrossRef]
82. Zhu, Y.; Olsen, S.F.; Mendola, P.; Halldorsson, T.I.; Rawal, S.; Hinkle, S.N.; Yeung, E.H.; Chavarro, J.E.; Grunnet, L.G.; Granström, C.; et al. Maternal Consumption of Artificially Sweetened Beverages during Pregnancy, and Offspring Growth through 7 Years of Age: A Prospective Cohort Study. *Int. J. Epidemiol.* **2017**, *46*, 1499–1508. [CrossRef]
83. Rasmussen, M.A.; Maslova, E.; Halldorsson, T.I.; Olsen, S.F. Characterization of Dietary Patterns in the Danish National Birth Cohort in Relation to Preterm Birth. *PLoS ONE* **2014**, *9*, e93644. [CrossRef]
84. Bärebring, L.; Brembeck, P.; Löf, M.; Brekke, H.K.; Winkvist, A.; Augustin, H. Food Intake and Gestational Weight Gain in Swedish Women. *SpringerPlus* **2016**, *5*, 377. [CrossRef]
85. Englund-Ögge, L.; Brantsæter, A.L.; Sengpiel, V.; Haugen, M.; Birgisdottir, B.E.; Myhre, R.; Meltzer, H.M.; Jacobsson, B. Maternal Dietary Patterns and Preterm Delivery: Results from Large Prospective Cohort Study. *BMJ* **2014**, *348*, g1446. [CrossRef] [PubMed]
86. Mikeš, O.; Brantsæter, A.L.; Knutsen, H.K.; Torheim, L.E.; Bienertová Vašků, J.; Pruša, T.; Čupr, P.; Janák, K.; Dušek, L.; Klánová, J. Dietary Patterns and Birth Outcomes in the ELSPAC Pregnancy Cohort. *J. Epidemiol. Community Health* **2021**, *76*, 613–619. [CrossRef]
87. Wen, L.M.; Simpson, J.M.; Rissel, C.; Baur, L.A. Maternal "Junk Food" Diet During Pregnancy as a Predictor of High Birthweight: Findings from the Healthy Beginnings Trial. *Birth* **2013**, *40*, 46–51. [CrossRef]
88. Baskin, R.; Hill, B.; Jacka, F.N.; O'Neil, A.; Skouteris, H. Antenatal Dietary Patterns and Depressive Symptoms during Pregnancy and Early Post-Partum. *Matern. Child Nutr.* **2017**, *13*, e12218. [CrossRef]
89. Ferreira, L.B.; Lobo, C.V.; do Carmo, A.S.; Souza, R.C.V.E.; Dos Santos, L.C. Dietary Patterns During Pregnancy and Their Association with Gestational Weight Gain and Anthropometric Measurements at Birth. *Matern. Child Health J.* **2022**, *26*, 1464–1472. [CrossRef] [PubMed]
90. De Silva, C.O.; de Souza, J. Diet during Pregnancy: Ultra-Processed Foods and the Inflammatory Potential of Diet. *Nutrition* **2022**, *97*, 111603. [CrossRef]
91. Pagliai, G.; Dinu, M.; Madarena, M.P.; Bonaccio, M.; Iacoviello, L.; Sofi, F. Consumption of Ultra-Processed Foods and Health Status: A Systematic Review and Meta-Analysis. *Br. J. Nutr.* **2021**, *125*, 308–318. [CrossRef]
92. Chen, X.; Zhang, Z.; Yang, H.; Qiu, P.; Wang, H.; Wang, F.; Zhao, Q.; Fang, J.; Nie, J. Consumption of Ultra-Processed Foods and Health Outcomes: A Systematic Review of Epidemiological Studies. *Nutr. J.* **2020**, *19*, 86. [CrossRef] [PubMed]
93. Barbosa, S.S.; Sousa, L.C.M.; de Oliveira Silva, D.F.; Pimentel, J.B.; de Evangelista, K.C.M.S.; de Lyra, C.O.; Lopes, M.M.G.D.; Lima, S.C.V.C. A Systematic Review on Processed/Ultra-Processed Foods and Arterial Hypertension in Adults and Older People. *Nutrients* **2022**, *14*, 1215. [CrossRef] [PubMed]
94. Hassani Zadeh, S.; Boffetta, P.; Hosseinzadeh, M. Dietary Patterns and Risk of Gestational Diabetes Mellitus: A Systematic Review and Meta-Analysis of Cohort Studies. *Clin. Nutr. ESPEN* **2020**, *36*, 1–9. [CrossRef] [PubMed]

95. Quan, W.; Zeng, M.; Jiao, Y.; Li, Y.; Xue, C.; Liu, G.; Wang, Z.; Qin, F.; He, Z.; Chen, J. Western Dietary Patterns, Foods, and Risk of Gestational Diabetes Mellitus: A Systematic Review and Meta-Analysis of Prospective Cohort Studies. *Adv. Nutr.* **2021**, *12*, 1353–1364. [CrossRef]
96. Kibret, K.T.; Chojenta, C.; Gresham, E.; Tegegne, T.K.; Loxton, D. Maternal Dietary Patterns and Risk of Adverse Pregnancy (Hypertensive Disorders of Pregnancy and Gestational Diabetes Mellitus) and Birth (Preterm Birth and Low Birth Weight) Outcomes: A Systematic Review and Meta-Analysis. *Public Health Nutr.* **2019**, *22*, 506–520. [CrossRef] [PubMed]
97. Abdollahi, S.; Soltani, S.; De Souza, R.J.; Forbes, S.C.; Toupchian, O.; Salehi-Abargouei, A. Associations between Maternal Dietary Patterns and Perinatal Outcomes: A Systematic Review and Meta-Analysis of Cohort Studies. *Adv. Nutr.* **2021**, *12*, 1332–1352. [CrossRef] [PubMed]
98. Traore, S.S.; Bo, Y.; Amoah, A.N.; Khatun, P.; Kou, G.; Hu, Y.; Lyu, Q. A Meta-Analysis of Maternal Dietary Patterns and Preeclampsia. *Clin. Nutr. Open Sci.* **2021**, *40*, 15–29. [CrossRef]
99. Rana, S.; Lemoine, E.; Granger, J.; Karumanchi, S.A. Preeclampsia: Pathophysiology, Challenges, and Perspectives. *Circ. Res.* **2019**, *124*, 1094–1112. [CrossRef] [PubMed]
100. Garovic, V.D.; Dechend, R.; Easterling, T.; Karumanchi, S.A.; Baird, S.M.M.; Magee, L.A.; Rana, S.; Vermunt, J.V.; August, P. Hypertension in Pregnancy: Diagnosis, Blood Pressure Goals, and Pharmacotherapy: A Scientific Statement From the American Heart Association. *Hypertension* **2022**, *79*, E21–E41. [CrossRef] [PubMed]
101. Teng, H.; Wang, Y.; Han, B.; Liu, J.; Cao, Y.; Wang, J.; Zhu, X.; Fu, J.; Ling, Q.; Xiao, C.; et al. Gestational Systolic Blood Pressure Trajectories and Risk of Adverse Maternal and Perinatal Outcomes in Chinese Women. *BMC Pregnancy Childbirth* **2021**, *21*, 155. [CrossRef] [PubMed]
102. Diemert, A.; Lezius, S.; Pagenkemper, M.; Hansen, G.; Drozdowska, A.; Hecher, K.; Arck, P.; Zyriax, B.C. Maternal Nutrition, Inadequate Gestational Weight Gain and Birth Weight: Results from a Prospective Birth Cohort. *BMC Pregnancy Childbirth* **2016**, *16*, 224. [CrossRef] [PubMed]
103. Mishra, K.G.; Bhatia, V.; Nayak, R. Maternal Nutrition and Inadequate Gestational Weight Gain in Relation to Birth Weight: Results from a Prospective Cohort Study in India. *Clin. Nutr. Res.* **2020**, *9*, 213. [CrossRef]
104. Lagiou, P.; Tamimi, R.M.; Mucci, L.A.; Adami, H.-O.; Hsieh, C.-C.; Trichopoulos, D. Diet during Pregnancy in Relation to Maternal Weight Gain and Birth Size. *Eur. J. Clin. Nutr.* **2004**, *58*, 231–237. [CrossRef]
105. Dude, A.M.; Grobman, W.; Haas, D.; Mercer, B.M.; Parry, S.; Silver, R.M.; Wapner, R.; Wing, D.; Saade, G.; Reddy, U.; et al. Gestational Weight Gain and Pregnancy Outcomes among Nulliparous Women. *Am. J. Perinatol.* **2021**, *38*, 182–190. [CrossRef]
106. Sun, Y.; Shen, Z.; Zhan, Y.; Wang, Y.; Ma, S.; Zhang, S.; Liu, J.; Wu, S.; Feng, Y.; Chen, Y.; et al. Effects of Pre-Pregnancy Body Mass Index and Gestational Weight Gain on Maternal and Infant Complications. *BMC Pregnancy Childbirth* **2020**, *20*, 390. [CrossRef] [PubMed]
107. Mbu, R.E.; Fouedjio, H.J.; Tabot, M.; Fouelifack, F.Y.; Tumasang, F.N.; Tonye, R.N.; Leke, R.J.I. Effects of Gestational Weight Gain on the Outcome of Labor at the Yaounde Central Hospital Maternity, Cameroon. *Open J. Obstet. Gynecol.* **2013**, *3*, 648–652. [CrossRef]
108. Ashley-Martin, J.; Woolcott, C. Gestational Weight Gain and Postpartum Weight Retention in a Cohort of Nova Scotian Women. *Matern. Child Health J.* **2014**, *18*, 1927–1935. [CrossRef] [PubMed]
109. Widen, E.M.; Whyatt, R.M.; Hoepner, L.A.; Ramirez-Carvey, J.; Oberfield, S.E.; Hassoun, A.; Perera, F.P.; Gallagher, D.; Rundle, A.G. Excessive Gestational Weight Gain Is Associated with Long-Term Body Fat and Weight Retention at 7 y Postpartum in African American and Dominican Mothers with Underweight, Normal, and Overweight Prepregnancy BMI. *Am. J. Clin. Nutr.* **2015**, *102*, 1460–1467. [CrossRef] [PubMed]
110. Shin, D.; Lee, K.W.; Song, W.O. Dietary Patterns during Pregnancy Are Associated with Risk of Gestational Diabetes Mellitus. *Nutrients* **2015**, *7*, 9369–9382. [CrossRef] [PubMed]
111. Silva, C.F.M.; Saunders, C.; Peres, W.; Folino, B.; Kamel, T.; dos Santos, M.S.; Padilha, P. Effect of Ultra-Processed Foods Consumption on Glycemic Control and Gestational Weight Gain in Pregnant with Pregestational Diabetes Mellitus Using Carbohydrate Counting. *PeerJ* **2021**, *9*, e10514. [CrossRef]
112. Hsieh, T.-T.; Chen, S.-F.; Lo, L.-M.; Li, M.-J.; Yeh, Y.-L.; Hung, T.-H. The Association Between Maternal Oxidative Stress at Mid-Gestation and Subsequent Pregnancy Complications. *Reprod. Sci.* **2012**, *19*, 505–512. [CrossRef]
113. Kim, H.; Hwang, J.-Y.; Ha, E.-H.; Park, H.; Ha, M.; Lee, S.-H.; Hong, Y.-C.; Chang, N. Fruit and Vegetable Intake Influences the Association between Exposure to Polycyclic Aromatic Hydrocarbons and a Marker of Oxidative Stress in Pregnant Women. *Eur. J. Clin. Nutr.* **2011**, *65*, 1118–1125. [CrossRef]
114. Pistollato, F.; Sumalla Cano, S.; Elio, I.; Masias Vergara, M.; Giampieri, F.; Battino, M. Plant-Based and Plant-Rich Diet Patterns during Gestation: Beneficial Effects and Possible Shortcomings. *Adv. Nutr.* **2015**, *6*, 581–591. [CrossRef] [PubMed]
115. WHO. Low Birth Weight. Available online: https://www.who.int/data/nutrition/nlis/info/low-birth-weight (accessed on 4 March 2022).

nutrients

MDPI

Article

Higher Ultra-Processed Food Consumption Is Associated with Greater High-Sensitivity C-Reactive Protein Concentration in Adults: Cross-Sectional Results from the Melbourne Collaborative Cohort Study

Melissa M. Lane [1,*], Mojtaba Lotfaliany [1], Malcolm Forbes [1,2,3], Amy Loughman [1], Tetyana Rocks [1], Adrienne O'Neil [1], Priscila Machado [4,5], Felice N. Jacka [1,6,7], Allison Hodge [8,9] and Wolfgang Marx [1]

1 Food & Mood Centre, The Institute for Mental and Physical Health and Clinical Translation (IMPACT), School of Medicine, Deakin University, Geelong, VIC 3220, Australia
2 Mental Health, Drugs & Alcohol Service, University Hospital Geelong, Barwon Health, Geelong, VIC 3220, Australia
3 Department of Psychiatry, University of Melbourne, Parkville, VIC 3050, Australia
4 Institute for Physical Activity and Nutrition, School of Exercise and Nutrition Sciences, Deakin University, Melbourne, VIC 3125, Australia
5 Center for Epidemiological Research in Nutrition and Health, University of Sao Paulo, Sao Paulo 01246-904, Brazil
6 Centre for Adolescent Health, Murdoch Children's Research Institute, Parkville, VIC 3052, Australia
7 College of Public Health, Medical & Veterinary Sciences, James Cook University, Townsville, QLD 4811, Australia
8 Cancer Epidemiology Division, Cancer Council Victoria, Melbourne, VIC 3004, Australia
9 Centre for Epidemiology and Biostatistics, Melbourne School of Population and Global Health, University of Melbourne, Melbourne, VIC 3010, Australia
* Correspondence: laneme@deakin.edu.au

Citation: Lane, M.M.; Lotfaliany, M.; Forbes, M.; Loughman, A.; Rocks, T.; O'Neil, A.; Machado, P.; Jacka, F.N.; Hodge, A.; Marx, W. Higher Ultra-Processed Food Consumption Is Associated with Greater High-Sensitivity C-Reactive Protein Concentration in Adults: Cross-Sectional Results from the Melbourne Collaborative Cohort Study. *Nutrients* **2022**, *14*, 3309. https://doi.org/10.3390/nu14163309

Academic Editors: Monica Dinu and Daniela Martini

Received: 15 July 2022
Accepted: 10 August 2022
Published: 12 August 2022

Abstract: Background: Few studies have examined associations between ultra-processed food intake and biomarkers of inflammation, and inconsistent results have been reported in the small number of studies that do exist. As such, further investigation is required. Methods: Cross-sectional baseline data from the Melbourne Collaborative Cohort Study (MCCS) were analysed (n = 2018). We applied the NOVA food classification system to data from a food frequency questionnaire (FFQ) to determine ultra-processed food intake (g/day). The outcome was high-sensitivity C-reactive protein concentration (hsCRP; mg/L). We fitted unadjusted and adjusted linear regression analyses, with sociodemographic characteristics and lifestyle- and health-related behaviours as covariates. Supplementary analyses further adjusted for body mass index (kg/m^2). Sex was assessed as a possible effect modifier. Ultra-processed food intake was modelled as 100 g increments and the magnitude of associations expressed as estimated relative change in hsCRP concentration with accompanying 95% confidence intervals (95%CIs). Results: After adjustment, every 100 g increase in ultra-processed food intake was associated with a 4.0% increase in hsCRP concentration (95%CIs: 2.1–5.9%, $p < 0.001$). Supplementary analyses showed that part of this association was independent of body mass index (estimated relative change in hsCRP: 2.5%; 95%CIs: 0.8–4.3%, $p = 0.004$). No interaction was observed between sex and ultra-processed food intake. Conclusion: Higher ultra-processed food intake was cross-sectionally associated with elevated hsCRP, which appeared to occur independent of body mass index. Future prospective and intervention studies are necessary to confirm directionality and whether the observed association is causal.

Keywords: ultra-processed food; NOVA; diet; inflammation; high-sensitivity C-reactive protein; non-communicable diseases; cross-sectional

1. Introduction

Nutrition science has long sought to understand the effects of diet on human health. This has largely been done by classifying foods based on their nutrient composition. The impacts on health of individual macro- and micro-nutrients as well as kilojoules have typically been considered independent of different foods and food group sources [1]. Excess intakes of sugar, salt, saturated fat, and kilojoules have been previously linked with increased risk of cardiometabolic conditions [2–4]. Such research has been beneficial for understanding nutritional physiology and has subsequently informed dietary recommendations [5]. However, this nutrient-centric perspective does not capture the effect of complex food matrices. A food matrix is characterised as the molecular interactions between nutrient and non-nutrient components of food [6]. Indeed, emerging experimental [7] and epidemiological [8–10] evidence implicates the extent to which a food has been processed (or undergone food matrix alterations) as a risk factor for chronic non-communicable diseases, morbidity, and mortality.

The NOVA (name, not acronym) food classification system was recently developed to allow for the categorisation of food items based on their level of processing: from unprocessed or minimally processed food, processed culinary ingredients, processed food, to extensively processed food termed "ultra-processed" [11]. Ultra-processed foods, are defined by NOVA as industrial formulations created from compounds extracted, derived, or synthesised from food or food substrates. Ultra-processed foods also typically contain five or more ingredients including artificial food additives (e.g., colours, texturising agents, and olfactory and taste enhancers) and are commonly inexpensive, virtually imperishable, easily consumed, and highly palatable [12]. Time-series country-level sales data from 2006 to 2019 show a substantial growth in the types and quantities of ultra-processed foods sold worldwide, with projected increases to 2024 [13,14]. This suggests a transition away from non-ultra-processed food and toward a more processed global diet [13,14].

Chronic low-grade inflammation, marked by the presence of elevated inflammatory cytokines, is both a driver of chronic diseases and a characteristic of an established diseased state [15]. These diseases include cancers [16], cardiometabolic conditions [17], and mental disorders [18,19]. The shared link between chronic low-grade inflammation and diseased states exists despite the different organs and systems involved in their onset, prognosis, and morbidity [20]. Hence, better understanding and addressing possible drivers of inflammation is of significant public health interest. However, little data are available that have directly linked ultra-processed food intake to inflammation.

In the three epidemiological studies that do exist [21–23], inconsistent results have been observed. These included sex- and cohort-specific differences within and between studies [21,22] as well as associations of ultra-processed food with some inflammatory biomarkers but not others [23]. Importantly, each of these three studies included samples from Brazil, where the concept of avoiding ultra-processed food has received recognition in official dietary guidelines since 2014 [24], and where consumption of ultra-processed food is estimated to be lower than higher-income countries [11,14]. Thus, there is a need for further investigation of the association between ultra-processed food intake and inflammation in other regions, particularly in settings where the substitution of non-ultra-processed foods for those that are ultra-processed is increasingly common. The current study aimed to address this gap by using data from the Melbourne Collaborative Cohort Study (MCCS) to investigate cross-sectional associations between ultra-processed food intake and plasma concentrations of the inflammatory cytokine, high-sensitivity C-reactive protein (hsCRP).

2. Methods

A full description of methods for data collection in the MCCS was published elsewhere [25]. In brief, the MCCS is a study that aimed to assess prospective associations between diet and lifestyle and chronic non-communicable diseases [25]. Between 1990 and 1994 (baseline), 41,513 people (24,469 women) aged between 27 and 76 (99% 40–69) years were recruited from the Melbourne metropolitan area, with migrants from Southern

Europe included and deliberately recruited to expand the span of diet and lifestyle exposures. At baseline, participants completed surveys and anthropometric measurements, and blood samples were collected. A case-cohort sub-study was undertaken to provide a more economical design within which to perform assays on blood samples collected at baseline (as part of the original MCCS project) and analyse associations of various molecules in the stored plasma with selected disease outcomes; cases were included in the case-cohort if they were identified as having the outcome of interest by 30 June 2002. High-sensitivity C-reactive protein as a biomarker of inflammation was measured in incident cardiovascular death cases together with a random sample (sub-cohort) of all participants in the MCCS.

The current cross-sectional study thus comprised a sample of participants from the case-cohort sub-study for whom valid baseline dietary data and plasma hsCRP measurements were available (see Figure 1). We excluded participants who had missing hsCRP data (n = 161), total energy intake (kJ/day) below the 1st and above the 99th percentiles (n = 40), or hsCRP concentration above the 99th percentile (n = 23). Two thousand and eighteen participants remained for analysis, including both the cardiovascular disease death group as cases (n = 567) and the random sample of all participants from the original MCCS project as the sub-cohort (n = 1451). Cardiovascular disease death cases were identified from notifications to the Victorian Registry of Births, Deaths and Marriages, and the Australian National Death Index (codes 390–459 and I00–I99 of the International Classification of Diseases, ICD-9 and ICD-10, respectively).

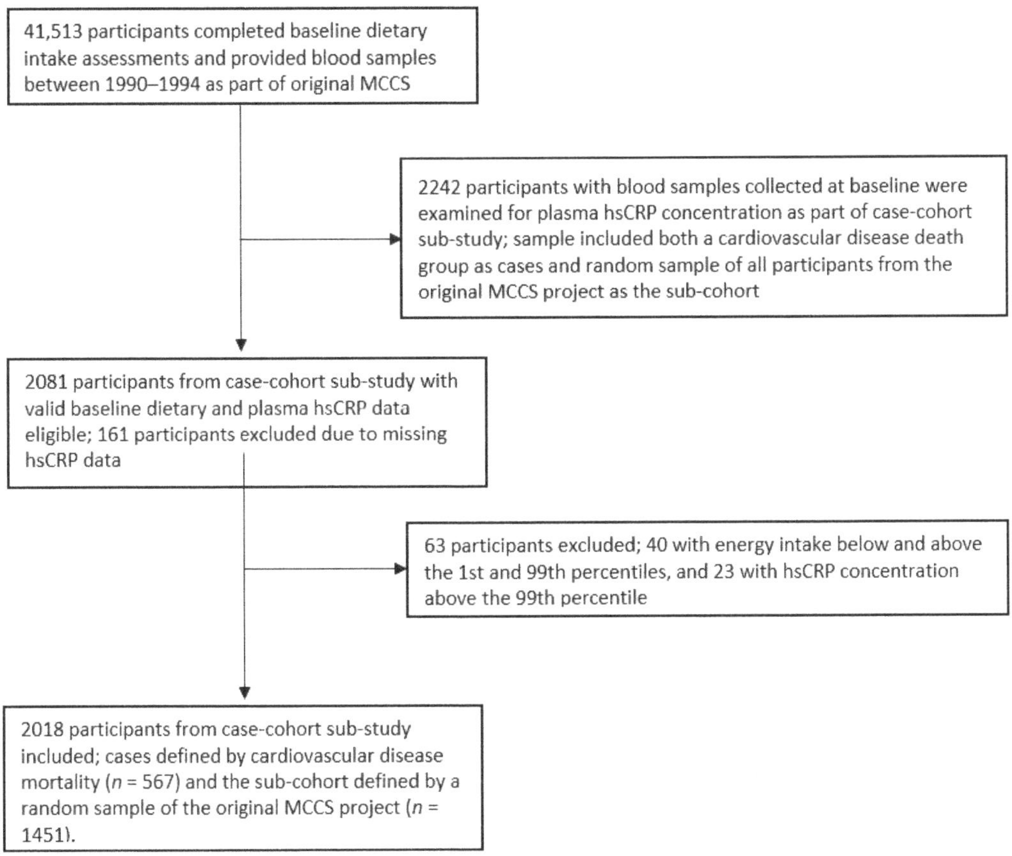

Figure 1. Flow-chart of participant selection. MCCS—Melbourne Collaborative Cohort Study, hsCRP—high-sensitivity C-reactive protein.

The current study was reported in accordance with the Strengthening the Reporting of Observational Studies in Epidemiology (STROBE) statement and checklist for cross-sectional studies [25]. The current study was prospectively registered with Open Science Framework (OSF) registry (registration DOI: 10.17605/OSF.IO/EHFXD) and was approved for exemption from ethical review in accordance with the National Statement on Ethical Conduct in Human Research (2007, updated 2018) Section 5.1.22 by the Deakin University Human Research Ethics Committee (project number: 2020-413, received 18th of November 2020). The study protocol for the original MCCS project was approved by the Cancer Council Victoria's Human Research Ethics Committee (project number: IEC 9001, received 23rd of August 1990). Participants provided written consent to participate including researcher access to their medical records [26].

2.1. Exposure: Dietary Assessment

At baseline of the original MCCS project, participants attended clinics where dietary data were collected using a self-administered 121-item food frequency questionnaire (FFQ) specifically developed for use in this multiethnic cohort [27]. This FFQ was based on weighed food records in 810 Melbournians of similar demographics to the cohort [27]. For this study, all FFQ food items were classified according to the NOVA food classification system as ultra-processed foods and non-ultra-processed foods by two experts with Australian food and dietary intake knowledge. Examples of NOVA's classification of ultra-processed food (NOVA group 4) include soft drinks, sweet or savoury packaged snacks, confectionery, packaged breads and buns, margarine, reconstituted meat products, and pre-prepared frozen or shelf-stable dishes. Examples of NOVA's classification of non-ultra-processed food include unprocessed or minimally processed foods (NOVA group 1) such as rice and other cereals, meat, fish, milk, eggs, fruit, roots and tubers, vegetables, and nuts and seeds; processed culinary ingredients (NOVA group 2) such as sugar, plant oils, and butter; and processed foods (NOVA group 3) such as processed breads and cheese, canned fruit and fish, and salted and smoked meats. More information regarding the NOVA food classification system can be found elsewhere (1). When it was not possible to discriminate, (e.g., food items such as 'bread', 'pasta or noodles', 'low fat cheese', 'yoghurt', and 'fruit juice'), cross-sectional data from the National Nutrition Survey 1995–1996 (data not published) and Australian National Nutrition and Physical Activity Survey (NNPAS) 2011–2012 were used for comparison and decision making [28].

As per previous research [27,29,30], the mean daily contribution of ultra-processed foods to intake of total energy (kJ) and weight (g) was calculated by transforming frequencies into grams based on sex-specific portion sizes of each food multiplied by the daily equivalent frequency. Energy was estimated based on the Nutrient Data Table for Use in Australia 1995 (NUTTAB 95). The NUTTAB 95 is a food composition database that contains information for 1800 foods and beverages available in Australia [31].

2.2. Outcome: Inflammatory Cytokine Assessment

Blood samples were also collected at baseline. Venous blood samples were drawn in lithium-heparin tubes, and plasma was subsequently stored at −180 °C (liquid nitrogen) until assayed as part of the case-cohort sub-study. HsCRP concentration expressed as mg/L was measured by a high-sensitivity immunonephelometric assay. The assay was a Dade Behring nephelometric assay done on a BNII Nehphelometric Analyser, Dade-Behring Diagnostics, Lane Cove, NSW, Australia.

2.3. Assessment of Covariates

Potential covariates were identified based on previous literature [7,21,32–35] and included in a directed acyclic graph to map hypothesised causal relationships between all relevant variables (Supplementary Materials Figure S1). These covariates were assessed through a structured interview that was administered at baseline providing data on sociodemographic characteristics and lifestyle- and health-related factors.

In particular, the sociodemographic characteristics included: age (continuous), sex (men and women), country of birth (Australia/New Zealand, United Kingdom/Malta, Italy, and Greece), marital status (married, de facto, single, divorced, separated, and widow), highest level of education (primary school, high/technical school, and tertiary degree or diploma) and Socio-Economic Indexes for Areas (SEIFA)—Index of Relative Socio-Economic Disadvantage [36]. SEIFA scores are recorded by the Australian Bureau of Statistics and refer to the relative socioeconomic advantage and disadvantage of defined geographical areas such as postal code [37] (we divided these scores into quintiles, with the lowest and highest representing the greatest and least disadvantaged, respectively).

The lifestyle- and health-related factors included: smoking status (never smoked, current smoker, and former smoker), leisure time physical activity over the last 6 months (a score was calculated ranging from 0–16 based on the frequency of walking and less vigorous and vigorous activity multiplied by two, which was then divided into categories, namely: 0 (none), >0 and <4 (low), ≥4 and <6 (moderate), and ≥6 (high) [32,34]), and alcohol intake using beverage-specific quantity frequency questions (lifetime abstainers, ex-drinkers, and current drinkers (further categorised as up to 19, 20–29, 30–39, and 40+ g/day)) [26,32,33]. Height and weight were measured, and body mass index was calculated as kg/m^2 [26]. These sociodemographic characteristics and lifestyle- and health-related variables were used as covariates.

2.4. Statistical Analyses

An inverse probability weighting method was applied to address the case-cohort design and adjust for the possibility of oversampling cases versus participants from the sub-cohort [26]. Characteristics of participants were summarised using mean and standard deviation (SD) or median and interquartile range (IQR) for continuous variables and frequency and percentage for categorical variables.

Linear regression analysis was used to examine associations between the consumption of ultra-processed food and hsCRP concentration. We aimed to better account for ultra-processed formulations that did not provide energy or were low in energy (e.g., artificially sweetened beverages) as per [38–40]. Thus, the total weight of ultra-processed foods in grams per day (g/day) was adjusted for energy using Willet's residual method [41] and used to model our exposure in 100 g increments. We chose 100 g increments to aid in reporting and interpretation. We assessed our outcome variable, hsCRP concentration, continuously; however, because the variance of residuals for hsCRP concentration was not homogeneous along all of the fitted values, hsCRP was log (to base e) transformed; the exponentiated coefficients represent the percent change from the geometric mean (anti-log). There were no zero values for hsCRP concentration. We verified the assumptions for a linear model with graphical and statistical tests of the associations between ultra-processed food intake and hsCRP concentration as well as between the fitted values and residuals. We further added the exposure value squared in the model to assess whether there was curvature in the association between ultra-processed food intake and hsCRP concentration. We used locally weighted scatterplot smoothing (LOWESS) models to examine whether there was a threshold for the association between ultra-processed food intake and hsCRP concentration. We also tested with graphical and statistical tests the normality of residuals and homoscedasticity (homogeneity of variance). We used the variance inflation factor to assess collinearity between the potential confounders included in our models. Lastly, to allow for comparison with the three previously published ultra-processed inflammation studies [21–23], and given that the assumptions associated with a linear model were not violated, we used linear models. The estimated relative change in hsCRP concentration (mg/L) for each energy-adjusted 100 g increase in ultra-processed food consumption was thus calculated along with robust standard 95% confidence intervals (95%CIs) and p-values. We also estimated the variance explained (Pseudo-R^2) in hsCRP by ultra-processed food consumption via the Cragg-Uhler method [42].

Four different sequential models were fitted: energy-adjusted ultra-processed food as the exposure variable and otherwise unadjusted (model 1), additionally adjusted for sociodemographic characteristics (model 2), and a fully adjusted model that further controlled for lifestyle- and health-related behaviours (model 3). Since previous studies have highlighted body mass index as a potential mediator in the association between ultra-processed food consumption and inflammation [7,21], supplementary analyses were performed by additionally adjusting for body mass index (model 4). It is important to highlight here that we were interested in assessing the "total effect" of ultra-processed food consumption on hsCRP concentration. As such, and given that body mass index was a prespecified mediator, we assessed its possible impact as part of our supplementary analyses (model 4). However, given the cross-sectional nature of our study, we refrained from referring to and did not formally test for mediation [43]. Because body mass index did not qualify as a confounder (see [44]), model 3 was considered the main model. Other studies have also reported differences between sexes in the association between ultra-processed food consumption and inflammation [21,22]. We thus undertook further supplementary analyses to (a) stratify by sex and (b) assess the potential effect modification of sex with ultra-processed food consumption. To explore sex as a possible effect modifier, we added interaction terms between sex and ultra-processed food consumption into the main effects model.

To ensure the sampling methods did not affect the results, sensitivity analyses were performed across all models with the following exclusions: (1) people with hsCRP > 10 mg/L (n = 122), which may indicate acute inflammation [45], although these values can also be seen in cases of chronic inflammation [46]; and (2) cases defined by cardiovascular disease mortality (n = 567). We conducted further sensitivity analyses on our main model 3 by excluding people with history of non-communicable diseases, such as hypertension (n = 555), stroke (n = 44), heart attack (n = 129), cancer (n = 166), diabetes mellitus (n = 100), and body mass index $\geq$30 (n = 520).

Lastly, we conducted post hoc analyses by fitting a logistic regression on our main model 3 (adjusted for sociodemographic characteristics and lifestyle- and health-related behaviours) to assess associations between each energy-adjusted 100 g increase in ultra-processed food consumption and the odds of hsCRP at or above 3 mg/L, which is considered a risk factor for cardiovascular events [45].

The analyses were undertaken using R version 3.6.3 (R Development Core Team, Vienna, Austria) [47].

3. Results

The current study included 1261 men and 757 women. Table 1 shows the sociodemographic and lifestyle characteristics of participants. The mean age of participants was 57 years. Most people reported that they were married or in a de facto relationship (75.6%) as well as reporting their country of birth as Australia or New Zealand (64.2%). Approximately one quarter of participants were in the top quintile of SEFIA (least disadvantaged; 25.1%) and reported that they had either some study towards or had completed a tertiary degree or diploma (24.0%). Less than a fifth of participants reported that they were current smokers (14.2%) and over one fifth (21.7%) reported that they had engaged in high physical activity over the last six months. Most participants (40.8%) had an average alcohol intake of less than 19 (g/day). The mean body mass index for all participants was 27.8 (kg/m^2), and the mean proportion of ultra-processed food in the overall diet by weight and energy was 26% (g/day) and 40% (kJ/day), respectively. In terms of ultra-processed food intake in absolute weight and energy, the median was 364.4 (g/day) and 2975.1 (kJ/day), respectively. The median hsCRP concentration for participants was 1.6 (mg/L).

Table 1. Descriptive characteristics of the study population.

n	**Total = 2018**
n for cardiovascular death cases	Total = 632
n for sub-cohort (random sample of all MCCS participants)	Total = 1386
Age (years)—mean (SD)	57.0 (8.8)
Women	757 (37.5%)
Married/de facto	1431 (75.6%)
(In)complete tertiary degree or diploma [a]	485 (24.0%)
Top quintile of SEIFA [b] index (Q5: least disadvantaged)	504 (25.1%)
Born in Australia/New Zealand	1296 (64.2%)
Current smoker	287 (14.2%)
High physical activity score [c] (≥6)	438 (21.7%)
Alcohol intake of up to 19 g/day	801 (40.8%)
Body mass index (kg/m^2)—mean (SD)	27.8 (4.7)
Proportion (%) of ultra-processed food (g/day)—mean (SD)	26.0 (11.4)
Proportion (%) of ultra-processed food (kJ/day)—mean (SD)	40.0 (13.0)
Total ultra-processed food (g/day)—median (interquartile range: Q1, Q3)	364 (248, 518)
Total ultra-processed food (kJ/day)—median (interquartile range: Q1, Q3)	2975 (2091, 4244)
hsCRP concentration (mg/L)—median (interquartile range: Q1, Q3)	1.6 (0.8, 3.6)

[a] (In)complete tertiary degree or diploma referred to participants who had some study towards a tertiary degree or diploma as well as participants who had completed a tertiary degree or diploma. [b] SEIFA Socio-Economic Indexes for Areas [37]. [c] Ordinal score based on frequency of walking plus frequency of less vigorous activity plus twice the frequency of vigorous activity, and ranging from 0 to 16 [32,34]. MCCS—Melbourne Collaborative Cohort Study, hsCRP—high-sensitivity C-reactive protein, SD—standard deviation.

Table 2 details the results of the multivariable adjusted models. In model 1, every 100 g increase in ultra-processed food intake was associated with a 3.6% increase in hsCRP concentration (95%CIs: 1.7–5.5%, $p < 0.001$). After accounting for sociodemographic characteristics and lifestyle- and health-related behaviours in the main multivariable analysis (model 3), the association remained robust (expected relative change in hsCRP: 4.0%; 95%CIs: 2.1–5.9%, $p < 0.001$). The supplementary analyses including all participants and further adjustment for body mass index are also shown in Table 2 (model 4). Part of the association between ultra-processed food intake and hsCRP concentration was independent of body mass index, where every 100 g increase in ultra-processed food intake was associated with a 2.5% increase in hsCRP concentration (95%CIs: 0.8–4.3%, $p = 0.004$). Results remained relatively stable in our sensitivity analyses that excluded people with hsCRP concentrations above 10 mg/L, cardiovascular disease mortality and history of cardiovascular diseases, cancer, diabetes mellitus, and body mass index ≥30 (see Supplementary Materials Tables S1 and S2).

Table 2. Cross-sectional associations between ultra-processed food intake and hsCRP concentration (MCCS, 1990–1994).

		Main Analyses		
Variable	*n*	Estimated Relative Change in hsCRP Concentration (mg/L) for Each Energy-Adjusted 100 (g) Increase in Ultra-Processed Food Intake (95%CIs)	*p*-Value	R^2
Model 1 [a]	2018	3.6% (1.7–5.5%)	<0.001	6%
Model 2 [b]	1899	4.2% (2.3–6.0%)	<0.001	11.3%
*Model 3 [c]	1852	4.0% (2.1–5.9%)	<0.001	15.1%
**Model 4 [d]	1850	2.5% (0.8–4.3%)	0.004	27.7%

Regressions performed with hsCRP on a logarithmic scale. [a] Model 1 = unadjusted. [b] Model 2 = additionally adjusted for sociodemographic characteristics: sex (men and women), age (continuous), education ((in)complete tertiary degree or diploma, completed high/technical school, (in)complete high/technical school, completed primary school, and (in)complete primary school), country of birth (Australia/New Zealand/Other, United Kingdom/Malta, Italy, and Greece), marital status (married, de facto, divorced, separated, and widow), and SEIFA quintiles (Q1–Q5). Change to *n* due missing values for confounders marital status and SEIFA quintiles. [c] *Model 3 = main model additionally adjusted for lifestyle- and health-related behaviours: smoking status (never smoked, current smoker, and former smoker), physical activity over the last 6 months (0 (none), >0 and <4 (low), ≥4 and <6 (moderate), and ≥6 (high)), and alcohol intake (g/day) (lifetime abstainers, ex-drinkers, and up to 19, 20–29, 30–39, and 40+). Change to *n* due missing values for confounder alcohol intake. [d] **Model 4 = supplementary analyses additionally adjusted for body mass index (kg/m^2). Change to *n* due missing values for confounders alcohol intake and body mass index. SEIFA—Socio-Economic Indexes for Areas, 95%CIs—95% confidence intervals.

There was no evidence of sex interactions (all *p*-values > 0.05 and estimates of interaction range: 0.0–2.6%). The supplementary analyses stratified by sex are shown in Supplementary Materials Table S3. After accounting for potential confounders in our main model 3, every 100 g increase in ultra-processed food intake was associated with an increase in hsCRP concentration in both men (estimated relative change in hsCRP: 3.5%; 95%CIs: 1.3–5.7%, $p = 0.002$) and women (estimated relative change in hsCRP: 5.5%; 95%CIs: 0.5–10.5%, $p = 0.032$). However, after further adjustment for body mass index (model 4), the association remained robust in men only (estimated relative change in hsCRP for men: 2.8%; 95%CIs: 0.7–4.9%, $p = 0.010$ versus women: 2.4%, 95%CIs: −2.1–6.8%, $p = 0.296$). Post hoc analyses on our main model 3 showed that each 100 g increase in ultra-processed food consumption was associated with 1.08-fold increased odds of hsCRP concentration at or above 3 mg/L after adjusting for sociodemographic characteristics and lifestyle- and health-related behaviours (odds ratio: 1.080; 95%CIs: 1.034–1.128, $p < 0.001$).

4. Discussion

This study aimed to examine whether greater ultra-processed food intake was associated with higher hsCRP concentration in a sample of Australian adults. We found evidence of this association, and at least part of this association was independent of body mass index.

Three epidemiological studies have previously tested associations between ultra-processed food consumption and biomarkers of inflammation [21–23]. Overall associations with men and women combined were not tested in two of these ultra-processed food-inflammation studies [21,22]. This makes it challenging to compare these studies' results with the main results from our study. However, our results are partly consistent with another that assessed overall associations in male and female adolescents aged from 17 to 18 years [23]. That study demonstrated direct cross-sectional associations between the consumption of ultra-processed food and concentration of the inflammatory cytokine, interleukin-8 [23]. It also showed that participants with the highest intake of ultra-processed food had increased concentrations of leptin and C-reactive protein compared to participants with the lowest intake, but these associations were less certain given the 95% confidence intervals that crossed zero in both the unadjusted and fully adjusted linear models [23]. These less certain findings, particularly regarding C-reactive protein, may be partly explained by the included sample of adolescents who were exclusively from public schools in a lower socio-economic region of Brazil [23]. The authors of that study noted that these sociodemographic characteristics have been associated with lower consumption of ultra-processed food, with generalisability issues and underestimated effect estimates remaining possible [23]. Indeed, ultra-processed food contributed 26% to total daily energy intake in that Brazilian sample of adolescents compared to 40% in our sample.

While sex was not a significant effect modifier in our study, we conducted sex-stratified supplementary analyses to allow for comparison with previous literature [21,22]. One previous study reported direct prospective associations between the intake of ultra-processed food and interleukin-6 concentrations across two separate cohorts, with one cohort showing an association in women only and the other showing an association in men only [22]. Adiposity did not appear to explain these cohort- and sex-specific findings [22]. Results for the men-only analysis in our study support the data from the second cohort [22], where we also found associations between higher ultra-processed food intake and elevated hsCRP concentration across all models, including additional adjustment for body mass index.

In contrast, for the women in our study, observed associations were not independent of body mass index. These findings are somewhat concordant with another previous ultra-processed food-inflammation study [21], which reported direct cross-sectional associations between ultra-processed food intake and high-sensitivity C-reactive protein in women only and that appeared to be explained by body mass index [21]. These findings suggest that in women, adiposity is a possible intermediate on the causal pathway from ultra-processed food consumption to inflammation. This notion may be explained by the greater accumulation of adiposity, on average, in women compared to men; associations between body mass index and C-reactive protein concentration are suggested to be stronger for women than men [48]. However, cross-sectional studies do not allow for formal tests of mediation [43], and this notion requires further investigation in prospective analyses. However, it is important to reiterate that testing for effect modification by sex in our study showed no evidence of interaction. Given the underrepresentation of women compared to men in our study (37.5%), it is possible that we may not have had adequate power to detect this interaction. Further investigation with more appropriately designed studies is needed.

Our study is consistent with recent systematic reviews and meta-analyses [8–10] showing direct associations between intake of ultra-processed food and the prevalence and incidence of common chronic non-communicable diseases, morbidity, and mortality, all of which include inflammation as part of their pathophysiology [49]. Our results are also consistent with a recent systematic review of observational studies and broader whole of diet or dietary pattern analyses [50]. Not unlike the NOVA food classification system, dietary patterns expand beyond isolated nutrients and account for the fact that foods are consumed in complex combinations [50]. This systematic review reported that indices and scores used to assess the inflammatory potential of diets (e.g., Dietary Inflammatory Index) were directly and cross-sectionally associated with inflammatory biomarkers, including C-reactive protein, interleukin-6, tumour necrosis factor-α, and fibrinogen [50]. Pro-inflammatory dietary patterns were characterised by, for example, excess consumption of kilojoule-dense Western-style foods, including red and processed meats, sweets, desserts, fried foods, and refined grains [51]. Similarly, diet scores measuring adherence to healthy or Mediterranean-style diets—rich in fruits, vegetables, fatty fish, poultry, extra virgin olive oil, and whole grains—appeared to be inversely associated with inflammatory biomarkers in cross-sectional analyses [50]. In terms of experimental evidence, our results are also consistent with an earlier meta-analysis of intervention studies showing that Mediterranean diets higher in unprocessed or minimally processed foods were anti-inflammatory [52].

A potential role of the gut microbiota in the link between ultra-processed food intake and inflammation was hypothesised [53]. Preliminary theory posits that extensive food processing leading to the degradation of cell walls within food and acellular compounds (i.e., deconstruction of the food matrix and nutrients not contained within cells, respectively [54]) may impact abnormal absorption and signalling from the gastrointestinal tract as well as its interactions with gastrointestinal microbiota [53–55]. Both may in turn promote microbe encroachment on the gastrointestinal wall and a cascade of inflammatory processes [53].

Though not extensively demonstrated in humans, several pre-clinical rodent studies have indicated an effect of advanced glycation end-products (AGEs) formed during the thermal treatment of food products [56,57] and artificial additives common to ultra-processed food (e.g., carboxymethylcellulose [58,59], polysorbate-80 [59,60], saccharin [61], and sucralose [62]) on the gut microbiota composition and activity together with host physiology including pro-inflammatory states. One recent randomised controlled-feeding study in humans reported a detrimental effect of the emulsifier carboxymethylcellulose on the gut microbiota and metabolome, with the authors of that study surmising that carboxymethylcellulose may be contributing to an array of chronic inflammatory diseases [63]. This emerging evidence is certainly suggestive and warrants further investigation to determine the precise features of ultra-processed food that elicit their unhealthful effects.

4.1. Limitations and Future Research

Our results should be interpreted with consideration of the following limitations. First, the possible temporality of these associations cannot be established from this single cross-sectional study, and residual confounding cannot be excluded. However, our results were unchanged after adjusting for common confounders and after various sensitivity analyses, which showed that the associations remained relatively stable with or without the inclusion of people with markedly elevated hsCRP concentration. Sensitivity analyses also highlighted that our sampling methods may not have biased results given the stability of effect estimates no matter whether we included cases defined by cardiovascular disease mortality (which occurred after data and sample collection) or participants with a history of non-communicable diseases, such as cardiovascular diseases, cancer, diabetes mellitus, and body mass index $\geq$30 at baseline.

Second, although the FFQ was not specifically designed to identify ultra-processed food, there is some evidence in certain populations (e.g., New Zealand children [64] and adults from Italy [65] and Mexico [66]) that FFQs have acceptable validity to assess food consumption based on the NOVA food classification system. FFQs have also been reported as the most frequently used dietary data collection tool in reviews investigating ultra-processed food–chronic disease relationships [8]. Nonetheless, some degree of misclassification bias may exist.

Lastly, while the FFQ dietary data used to investigate ultra-processed food intake in the current study were captured over 20 years ago, this may not limit the generalisability of our results since both the participant characteristics and level of ultra-processed food consumption reported in the current study are comparable to current estimates globally. As specified in Table 1, ultra-processed food contributed 40% of total energy intake, which is consistent with the most recent analysis of ultra-processed food intake in a nationally representative sample of Australians taken from the National Nutrition and Physical Activity Survey (2011–2012) [28], where ultra-processed food contributed 42% of total energy. This is also comparable with other estimates from Western countries such as Canada (42%), the United Kingdom (54%), and the United States (56%) [8].

4.2. Implications

Historically, and as previously noted, nutrition research has focused on the effect of dietary intakes of energy and macro- and micro-nutrients on human physiology and health, including inflammatory processes together with the mechanistic link between inflammation and chronic diseases. The relevance and novelty of the NOVA food classification system becomes prominent, however, when considering emerging evidence for differential health outcomes that depend on the extent and level of food processing [67–70]. Indeed, NOVA largely ignores the nutrient profiles of ultra-processed food, instead focusing on the extent and purpose of food processing [11]. One tightly controlled randomised trial in humans that specifically applied the NOVA food classification system in its design [7] demonstrated a causal effect of an ultra-processed versus unprocessed diet on increased energy intake as well as adiposity (both of which have been associated with pro-inflammatory states [71]).

While this landmark study targeted different metabolic outcomes, it also showed a within-group reduction from baseline to endpoint in hsCRP concentration when participants were allocated to the unprocessed diet [7]. It also underscored the futility of focusing only on nutrient composition given that the two diets were matched for presented energy, sugar, fat, fibre, and macronutrients [7].

Given the association between ultra-processed food intake and morbidity and mortality [8], there were recent calls urging countries to adopt policy interventions that limit the production, distribution, and dietary intake of ultra-processed food [72]. The importance of our study includes its potential generalisability to other Anglo-European populations and ability to inform and encourage future research investigating the possible biological mechanisms of action involved in the observed associations between consumption of ultra-processed food, chronic non-communicable diseases, and all-cause mortality.

5. Conclusions

The current study showed a cross-sectional association between higher ultra-processed food intake and elevated hsCRP as a biomarker of inflammation. Part of the association between consumption of ultra-processed food and hsCRP was independent of body mass index. Further prospective and experimental studies in humans are needed to examine whether this association is causal. Such information will be key to appropriate health messages in the future.

Supplementary Materials: The following supporting information can be downloaded at: https://www.mdpi.com/article/10.3390/nu14163309/s1, Figure S1: A directed acyclic graph mapping hypothesised relationships between all relevant variables; Table S1: Sensitivity excluding individuals with hsCRP concentrations above 10 mg/L and cardiovascular disease mortality; Table S2: Sensitivity analyses excluding individuals with history of non-communicable diseases; Table S3: Sex-stratified cross-sectional associations between the ultra-processed food intake and hsCRP concentration (MCCS, 1990–1994).

Author Contributions: Conceptualization, M.M.L. and W.M.; formal analysis, M.M.L. and M.L.; data curation, M.M.L. and M.L.; writing—original draft preparation, M.M.L.; writing—review and editing, M.M.L., M.L., M.F., A.L., T.R., A.O., P.M., F.N.J., A.H., and W.M.; supervision, W.M., M.L., A.L., T.R., A.O. and F.N.J.; project administration, M.M.L. All authors have read and agreed to the published version of the manuscript.

Funding: This research received no external funding.

Institutional Review Board Statement: The current study was approved for exemption from ethical review in accordance with the National Statement on Ethical Conduct in Human Research (2007, updated 2018) Section 5.1.22 by the Deakin University Human Research Ethics Committee (project number: 2020-413, received 18th of November 2020). The study protocol for the original Melbourne Collaborative Cohort Study project was approved by the Cancer Council Victoria's Human Research Ethics Committee (project number: IEC 9001, received 23rd of August 1990).

Informed Consent Statement: Participants provided written consent to participate including researcher access to their medical records.

Acknowledgments: MCCS cohort recruitment was funded by Cancer Council Victoria and VicHealth. The MCCS was further supported by Australian National Health and Medical Research Council (NHMRC) grants 209057, 396414, and 1074383, and ongoing follow-up and data management has been funded by Cancer Council Victoria since 1995. Cases and their vital status were ascertained through the Victorian Cancer Registry and the Australian Institute of Health and Welfare, including the National Death Index and the Australian Cancer Database.

Conflicts of Interest: The Food & Mood Centre has received Grant/Research support from Fernwood Foundation, Wilson Foundation, the A2 Milk Company, and Be Fit Foods. M.M.L. is supported by a Deakin University Scholarship and has received research funding support from Be Fit Foods. M.L. and P.M. are currently funded by an Alfred Deakin Postdoctoral Research Fellowship. M.F. is supported by a Deakin University Scholarship. A.L. has received grants, fellowships and research support from the University of New South Wales, the University of Melbourne, RMIT University, Deakin University, the

National Health and Medical Research Council (NHMRC), Australian Academy of Science, National Institutes of Health (NIH), and The Jack Brockhoff Foundation. A.L. has received honoraria and travel funds from Sydney University, the University of Technology Sydney, American Epilepsy Society, Epilepsy Society of Australia, International Human Microbiome Congress, European Society of Neurogastroenterology, Australian and New Zealand College of Anaesthetists, Falk Foundation and Fonds de la Recherche Scientifique (FNRS). T.R. has received grants, fellowships and research support from University of the Sunshine Coast, Australian Postgraduate Awards, Fernwood Foundation, Roberts Family Foundation, Be Fit Food and Wilson Foundation. T.R. received consultancy, honoraria and travel funds from Oxford University Press, the University of Melbourne, the University of Sydney, Bond University, University of Southern Queensland, Dietitians Association of Australia, Nutrition Society of Australia, The Royal Australian and New Zealand College of Psychiatrists, Academy of Nutrition and Dietetics, Black Dog Institute, Australian Rotary Health, Australian Disease Management Association, Department of Health and Human Services, Primary Health Networks, Barwon Health, West Gippsland Healthcare Group, Central West Gippsland Primary Care Partnership, Parkdale College, City of Greater Geelong and Global Age. A.O. is supported by a Future Leader Fellowship (#101160) from the Heart Foundation Australia and Wilson Foundation. She has received research funding from National Health & Medical Research Council, Australian Research Council, University of Melbourne, Deakin University, Sanofi, Meat and Livestock Australia and Woolworths Limited and Honoraria from Novartis. F.J. has received Grant/Research support from the Brain and Behaviour Research Institute, the National Health and Medical Research Council (NHMRC), Australian Rotary Health, the Geelong Medical Research Foundation, the Ian Potter Foundation, Eli Lilly, Meat and Livestock Australia, Woolworths Limited, the Fernwood Foundation, Wilson Foundation, the A2 Milk Company, Be Fit Foods, and The University of Melbourne, and has received speakers honoraria from Sanofi-Synthelabo, Janssen Cilag, Servier, Pfizer, Health Ed, Network Nutrition, Angelini Farmaceutica, Eli Lilly and Metagenics. F.N.J. has written two books for commercial publication and has a personal belief that good diet quality is important for mental and brain health. W.M. is currently funded by an NHMRC Investigator Grant (#2008971) and a Multiple Sclerosis Research Australia early-career fellowship. Wolfgang has previously received funding from the Cancer Council Queensland and university grants/fellowships from La Trobe University, Deakin University, University of Queensland, and Bond University. Wolfgang has received industry funding and/or has attended events funded by Cobram Estate Pty. Ltd. and Bega Dairy and Drinks Pty Ltd. Wolfgang has received travel funding from Nutrition Society of Australia. Wolfgang has received consultancy funding from Nutrition Research Australia and ParachuteBH. Wolfgang has received speakers honoraria from The Cancer Council Queensland and the Princess Alexandra Research Foundation.

References

1. Fardet, A.; Rock, E. Exclusive reductionism, chronic diseases and nutritional confusion: The degree of processing as a lever for improving public health. *Crit. Rev. Food Sci. Nutr.* **2020**, *62*, 2784–2799. [CrossRef] [PubMed]
2. Ruanpeng, D.; Thongprayoon, C.; Cheungpasitporn, W.; Harindhanavudhi, T. Sugar and artificially sweetened beverages linked to obesity: A systematic review and meta-analysis. *QJM Int. J. Med.* **2017**, *110*, 513–520. [CrossRef] [PubMed]
3. He, F.J.; Tan, M.; Ma, Y.; MacGregor, G.A. Salt Reduction to Prevent Hypertension and Cardiovascular Disease. *J. Am. Coll. Cardiol.* **2020**, *75*, 632–647. [CrossRef]
4. Hooper, L.; Martin, N.; Jimoh, O.F.; Kirk, C.; Foster, E.; Abdelhamid, A.S. Reduction in saturated fat intake for cardiovascular disease. *Cochrane Database Syst. Rev.* **2015**, *6*, CD011737. [CrossRef] [PubMed]
5. Fardet, A.; Rock, E.; Bassama, J.; Bohuon, P.; Prabhasankar, P.; Monteiro, C.B.D.M.; Moubarac, J.-C.; Achir, N. Current Food Classifications in Epidemiological Studies Do Not Enable Solid Nutritional Recommendations for Preventing Diet-Related Chronic Diseases: The Impact of Food Processing. *Adv. Nutr. Int. Rev. J.* **2015**, *6*, 629–638. [CrossRef] [PubMed]
6. Fardet, A.; Rock, E. Perspective: Reductionist Nutrition Research Has Meaning Only within the Framework of Holistic and Ethical Thinking. *Adv. Nutr. Int. Rev. J.* **2018**, *9*, 655–670. [CrossRef] [PubMed]
7. Hall, K.D.; Ayuketah, A.; Brychta, R.; Cai, H.; Cassimatis, T.; Chen, K.Y.; Chung, S.T.; Costa, E.; Courville, A.; Darcey, V.; et al. Ultra-Processed Diets Cause Excess Calorie Intake and Weight Gain: An Inpatient Randomized Controlled Trial of Ad Libitum Food Intake. *Cell Metab.* **2019**, *30*, 67–77.e3. [CrossRef] [PubMed]
8. Lane, M.M.; Davis, J.A.; Beattie, S.; Gómez-Donoso, C.; Loughman, A.; O'Neil, A.; Jacka, F.; Berk, M.; Page, R.; Marx, W.; et al. Ultraprocessed Food and Chronic Noncommunicable Diseases: A Systematic Review and Meta-Analysis of 43 Observational Studies. *Obes. Rev.* **2021**, *22*, e13146. [CrossRef] [PubMed]

9. Lane, M.M.; Gamage, E.; Travica, N.; Dissanayaka, T.; Ashtree, D.N.; Gauci, S.; Lotfaliany, M.; O'Neil, A.; Jacka, F.N.; Marx, W. Ultra-Processed Food Consumption and Mental Health: A Systematic Review and Meta-Analysis of Observational Studies. *Nutrients* **2022**, *14*, 2568. [CrossRef] [PubMed]
10. Moradi, S.; Ma, H.K.; Bagheri, R.; Mohammadi, H.; Jayedi, A.; Lane, M.M.; Asbaghi, O.; Mehrabani, S.; Suzuki, K. Ultra-Processed Food Consumption and Adult Diabetes Risk: A Systematic Review and Dose-Response Meta-Analysis. *Nutrients* **2021**, *13*, 4410. [CrossRef] [PubMed]
11. Monteiro, C.A.; Cannon, G.; Levy, R.B.; Moubarac, J.-C.; Louzada, M.L.C.; Rauber, F.; Khandpur, N.; Cediel, G.; Neri, D.; Martinez-Steele, E.; et al. Ultra-Processed Foods: What They Are and How to Identify Them. *Public Health Nutr.* **2019**, *22*, 936–941. [CrossRef] [PubMed]
12. Monteiro, C.A.; Cannon, G.; Moubarac, J.-C.; Levy, R.B.; Louzada, M.L.C.; Jaime, P.C. The UN Decade of Nutrition, the NOVA Food Classification and the Trouble with Ultra-Processing. *Public Health Nutr.* **2018**, *21*, 5–17. [CrossRef] [PubMed]
13. Baker, P.; Machado, P.; Santos, T.; Sievert, K.; Backholer, K.; Hadjikakou, M.; Russell, C.; Huse, O.; Bell, C.; Scrinis, G.; et al. Ultra-processed foods and the nutrition transition: Global, regional and national trends, food systems transformations and political economy drivers. *Obes. Rev.* **2020**, *21*, e13126. [CrossRef] [PubMed]
14. Monteiro, C.A.; Moubarac, J.-C.; Cannon, G.; Ng, S.W.; Popkin, B. Ultra-Processed Products Are Becoming Dominant in the Global Food System. *Obes. Reviews Obes Rev.* **2013**, *14*, 21–28. [CrossRef]
15. Marx, W.; Veronese, N.; Kelly, J.T.; Smith, L.; Hockey, M.; Collins, S.; Trakman, G.L.; Hoare, E.; Teasdale, S.B.; Wade, A.; et al. The Dietary Inflammatory Index and Human Health: An Umbrella Review of Meta-Analyses of Observational Studies. *Adv. Nutr. Int. Rev. J.* **2021**, *12*, 1681–1690. [CrossRef] [PubMed]
16. Mantovani, A.; Allavena, P.; Sica, A.; Balkwill, F. Cancer-related inflammation. *Nature* **2008**, *454*, 436–444. [CrossRef] [PubMed]
17. Kaptoge, S.; Seshasai, S.R.K.; Gao, P.; Freitag, D.F.; Butterworth, A.S.; Borglykke, A.; Angelantonio, E.D.; Gudnason, V.; Rumley, A.; Lowe, G.D. Inflammatory cytokines and risk of coronary heart disease: New prospective study and updated meta-analysis. *Eur. Heart J.* **2014**, *35*, 578–589. [CrossRef] [PubMed]
18. Pasco, J.A.; Nicholson, G.C.; Williams, L.J.; Jacka, F.N.; Henry, M.J.; Kotowicz, M.A.; Schneider, H.G.; Leonard, B.E.; Berk, M. Association of high-sensitivity C-reactive protein with de novo major depression. *Br. J. Psychiatry* **2010**, *197*, 372–377. [CrossRef] [PubMed]
19. Berk, M.; Williams, L.J.; Jacka, F.N.; O'Neil, A.; Pasco, J.A.; Moylan, S.; Allen, N.B.; Stuart, A.L.; Hayley, A.; Byrne, M.L.; et al. So depression is an inflammatory disease, but where does the inflammation come from? *BMC Med.* **2013**, *11*, 200. [CrossRef] [PubMed]
20. Camps, J.; García-Heredia, A. Introduction: Oxidation and Inflammation, A Molecular Link between Non-communicable Diseases. In *Oxidative Stress and Inflammation in Non-communicable Diseases—Molecular Mechanisms and Perspectives in Therapeutics;* Camps, J., Ed.; Springer: Cham, Switzerland, 2014; pp. 1–4.
21. Lopes, A.E.D.S.C.; Araújo, L.F.; Levy, R.B.; Barreto, S.M.; Giatti, L. Association between consumption of ultra-processed foods and serum C-reactive protein levels: Cross-sectional results from the ELSA-Brasil study. *Sao Paulo Med. J.* **2019**, *137*, 169–176. [CrossRef]
22. dos Santos, F.S.; Mintem, G.C.; de Oliveira, I.O.; Horta, B.L.; Ramos, E.; Lopes, C.; Gigante, D.P. Consumption of ultra-processed foods and interleukin-6 in two cohorts from high- and middle-income countries. *Br. J. Nutr.* **2022**, 1–28. [CrossRef] [PubMed]
23. Martins, G.M.d.S.; França, A.K.T.D.C.; Viola, P.C.D.A.F.; de Carvalho, C.A.; Marques, K.D.S.; dos Santos, A.M.; Batalha, M.A.; Alves, J.D.d.A.; Ribeiro, C.C.C. Intake of ultra-processed foods is associated with inflammatory markers in Brazilian adolescents. *Public Health Nutr.* **2021**, *25*, 591–599. [CrossRef] [PubMed]
24. Monteiro, C.A.; Astrup, A. Does the concept of "ultra-processed foods" help inform dietary guidelines, beyond conventional classification systems? YES. *Am. J. Clin. Nutr.* **2022**, *00*, 1–6. [CrossRef] [PubMed]
25. Vandenbroucke, J.P.; von Elm, E.; Altman, D.G.; Gøtzsche, P.C.; Mulrow, C.D.; Pocock, S.J.; Poole, C.; Schlesselman, J.J.; Egger, M. Strengthening the Reporting of Observational Studies in Epidemiology (STROBE): Explanation and Elaboration. *PLOS Medicine* **2007**, *4*, e297. [CrossRef] [PubMed]
26. Milne, R.L.; Fletcher, A.S.; MacInnis, R.J.; Hodge, A.M.; Hopkins, A.H.; Bassett, J.K.; Bruinsma, F.J.; Lynch, B.M.; Dugué, P.A.; Jayasekara, H.; et al. Cohort Profile: The Melbourne Collaborative Cohort Study (Health 2020). *Int. J. Epidemiology* **2017**, *46*, 1757–1757i. [CrossRef] [PubMed]
27. Bassett, J.K.; English, D.; Fahey, M.; Forbes, A.; Gurrin, L.; Simpson, J.A.; Brinkman, M.T.; Giles, G.; Hodge, A. Validity and calibration of the FFQ used in the Melbourne Collaborative Cohort Study. *Public Health Nutr.* **2016**, *19*, 2357–2368. [CrossRef]
28. Machado, P.P.; Steele, E.M.; Levy, R.B.; Sui, Z.; Rangan, A.; Woods, J.; Gill, T.; Scrinis, G.; Monteiro, C.A. Ultra-processed foods and recommended intake levels of nutrients linked to non-communicable diseases in Australia: Evidence from a nationally representative cross-sectional study. *BMJ Open* **2019**, *9*, e029544. [CrossRef]
29. The Cancer Council Victoria Epidemiology Centre. Melbourne Collaborative Cohort Study Databook Vol. 3: Diet & Alcohol. Available online: https://www.cancervic.org.au/research/epidemiology/health_2020/health2020-databook3-diet (accessed on 1 September 2021).
30. Ireland, P.; Jolley, D.; Giles, G.; O'Dea, K.; Powles, J.; Rutishauser, I.; Wahlqvist, M.L.; Williams, J. Development of the Melbourne FFQ: A food frequency questionnaire for use in an Australian prospective study involving an ethnically diverse cohort. *Asia. Pac. J. Clin. Nutr.* **1994**, *3*, 19–31. [PubMed]

31. Lewis, J.; Milligan, G.; Hunt, A. NUTTAB95: Nutrient data table for use in Australia. Commonwealth of Australia: Canberra, Australia, 1995.
32. Hodge, A.M.; Bassett, J.K.; Shivappa, N.; Hébert, J.R.; English, D.R.; Giles, G.G.; Severi, G. Dietary inflammatory index, Mediterranean diet score, and lung cancer: A prospective study. *Cancer Causes Control* **2016**, *27*, 907–917. [CrossRef]
33. Jayasekara, H.; MacInnis, R.J.; Williamson, E.J.; Hodge, A.M.; Clendenning, M.; Rosty, C.; Walters, R.; Room, R.; Southey, M.C.; Jenkins, M.A.; et al. Lifetime alcohol intake is associated with an increased risk of *KRAS*+ and *BRAF-/KRAS-* but not *BRAF*+ colorectal cancer. *Int. J. Cancer* **2016**, *140*, 1485–1493. [CrossRef]
34. MacInnis, R.J.; English, D.R.; Hopper, J.L.; Haydon, A.M.; Gertig, D.M.; Giles, G.G. Body Size and Composition and Colon Cancer Risk in Men. *Cancer Epidemiol. Biomark. Prev.* **2004**, *13*, 553–559. [CrossRef]
35. Machado, P.P.; Steele, E.M.; Levy, R.B.; da Costa Louzada, M.L.; Rangan, A.; Woods, J.; Gill, T.; Scrinis, G.; Monteiro, C.A. Ultra-processed food consumption and obesity in the Australian adult population. *Nutr. Diabetes* **2020**, *10*, 39. [CrossRef] [PubMed]
36. Pink, B. *An Introduction to Socio-Economic Indexes for Areas (SEIFA) In Australian Bureau of Statistics*; Australian Bureau of Statistics: Canberra, Australia, 2008.
37. Australian Bureau of Statistics Socio-Economic Indexes for Areas. 2022. Available online: https://www.abs.gov.au/websitedbs/censushome.nsf/home/seifa. (accessed on 2 July 2022).
38. Srour, B.; Fezeu, L.K.; Kesse-Guyot, E.; Allès, B.; Méjean, C.; Andrianasolo, R.M.; Chazelas, E.; Deschasaux, M.; Hercberg, S.; Galan, P.; et al. Ultra-processed food intake and risk of cardiovascular disease: Prospective cohort study (NutriNet-Santé). *BMJ* **2019**, *365*, l1451. [CrossRef] [PubMed]
39. Schnabel, L.; Kesse-Guyot, E.; Allès, B.; Touvier, M.; Srour, B.; Hercberg, S.; Buscail, C.; Julia, C. Association Between Ultraprocessed Food Consumption and Risk of Mortality Among Middle-aged Adults in France. *JAMA Intern. Med.* **2019**, *179*, 490–498. [CrossRef] [PubMed]
40. Julia, C.; Martinez, L.; Allès, B.; Touvier, M.; Hercberg, S.; Méjean, C.; Kesse-Guyot, E. Contribution of ultra-processed foods in the diet of adults from the French NutriNet-Santé study. *Public Health Nutr.* **2018**, *21*, 27–37. [CrossRef] [PubMed]
41. Willett, W.C.; Howe, G.R.; Kushi, L.H. Adjustment for total energy intake in epidemiologic studies. *Am. J. Clin. Nutr.* **1997**, *65*, 1220S–1228S. [CrossRef]
42. Cragg, J.G.; Uhler, R.S. The demand for automobiles. *Can. J. Econ. Rev. Can. D'economique* **1970**, *3*, 386–406. [CrossRef]
43. Fairchild, A.J.; McDaniel, H.L. Best (but oft-forgotten) practices: Mediation analysis. *Am. J. Clin. Nutr.* **2017**, *105*, 1259–1271. [CrossRef]
44. Ananth, C.V.; Schisterman, E.F. Confounding, causality, and confusion: The role of intermediate variables in interpreting observational studies in obstetrics. *Am. J. Obstet. Gynecol.* **2017**, *217*, 167–175. [CrossRef]
45. Ridker, P.M.; Cook, N. Clinical Usefulness of Very High and Very Low Levels of C-Reactive Protein Across the Full Range of Framingham Risk Scores. *Circulation* **2004**, *109*, 1955–1959. [CrossRef]
46. Pepys, M.B.; Hirschfield, G.M. C-reactive protein: A critical update. *J. Clin. Investig.* **2003**, *111*, 1805–1812. [CrossRef] [PubMed]
47. R Development Core Team. R: A Language and Environment for Statistical Computing. R Foundation for Statistical Computing, Vienna, Austria. 2020. Available online: https://www.R-project.org/ (accessed on 11 August 2021).
48. Khera, A.; McGuire, D.K.; Murphy, S.A.; Stanek, H.G.; Das, S.R.; Vongpatanasin, W.; Wians, F.H.; Grundy, S.M.; de Lemos, J.A. Race and Gender Differences in C-Reactive Protein Levels. *J. Am. Coll. Cardiol.* **2005**, *46*, 464–469. [CrossRef] [PubMed]
49. Furman, D.; Campisi, J.; Verdin, E.; Carrera-Bastos, P.; Targ, S.; Franceschi, C.; Ferrucci, L.; Gilroy, D.W.; Fasano, A.; Miller, G.W.; et al. Chronic inflammation in the etiology of disease across the life span. *Nat. Med.* **2019**, *25*, 1822–1832. [CrossRef] [PubMed]
50. Hart, M.J.; Torres, S.J.; McNaughton, S.A.; Milte, C.M. Dietary patterns and associations with biomarkers of inflammation in adults: A systematic review of observational studies. *Nutr. J.* **2021**, *20*, 1–14. [CrossRef] [PubMed]
51. Lopez-Garcia, E.; Schulze, M.B.; Fung, T.T.; Meigs, J.B.; Rifai, N.; Manson, J.E.; Hu, F.B. Major dietary patterns are related to plasma concentrations of markers of inflammation and endothelial dysfunction. *Am. J. Clin. Nutr.* **2004**, *80*, 1029–1035. [CrossRef]
52. Schwingshackl, L.; Hoffmann, G. Mediterranean dietary pattern, inflammation and endothelial function: A systematic review and meta-analysis of intervention trials. *Nutr. Metab. Cardiovasc. Dis.* **2014**, *24*, 929–939. [CrossRef]
53. Zinöcker, M.K.; Lindseth, I.A. The Western Diet–Microbiome-Host Interaction and Its Role in Metabolic Disease. *Nutrients* **2018**, *10*, 365. [CrossRef]
54. Spreadbury, I. Comparison with ancestral diets suggests dense acellular carbohydrates promote an inflammatory microbiota, and may be the primary dietary cause of leptin resistance and obesity. *Diabetes Metab. Syndr. Obesity Targets Ther.* **2012**, *5*, 175–189. [CrossRef]
55. Jang, C.; Hui, S.; Lu, W.; Cowan, A.J.; Morscher, R.J.; Lee, G.; Liu, W.; Tesz, G.J.; Birnbaum, M.J.; Rabinowitz, J.D. The Small Intestine Converts Dietary Fructose into Glucose and Organic Acids. *Cell Metab.* **2018**, *27*, 351–361.e3. [CrossRef]
56. Miclotte, L.; Van De Wiele, T. Food processing, gut microbiota and the globesity problem. *Crit. Rev. Food Sci. Nutr.* **2019**, *60*, 1769–1782. [CrossRef]
57. D'Cunha, N.M.; Sergi, D.; Lane, M.M.; Naumovski, N.; Gamage, E.; Rajendran, A.; Kouvari, M.; Gauci, S.; Dissanayka, T.; Marx, W.; et al. The Effects of Dietary Advanced Glycation End-Products on Neurocognitive and Mental Disorders. *Nutrients* **2022**, *14*, 2421. [CrossRef] [PubMed]

58. Swidsinski, A.; Ung, V.; Sydora, B.C.; Loening-Baucke, V.; Doerffel, Y.; Verstraelen, H.; Fedorak, R.N. Bacterial Overgrowth and Inflammation of Small Intestine After Carboxymethylcellulose Ingestion in Genetically Susceptible Mice. *Inflamm. Bowel Dis.* **2009**, *15*, 359–364. [CrossRef] [PubMed]
59. Chassaing, B.; Koren, O.; Goodrich, J.K.; Poole, A.C.; Srinivasan, S.; Ley, R.E.; Gewirtz, A.T. Dietary emulsifiers impact the mouse gut microbiota promoting colitis and metabolic syndrome. *Nature* **2015**, *519*, 92–96. [CrossRef] [PubMed]
60. Singh, R.K.; Wheildon, N.; Ishikawa, S. Food Additive P-80 Impacts Mouse Gut Microbiota Promoting Intestinal Inflammation, Obesity and Liver Dysfunction. *SOJ Microbiol. Infect. Dis.* **2016**, *4*, 10. [CrossRef]
61. Bian, X.; Tu, P.; Chi, L.; Gao, B.; Ru, H.; Lu, K. Saccharin induced liver inflammation in mice by altering the gut microbiota and its metabolic functions. *Food Chem. Toxicol.* **2017**, *107*, 530–539. [CrossRef]
62. Bian, X.; Chi, L.; Gao, B.; Tu, P.; Ru, H.; Lu, K. Gut Microbiome Response to Sucralose and Its Potential Role in Inducing Liver Inflammation in Mice. *Front. Physiol.* **2017**, *8*, 487. [CrossRef]
63. Chassaing, B.; Compher, C.; Bonhomme, B.; Liu, Q.; Tian, Y.; Walters, W.; Nessel, L.; Delaroque, C.; Hao, F.; Gershuni, V.; et al. Randomized Controlled-Feeding Study of Dietary Emulsifier Carboxymethylcellulose Reveals Detrimental Impacts on the Gut Microbiota and Metabolome. *Gastroenterology* **2022**, *162*, 743–756. [CrossRef]
64. Fangupo, L.J.; Haszard, J.J.; Leong, C.; Heath, A.-L.M.; Fleming, E.A.; Taylor, R.W. Relative Validity and Reproducibility of a Food Frequency Questionnaire to Assess Energy Intake from Minimally Processed and Ultra-Processed Foods in Young Children. *Nutrients* **2019**, *11*, 1290. [CrossRef]
65. Dinu, M.; Bonaccio, M.; Martini, D.; Madarena, M.P.; Vitale, M.; Pagliai, G.; Esposito, S.; Ferraris, C.; Guglielmetti, M.; Rosi, A.; et al. Reproducibility and validity of a food-frequency questionnaire (NFFQ) to assess food consumption based on the NOVA classification in adults. *Int. J. Food Sci. Nutr.* **2021**, *72*, 861–869. [CrossRef]
66. Oviedo-Solís, C.I.; Monterrubio-Flores, E.A.; Rodríguez-Ramírez, S.; Cediel, G.; Denova-Gutiérrez, E.; Barquera, S. A Semi-quantitative Food Frequency Questionnaire Has Relative Validity to Identify Groups of NOVA Food Classification System Among Mexican Adults. *Front. Nutr.* **2022**, *9*, 737432. [CrossRef]
67. Kaluza, J.; Wolk, A.; Larsson, S. Red Meat Consumption and Risk of Stroke. *Stroke* **2012**, *43*, 2556–2560. [CrossRef] [PubMed]
68. Micha, R.; Michas, G.; Mozaffarian, D. Unprocessed Red and Processed Meats and Risk of Coronary Artery Disease and Type 2 Diabetes–An Updated Review of the Evidence. *Curr. Atheroscler. Rep.* **2012**, *14*, 515–524. [CrossRef] [PubMed]
69. Chan, D.S.M.; Lau, R.; Aune, D.; Vieira, R.; Greenwood, D.C.; Kampman, E.; Norat, T. Red and Processed Meat and Colorectal Cancer Incidence: Meta-Analysis of Prospective Studies. *PLoS ONE* **2011**, *6*, e20456. [CrossRef] [PubMed]
70. Tørris, C.; Molin, M.; Småstuen, M.C. Lean Fish Consumption Is Associated with Beneficial Changes in the Metabolic Syndrome Components: A 13-Year Follow-Up Study from the Norwegian Tromsø Study. *Nutrients* **2017**, *9*, 247. [CrossRef] [PubMed]
71. Ellulu, M.S.; Patimah, I.; KhazáAi, H.; Rahmat, A.; Abed, Y. Obesity and inflammation: The linking mechanism and the complications. *Arch. Med. Sci.* **2017**, *13*, 851–863. [CrossRef] [PubMed]
72. Monteiro, C.A.; Lawrence, M.; Millett, C.; Nestle, M.; Popkin, B.M.; Scrinis, G.; Swinburn, B. The need to reshape global food processing: A call to the United Nations Food Systems Summit. *BMJ Glob. Health* **2021**, *6*, e006885. [CrossRef] [PubMed]

nutrients

MDPI

Article

Ultra-Processed Foods as Ingredients of Culinary Recipes Shared on Popular Brazilian YouTube Cooking Channels

Anice Milbratz de Camargo [1], Alyne Michelle Botelho [1], Állan Milbratz de Camargo [2], Moira Dean [3] and Giovanna Medeiros Rataichesck Fiates [1,*]

1 Nutrition in Foodservice Research Centre, Nutrition Post Graduation Program, Universidade Federal de Santa Catarina, Campus Universitário João David Ferreira Lima—Trindade, Florianópolis 88040-900, SC, Brazil
2 Computer Science Undergraduate Course, Universidade Regional de Blumenau, Rua Antônio da Veiga, 140, Blumenau 89012-900, SC, Brazil
3 Institute for Global Food Security, School of Biological Sciences, Queen's University Belfast, Belfast BT9 5DL, UK
* Correspondence: giovanna.fiates@ufsc.br

Abstract: Social media platforms are readily accessible sources of information about cooking, an activity deemed crucial for the improvement of a population's diet. Previous research focused on the healthiness of the content shared on websites and blogs, but not on social media such as YouTube®. This paper analysed the healthiness of 823 culinary recipes retrieved from 755 videos shared during a six-month period on ten popular Brazilian YouTube® cooking channels. Recipes were categorized by type of preparation. To assess recipes' healthiness, ingredients were classified according to the extension and purpose of industrial processing, in order to identify the use of ultra-processed foods. Additionally, a validated framework developed from criteria established in both editions of the Dietary Guidelines for the Brazilian Population was employed. Recipes for cakes and baked goods, puddings, snacks and homemade fast foods, which were among the most frequently posted, contained the lowest proportion of unprocessed/minimally processed ingredients and the highest proportion of ultra-processed ingredients. Recipes containing whole cereals, fruits, legumes, nuts, and seeds were scarce. Results indicate that users should be critical about the quality of recipes shared on YouTube® videos, also indicating a need for strategies aimed at informing individuals on how to choose healthier recipes or adapt them to become healthier

Keywords: social media; social network site; Internet; cookery channels; recipe quality; cooking instruction; ultra-processed foods

Citation: Camargo, A.M.d.; Botelho, A.M.; Camargo, Á.M.d.; Dean, M.; Fiates, G.M.R. Ultra-Processed Foods as Ingredients of Culinary Recipes Shared on Popular Brazilian YouTube Cooking Channels. *Nutrients* **2022**, *14*, 3689. https://doi.org/10.3390/nu14183689

Academic Editors: Monica Dinu and Daniela Martini

Received: 5 July 2022
Accepted: 7 August 2022
Published: 7 September 2022

1. Introduction

Public health initiatives from many countries encourage home cooking as a health-promoting strategy [1]. This is also true for both editions of the Dietary Guidelines for the Brazilian Population [2,3], which adopt distinct but complementary approaches for the promotion of healthy eating. The first edition of the Guidelines valued the act of eating at home and provided information on how to prepare food in a healthy way. Its directives were based on the intake of adequate amounts of foods, classified into food groups, to prevent nutritional deficiencies and chronic non-communicable diseases [2].

Aside from stressing the importance of home cooking, the Dietary Guidelines for the Brazilian Population published in 2015 focused on categorizing foods according to the extension and purpose of industrial processing [3]. Individuals should base their diets on unprocessed/minimally-processed foods (U/MP) and avoid ingesting ultra-processed foods (UP) as much as possible [3]. UP foods are formulations of ingredients that are usually nutritionally unbalanced, being rich in fats and sugars while poor in fibre and micronutrients [3]. Carbonated soft drinks, packaged snacks, mass-produced breads, margarines, candies, cake mixes, and many ready-to-heat frozen products (pies, pizza, sausages, burgers) are examples of UP foods [3,4].

High consumption of UP foods has been associated with chronic non-communicable diseases and all-cause mortality [5–7]. Conversely, cooking at home more often has been associated with a lower risk of developing chronic non-communicable diseases [8,9], possibly as a result of a better diet quality [9,10] due to the use of fresh ingredients. A pattern of healthy cooking practices, where individuals can confidently cook several meals using fresh foods and natural seasonings, and use healthier cooking techniques, was inversely associated with ultra-processed food consumption [11]. A diet composed mostly of U/MP foods, however, can only be achieved if individuals master a certain number of cooking skills [3].

Informal cooking education happens through culinary socialization over the course of a person's life, a process in which individuals acquire patterns of practices and perceptions related to cooking, from socializing agents [12]. The first culinary socializing agents are family members; later in life, different agents start to influence cooking practices, such as friends, partners, cookbooks, culinary television programs, and more recently, the Internet [9,13–15]. Individuals report favouring Internet searches and digital sources when looking for recipes, instead of printed sources such as books, for the convenience of being 'at hand' [13].

Brazilians spend an average of 3.5 h daily on the Internet [16], mainly accessing social media [17]. Social media platforms have become accessible sources of information regarding cooking-related matters—people use Facebook®, Instagram®, Pinterest® and YouTube® to share and search for recipes, and to find meal suggestions and inspiration [13,18–20].

YouTube® was created in 2005 and works as a video sharing platform, which is accessible via personal computers or smartphones through an Internet browser or application [21]. On the platform's homepage, an algorithm suggests videos based on visualization history and the popularity of the content, among other information. Users can also actively search for videos using keywords or browsing channels. A user can interact with a video by watching it, liking, sharing with others, and/or publicly commenting, all of which are important social media features [22–24].

Previous research mentions that YouTube® is one of many people's favourite ways to learn how to cook [20]. Understandably, when compared to just text and images, recipes shared through video technology can favour user engagement, increase the motivation to cook, and reduce the perception of time, skills, and cost barriers [25]. Video recipes also potentially assist with the development of new skills, increase the pleasure of cooking, provide real-time assurance during the cooking process, help people remember the steps, and improve the understanding of the process [26]. In Brazil, YouTube® is the most popular social media platform among individuals aged between 16 and 64 years [16].

Accessing the Internet to search for recipes, learn how to cook, and develop cooking skills is recommended by the Dietary Guidelines for the Brazilian Population [3], but the healthiness of recipes obviously depends on the ingredients and preparation methods employed [11]. In this sense, exploring the sources of knowledge and inspiration to cook is as key as getting people to cook more often. As tools that guide the preparation of dishes [27], culinary recipes can potentially promote health if aligned with recommendations for healthy eating, expanding and encouraging individuals' decision-making autonomy regarding the adoption of healthy eating practices [2,18]. However, in the context of social media, content can be produced and shared by anyone, including lay people not qualified to give nutritional advice or create content that promotes healthy eating.

Previous studies assessed the healthiness of Internet recipes on websites and blogs (which are not social media), and concluded that users tend to interact more often with the least healthy recipes [28]. Authors concluded that even recipes tagged as 'healthy' are often quite unhealthy [28,29]. We identified only one paper on the healthiness of culinary recipes on social media, which used Pinterest® as a data source [30]. The paper reported that recipes using seafood or vegetables as main ingredients had fewer calories, sodium, sugar, and cholesterol than meat- or poultry-based recipes. However, the study's sample was small due to the adoption of many exclusion criteria [30]. No research investigating the healthiness of culinary recipes shared on other social media was found.

To address this gap, this descriptive and exploratory study analyses the healthiness of culinary recipes shared on popular YouTube® cooking channels from Brazil, using both national dietary guidelines as references. We adopted complementary approaches to assess recipes' healthiness, the first being the analysis of recipes' ingredients according to the extension and purpose of industrial processing, an important and widely used approach to categorize foods. Subsequently, a specially designed qualitative framework was used to characterize recipes according to cooking method, and by the presence of healthy or unhealthy ingredients. We believe this study has the potential to inform the design of public health initiatives that guide individuals and inform dietitians on how to select and critically evaluate sources of cooking information, and improve the quality of homecooked meals.

2. Materials and Methods

Considering the scarcity of literature on the research topic, a pilot study was carried out to inform the data collection protocol, which included aspects of various channels' eligibility criteria and database layout, introduced different video characterization variables, validated recipes' assessment method, and determined the data collection period, taking into account the temporal feasibility of the study and the amount of content to be analysed [31].

2.1. Selection of YouTube® Cooking Channels

Cooking channels were purposely selected by taking into consideration that popularity (number of subscribers) can promote a greater reach and be a proxy for users' preference. Channels were selected according to the number of subscribers in February 2020 using The YouTube Channel Crawler page (https://www.channelcrawler.com/, accessed on 10 February 2020), which classifies channels according to criteria established by the researcher (in this study: category, language, country of origin, and number of subscribers). During the pilot study, it was observed that cooking channels belonged to the 'How to and Style' category on YouTube®, thus, all channels of the platform within that category were accessed in decreasing order of subscribers to identify which ones best fit the eligibility criteria. The ten biggest channels which (1) presented audio-visual content in Portuguese and were Brazilian based; (2) were a cooking channel; (3) posted culinary videos at least once a week; and (4) were not an advertising channel or reproduced television cooking programs were selected (Figure 1).

With the aim of having a high number of videos to be analysed during data collection, it was established that channels that posted videos less than once a week would not be included. The pilot study also revealed that some channels which were among the most popular in terms of number of subscribers had suddenly stopped producing content in the weeks preceding the selection of channels. They were not included to avoid the possibility of not having enough content to analyse in the following months. Another reason for adopting this criterion was to try to standardize the number of videos per channel. Novelty was another important factor, as channels need not only to attract, but also maintain users' interest and engagement with content [23].

Eighty-two channels were excluded from the sample because they were not cooking channels, five were excluded because they did not post videos with the desired frequency, and one was excluded for being an advertising channel.

The included channels were mostly presented by women (n = 7), two by men, and one by a couple; none of them were popularly known chefs or food celebrities. Subscribers ranged from 514 thousand to 4.25 million; channels' time of existence ranged from 4 to 9 years, and posting frequency varied from 2 to 7 videos per week.

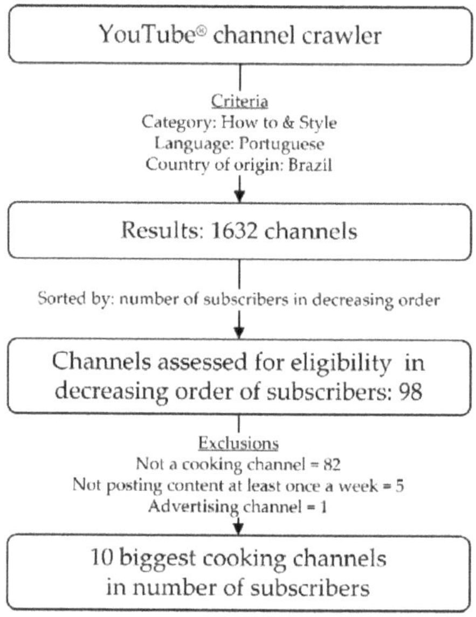

Figure 1. Brazilian YouTube® Cooking channels' selection flowchart, February 2020.

2.2. Selection of Recipes

A sample of 823 recipes presented in 755 videos (104 h and 21 min in total) posted during a six-month period (from February to August 2020) on ten different cooking channels was selected. Considering that this is a recent field of study and there is no specific recommendation in the literature for how long data collection on YouTube® should take place, the pilot study also informed the choice of an appropriate data collection period. With the pilot study, we were able to project that a 6-month data collection period would capture a high number of videos from each channel, carefully accounting for at least three seasons of the year. At the same time, the amount of content collected would meet the temporal and operational feasibility criteria of the study.

All videos with recipes posted within the period were watched in full (first author) to determine if they contained all the ingredients needed, as well as the preparation method. A total of 106 videos were excluded from analysis because they did not meet the eligibility criteria: (1) were recorded live transmissions (n = 35), (2) presented a festive recipe (Easter n = 20; Mother's day or father's day n = 6; Valentine's day = 4; June festivities in Brazil n = 12; Channel's subscribers milestone celebration n = 4; total n = 46), (3) were sponsored by the food industry (n = 9), (4) were a repost (n = 7), (5) presented recipes linked to the COVID-19 pandemic (with connotations of treatment for the virus, for improving immunity or with tips for food sales during the period of social isolation; n = 9).

Reasons for not including recorded live transmissions were: (1) during the pilot study, we observed that those kinds of videos were usually presented as 'extra' content and were produced by only four of the ten channels. They were not included for the sake of standardization (type and number of videos per channel). (2) 'Live' transmissions lasted more than one hour each, as the recipe-related content was diluted among various other content during the video. This affected both the practical relevance of the recipe and the temporal feasibility of the research.

2.3. Data Collection

Weekly, from February to August 2020, each selected channel was accessed via computer and all videos posted during the previous week were registered. A database in Microsoft Excel 2016® was created to include the following information for each video: title, access link, ID provided by YouTube®, video description, date of posting, date of access, duration in seconds, number of likes, dislikes, and views. To obtain the number of comments posted by users in each video, a command line application was developed in Python 3.0 (third author). Using the public and free Google Data application programming interface (API) service® as the data source, the application generated automated reports from the video ID and the period determined by the researcher (freely available at https://bitbucket.org/amcamargo/healthy-recipe-youtube-br.git, last accessed on 11 August 2020).

Next, the first author watched each video to register the ingredients and the cooking method in the database. If further details about ingredients were needed, the researcher consulted the recipe's ingredient list provided in the video description, or, in case of industrialized products, the packaging, when information was clearly visible on screen. Steps or ingredients mentioned by the youtuber as 'optional' and not shown in the video were not assessed.

2.4. Data Analysis

2.4.1. Videos' Characteristics

Variables assessed to characterize the videos were duration in minutes, day of the week of posting, and interaction measures including popularity as daily views in the first week, approval as daily likes and dislikes in the first week, direct interaction of users with content through daily comments in the first week, and total comments in the first week and in the first month after the video was posted.

To classify the recipe into a category (e.g., salad, pudding, etc.) a content analysis was carried out based on the video's title, description, and list of ingredients used. This analysis was manually organized in Microsoft Excel 2016® by determining the degree of similarity of the words and phrases used and the characteristics of the recipes, starting at coding recipes' names in videos' titles (first author). After coding, data was categorized until strong or terminal categories appeared [32].

2.4.2. Recipes' Healthiness

Recipes had their ingredients classified according to the extension and purpose of industrial processing, as unprocessed/minimally processed (U/MP), processed culinary ingredient (PCI), processed (P), or ultra-processed (UP) (first author) [3,4,33]. Ingredients that did not have their preparation described in the recipe but are available for purchase as an industrialized version were classified as P or UP (e.g., sweetened condensed milk, mayonnaise), according to the predominant characteristic of products available in Brazilian retail outlets. Whenever agreement about the extension and purpose of industrial processing was not achieved, a conservative criterion was applied, meaning that a lower extension of processing was adopted for the ingredient [34]. Ingredients used twice in the same recipe counted as one (e.g., sugar used in a cake's batter and icing).

Subsequently, the Qualitative Framework for the Assessment of Culinary Recipes' Healthiness [31] was applied to evaluate recipes' cooking methods and presence of key healthy and unhealthy ingredients (first author). The framework was specifically developed and validated to assess culinary recipes' healthiness, and was based on recommendations for healthy eating retrieved from both Dietary Guidelines for the Brazilian Population [2,3].

2.4.3. Data Treatment

To ensure data quality control, the second author independently analysed 10% of the recipes from the dataset. Weighted kappa of agreement between raters for the assessment of ingredients' extension and purpose of industrial processing was 0.96, and ranged between 0.90 and 1.00 (kappa and weighted kappa) for the application of the Qualitative Framework for the Assessment of Culinary Recipes' Healthiness, indicating almost perfect agreement in both analyses [35]. Content analysis for the categorization of recipes was firstly discussed between the first two authors, and divergences were resolved with the participation of the last author.

2.4.4. Statistical Analysis

Qualitative dichotomous and polytomous variables are presented in absolute and relative frequencies. Quantitative variables are presented as median and interquartile range (IQR), considering the non-normality in data distribution when assessed by Shapiro–Wilk test, histogram, kurtosis value, and mean/median proximity.

Variables of videos' characteristics and recipes' healthiness among the categories of recipes were compared. Also, as data collection took place mostly during a social distancing period due to the COVID-19 pandemic, when searches for recipes online increased [36], we also checked for differences in videos' interaction measures (popularity, approval, interaction through comments), and recipes' healthiness in the periods preceding (n = 141) vs. during social isolation (n = 614) (which, in Brazil, started around 15 March). Mann–Whitney and Kruskall–Wallis tests were used for quantitative variables. For qualitative variables, Pearson's chi-square test was employed. Stata 13.0® (StataCorp LLC, College Station, TX, USA) was used for analysis and a post-hoc power analysis was applied on G*power 3.1.9.2 whenever necessary, considering a two-tailed test. An alpha of 0.05 was established as the significance level for all analyses.

3. Results

3.1. Videos' Characteristics

The videos' durations ranged from 45 s to 27.33 min (n = 755). The number of daily likes in the first week was superior to daily dislikes. The option of liking or disliking a video was not enabled by the youtubers for all videos (only n = 611), therefore, even if users wanted to give a particular video a thumbs-up or down, they could not. Direct interaction through comments was concentrated in the first week after the videos were posted, as the median of total comments in the first month was close to the median in the first week. Sunday was the day of the week with the lowest number of videos posted, nevertheless, the distribution of videos was similar among the other days (Table 1).

Table 1. Videos' characterization variables (n = 755).

Variable	Median (IQR)
Duration (minutes)	7.8 (5.1; 10.8)
Popularity	
Daily views in the first week (n)	5194 (2094; 10,827)
Approval	
Daily likes in the first week (n)	995 (322; 1805) [1]
Daily dislikes in the first week (n)	12 (3; 23) [1]
Direct interaction of users	
Daily comments in the first week (n)	37 (14; 68)
Total comments in the first week (n)	150 (61; 272)
Total comments in the first month (n)	161 (67; 291)
Day of posting	**% (n)**
Monday	18 (136)
Tuesday	14 (103)
Wednesday	18 (138)
Thursday	16 (118)
Friday	15 (113)
Saturday	13 (97)
Sunday	6 (50)

Footnote: [1] n = 611 videos.

The only observed difference between videos collected in the period preceding vs. during social isolation was in the total of comments in the first week, which was higher during the social isolation period (median = 154, IQR = 64; 280) than before the pandemic (median = 140; IQR = 47; 234) (Mann–Whitney's p = 0.04, power = 0.19).

More than two thirds of all recipes (68.1%) comprised preparations from only four categories, namely: meat or egg main dishes; cakes and baked goods; snacks and home-made fast foods; and puddings (Table 2). The sixteen different categories of recipes had comparable video characteristics (all Kruskall–Wallis $p > 0.10$; $\chi^2 = 91.19$, $p = 0.445$). The frequency of categories observed in the period preceding vs. during social isolation was statistically the same ($\chi^2 = 18.25$; $p = 0.07$).

Table 2. Recipes' healthiness according to ingredients' extension and purpose of industrial processing.

Recipes' Categories	Examples	Recipes % (*n*)	Ingredients Median (IQR)	Ingredients Distribution according to the Extension and Purpose of Industrial Processing			
				U/MP % (*n*)	PCI % (*n*)	P % (*n*)	UP % (*n*)
Meat or egg main dishes	Stroganoff, meat stew, omelette, chicken lasagne, one pan pepperoni pasta	22.8 (185)	12.0 (9.0; 15.0)	63.6 (1428)	16.1 (361)	9.1 (205)	11.2 (251)
Cakes and baked goods	Banana cake, chocolate cake, pies, biscuits, Brazilian cornbread, pancakes	18.2 (148)	8.0 (7.0; 10.0)	40.8 (530)	37.8 (491)	5.5 (72)	15.9 (206)
Snacks and homemade fast foods	Fried snacks, hotdog, pizza, sandwiches, *pão de queijo* [1], sweet popcorn	15.1 (123)	11.0 (7.0; 13.0)	49.6 (624)	23.5 (295)	13.4 (168)	13.9 (171)
Puddings	Mousses, sweetened condensed milk trifles, ice cream, rice pudding	12.1 (98)	6.0 (5.0; 7.0)	37.2 (225)	21.5 (130)	4.6 (28)	36.7 (222)
Side dishes	Cooked rice, *farofa* [2], cooked beans, roasted potatoes	9.2 (75)	10.0 (7.0; 12.0)	61.9 (440)	19.8 (141)	9.0 (64)	9.3 (66)
Breads	Basic homemade bread, homemade sliced bread, whole wheat bread, onion bread, aussie bread	5.0 (41)	8.0 (6.0; 10.0)	45.7 (148)	46.6 (151)	2.5 (8)	5.2 (17)
Savoury cakes and pies	Vegetable and cheese pie, sardine pie, quiche	4.8 (39)	14.0 (12.0; 18.0)	55.3 (306)	20.1 (111)	14.6 (81)	10.0 (55)
Salads	Raw vegetables with legumes salad, sautéed vegetables	2.6 (21)	9.0 (6.0; 13.0)	70.4 (143)	19.7 (40)	4.4 (9)	5.4 (11)
Soups and creams	Vegetables and/or chicken soups	2.2 (18)	13.0 (12.0; 14.0)	74.4 (169)	13.7 (31)	5.7 (13)	6.2 (14)
Appetizers	Onion toast, French fries, rice balls, fried beans	2.1 (17)	9.0 (6.0; 10.0)	61.8 (84)	25.0 (34)	8.1 (11)	5.1 (7)
Non-alcoholic beverages	Creamy coffee, hot-chocolate, juices	2.0 (16)	4.0 (4.0; 4.0)	63.8 (44)	17.4 (12)	5.8 (4)	13.0 (9)
Homemade ingredients	Butter, stock, homemade seasoning mix, pastry dough	1.3 (11)	5.0 (2.0; 9.0)	62.5 (40)	26.6 (17)	6.2 (4)	4.7 (3)

Table 2. *Cont.*

Recipes' Categories	Examples	Recipes % (*n*)	Ingredients Median (IQR)	Ingredients Distribution according to the Extension and Purpose of Industrial Processing			
				U/MP % (*n*)	PCI % (*n*)	P % (*n*)	UP % (*n*)
Savoury spreads and pâtés	Cheese pâté, dried tomato pâté, olive pâté	1.0 (8)	4.5 (4.0; 6.5)	45.8 (27)	27.1 (16)	15.2 (9)	11.9 (7)
Sauces	Bechamel sauce, pepper sauce, yoghurt sauce, rosé sauce	0.7 (6)	6.0 (5.0; 8.0)	51.6 (16)	25.8 (8)	-	22.6 (7)
Preserves	Onion preserve, beans preserve	0.5 (4)	5.5 (2.0; 9.5)	69.6 (16)	30.4 (7)	-	-
Sweet spreads	Dulce de leche	0.4 (3)	2.0 (2.0; 3.0)	28.6 (2)	28.6 (2)	-	42.8 (3)
Total		100.0 (813)	9.0 (6.0; 12.0)	54.3 (4242)	23.6 (1844)	8.6 (676)	13.5 (1052)

Footnote: U/MP—unprocessed/minimally processed foods. PCI—processed culinary ingredients. P—processed foods. UP—ultra-processed foods. [1] Traditional Brazilian recipe of small cheese bread made of fermented tapioca flour. [2] Traditional Brazilian dish made of manioc flour fried in fat, which can be enriched with other ingredients.

3.2. Recipes' Healthiness

Of the total 7814 ingredients analysed, the majority were U/MP (54.3%, n = 4242) and PCI (23.6%, n = 1844). Ingredients classified as P (8.6%, n = 676) and UP (13.5%, n = 1052) were less frequent. The categories of recipes differed in terms of the ingredients' distinct extension and purpose of industrial processing (χ^2 = 859.22; p < 0.001). As Table 2 shows, in many categories, less than half of the ingredients were U/MP, i.e., cakes and baked goods, snacks and homemade fast foods, puddings, breads, sweet and savoury spreads, and pâtés. The ten most frequent U/MP ingredients in the sample were, in decreasing order: water, eggs, onion, all-purpose flour, garlic, milk, black pepper, oregano, spring onions, and tomatoes. Some of the categories with the lowest frequency of U/MP foods also had the highest frequencies of UP foods in the sample, i.e., puddings, cakes and baked goods, snacks and homemade fast foods, sauces, sweet and savoury spreads, and pâtés. The ten most frequent UP ingredients in the sample were, in decreasing order: UHT cream, sweetened condensed milk, Brazilian cheese spread, margarine, ham, industrialized tomato sauce, spicy sausage, vanilla essence, industrialized seasoning mix, and semi-sweet chocolate. The frequency of ingredients with distinct extension and purpose of industrial processing observed in the period preceding vs. during social isolation was not statistically different (χ^2 = 0.68; p = 0.877).

Application of the Qualitative Framework for the Assessment of Culinary Recipes' Healthiness (Table 3) identified positive and negative aspects of the recipes. Positively, most recipes that mentioned some type of fat as an ingredient did not suggest the use of margarine (88.4%, n = 518). Mentions of tomato sauce with herbs (bottled or freshly made) were more frequent than exclusive mentions of white sauce with mayonnaise or cheese (69.6%, n = 131). Exclusive use of industrialized seasonings (1.5%, n = 7) and of frying as a cooking method (7.9%, n = 60) was also not frequently mentioned. On the other hand, the presence of whole cereals, breads and/or pasta, either exclusively or mixed with refined cereals was low in the recipes (7.1%, n = 34), as well as were the presence of fruits (13.7%, n = 111), legumes (4.5%, n = 37), and nuts and seeds (3.5%, n = 28). The categories that presented the most evenly distributed positive and negative criteria were types of meats, presence of foods with high sugar concentration, and presence of vegetables. All results from the framework analysis were statistically the same regarding the period of data collection (preceding vs. during social distancing; all 0.01 < χ^2 > 4.73 and p > 0.07).

Table 3. Recipes' healthiness according to the Qualitative Framework for the Assessment of Culinary Recipes' Healthiness.

Category	Description of Components	Criteria	% (n)
Foods with high starch content	Exclusive presence of whole cereals, breads and/or pasta	+	5.0 (24)
	Mixed presence of whole and refined cereals, breads and/or pasta	+	2.1 (10)
	Exclusive presence of refined cereals, breads and/or pasta	−	92.9 (446)
Fruits, vegetables and legumes	Presence of vegetables	+	43.3 (353)
	Absence of vegetables	−	56.6 (460)
	Presence of legumes	+	4.5 (37)
	Absence of legumes	−	95.5 (776)
	Presence of fresh, frozen or dried fruits	+	13.7 (111)
	Absence of fresh, frozen or dried fruits	−	86.3 (700)
Nuts and seeds	Presence of nuts and seeds	+	3.5 (28)
	Absence of nuts and seeds	−	96.5 (784)

Table 3. *Cont.*

Category	Description of Components	Criteria	% (*n*)
Meats and eggs	Exclusive presence of lean cuts of meat, poultry cuts without skin, fish, seafood and/or eggs	+	32.9 (109)
	Mixed presence of lean cuts of meat, poultry cuts without skin, fish, seafood and/or eggs and non-lean cuts of meat, poultry cuts with skin and/or processed meats	+	19.6 (65)
	Exclusive presence of non-lean cuts of meat, poultry cuts with skin and/or processed meats	−	47.4 (157)
Fats	Exclusive use of vegetable oils, butter and/or lard in place of margarine	+	88.4 (518)
	Presence of margarine	−	11.6 (68)
Sauces	Exclusive presence of tomato sauce with herbs	+	52.1 (98)
	Mixed presence of tomato sauce with herbs and white sauce, with mayonnaise or cheese	+	17.5 (33)
	Exclusive presence of white sauce, with mayonnaise or cheese	−	30.3 (57)
Seasonings	Exclusive presence of olive oil, lemon and/or fresh or dried herbs	+	68.9 (333)
	Mixed presence of olive oil, lemon and/or fresh or dried herbs, and industrialized spices, sauces and/or broths	+	29.6 (143)
	Exclusive presence of industrialized spices, sauces and/or broths	−	1.5 (7)
Sugars	Presence of foods with high sugar concentration	−	41.8 (338)
	Absence of foods with high sugar concentration	+	58.2 (470)
Cooking method	Use of steam, cooking in water without or with little fat, stewing, roasting, broiling, sautéing	+	92.1 (696)
	Use of steam, cooking in water without or with little fat, stewing, roasting, broiling, sautéing—and/or frying	−	7.9 (60)

Footnote: Fruits, vegetables, and legumes; nuts and seeds, and sugars categories are mandatorily assessed in all recipes. The remaining categories are assessed only when applicable. Criteria: + and − indicate recommended and not recommended components for healthy recipes, respectively [31].

4. Discussion

This study analysed the healthiness of recipes shared on popular YouTube® cooking channels from Brazil using the Dietary Guidelines for the Brazilian Populationas references. Recipes posted during a six-month period were retrieved and categorized into sixteen different groups. The most frequently posted recipes were of meat/egg-based main dishes; cakes/baked goods; snacks/homemade fast foods; and puddings. This means that recipes for salads and side dishes, which usually contain vegetables, fruits, and legumes, were shared less often than recipes with animal sources of protein, all-purpose flour, fats, and sugar as the main ingredients. This result is not favourable from a health standpoint, as individuals are possibly being led to prepare fewer recipes with fruits, vegetables, and legumes, which are linked to a lower risk of chronic non-communicable diseases, and are largely present in most healthy eating patterns [37,38]. Interestingly, the study by Trattner and Elsweiler (2017) identified different results—in their study, which evaluated content from a recipes' website, the category 'fruits and vegetables' was much more prevalent than 'main dishes,' 'meat and poultry,' 'desserts,' and 'salads.' This disparity may be attributed to differences in the process of categorizing recipes, as in food blogs [29,39] and websites [28], recipes are usually pre-categorized, while we conducted our own categorization. Because YouTube® is a multi-content platform not specifically focused on recipes, our recipe categories were qualitatively and inductively generated from recipes' titles, descriptions, and ingredients. Additionally, several studies only assessed specific categories of recipes [28–30,39], since their aim was not to have an overall picture of what is shared.

Another possible explanation for the low prevalence of fruit- and vegetable-based recipes in our sample may be that content producers expect users to interact with the postings through comments and shares, as interaction is fundamental for a channel's engagement and sustainability [23]. It has been reported by previous studies on a recipes' website [28] and on Pinterest® [30] that interaction is more frequent with posts of highly palatable recipes. In our study conducted on YouTube®, every culinary preparation had statistically equal measures of interaction (popularity, approval, and direct interaction through comments), possibly due to differences between the profiles of users from recipe websites [28] and even between different social media apps [30]. YouTube®, as a video platform, enables a kind of interaction that gives users a feeling of being connected not only to a video, but to a person who shares their beliefs and interests. This feature can promote a certain measure of social bonding in which people feel connected with one another and start following the channel for further communication. For user-created content such as the videos analysed, a sense of community is fundamental; so it is possible that subscribers give the same attention to a recipe, regardless of whether it is a salad or a cake, in order to provide support through constancy of viewership and interaction [21]. The reasoning behind youtubers' choices of categories of recipes for cooking videos, the channels' features that promote connection with users, as well as subscribers' motivations for interaction with content, deserve to be further explored in future research. Nevertheless, health professionals should be aware that, in order to expose individuals to more recipes based on vegetables, fruits, and legumes (such as salads and side dishes), active searching is preferable to just following content from popular cooking channels. Nutritionists and other health professionals can also search for cooking channels whose content is more in line with the healthy eating recommendations of national guidelines to suggest to patients.

Recipes' ingredients were mainly U/MP foods and PCI, so one can argue that from a wide perspective, the recipes could lead individuals to cook recipes that are aligned with the recommendations of the Dietary Guidelines for the Brazilian Population [3]. Nevertheless, this is not true when different categories of culinary recipes are considered. Some categories of recipes had lower frequencies of U/MP foods as ingredients, and a few of them had, in addition to this, higher frequencies of the UP foods in the sample (more than 10%)—i.e., puddings, cakes and baked goods, snacks and homemade fast foods, sauces, sweet and savoury spreads, and pâtés. This result is cause for concern, as some of these were among the most frequently posted recipes. To cook healthily, the Dietary Guidelines for the Brazilian Population recommends the avoidance of UP foods [3], as high consumption of UP foods has been associated with chronic non-communicable diseases and all-cause mortality [5–7]. UP food consumption has been associated with a poor dietary intake (excess calories from free sugars and unhealthy saturated fats, poor in fibre, and an intake of many micronutrients) [40]. Additionally, recent research shows that the majority of the associations between UP food consumption, obesity, and health-related outcomes can be attributed to UP foods on their own, regardless of diet quality or pattern [41].

The presence of UP ingredients in the recipes may be explained by their convenience appeal [4,33]. It is rather common for UP foods to replace U/MP foods in recipes (e.g., sausage vs. U/MP meat seasoned with spices and herbs). Generations have learned to cook using recipes combining UP and U/MP foods through teaching investments by the food industry (leaflets, books, courses, recipes on packaging) [42]. Nowadays, the ongoing increase of options of UP foods and of social media marketing play an important incentivizing role [43]. We observed with our framework analysis (Table 3) that the mixed use of industrialized seasonings with fresh or dried herbs and spices was frequent, indicating an attachment to this type of UP product. To mitigate this effect, strategies involving the promotion of healthy eating through cooking (such as workshops, intervention programs, creation of content for social media, health professionals' advice, etc.) need to consider that people must be taught how to identify UP foods so they can choose recipes in which they are not included. People must also be taught how to substitute UP foods for healthier ingredients, so they can use U/MP foods practically when cooking. For instance, instead of relying on UP foods as seasoning in puddings, snacks, and homemade fast foods as observed in this sample of recipes, one can substitute such ingredients for fruit zest and juices, fresh or dried herbs, and spices. Another valuable strategy is to rescue and promote the sharing of traditional recipes that do not contain UP foods as ingredients.

While the majority of recipes were healthy with respect to avoiding the use of margarine, avoiding frying, and opting for sauces with lower fat content, other aspects such as incorporating whole cereals, fruits, legumes, nuts, and seeds in preparation were not frequently present. Considering that social media platforms such as YouTube® reach a wide audience, this finding reinforces the need to not only encourage people to look for recipes online [3], but also to teach them how to choose or adapt these recipes by evaluating their healthiness. One strategy is to use the same medium to do this, as video technology can help individuals overcome barriers to cook and incorporate healthier foods in recipes [44], while reducing the perception of barriers to cook with vegetables [25]. We are aware that, for some recipes, whole cereals, fruits, legumes, nuts, and seeds may not be all traditionally present (e.g., a basic homemade bread), but different 'improved' versions of recipes can be proposed and shared. As a matter of fact, many channels assessed in this study adapted recipes to keep producing new content weekly. As a practical implication, we argue that many categories of recipes can be adapted to become healthier—for examples and suggestions, see [31]. Members of academia, health professionals, and social media content creators can also work together and establish partnerships to promote healthier content on the Internet.

Limitations and Strong Points

The adoption of a conservative criterion for classifying ingredients by the extension and purpose of industrial processing may have led to an underestimation of the number of UP ingredients. Nevertheless, this approach mirrors how information reaches users—they also do not necessarily have access to information on labels when watching videos.

Data collection took place during months of social isolation due to the COVID-19 pandemic, when searches for recipes online increased [36]. This was handled by avoiding the inclusion of videos linked to the COVID-19 pandemic in the sample. Our post-hoc analysis found an underpowered difference in the number of comments in the first week after videos were posted (power = 0.19) [45]; no change in the categories of recipes shared, nor in their healthiness compared to videos from before the pandemic.

YouTube® channels' popularity oscillates constantly. To handle this, we repeated the channel selection step at the end of data collection, and verified that they remained as the ten most popular in the period, despite some outperforming others in the number of subscribers.

As the number of views is validated by YouTube®'s own algorithms, a view might not indicate a user who has watched the content in its entirety. Content approval (likes and dislikes) also does not indicate whether or not an individual fully watched the content before giving a positive or negative rating. Although views made by computer programs rather than by humans are not counted [46], those interaction measures should be cautiously interpreted [47].

In the context of television cooking shows, some researchers argue that the consumption of this content is unlikely to impact habitual dietary intake, because entertainment and leisure are the main reasons people watch those programs [48,49]. Notwithstanding, social media, through its networked nature, provides an additional layer of complexity not experienced by those earlier media scholars [47]. Through observation, people indeed acquire behaviours, knowledge, values, and skills, including those related to cooking [50].

We understand that it is not possible within the confines of the present study to account for variations in the reproduction of recipes at home, such as instances when people do not follow all the steps, or when ingredients are exchanged, which may result in a different assessment of their healthiness.

As positive points, we highlight the investigation of culinary recipes posted in the most used social media in Brazil, by adults [16]; the rigorous quality control; the long period of data collection throughout three seasons of the year, and, therefore, the large sample size. Recipes were also very diverse in terms of categories, video duration, and days of posting, probably reaching different types of audiences. Additionally, collecting measures of interaction (views, comments, likes, and dislikes) reinforces the wide reach that this type of content has. Finally, using a validated framework for the assessment of recipes' healthiness, we were able to deliver a more specific picture of the research problem.

5. Conclusions

This study provides a comprehensive overview of the healthiness of culinary recipes shared on a social media platform, one of the favoured avenues for the search of cooking-related content. On a professional practice and health promotion note, although it is praiseworthy that people are cooking and sharing their knowledge on platforms such as YouTube®, users and subscribers to popular cooking channels should be aware that most recipes are based on ingredients such as meats, eggs, all-purpose flour, fats and sugar, and only a few have whole cereals, fruits, legumes, nuts, and seeds. Recipes for puddings, cakes and baked goods, snacks and homemade fast foods, sauces, sweet and savoury spreads, and pâtés had, in addition to low numbers of U/MP food ingredients, higher numbers of UP foods as ingredients. Our findings can inform health professionals and policymakers on how to promote healthier culinary recipes, how to interact with content creators, and how to advise individuals about the quality of the recipes shared on YouTube® videos, and hence, can help them choose healthier recipes or teach them how to modify the recipes into healthier versions. Future research exploring how users from different populational groups interact with culinary content on distinct social media platforms will be relevant for advancing this field of study.

Author Contributions: Conceptualization, A.M.d.C.; data curation, A.M.d.C. and Á.M.d.C.; formal analysis, A.M.d.C., A.M.B. and G.M.R.F.; methodology, A.M.d.C. and G.M.R.F.; software, Á.M.d.C.; validation, A.M.d.C., A.M.B. and G.M.R.F.; writing—original draft preparation, A.M.d.C.; writing—review and editing, A.M.d.C., A.M.B., Á.M.d.C., M.D. and G.M.R.F.; supervision, G.M.R.F.; funding acquisition, M.D. All authors have read and agreed to the published version of the manuscript.

Funding: This work was supported by Newton Mobility Grant Scheme 2015 [Award Reference: NG150026. 2 to M.D. and G.M.R.F.], the UK Academies Fellowships Research Mobility, and Young Investigator Awards for UK Researchers in Brazil FAPESC/CONFAP/FUNDO NEWTON [Call N° 02/2017; 3 to M.D. and G.M.R.F.]; Coordination for the Improvement of Higher Education Personnel (CAPES) in the form of scholarships [Finance code 001] to [A.M.d.C. and A.M.B.]. Founding sources had no role in study design, in the collection, analysis and interpretation of data; in the writing of the report; and in the decision to submit the article for publication.

Institutional Review Board Statement: Not applicable.

Informed Consent Statement: Not applicable.

Data Availability Statement: Not applicable.

Conflicts of Interest: The authors declare no conflict of interest.

References

1. Jomori, M.M.; Vasconcelos, F.d.A.G.d.; Bernardo, G.L.; Uggioni, P.L.; Proença, R.P.d.C. The Concept of Cooking Skills: A Review with Contributions to the Scientific Debate. *Rev. Nutr.* **2018**, *31*, 119–135. [CrossRef]
2. Ministério da Saúde do Brasil. *Guia Alimentar Para a População Brasileira*; Ministério da Saúde: Brasília, Brazil, 2006; 210p. (In Portuguese)
3. Ministry of Health of Brazil. *Dietary Guidelines for the Brazilian Population*; Ministry of Health of Brazil: Brasília, Brazil, 2015; 150p.
4. Monteiro, C.A.; Cannon, G.; Levy, R.B.; Moubarac, J.-C.; Louzada, M.L.; Rauber, F.; Khandpur, N.; Cediel, G.; Neri, D.; Martinez-Steele, E.; et al. Ultra-Processed Foods: What They Are and How to Identify Them. *Public Health Nutr.* **2019**, *22*, 936–941. [CrossRef] [PubMed]

5. Chen, X.; Zhang, Z.; Yang, H.; Qiu, P.; Wang, H.; Wang, F.; Zhao, Q.; Fang, J.; Nie, J. Consumption of Ultra-Processed Foods and Health Outcomes: A Systematic Review of Epidemiological Studies. *Nutr. J.* **2020**, *19*, 86. [CrossRef]
6. Lane, M.M.; Davis, J.A.; Beattie, S.; Gómez-Donoso, C.; Loughman, A.; O'Neil, A.; Jacka, F.; Berk, M.; Page, R.; Marx, W.; et al. Ultraprocessed Food and Chronic Noncommunicable Diseases: A Systematic Review and Meta-Analysis of 43 Observational Studies. *Obes. Rev.* **2020**, *22*, e13146. [CrossRef] [PubMed]
7. Matos, R.A.; Adams, M.; Sabaté, J. Review: The Consumption of Ultra-Processed Foods and Non-Communicable Diseases in Latin America. *Front. Nutr.* **2021**, *8*, 110. [CrossRef]
8. Méjean, C.; Lampuré, A.; Si Hassen, W.; Gojard, S.; Péneau, S.; Hercberg, S.; Castetbon, K. Influence of Food Preparation Behaviors on 5-Year Weight Change and Obesity Risk in a French Prospective Cohort. *Int. J. Behav. Nutr. Phys. Act.* **2018**, *15*, 120. [CrossRef]
9. Mills, S.; White, M.; Brown, H.; Wrieden, W.; Kwasnicka, D.; Halligan, J.; Robalino, S.; Adams, J. Health and Social Determinants and Outcomes of Home Cooking: A Systematic Review of Observational Studies. *Appetite* **2017**, *111*, 116–134. [CrossRef]
10. Wolfson, J.A.; Leung, C.W.; Richardson, C.R. More Frequent Cooking at Home Is Associated with Higher Healthy Eating Index-2015 Score. *Public Health Nutr.* **2020**, *23*, 2384–2394. [CrossRef]
11. Martins, C.A.; Andrade, G.C.; de Oliveira, M.F.B.; Rauber, F.; de Castro, I.R.R.; Couto, M.T.; Levy, R.B. "Healthy", "Usual" and "Convenience" Cooking Practices Patterns: How Do They Influence Children's Food Consumption? *Appetite* **2021**, *158*, 105018. [CrossRef]
12. Urdapilleta, I.; Dany, L.; Boussoco, J.; Schwartz, C.; Giboreau, A. Culinary Choices: A Sociopsychological Perspective Based on the Concept of Distance to the Object. *Food Qual. Prefer.* **2016**, *48*, 50–58. [CrossRef]
13. Lavelle, F.; Spence, M.; Hollywood, L.; McGowan, L.; Surgenor, D.; McCloat, A.; Mooney, E.; Caraher, M.; Raats, M.; Dean, M. Learning Cooking Skills at Different Ages: A Cross-Sectional Study. *Int. J. Behav. Nutr. Phys. Act.* **2016**, *13*, 119. [CrossRef] [PubMed]
14. Tobey, L.N.; Mouzong, C.; Angulo, J.S.; Bowman, S.; Manore, M.M. How Low-Income Mothers Select and Adapt Recipes and Implications for Promoting Healthy Recipes Online. *Nutrients* **2019**, *11*, 339. [CrossRef] [PubMed]
15. Romero, M.Y.M.; Francis, L.A. Youth Involvement in Food Preparation Practices at Home: A Multi-Method Exploration of Latinx Youth Experiences and Perspectives. *Appetite* **2020**, *144*, 104439. [CrossRef] [PubMed]
16. Kemp, S. Digital Brazil: 2020. Available online: https://datareportal.com/reports/digital-2020-brazil (accessed on 1 June 2021).
17. Instituto Brasileiro de Geografia e Estatística (IBGE). Pesquisa Nacional por Amostra de Domicílios Contínua. 2018. Available online: https://www.ibge.gov.br/estatisticas/sociais/habitacao/17270-pnad-continua.html?=&t=o-que-e (accessed on 10 October 2021). (In Portuguese)
18. Doub, A.E.; Small, M.L.; Levin, A.; LeVangie, K.; Brick, T.R. Identifying Users of Traditional and Internet-Based Resources for Meal Ideas: An Association Rule Learning Approach. *Appetite* **2016**, *103*, 128–136. [CrossRef]
19. Nour, M.M.; Rouf, A.S.; Allman-Farinelli, M. Exploring Young Adult Perspectives on the Use of Gamification and Social Media in a Smartphone Platform for Improving Vegetable Intake. *Appetite* **2018**, *120*, 547–556. [CrossRef]
20. Worsley, A.; Wang, W.; Ismail, S.; Ridley, S. Consumers' Interest in Learning about Cooking: The Influence of Age, Gender and Education. *Int. J. Consum. Stud.* **2014**, *38*, 258–264. [CrossRef]
21. Snickars, P.; Vonderau, P. *The YouTube Reader*, 1st ed.; Mediehistoriskt Arkiv: Estocolmo, Sweden, 2009.
22. Carr, C.T.; Hayes, R.A. Social Media: Defining, Developing, and Divining. *Atl. J. Commun.* **2015**, *23*, 46–65. [CrossRef]
23. O'Brien, H.L.; Toms, E.G. What Is User Engagement? A Conceptual Framework for Defining User Engagement with Technology. *J. Am. Soc. Inf. Sci. Technol.* **2008**, *59*, 938–955. [CrossRef]
24. YouTube. YouTube Terms of Service 2021. Available online: https://www.youtube.com/static?template=terms#3f63e0eb0d (accessed on 25 May 2021).
25. Nour, M.; Cheng, Z.G.Y.; Farrow, J.L.; Allman-Farinelli, M. Short Videos Addressing Barriers to Cooking with Vegetables in Young Adults: Pilot Testing. *J. Am. Coll. Nutr.* **2018**, *37*, 724–730. [CrossRef]
26. Surgenor, D.; Hollywood, L.; Furey, S.; Lavelle, F.; McGowan, L.; Spence, M.; Raats, M.; McCloat, A.; Mooney, E.; Caraher, M.; et al. The Impact of Video Technology on Learning: A Cooking Skills Experiment. *Appetite* **2017**, *114*, 306–312. [CrossRef]
27. Milică, I.; Guia, S. Culinary Recipes: Orality and Scripturality (I). *Diacronia* **2017**, *5*, A71. [CrossRef]
28. Trattner, C.; Elsweiler, D. Investigating the Healthiness of Internet-Sourced Recipes Implications for Meal Planning and Recommender Systems. In Proceedings of the 26th International Conference on World Wide Web, Perth, Autralia, 3–7 April 2017; pp. 489–498. [CrossRef]
29. Dickinson, K.M.; Watson, M.S.; Prichard, I. Are Clean Eating Blogs a Source of Healthy Recipes? A Comparative Study of the Nutrient Composition of Foods with and without Clean Eating Claims. *Nutrients* **2018**, *10*, 1440. [CrossRef]
30. Cheng, X.; Lin, S.-Y.; Wang, K.; Hong, Y.A.; Zhao, X.; Gress, D.; Wojtusiak, J.; Cheskin, L.J.; Xue, H. Healthfulness Assessment of Recipes Shared on Pinterest: Natural Language Processing and Content Analysis. *J. Med. Internet Res.* **2021**, *23*, e25757. [CrossRef] [PubMed]
31. Camargo, A.M.; Botelho, A.M.; Dean, M.; Fiates, G.M.R. A Qualitative Framework for the Assessment of Culinary Recipes' Healthiness. *Int. J. Gastron. Food Sci.* **2021**, *28*, 100391. [CrossRef]
32. Bardin, L. *Análise de Conteúdo*, 1st ed.; Edições 70: São Paulo, Brazil, 2015.
33. Monteiro, C.A.; Cannon, G.; Levy, R.B.; Moubarac, J.-C.; Jaime, P.; Martins, A.P.B.; Canella, D.; Louzada, M.; Parra, D. NOVA The Star Shines Bright. *World Nutr.* **2016**, *7*, 28–38.

34. Steele, E.M.; Baraldi, L.G.; da Costa Louzada, M.L.; Moubarac, J.-C.; Mozaffarian, D.; Monteiro, C.A. Ultra-Processed Foods and Added Sugars in the US Diet: Evidence from a Nationally Representative Cross-Sectional Study. *BMJ Open* **2016**, *6*, e009892. [CrossRef]
35. Landis, J.R.; Koch, G.G. The Measurement of Observer Agreement for Categorical Data. *Biometrics* **1977**, *33*, 159–174. [CrossRef] [PubMed]
36. Laguna, L.; Fiszman, S.; Puerta, P.; Chaya, C.; Tárrega, A. The Impact of COVID-19 Lockdown on Food Priorities. Results from a Preliminary Study Using Social Media and an Online Survey with Spanish Consumers. *Food Qual. Prefer.* **2020**, *86*, 104028. [CrossRef]
37. Miller, V.; Mente, A.; Dehghan, M.; Rangarajan, S.; Zhang, X.; Swaminathan, S.; Dagenais, G.; Gupta, R.; Mohan, V.; Lear, S.; et al. Fruit, Vegetable, and Legume Intake, and Cardiovascular Disease and Deaths in 18 Countries (PURE): A Prospective Cohort Study. *Lancet* **2017**, *390*, 2037–2049. [CrossRef]
38. Wallace, T.C.; Bailey, R.L.; Blumberg, J.B.; Burton-Freeman, B.; Chen, C.-Y.O.; Crowe-White, K.M.; Drewnowski, A.; Hooshmand, S.; Johnson, E.; Lewis, R.; et al. Fruits, Vegetables, and Health: A Comprehensive Narrative, Umbrella Review of the Science and Recommendations for Enhanced Public Policy to Improve Intake. *Crit. Rev. Food Sci. Nutr.* **2020**, *60*, 2174–2211. [CrossRef]
39. Schneider, E.P.; McGovern, E.E.; Lynch, C.L.; Brown, L.S. Do Food Blogs Serve as a Source of Nutritionally Balanced Recipes? An Analysis of 6 Popular Food Blogs. *J. Nutr. Educ. Behav.* **2013**, *45*, 696–700. [CrossRef] [PubMed]
40. Martini, D.; Godos, J.; Bonaccio, M.; Vitaglione, P.; Grosso, G. Ultra-Processed Foods and Nutritional Dietary Profile: A Meta-Analysis of Nationally Representative Samples. *Nutrients* **2021**, *13*, 3390. [CrossRef] [PubMed]
41. Dicken, S.J.; Batterham, R.L. The Role of Diet Quality in Mediating the Association between Ultra-Processed Food Intake, Obesity and Health-Related Outcomes: A Review of Prospective Cohort Studies. *Nutrients* **2022**, *14*, 23. [CrossRef] [PubMed]
42. Oliveira, D. *Dos Cadernos de Receitas as Receitas de Latinha: Indústria e Tradição Culinária No Brasil*, 1st ed.; Senac: São Paulo, Brazil, 2013. (In Portuguese)
43. Sato, P.M.; Couto, M.T.; Wells, J.; Cardoso, M.A.; Devakumar, D.; Scagliusi, F.B. Mothers' Food Choices and Consumption of Ultra-Processed Foods in the Brazilian Amazon: A Grounded Theory Study. *Appetite* **2020**, *148*, 104602. [CrossRef]
44. Mc Morrow, L.; Ludbrook, A.; Macdiarmid, J.I.; Olajide, D. Perceived Barriers towards Healthy Eating and Their Association with Fruit and Vegetable Consumption. *J. Public Health* **2017**, *39*, 330–338. [CrossRef] [PubMed]
45. Cohen, J. *Statistical Power Analysis for the Behavioral Sciences*, 2nd ed.; Routledge: Hillsdale, NJ, USA, 1988.
46. Google. How Video Views Are Counted 2020. Available online: https://support.google.com/youtube/answer/2991785?hl=en (accessed on 17 February 2020).
47. Dutton, W.H. *The Oxford Handbook of Internet Studies*, 1st ed.; Oxford University Press: Oxford, UK, 2013.
48. Backer, C.J.S.; Hudders, L. Look Who's Cooking. Investigating the Relationship between Watching Educational and Edutainment TV Cooking Shows, Eating Habits and Everyday Cooking Practices among Men and Women in Belgium. *Appetite* **2016**, *96*, 494–501. [CrossRef]
49. Villani, A.M.; Egan, T.; Keogh, J.B.; Clifton, P.M. Attitudes and Beliefs of Australian Adults on Reality Television Cooking Programmes and Celebrity Chefs. Is There Cause for Concern? Descriptive Analysis Presented from a Consumer Survey. *Appetite* **2015**, *91*, 7–12. [CrossRef]
50. Bandura, A. *Social Foundations of thought and Action: A Social Cognitive Theory*, 1st ed.; Prentice-Hall, Inc.: Englewood Cliffs, NJ, USA, 1986; ISBN 978-0-13-815614-5.

nutrients

MDPI

Article

Zinc Supplementation Partially Decreases the Harmful Effects of a Cafeteria Diet in Rats but Does Not Prevent Intestinal Dysbiosis

Samia Squizani [1], Jeferson Jantsch [1], Fernanda da Silva Rodrigues [1], Matheus Filipe Braga [2], Sarah Eller [3], Tiago Franco de Oliveira [3], Alexandre Kleber Silveira [4], José Cláudio Fonseca Moreira [1,4], Marcia Giovenardi [1,3], Marilene Porawski [1,5] and Renata Padilha Guedes [1,3,*]

1 Programa de Pós-Graduação em Biociências, Universidade Federal de Ciências da Saúde de Porto Alegre (UFCSPA), Porto Alegre 90050-170, Brazil
2 Acadêmico do Curso de Biomedicina, Universidade Federal de Ciências da Saúde de Porto Alegre (UFCSPA), Porto Alegre 90050-170, Brazil
3 Programa de Pós-Graduação em Ciências da Saúde, Universidade Federal de Ciências da Saúde de Porto Alegre (UFCSPA), Porto Alegre 90050-170, Brazil
4 Departamento de Bioquímica, Universidade Federal do Rio Grande do Sul (UFRGS), Porto Alegre 90035-003, Brazil
5 Programa de Pós-Graduação em Medicina: Hepatologia, Universidade Federal de Ciências da Saúde de Porto Alegre (UFCSPA), Porto Alegre 90050-170, Brazil
* Correspondence: renata.guedes@ufcspa.edu.br; Tel.: +55-51-3303-8751

Abstract: Zinc (Zn) plays an important role in metabolic homeostasis and may modulate neurological impairment related to obesity. The present study aimed to evaluate the effect of Zn supplementation on the intestinal microbiota, fatty acid profile, and neurofunctional parameters in obese male Wistar rats. Rats were fed a cafeteria diet (CAF), composed of ultra-processed and highly caloric and palatable foods, for 20 weeks to induce obesity. From week 16, Zn supplementation was started (10 mg/kg/day). At the end of the experiment, we evaluated the colon morphology, composition of gut microbiota, intestinal fatty acids, integrity of the intestinal barrier and blood–brain barrier (BBB), and neuroplasticity markers in the cerebral cortex and hippocampus. Obese rats showed dysbiosis, morphological changes, short-chain fatty acid (SCFA) reduction, and increased saturated fatty acids in the colon. BBB may also be compromised in CAF-fed animals, as claudin-5 expression is reduced in the cerebral cortex. In addition, synaptophysin was decreased in the hippocampus, which may affect synaptic function. Our findings showed that Zn could not protect obese animals from intestinal dysbiosis. However, an increase in acetate levels was observed, which suggests a partial beneficial effect of Zn. Thus, Zn supplementation may not be sufficient to protect from obesity-related dysfunctions.

Keywords: obesity; cafeteria diet (CAF); inflammation; gut microbiota; short-chain fatty acid (SCFA); zinc (Zn)

Citation: Squizani, S.; Jantsch, J.; Rodrigues, F.d.S.; Braga, M.F.; Eller, S.; de Oliveira, T.F.; Silveira, A.K.; Moreira, J.C.F.; Giovenardi, M.; Porawski, M.; et al. Zinc Supplementation Partially Decreases the Harmful Effects of a Cafeteria Diet in Rats but Does Not Prevent Intestinal Dysbiosis. *Nutrients* **2022**, *14*, 3921. https://doi.org/10.3390/nu14193921

Academic Editors: Monica Dinu and Daniela Martini

Received: 29 August 2022
Accepted: 15 September 2022
Published: 22 September 2022

1. Introduction

Zinc (Zn) is a mineral widely distributed throughout the human body in small concentrations, and is involved in several biochemical and enzymatic processes [1–3]. It is an enzymatic cofactor for several enzymes that regulate the metabolism of carbohydrates, proteins, and lipids [1,3]. In addition, it exerts beneficial effects on immune function and inflammatory response, and contributes to barrier integrity and cognitive processing [2–7]. Moreover, it has already been established that obesity can affect the distribution of Zn in the body, reducing its systemic availability in obese individuals [8,9]. Previous studies have shown that Zn supplementation may improve anthropometric measurements and reduce inflammatory markers, insulin resistance, oxidative stress, and obesity-related neuroinflammation. Thus, Zn supplementation reduces metabolic dysfunction and may decrease neurological impairment related to obesity [10–12].

The consumption of processed energy-dense food, combined with overeating, urbanization, and a sedentary lifestyle in modern Western societies, is likely a major contributor to the obesity epidemic [13–15]. Obesity is associated with low-grade inflammation characterized by an increased production of proinflammatory cytokines, such as tumor necrosis factor-α (TNF-α), interleukin-1β (IL-1β), and interleukin-6 (IL-6) [16–18]. As a result of this inflammatory state, this complex and multifactorial disease is implicated in a higher prevalence of metabolic disorders, such as type 2 diabetes, cardio-metabolic disease, non-alcoholic liver disease, and also dysbiosis of the gut microbiota and neurological impairment [15,19].

Another consequence of low-grade systemic inflammation is the disruption of the intestinal epithelial barrier and blood–brain barrier (BBB), which leads to neuroinflammation [20,21]. The BBB comprises a tight junction complex and is assembled by several proteins, such as transmembrane, cytoplasmic attachment, and cytoskeleton proteins that form epithelial barriers [22,23]. Zonula occludens-1 (ZO-1), claudin-5, and occludin are recognized as markers of the integrity of the BBB due to their critical role in maintaining the integrity of tight junctions [24,25].

In addition to BBB dysfunction and neuroinflammation, neuroplasticity might be affected following inflammatory conditions, such as obesity. Brain-derived neurotrophic factor (BDNF) is a protein of the neurotrophin family that is essential for neuronal development, maintenance, survival, cognitive function, and synaptic plasticity. In addition, it is considered a key regulator of synaptic transmission, mainly in the hippocampus and neocortex [6,26]. Synaptophysin is a protein present in presynaptic vesicles. It participates in synaptic formation and neurotransmitter release. Additionally, it is used as a marker of synapsis distribution and density [27–29]. Thus, in the present study, we investigated BDNF and synaptophysin to assess whether consuming ultra-processed and hypercaloric food might affect neuroplasticity.

Regarding gut microbiota, there is a high variability of microorganisms in their composition depending on dietary habits [30,31]. Additionally, several diseases can be associated with a certain microbiota profile [32–34]. Previous studies have indicated that obesity is linked to disruptions in the gut microbiota composition, creating an imbalance in the microbial ecosystem defined as dysbiosis [35–37]. In addition, in healthy conditions, commensal anaerobic colonic bacteria produce functional metabolites by highly fermentable dietary fibers and resistant starch that benefit their host [38]. Short-chain fatty acids (SCFA) are functional metabolites represented mainly by acetate, butyrate, and propionate. These metabolites promote intestinal epithelium integrity, regulate immune function, modulate neurotransmission, and appear involved in neuroimmunoendocrine regulation [39,40]. Unbalanced SCFA concentrations due to consuming a Western diet can lead to dysbiosis. Subsequently, this leads to an increased intestinal permeability and induces low-grade systemic inflammation [19,41].

Although Zn's role in decreasing inflammation and improving cognition is already described, its effect on reversing the harmful consequences of consuming ultra-processed foods on microbiota composition and neuroplasticity is still unclear. The main aim of this study was to investigate the effects of Zn supplementation on the intestinal microbiota, intestinal barrier, and BBB in Wistar rats fed with a cafeteria diet (CAF). The CAF mimics the foods consumed by Western civilization by exposing the animals to highly palatable energy-dense foods with a high content of saturated fat and refined sugars and a large amount of food additives [42,43]. This model elicits obesity-related disorders, such as dysbiosis, leaky gut, metabolic disorders, low-grade systemic inflammation, neuroinflammation, and behavioral dysfunction in rodents [42,44]. As expected, the chronic consumption of the CAF for 20 weeks increased body weight, as shown in a previous study [10]. Thus, the present work also aimed to evaluate Zn's ability to reverse the mentioned harmful effects of obesity, since its supplementation started after obesity had already been developed.

2. Materials and Methods

2.1. Animals

Male three-month-old Wistar rats were obtained from the facility of the Federal University of Health Sciences of Porto Alegre (UFCSPA). Only male rats were used to avoid hormonal fluctuation and a possible impact on the results of the study. During the period of diet administration, one or two animals were housed per cage in a temperature-controlled environment (21 °C ± 2 °C) under a 12 h light/dark cycle and with ad libitum diet and water. The study was approved by the Institutional Animal Care and Use Committee of UFCSPA under the protocol number 570/18. All research procedures were designed to minimize the number of animals and suffering.

2.2. Experimental Groups and Diet

Twenty-eight animals were divided into four groups (n = 7/group), namely: (i) the control group (CT); (ii) control group + zinc (CT Zn); (iii) the cafeteria group (CAF), and (iv) the cafeteria + zinc group (CAF Zn). Standard chow (CT) or cafeteria (CAF, high fat and high calorie) diets were administered for 20 weeks. The cafeteria diet consisted of the alternating offer of three menus per week, containing bacon mortadella (Perdigão®, Itajaí, Brazil), strawberry-flavored biscuits (Isabela®, Bento Gonçalves, Brazil), chocolate biscuits (Isabela®, Bento Gonçalves, Brazil), pizza-flavored hot crackers (Parati®, São Lourenço do Oeste, Brazil), white chocolate (Harald®, Santana de Parnaíba, Brazil), orange-flavored soda (Sukita®, Sapucaia do Sul, Brazil), sausages (Alibem®, Porto Alegre, Brazil) offered concomitantly with ad libitum standard chow, and water. The alternation between cafeteria diet menus was performed every two days in order to maintain novelty. This diet's average energy supply was 4.5 Kcal/g (42% carbohydrates, 16% proteins, 42% lipids). The standard diet consisted of the standard chow Nuvilab CR-1 provided by the animal facility of UFCSPA (Nuvital®, Colombo, Brazil), which had an average energy supply of 3.4 Kcal/g (63% carbohydrates, 26% proteins, 11% lipids). The total energy content of the CAF was calculated based on manufacturer's information. Leftovers were evaluated, and food consumption was calculated at the end of the experiment. The average energy consumption of CT groups was 771.5 Kcal/day, while CAF groups consumed 963 Kcal/day [10].

2.3. Zn Treatment

Chelated zinc bisglycinate (Deconto Farma®, Porto Alegre, Brazil) was administered at a dose of 10 mg/Kg/day by gavage for 4 weeks from the 16th week of the diets. Gavage was performed daily between 2 and 3 pm, during the light cycle. Animals that did not receive zinc supplementation were also subjected to gavage with vehicle solution (water, 0.5 mL per animal). Animals were weighed weekly to determine weight gain. It should be noted that the treatment with zinc was started after obesity had already been induced. Zn content in the liver and cerebellum was determined by flame atomic absorption spectrometry (model AA 7000F; Shimadzu, Kyoto, Japan) equipped with a hollow cathode lamp and a deuterium lamp as a background corrector, and dosage was previously published [10].

2.4. Tissue, Blood and Feces Collection

At the end of the experimental period of 20 weeks, the animals were euthanized by decapitation without anesthesia. After that, feces were collected directly from the proximal colon. The proximal portion of the colon and brain were also collected. The brain was dissected out and the hippocampus and cerebral cortex were stored at −80 °C for further homogenization and analysis.

2.5. Histological Analysis

Histology of the proximal colon was performed for the general morphological analysis. Three to five animals per group were used for histology. For this, the tissue was fixed in paraformaldehyde, posteriorly embedded in a paraffin block, and cut transversely to a thickness of 4 μm. After this, the sections were stained with hematoxylin and eosin (HE). The slides were then analyzed under an EVOS microscope (EVOS Cell Imaging Systems, Thermo Fisher Scientific, Waltham, MA, USA). At least six measurements were taken per animal at 10× magnification. The crypt depth was measured using ImageJ software, and the size was normalized to a 500 μm scale [45].

2.6. Next-Generation Sequencing 16s rRNA

Stool samples were collected in sterile tubes and immediately stored at −20 °C. We also used samples from CT and CAF groups provided from a previous experiment using the same diet protocol, and the sample size was 8 animals for these groups. The genomic material of microbial DNA was obtained from approximately 200 mg of a fecal sample with a specific extraction kit (Microbiome DNA Purification kit, Invitrogen®, Waltham, MA, USA) following the manufacturer's instructions. After DNA extraction, the quantification of the DNA of each sample was performed by a NanoDrop spectrophotometer (Shimadzu, Japan). The libraries were prepared from an average of 5 to 10 μg of genomic material.

The hypervariable V3-V4 region from the 16S ribosomal RNA (rRNA) gene was amplified through PCR using genomic DNA (approximately 50 ng per reaction) and the following primer pair: 515F (5′-GTGCCAGCMGCCGCGGTAA-3′) and 806R (5′-GGACTACHVGGGT WTCTAAT-3′). In order to pool different samples in the same reaction, the primer-fusion method and each sample had a distinct barcode attached to the corresponding PCR product. The amplification was performed using Platinum™ PCR SuperMix High Fidelity (Invitrogen®, Waltham, MA, USA). The products were verified through electrophoresis in an agarose gel, purified with the AMPure XP PCR Purification Kit (Cat# A63881, Beckman Coulter Inc. Life Sciences, Indianapolis, IN, USA), quantified using Qubit™ dsDNA HS Assay Kit (Invitrogen®, Waltham, MA, USA) and subjected to emulsion PCR using the Ion PGM™ Hi-Q™ View OT2 Kit (Cat# A29900, Thermo Fisher Scientific, Waltham, MA, USA). Afterwards, the resulting enriched beads were sequenced in a next-generation sequencing (NGS) machine (Ion Torrent PGM™, Life Technologies, Carlsbad, CA, USA) using the Ion PGM™ Hi-Q™ View Sequencing Kit (Cat# A30044, Thermo Fisher Scientific, Waltham, MA, USA).

The 16S rRNA reads generated by high-throughput sequencing were submitted to a quality control analysis that retained sequences with a minimum length of 100 base pairs and trimmed the sequences to remove low-quality bases to obtain a minimum Phred score of 30 using PRINSEQ [46]. The remaining sequences were dereplicated and sorted by decreasing read abundance and filtered to exclude singletons using USEARCH v7.0.1090. Clusters were assembled using a minimum identity of 99% and chimeras were removed using the Ribosomal Database Project (RDP) reference database [47].

The taxonomic assignment was obtained using QIIME v1.7 [48]. Operational taxonomical units (OTUs) were selected based on 97% sequence similarity and taxonomic data were obtained using a classification algorithm with the 97% OTUs version of GreenGenes 13.8 [49]. Diversity index was calculated utilizing the vegan:ecological diversity package in R.

2.7. Determination of Fatty Acid Profile

The levels of saturated fatty acid (SFA) and SCFA were measured in the large intestine of the animals. The determination of SFA was performed by gas chromatography coupled to mass spectrometry (GC-MS). For this, 100 mg of intestinal tissue samples were homogenized in 1 mL of deionized water. After that, an aliquot of 50 μL of the homogenate was used for the extraction of the fatty acids with the addition of 950 μL of a chloroform/methanol (2:1, v/v) solution and one drop of concentrated hydrochloric acid

(HCl) was added to a plastic tubes, forming a single liquid phase. Subsequently, 200 μL of deionized water was added, promoting a separation between the chloroform phase (rich in lipids) and the methanol/aqueous phase. The samples were decanted for 5 min at room temperature. Then, the entire fraction containing chloroform was collected, dried under nitrogen flow at room temperature, and stored at −20 °C. Thereafter, the samples were finally resuspended in 30 μL of methanol and 2 μL aliquots and then injected into an analytical system consisting of a gas chromatograph model GC-2010 A PLUS coupled to the QP-2010 Ultra mass spectrometer (Shimadzu, Japan) adapted from G. DEMERS et al. [50]. The spectrometer was operated in scan mode (50–700 m/z) and the results obtained were verified by comparing the findings with those described in the mass spectral reference library of the National Institute of Standards and Technology (NIST).

The determination of SCFA in the large intestine was performed by liquid chromatography with tandem mass spectrometry (LC-MS/MS), according to a protocol developed in-house. For this, 50 μL of the homogenate, 20 μL NaOH (2 M) and 175 μL of HCl (2 M) were added to plastic tubes followed by vortexing for 30 s. The samples were then centrifuged for 6 min at 12,000× g. After that, an aliquot of 175 μL of the supernatant was collected and transferred to a glass vial. In order to promote fatty acid derivatization, 25 μL of 2,4-Dinitrophenylhydrazine (DNPH) was added. The flask was closed and incubated for 30 min at 40 °C. Then, 100 μL of the mixture was transferred to a vial and 100 μL of acetonitrile was added. Thereafter, an aliquot of 10 μL was injected into the analytical system equipment. The LC-MS/MS equipment consisted of a chromatographic system Nexera UFLC (Shimadzu, Japan) equipped with two binary pumps (LC-30AD), a column oven (CTO-30A), a diode array absorbance detector (SPD-M20A), and an automatic injector (SIL-30AMP) coupled with a quadrupole mass spectrometer model LCMS-8045 (Shimadzu, Japan). The results obtained were processed and evaluated using LabSolutions software (Shimadzu, Japan).

2.8. Western Blotting

The protein expression of ZO-1, claudin-5, BDNF, and synaptophysin was analyzed in the cerebral cortex, hippocampus and proximal colon. Tissues were processed and homogenized in a lysis buffer, and then samples were centrifuged for 10 min at 8000 rpm. After the protein quantification by Bradford protein assay, Laemmli buffer was mixed with 30 μg of proteins and heated at 90 °C for 2 min. Proteins were loaded and separated by SDS-PAGE (sodium dodecyl sulfate polyacrylamide gel electrophoresis) gel and later transferred to nitrocellulose membranes using a transfer system with semi-dry equipment (mini Trans-blot Electrophoretic Transfer Cell, BioRad, Hercules, CA, USA) at 110 V for 1–2 h. Membranes were incubated with 8% powdered milk in saline Tris buffer containing 0.1% tween 20 (T-TBS) for 90 min in order to block nonspecific binding. Membranes were incubated overnight (at 4 °C) with the primary antibody. Primary antibodies to TLR-4 (1:500, Cat# sc-293072, Santa Cruz Biotechnology®, Dallas, TX, USA), ZO-1 (1:500, Cat# 61-7300, Invitrogen®, Waltham, MA, USA), claudin-5 (1:1000, Cat# ABT45, Merk®, Kenilworth, NJ, USA), BDNF (1:1000, Cat# BS-4989R, Thermo Fisher Scientific, Waltham, MA, USA), and synaptophysin (1:500, Cat# MA5-14532, Thermo Fisher Scientific, Waltham, MA, USA) were used. Then, the membranes were incubated with secondary anti-mouse (Cat# A9044, Sigma-Aldrich, St. Louis, MI, USA) or anti-rabbit (Cat# AP132P, Merk, Kenilworth, NJ, USA) antibodies for 2 h at room temperature. All incubations were performed under constant agitation, and between each incubation membranes were washed with T-TBS. A chemiluminescence reaction was performed to detect the labeled proteins and the images were obtained using the ChemiDoc MP photodocumenter (Bio-rad Laboratories, Hercules, CA, USA). The results of each membrane were relative to the values found by incubating them with the primary antibody anti-β-actin (1:500, Cat# sc-47778 horseradish peroxidase, Santa Cruz Biotechnology®, Dallas, TX, USA) or with nonspecific bands [51]. To avoid inter-assay variations, samples from all experimental groups were processed in parallel.

2.9. Statistical Analysis

Data are expressed as the mean and standard error of the mean (SEM). Two-way analysis of variance (ANOVA) was performed. The main factors were diet (CT or CAF) and supplementation (vehicle or zinc). The Bonferroni test was used for post hoc analysis. The value of $p < 0.05$ was considered as indicative of statistical significance. All analyses were performed using the GraphPad Prism® 9.0 program (San Diego, CA, USA).

3. Results

3.1. Colon Morphology

The analysis of the crypt depth of the colon is shown in Figure 1. Crypt depth was lower in CAF-fed animals when compared with CT-fed animals (diet effect: F1,146 = 20.11, $p < 0.0001$). Zn supplementation did not exert any effect on crypt depth measurements.

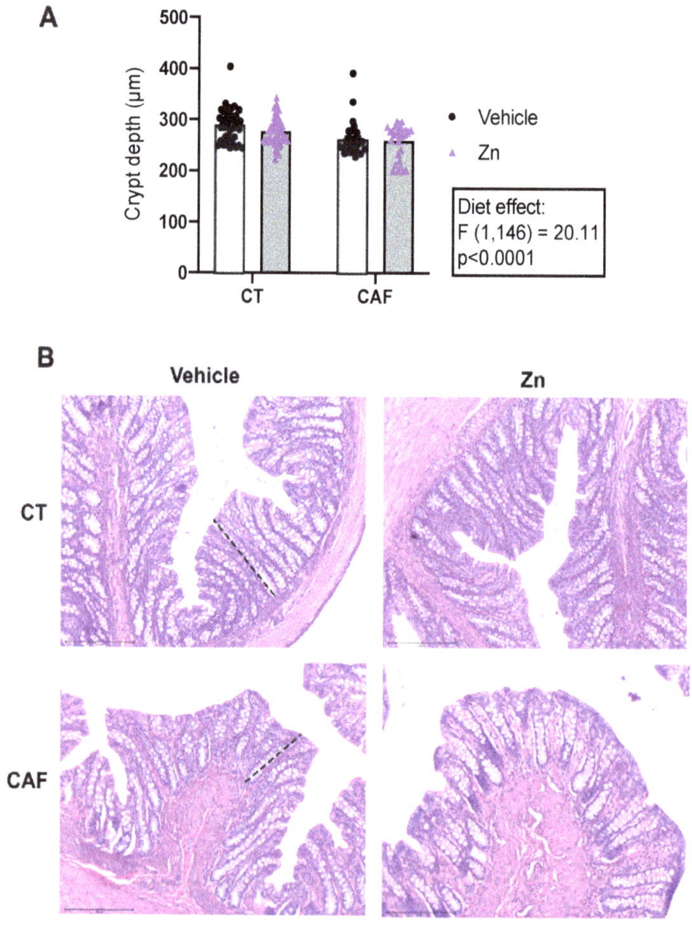

Figure 1. Crypt depth measurement in the proximal portion of the colon of control (CT) and cafeteria diet (CAF)-fed rats. (**A**) Crypt depth was reduced following CAF. (**B**) Representative images of each group of the colon stained with hematoxylin and eosin. The dashed lines indicate crypt length. Scale bar: 200 µm. The text box indicates significant differences shown by two-way ANOVA regarding the effects of the diet (CT and CT + Zn vs. CAF and CAF + Zn) and Zn treatment (CT and CAF vs. CT + Zn and CAF + Zn). n = 3–5 animals/group, 10 measurements per animal.

3.2. Composition of the Gut Microbiota

We performed the next-generation sequencing of the V4 16S rRNA region for the taxonomic identification of intestinal bacterial composition. The heatmap (Figure 2A) shows the number of reads of the main phyla and genera. Significant differences regarding diet and Zn supplementation (two-way ANOVA main effects) and the multiple comparisons (post hoc Bonferroni test) are described below.

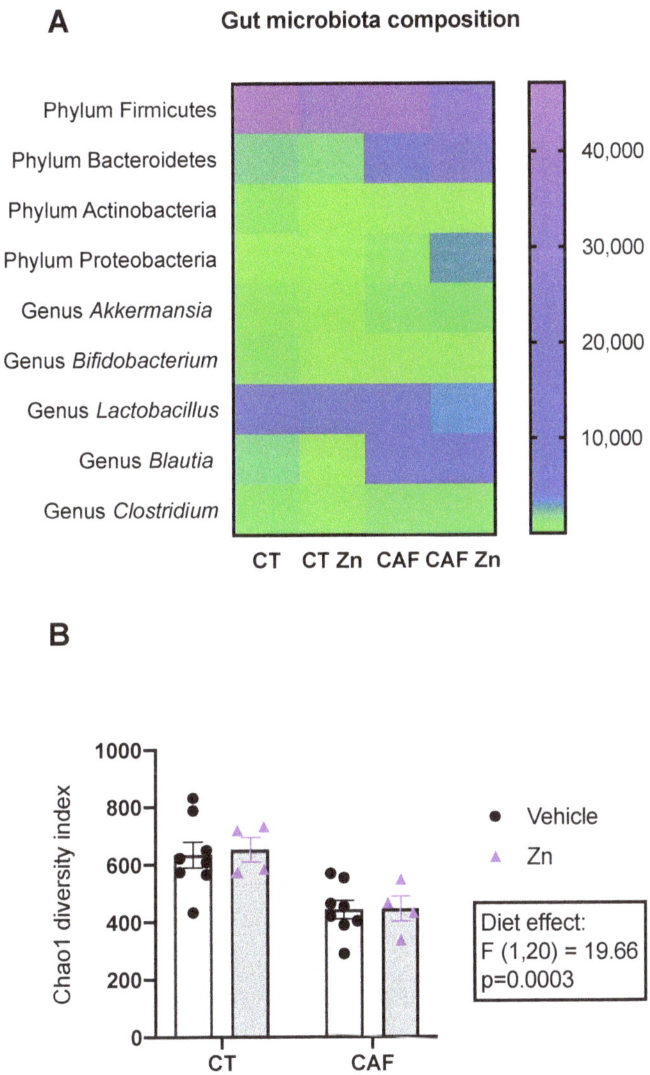

Figure 2. Composition of the gut microbiota of control (CT) and cafeteria diet (CAF)-fed animals. (**A**) Heatmap shows the phylum and genera levels of the main bacterial community found in the fecal samples of the intestinal colon. The range of colors, from green to lilac, indicates the abundance of each phylum or genera per group. (**B**) Chao1 diversity index. The text box indicates significant differences shown by two-way ANOVA regarding the effects of the diet (CT and CT + Zn vs. CAF and CAF + Zn). n = 4–8 animals/group.

We found a decrease in the Firmicutes phylum in CAF-fed rats (diet effect, $F1,9 = 13.55$, $p = 0.0016$). Additionally, the post hoc test showed a significant decrease in Firmicutes bacteria in CAF + Zn compared to the CAF group ($p = 0.0307$). On the other hand, Bacteroidetes phylum was significantly increased following CAF (diet effect, $F1,16 = 40.83$, $p < 0.0001$), but it also showed a Zn supplementation effect ($F1,16 = 5.596$, $p = 0.0310$) and an interaction between diet and Zn ($F1,16 = 7.121$, $p = 0.0168$). The Bonferroni post hoc test also showed an increase in CAF + Zn compared to the CAF group ($p = 0.0079$). These findings show that Zn supplementation was able to decrease the abundance of Firmicutes bacteria while increasing Bacteroidetes phylum in CAF + Zn group. There was no difference in the abundance of Actinobacteria phylum among the groups. The Proteobacteria phylum showed an interaction between diet and Zn ($F1,17 = 5.991$, $p = 0.0255$), a diet effect ($F1,17 = 65.03$, $p < 0.0001$), and a Zn effect ($F1,17 = 5.838$, $p = 0.0272$). Additionally, CAF + Zn showed a significantly higher abundance of Proteobacteria when compared to the CAF group ($p = 0.0101$). These findings show, once again, that Zn supplementation was able to increase the abundance of Proteobacteria phylum only in CAF-fed rats.

Regarding the genus taxonomic level, we analyzed the *Akkermansia*, *Bifidobacterium*, *Blautia*, *Clostridium* and *Lactobacillus*. There was an increase in the abundance of *Akkermansia* (diet effect, $F1, 18 = 6.520$, $p = 0.0200$) and *Blautia* (diet effect, $F1,18 = 36.07$, $p < 0.0001$) in CAF-fed rats, with no effect of Zn supplementation for either genus. Concerning *Lactobacillus*, we found an interaction between diet and Zn ($F1,17 = 6.583$, $p = 0.0201$), diet effect ($F1,17 = 37.34$, $p < 0.0001$) and Zn effect ($F1,17 = 13.80$, $p = 0.0017$). A significant decrease was also observed in *Lactobacillus* abundance in CT + Zn compared to the CT group ($p = 0.0008$). These findings show that Zn supplementation was able to reduce the abundance of *Lactobacillus* only in lean rats. There was no difference in the abundance of *Bifidobacterium* and *Clostridium* genera among the groups.

The diversity of the gut microbiota based on the Chao1 index (Figure 2B) was also analyzed, which showed a reduction in the CAF-fed groups (diet effect, $F1,20 = 19.66$, $p = 0.0003$) with no effect of Zn supplementation.

3.3. SCFA, MCFA, LCFA and VLCFA Profile in the Intestine

The determination of SCFA in the intestine is shown in Figure 3. The acetate concentration (Figure 3A) showed an interaction between diet and Zn ($F1,19 = 4.819$, $p = 0.0408$) and a Zn supplementation effect ($F1,19 = 4.753$, $p = 0.0420$). In the post hoc comparisons, an increase in acetate levels was observed in the CAF + Zn group compared to CAF group ($p = 0.0188$). These findings demonstrate that Zn supplementation in obese rats could return the acetate concentration to the level of the lean animals. However, there was a diet effect ($F1,19 = 5.238$, $p = 0.0337$) in butyrate levels (Figure 3B) which was decreased in CAF-fed rats with no effect of Zn supplementation. We noticed no difference in the concentration of propionate (Figure 3C) and isobutyrate (Figure 3D) among the groups. The valerate concentration (Figure 3E) increased in Zn-supplemented rats (Zn effect, $F1,21 = 4.754$, $p = 0.0407$), with no effect of diet. Furthermore, we found a decreased concentration of isovalerate (Figure 3F) (diet effect, $F1,23 = 6.526$, $p = 0.0177$) in CAF-fed rats, with no effect of Zn supplementation.

The determination of medium-chain fatty acid (MCFA), long-chain fatty acid (LCFA) and very long-chain fatty acid (VLCFA) concentration in the intestinal tissue (μg/mg) is shown in Table 1. Regarding the MCFAs, we analyzed the concentration of caprylic, decanoic, octanoic, lauric, and undecanoic acids. There was an increase in decanoic (diet effect, $F1,23 = 8.969$, $p = 0.0065$), octanoic (diet effect, $F1,23 = 5.175$, $p = 0.0326$) and lauric (diet effect, $F1,23 = 11.49$, $p = 0.0025$) fatty acids in CAF-fed rats, with no effect of Zn supplementation. The undecanoic fatty acid showed an interaction effect between diet and Zn ($F1,23 = 5.716$, $p = 0.0254$); in this case, post hoc analysis showed that the CT + Zn group was higher than the CT group ($p = 0.0285$). Additionally, we observed a decreased concentration of caprylic acid (diet effect, $F1,22 = 4.387$, $p = 0.0479$) in CAF-fed rats, with no effect of Zn supplementation. Moreover, we determined the concentration of the LCFAs such as elaidic, heptadecanoic, linoleic, myristic, myristoleic, palmitic, pentadecanoic, stearic, and tridecanoic. Similarly to the findings of the MCFAs, we noticed an increase in the concentration of elaidic (diet effect, F1,23 = 34.12, $p < 0.0001$), linoleic (diet effect, F1,23 = 14.46, $p = 0.0009$), myristic (diet effect, F1,23 = 32.76, $p < 0.0001$), palmitic (diet effect, F1,19 = 10.15, $p = 0.0049$), and stearic (diet effect, F1,21 = 6.600, $p = 0.0179$) acids in response to CAF, with no effect of Zn. Additionally, the myristoleic acid showed an interaction effect between diet and Zn (F1,20 = 6.217, $p = 0.0215$). There was no difference in the concentration of heptadecanoic, pentadecanoic, and tridecanoic acids among the groups. As for VLCFA, we analyzed the concentration of behenic, heneicosanoic, lignoceric, and tricosanoic acids. Again, an increase in behenic (diet effect, F1,23 = 26.61, $p < 0.0001$) and heneicosanoic (diet effect, F1,23 = 13.62, $p = 0.0012$) acids was observed in CAF-fed animals, with no effect of Zn supplementation. No differences were seen in the concentration of lignoceric and tricosanoic acids among the groups.

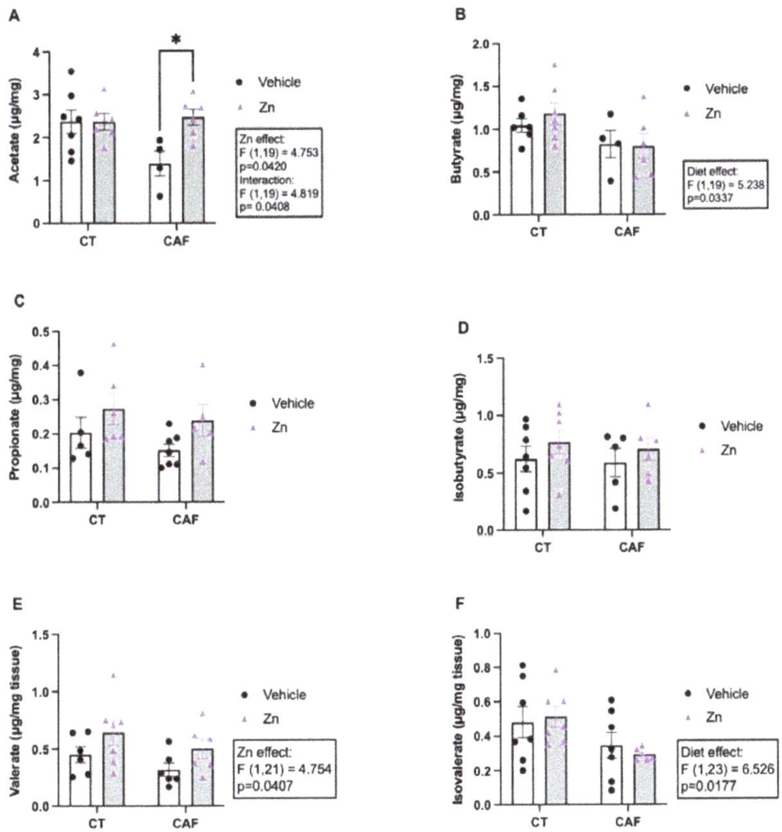

Figure 3. Short-chain fatty acid (SCFA) concentration in the proximal portion of the colon of control (CT) and cafeteria diet (CAF)-fed rats. (**A**) Acetate, (**B**) butyrate, (**C**) propionate, (**D**) isobutyrate, (**E**) valerate, and (**F**) isovalerate levels. The text box indicates significant differences shown by two-way ANOVA regarding the effects of diet (CT and CT + Zn vs. CAF and CAF + Zn), Zn supplementation (CT and CAF vs. CT + Zn and CAF + Zn), and the interaction between diet and Zn supplementation. Multiple comparisons were performed by Bonferroni post hoc test and significant differences are shown by the asterisk (* $p < 0.05$). n = 4–7 animals/group.

Table 1. Medium-chain fatty acid (MCFA), long-chain fatty acid (LCFA) and very long-chain fatty acid (VLCFA) concentrations in the proximal portion of the colon (μg/mg).

Fatty Acid	Classification	CT	CT + Zn	CAF	CAF + Zn	Two-Way ANOVA Results			Bonferroni's Post Hoc Test
						Interaction	Diet Effect	Zn Effect	
MCFA (6–12 carbons)									
Caprylic	Saturated	9.36 ± 4.14	5.78 ± 1.00	3.36 ± 0.93	3.67 ± 0.90		0.0479		
Decanoic	Saturated	0.32 ± 0.03	0.23 ± 0.04	0.65 ± 0.17	0.51 ± 0.10	ns	0.0065	ns	ns
Octanoic	Saturated	0.11 ± 0.03	0.09 ± 0.03	0.29 ± 0.08	0.19 ± 0.09	ns	0.0326	ns	ns
Lauric	Saturated	1.86 ± 0.31	1.44 ± 0.36	8.207 ± 2.05	5.72 ± 2.54	ns	0.0025	ns	ns
Undecanoic	Saturated	19.35 ± 3.12	11.07 ± 2.25	16.99 ± 1.42	19.49 ± 1.62	0.0254	ns	ns	CT vs. CT + Zn (0.0285)
LCFA (13–21 carbons)									
Elaidic	Unsaturated trans fatty acid	61.41 ± 13.43	48.10 ± 15.02	244.9 ± 40.66	223.6 ± 44.51	ns	0.0001	ns	ns
Heptadecanoic	Saturated	3.71 ± 0.91	3.73 ± 0.99	3.82 ± 0.93	3.38 ± 1.24	ns	ns	ns	ns
Linoleic	Unsaturated	110.7 ± 18.36	76.81 ± 23.58	458.9 ± 114.2	322.6 ± 109.3	ns	0.0009	ns	ns
Myristic	Saturated	2.37 ± 0.85	2.50 ± 0.89	8.48 ± 0.89	5.77 ± 0.47	ns	0.0001	ns	ns
Palmitic	Saturated	28.83 ± 7.76	39.78 ± 9.52	47.52 ± 9.32	86.56 ± 27.04	ns	0.0049	ns	ns
Pentadecanoic	Saturated	14.21 ± 2.05	13.43 ± 1.07	18.61 ± 5.09	12.84 ± 1.22	ns	ns	ns	ns
Stearic	Saturated	59.69 ± 11.16	60.76 ± 10.34	34.22 ± 5.74	39.11 ± 4.99	ns	0.0179	ns	ns
Myristoleic	Unsaturated	29.96 ± 7.17	33.63 ± 7.39	52.10 ± 7.96	40.38 ± 10.15	0.0215	ns	ns	ns
Tridecanoic	Saturated	6.96 ± 1.52	4.98 ± 0.48	5.90 ± 1.03	5.65 ± 0.97	ns	ns	ns	ns
VLCFA (≥22 carbons)									
Behenic	Saturated	31.91 ± 7.10	32,14 ± 7.29	122.0 ± 33.77	182.6 ± 32.56	ns	0.0001	ns	ns
Tricosanoic	Saturated	25.90 ± 6.64	23.64 ± 3.88	43.57 ± 13.66	38.23 ± 13.53	ns	ns	ns	ns
Heneicosanoic	Saturated	11.31 ± 2.75	6.23 ± 2.05	27.86 ± 5.99	19.42 ± 4.10	ns	0.0012	ns	ns
Lignoceric	Saturated	12.68 ± 5.60	33.14 ± 7.72	19.64 ± 7.05	16.92 ± 5.35	ns	ns	ns	ns

Values are expressed as mean ± SEM. p values are described when significant differences were found. ns (non-significant); CT (control diet); CAF (cafeteria diet); Zn (zinc). n = 5/7 animals/group.

3.4. Blood–Brain Barrier (BBB) and Intestinal Barrier Integrity Components

The protein expressions of ZO-1 and claudin-5 in the cerebral cortex, hippocampus, and proximal portion of the intestinal colon are shown in Figure 4. There was no difference in the ZO-1 expression in the cerebral cortex (Figure 4A), hippocampus (Figure 4C), or intestinal colon (Figure 4E) among the groups. However, there was a decrease in claudin-5 protein expression in the cerebral cortex (Figure 4B) of CAF-fed animals (diet effect, $F1,17 = 11.16$, $p = 0.0039$), with no effect of Zn supplementation. Intriguingly, we also observed a diet effect ($F1,17 = 6.889$, $p = 0.0178$) in claudin-5 expression in the intestinal colon (Figure 4F) which was increased in CAF-fed rats, with no effect of Zn supplementation. Claudin-5 did not change among groups in the hippocampus (Figure 4D).

3.5. Synaptic and Neuroplasticity Markers

The protein expressions of synaptophysin and brain-derived neurotrophic factor (BDNF) in the cerebral cortex and hippocampus are shown in Figure 5. We found no difference in the protein expression of synaptophysin (Figure 5A) in the cerebral cortex among the groups. However, there was a lower expression of synaptophysin in the hippocampus (Figure 5B) of CAF-fed animals (diet effect, $F1,18 = 16.69$, $p = 0.0007$), with no effect of Zn supplementation. Moreover, BDNF expression in the hippocampus (Figure 5C) did not change among the groups.

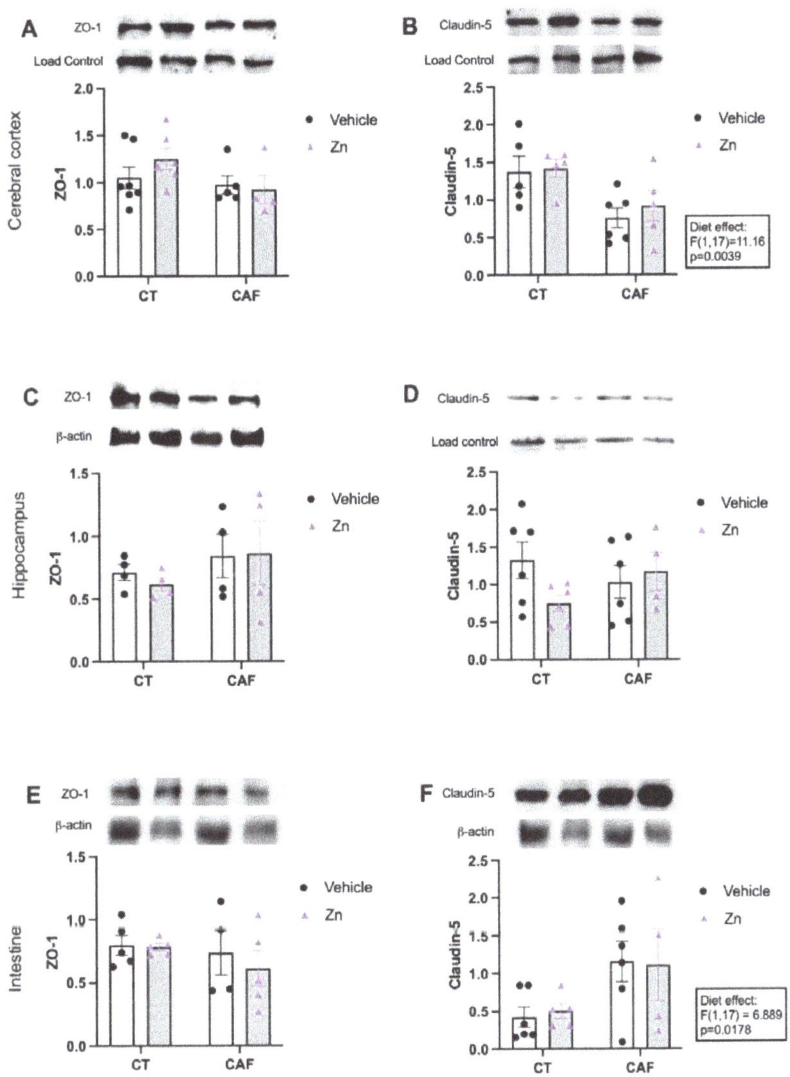

Figure 4. Protein expression of zonula occludens 1 (ZO-1) and claudin-5 in cerebral cortex (**A**,**B**), hippocampus (**C**,**D**), and proximal portion of the intestinal colon (**E**,**F**) in control (CT) and cafeteria diet (CAF)-fed rats. Representative bands of each group are shown on the top of the graphs. β-actin or nonspecific bands were used as a loading control. The text box indicates significant differences shown by two-way ANOVA regarding effects of the diet (CT and CT + Zn vs. CAF and CAF + Zn). n = 4–7 animals/group.

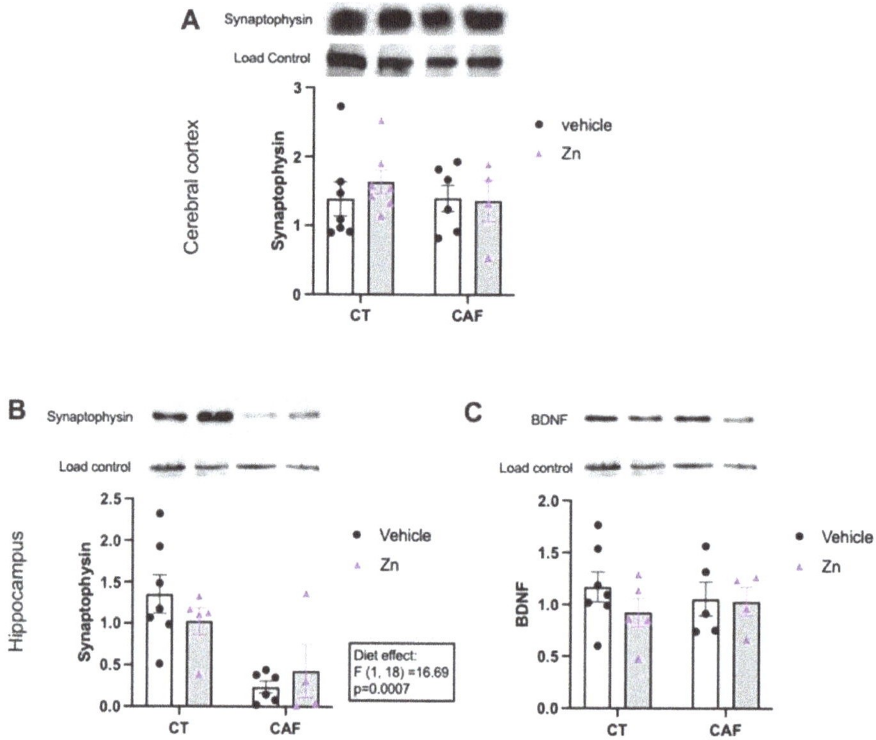

Figure 5. Protein expression of synaptophysin and brain-derived neurotrophic factor (BDNF) in the cerebral cortex (**A**) and hippocampus (**B**,**C**) of control (CT) and cafeteria diet (CAF)-fed rats. Representative bands of each group are shown on the top of the graphs. β-actin or nonspecific bands were used as a loading control. The text box indicates significant differences shown by two-way ANOVA regarding the effects of diet (CT and CT + Zn vs. CAF and CAF + Zn). n = 4–7 animals/group.

4. Discussion

We have previously shown that Zn supplementation could decrease metabolic dysfunction and neuroinflammation and improve memory in obese rats fed with a CAF [10]. Here, we showed that Zn changed the microbiota composition but did not enhance its diversity in obese animals. Additionally, Zn was able to increase acetate levels in obese rats. On the other hand, Zn did not affect the intestinal or cerebral barriers' integrity or the expression of neuroplasticity markers, such as synaptophysin and BDNF. Thus, we show that when obesity is severe, as in the CAF experimental model, the effects of Zn supplementation may not be as beneficial as in other populations, such as normal body weight or even overweight people. These findings might help to delimit the recommendation of Zn supplementation to these groups.

There are some controversial findings regarding intestinal morphology in diet-induced obesity (DIO) models. Some studies reported an increase in the crypt depth or villi height in obesity [52–54]. Zhou and collaborators found this result in the duodenum and jejunum but found no differences in the colons of mice [55]. It was also demonstrated that titanium dioxide, a coloring food additive, alters the functional absorption of nutrients by decreasing the number of microvilli and, consequently, reducing absorptive area. In addition, the same study showed that titanium dioxide significantly decreased zinc transport [56]. Our findings showed that chronic ingestion of the CAF exerts a decrease in the crypt depth of the colon in adult rats, with no effect of Zn supplementation. These findings may be related to the food additives present in the CAF.

In the present study, we also found that the CAF modulated the intestinal microbiota composition and induced changes on its profile. Although there is a lack of definition of a "healthy microbiota" due to the high variability of microorganisms among the microbiota composition, there are some patterns at taxonomic levels of phyla in all vertebrates [57,58]. In fact, several diseases are associated with a certain microbiota profile [32–34].

We demonstrated that CAF induced a decrease in Firmicutes and an increase in Bacteroidetes phyla abundance. In obese rats, Zn decreased Firmicutes while increasing Bacteroidetes. There is some divergence among studies regarding the abundance of these phyla. Although some reports demonstrated an increase in Firmicutes and a decrease in Bacteroidetes in obesity [42,59,60], others showed the opposite, or no difference in either phylum [44,61]. However, the divergence in the literature may be related to different diet protocols. It has already been shown that diet composition, including macronutrient and micronutrient distribution, or even the presence of food additives, can modulate the gut microbiota profile [42,62,63]. Furthermore, non-nutritive sweeteners, including acesulfame potassium, saccharin, and sucralose, exert a strong bacteriostatic effect. The consumption of these sweeteners can selectively inhibit the survival taxa of some bacterial populations, consequently changing gut microbiota homeostasis [64]. Since CAF consists of several ultra-processed foods and some of them have in their composition these sweeteners, it could have influenced our findings. In addition, the increase in the abundance of Bacteroidetes by Zn supplementation in CAF-fed rats seems interesting since it consists of several bacteria involved with SCFA production [30]. Additionally, the higher proportion of Firmicutes phyla in obese rodents fed with a Western diet may be associated with an increase in the abundance of *Clostridium ramosum*, which has been linked to metabolic syndrome [65]. Therefore, based on these findings, a protective effect of Zn supplementation could be suggested in CAF-fed rats.

Furthermore, we found an increase in the abundance of Proteobacteria phylum followed by CAF. Interestingly, obese animals that received Zn had an even higher abundance of this phylum. It is suggested that the increase in Proteobacteria is associated with an imbalance of the gut microbiota. Hence, it is a potential marker for dysbiosis [66]. It has already been reported that the increased colonization of Proteobacteria is associated with a high-fat diet (HFD) in both human and rodent models [60,67,68].

Among the analyzed genera, we found a decrease in the abundance of *Lactobacillus* following CAF and Zn supplementation, but Zn diminished *Lactobacillus* only in lean animals. *Lactobacillus* are beneficial bacteria with health-promoting properties which have been reported as contributors to host metabolism [68,69]. *Akkermansia* is a member of the Verrucomicrobia phylum that corroborates the maintenance of metabolic homeostasis. It increases goblet cell density, stimulates mucin production, and improves intestinal barrier integrity [70,71]. Indeed, studies have described the health-promoting effects of *Akkermansia* on energy metabolism and metabolic functions such as insulin sensitivity, dyslipidemia, and even cognition [72–74]. Interestingly, at the same time that *Akkermansia* improves mucin production, it appears to be also involved in mucin degradation [66,75]. *Akkermansia* is highly responsive to diet change. Previous DIO studies reported a higher prevalence of this genus [73,76] or even a decrease in its abundance [73,77]. Here, we observed an increase in *Akkermansia* abundance following CAF. Thus, the mechanisms related to alterations in the *Akkermansia* population warrant further investigation. As for *Blautia* genera, we found an increase in its abundance following CAF. This finding is in accordance with previous studies on obesogenic diets both in humans and rodents [78,79]. Again, while some studies reported deleterious effects linked with a higher abundance of *Blautia* [80,81], others reported beneficial effects [82].

Concerning the diversity taxa, it is consistently described that higher diversity characterizes healthy gut microbiota [58,83]. Indeed, lower microbiota richness is associated with an obesogenic diet in both humans and animals [42,79]. We found a reduction in alpha diversity in CAF-fed rats, in accordance with other studies [42,44,84]. However, we did not find an effect of Zn supplementation in this analysis. Previous studies showed the divergent effects of Zn on microbiota diversity. While some demonstrated that Zn supplementation tends to decrease overall bacterial richness [85], others showed that low dietary Zn supplementation had no effect [86]. In summary, our results suggest that Zn supplementation impacts bacterial communities by supporting or restricting the growth of selected taxa, corroborating with previous studies [75,85]. However, CAF is composed of several industrialized foods, and our study does not provide evidence of the impact of a particular diet component on the microbiota, which is a limitation. Future studies should address this point.

SCFA are microbiota metabolites that modulate several host functions. Butyrate interferes in host metabolism and immunity by regulating satiety [87,88], exerting an anti-inflammatory role [32,89], maintaining epithelial integrity in the intestine [39], and also providing an energy source for colonocytes [89]. Our findings showed a decrease in butyrate levels in CAF-fed animals, with no effect of Zn supplementation. The lower concentration of butyrate is in accordance with the decreased abundance of the Firmicutes phylum since butyrate-producing species belong to this phylum [62,90].

Acetate represents the most abundant SCFA in the body, being found in higher concentrations in blood and peripheral tissues [88,91]. It is a metabolite synthesized by the phylum Bacteroidetes, and is essential for the growth of several bacteria [62]. Acetate is the main SCFA synthesized by intestinal bacteria and exerts an important effect on the body's energy regulation [58,92]. Our findings demonstrated a significantly lower acetate concentration in the CAF group compared to the CAF + Zn group. Hence, Zn supplementation was capable of reestablishing the acetate concentration in CAF + Zn animals. Once again, a higher concentration of acetate is in accordance with the enhanced abundance of Bacteroidetes phylum observed in CAF + Zn rats. A previous pre-clinical study reported an increase in the presence of Gram-negative facultative anaerobic bacteria and the colonic concentration of SCFAs followed by Zn supplementation [93,94]. We also showed lower concentrations of isovalerate following CAF and an increase in valerate concentration in rats supplemented with Zn. A beneficial effect of valerate in intestinal epithelial integrity and an improvement in gastrointestinal function are reported [95]. Thus, Zn may exert a beneficial role by increasing valerate and acetate levels.

Regarding the concentration of SFAs in the colon, we observed an increase in several MCFAs, LCFAs, and VLCFAs following the CAF. The undecanoic and myristoleic fatty acids showed an interaction effect between diet and Zn supplementation. It has already been established that Zn is involved in several biochemical and enzymatic processes. Additionally, it is a part of desaturases and elongases and appears to influence the fatty acid profile. Changes in the amount of saturated fatty acids have already been reported following Zn supplementation [96]. Overall, we expected a higher concentration of these fatty acids due to CAF composition, characterized by a high saturated fatty acid content. Our findings are in accordance with previous studies concerning the quality and amount of dietary fat that modulates the gut microbiota composition and impacts metabolic health [97,98]. Moreover, it has already been reported that a high intake of dietary saturated fatty acids contributes to the establishment of low-grade systemic inflammation and enhances the risk of developing obesity and cancer-related diseases, and even impacts the onset of Alzheimer's and other dementias [78,99].

In addition, SFAs act as non-microbial agonists of TLR-4, sharing a common mechanism of action with lipopolysaccharide (LPS), a constituent of intestinal bacteria related to endotoxemia [100]. Moreover, it was demonstrated that saturated fatty acids stimulate the nuclear transcription factor kappa B (NF-κB) pathway in a TLR-4-dependent manner, leading to an increase in pro-inflammatory cytokine synthesis such as IL-6 and TNF-α [96,101]. Among the analyzed fatty acids in the present study, palmitate and stearate are the main representatives of saturated fatty acids in the human organism, and were both increased in CAF-fed rats [96].

The excessive consumption of SFAs has several consequences on the intestinal gut microbiota, including the dysbiosis and dysfunction of the gut barrier, enhancing the proinflammatory state [13,31]. Here, we analyzed the protein expression of ZO-1 and claudin-5 in the cerebral cortex, hippocampus, and intestine. There was no change in ZO-1 expression in these three regions. Mice that received an HFD for 14 weeks showed a decrease in the protein expression of claudin-5 and occludin in the frontal cortex, while ZO-1 was not affected by diet [102]. Regarding the claudin-5 protein expression in the hippocampus, we did not find differences between the groups. However, our findings showed a decrease in claudin-5 protein expression in the cerebral cortex of CAF-fed animals, with no effect of Zn supplementation. Intriguingly, we also observed an effect of diet in the claudin-5 expression in the intestine, which was increased in CAF-fed rats, with no effect of Zn supplementation. It has already been reported that the composition and distribution of epithelial claudins vary spatially along the length of the intestine. In intestinal inflammatory disorders, there is an overexpression of claudin-1, -2, and -18, simultaneously, with the downregulation of claudin-3, -4, -5, -7, -8, and -12. Such changes can modify epithelial barrier function and mucosal homeostasis [24,25]. Thus, we speculate that the damage caused by CAF may alter the expression of other proteins, such as occludins or other claudins, which were not investigated here. Claudin-5 may increase to compensate for other alterations in the tight junction complex. Although we found no effect of Zn supplementation on the expression of the tight junction proteins, it has already been described that Zn may have a protective effect on the intestinal barrier. The mechanisms are not fully elucidated, but zinc-mediated protection seems to be due to the stimulation of GPR39, a zinc-sensing receptor involved in barrier regulation [103]. Furthermore, Zn significantly enhanced the barrier function in an in vitro model [104].

The central nervous system is another important target of the harmful effects of obesity. It has already been described that there is a reciprocal interaction between gut microbiota and the central nervous system. This bidirectional communication, known as the gut–brain axis, is regulated via immune and neuroendocrine signals [105–107]. Several studies have reported the association between gut microbiota dysbiosis and neurologic disorders, such as cognitive impairments, Alzheimer's disease [99], Parkinson's disease [108], autism spectrum [109], and mood disorders [110]. Thus, we evaluated the effect of CAF on synaptophysin and BDNF as markers of neuroplasticity to further investigate the relationship between gut and brain in obesity.

Although the mechanisms are not fully elucidated, it appears that Zn implicates in the modulation of neurotrophic signaling [111–113]. In this study, we did not find differences in BDNF expression in the hippocampus. However, studies in rodents reported that Zn supplementation could increase the expression of BDNF in the hippocampus and prevent cognitive impairment. Additionally, it seems that the effectiveness of Zn supplementation in increasing BDNF in obese mice was dose-dependent [114,115]. We have previously shown that Zn diminished neuroinflammation and improved memory in obese rats. Thus, this mechanism of neuroprotection may be related to the reduction of inflammation without affecting BDNF expression. On the other hand, Zn supplementation in obese individuals increased the serum concentration of BDNF [116]. However, studies in humans have a limitation due to the circulating BDNF not necessarily reflecting its availability and concentration in encephalic structures [117,118].

We also evaluated synaptophysin as a marker of neuroplasticity. Although changes in synaptophysin expression in the cerebral cortex were not seen, the CAF reduced its expression in the hippocampus. A decreased synaptophysin expression in the hippocampus of rodents with an obesity-induced cognitive deficit has already been reported [119]. Cai and collaborators demonstrated that hyperglycemia and hyperlipidemia induced by obesity might enhance the hippocampal endoplasmic reticulum stress and impair the expression of BDNF and synaptophysin, consequently leading to memory and learning dysfunction in rats [120]. However, it was previously shown that a low dose of Zn was efficient in increasing synaptophysin following a high-fat diet in mice [115].

5. Conclusions

In summary, we showed that the chronic consumption of the CAF causes dysbiosis, morphological change, and a decrease in SCFA levels in the colon, along with increased saturated fatty acids. The BBB may also be compromised in CAF-fed animals since claudin-5 expression is reduced in the cerebral cortex. Additionally, synaptophysin was decreased in the hippocampus, which might affect synaptic function. Zinc supplementation did not protect from dysbiosis. However, it increased acetate levels, which suggests a partial beneficial effect of Zn. In the brain, Zn did not show a neuroprotective role in the present study. Nevertheless, in accordance with previous studies, we have demonstrated a beneficial role of Zn in neuroinflammation and cognitive function in obesity. Thus, the consumption of ultra-processed foods, as provided by the CAF, causes severe obesity. In this condition, Zn might not be sufficient to protect or revert obesity-related dysfunctions such as dysbiosis.

Author Contributions: Conceptualization, R.P.G. and M.P.; methodology, R.P.G., A.K.S., S.E., T.F.d.O. and J.C.F.M.; formal analysis, S.S., J.J., F.d.S.R. and M.G.; investigation, S.S., J.J., M.F.B. and F.d.S.R.; writing—original draft preparation, S.S.; writing—review and editing, S.S. and R.P.G.; supervision, R.P.G., M.G. and M.P.; project administration, R.P.G.; funding acquisition, R.P.G. All authors have read and agreed to the published version of the manuscript.

Funding: This research was funded by Research Support Foundation of Rio Grande do Sul (FAPERGS) funding number 17/2551-0001431-6, National Council for Scientific and Technological Development (CNPq) and Coordination for the Improvement of Higher Education Personnel (CAPES).

Institutional Review Board Statement: The animal study protocol was approved by the Institutional Animal Care and Use Committee of the Federal University of Health Sciences of Porto Alegre (protocol code 570/18 and date of approval 13/05/2018).

Informed Consent Statement: Not applicable.

Data Availability Statement: The data that support the findings of this study are available on reasonable request from the corresponding author.

Conflicts of Interest: The authors declare no conflict of interest.

References

1. Mangray, S.; Zweit, J.; Puri, P. Zinc Deficiency in Cirrhosis: Micronutrient for Thought? *Am. J. Dig. Dis.* **2015**, *60*, 2868–2870. [CrossRef] [PubMed]
2. Skalny, A.V.; Aschner, M.; Tinkov, A.A. Zinc. *Adv. Food Nutr. Res.* **2021**, *96*, 251–310. [CrossRef] [PubMed]
3. Mehri, A. Trace Elements in Human Nutrition (II)—An Update. *Int. J. Prev. Med.* **2020**, *11*, 2. [PubMed]
4. Prasad, A.S. Discovery of human zinc deficiency: 50 years later. *J. Trace Elem. Med. Biol.* **2012**, *26*, 66–69. [CrossRef] [PubMed]
5. Olechnowicz, J.; Tinkov, A.; Skalny, A.; Suliburska, J. Zinc status is associated with inflammation, oxidative stress, lipid, and glucose metabolism. *J. Physiol. Sci.* **2017**, *68*, 19–31. [CrossRef]
6. Travaglia, A.; La Mendola, D. *Zinc Interactions with Brain-Derived Neurotrophic Factor and Related Peptide Fragments*, 1st ed.; Vitamins and Hormones; Elsevier Inc.: Amsterdam, The Netherlands, 2017; Volume 104, pp. 29–56.
7. Mattei, D.; Pietrobelli, A. Micronutrients and Brain Development. *Curr. Nutr. Rep.* **2019**, *8*, 99–107. [CrossRef] [PubMed]
8. Gu, K.; Xiang, W.; Zhang, Y.; Sun, K.; Jiang, X. The association between serum zinc level and overweight/obesity: A meta-analysis. *Eur. J. Nutr.* **2020**, *11*, 398–411. [CrossRef]
9. Rios-Lugo, M.J.; Madrigal-Arellano, C.; Gaytán-Hernández, D.; Hernández-Mendoza, H.; Romero-Guzmán, E.T. Association of Serum Zinc Levels in Overweight and Obesity. *Biol. Trace Elem. Res.* **2020**, *198*, 51–57. [CrossRef]
10. De Oliveira, S.; Feijó, G.D.S.; Neto, J.; Jantsch, J.; Braga, M.F.; Castro, L.F.D.S.; Giovenardi, M.; Porawski, M.; Guedes, R.P. Zinc Supplementation Decreases Obesity-Related Neuroinflammation and Improves Metabolic Function and Memory in Rats. *Obesity* **2020**, *29*, 116–124. [CrossRef]
11. Fathi, M.; Alavinejad, P.; Haidari, Z.; Amani, R. The effects of zinc supplementation on metabolic profile and oxidative stress in overweight/obese patients with non-alcoholic fatty liver disease: A randomized, double-blind, placebo-controlled trial. *J. Trace Elem. Med. Biol.* **2020**, *62*, 126635. [CrossRef]
12. Khorsandi, H.; Nikpayam, O.; Yousefi, R.; Parandoosh, M.; Hosseinzadeh, N.; Saidpour, A.; Ghorbani, A. Zinc supplementation improves body weight management, inflammatory biomarkers and insulin resistance in individuals with obesity: A randomized, placebo-controlled, double-blind trial. *Diabetol. Metab. Syndr.* **2019**, *11*, 101. [CrossRef] [PubMed]
13. Malesza, I.J.; Malesza, M.; Walkowiak, J.; Mussin, N.; Walkowiak, D. High-Fat, Western-Style Diet, Systemic Inflammation, and Gut Microbiota: A Narrative Review. *Cells* **2021**, *10*, 3164. [CrossRef] [PubMed]

14. Bray, G.A.; Kim, K.K.; Wilding, J.P.H.; World Obesity Federation. Obesity: A chronic relapsing progressive disease process. A position statement of the World Obesity Federation. *Obes. Rev.* **2017**, *18*, 715–723. [CrossRef] [PubMed]
15. Blüher, M. Obesity: Global epidemiology and pathogenesis. *Nat. Rev. Endocrinol.* **2019**, *15*, 288–298. [CrossRef]
16. Pérez-Pérez, A.; Sánchez-Jiménez, F.; Vilariño-García, T.; Sánchez-Margalet, V. Role of Leptin in Inflammation and Vice Versa. *Int. J. Mol. Sci.* **2020**, *21*, 5887. [CrossRef]
17. Christ, A.; Lauterbach, M.; Latz, E. Western Diet and the Immune System: An Inflammatory Connection. *Immunity* **2019**, *51*, 794–811. [CrossRef]
18. Di Bonaventura, M.V.M.; Martinelli, I.; Moruzzi, M.; Di Bonaventura, E.M.; Giusepponi, M.E.; Polidori, C.; Lupidi, G.; Tayebati, S.K.; Amenta, F.; Cifani, C.; et al. Brain Alterations in High Fat Diet Induced Obesity: Effects of Tart Cherry Seeds and Juice. *Nutrients* **2020**, *12*, 623. [CrossRef]
19. López-Taboada, I.; González-Pardo, H.; Conejo, N.M. Western Diet: Implications for Brain Function and Behavior. *Front. Psychol.* **2020**, *11*, 564413. [CrossRef]
20. Rodríguez-Gómez, J.A.; Kavanagh, E.; Engskog-Vlachos, P.; Engskog, M.K.; Herrera, A.J.; Espinosa-Oliva, A.M.; Joseph, B.; Hajji, N.; Venero, J.L.; Burguillos, M.A. Microglia: Agents of the CNS Pro-Inflammatory Response. *Cells* **2020**, *9*, 1717. [CrossRef] [PubMed]
21. Alexaki, V. The Impact of Obesity on Microglial Function: Immune, Metabolic and Endocrine Perspectives. *Cells* **2021**, *10*, 1584. [CrossRef]
22. Rohr, M.W.; Narasimhulu, C.A.; Rudeski-Rohr, T.A.; Parthasarathy, S. Negative Effects of a High-Fat Diet on Intestinal Permeability: A Review. *Adv. Nutr.* **2020**, *11*, 77–91. [CrossRef]
23. Lochhead, J.J.; Yang, J.; Ronaldson, P.T.; Davis, T.P. Structure, Function, and Regulation of the Blood-Brain Barrier Tight Junction in Central Nervous System Disorders. *Front. Physiol.* **2020**, *11*, 914. [CrossRef]
24. Markov, A.G.; Aschenbach, J.R.; Amasheh, S. Claudin clusters as determinants of epithelial barrier function. *IUBMB Life* **2015**, *67*, 29–35. [CrossRef]
25. Garcia-Hernandez, V.; Quiros, M.; Nusrat, A. Intestinal epithelial claudins: Expression and regulation in homeostasis and inflammation. *Ann. N. Y. Acad. Sci.* **2017**, *1397*, 66–79. [CrossRef] [PubMed]
26. Park, H.; Poo, M.-M. Neurotrophin regulation of neural circuit development and function. *Nat. Rev. Neurosci.* **2012**, *14*, 7–23. [CrossRef]
27. Valtorta, F.; Pennuto, M.; Bonanomi, D.; Benfenati, F. Synaptophysin: Leading actor or walk-on role in synaptic vesicle exocytosis? *BioEssays* **2004**, *26*, 445–453. [CrossRef] [PubMed]
28. Arthur, C.P.; Stowell, M.H. Structure of Synaptophysin: A Hexameric MARVEL-Domain Channel Protein. *Structure* **2007**, *15*, 707–714. [CrossRef] [PubMed]
29. White, D.N.; Stowell, M.H.B. Room for Two: The Synaptophysin/Synaptobrevin Complex. *Front. Synaptic Neurosci.* **2021**, *13*, 740318. [CrossRef]
30. Rios-Covian, D.; Salazar, N.; Gueimonde, M.; de los Reyes-Gavilan, C.G. Shaping the Metabolism of Intestinal Bacteroides Population through Diet to Improve Human Health. *Front. Microbiol.* **2017**, *8*, 376. [CrossRef]
31. Rinninella, E.; Cintoni, M.; Raoul, P.; Lopetuso, L.R.; Scaldaferri, F.; Pulcini, G.; Miggiano, G.A.D.; Gasbarrini, A.; Mele, M.C. Food Components and Dietary Habits: Keys for a Healthy Gut Microbiota Composition. *Nutrients* **2019**, *11*, 2393. [CrossRef]
32. Scheithauer, T.P.M.; Rampanelli, E.; Nieuwdorp, M.; Vallance, B.A.; Verchere, C.B.; Van Raalte, D.H.; Herrema, H. Gut Microbiota as a Trigger for Metabolic Inflammation in Obesity and Type 2 Diabetes. *Front. Immunol.* **2020**, *11*, 571731. [CrossRef]
33. Scheithauer, T.P.; Dallinga-Thie, G.M.; de Vos, W.M.; Nieuwdorp, M.; van Raalte, D.H. Causality of small and large intestinal microbiota in weight regulation and insulin resistance. *Mol. Metab.* **2016**, *5*, 759–770. [CrossRef]
34. Meijnikman, A.S.; Gerdes, V.E.; Nieuwdorp, M.; Herrema, H. Evaluating Causality of Gut Microbiota in Obesity and Diabetes in Humans. *Endocr. Rev.* **2018**, *39*, 133–153. [CrossRef]
35. Cox, L.M.; Blaser, M.J. Pathways in Microbe-Induced Obesity. *Cell Metab.* **2013**, *17*, 883–894. [CrossRef]
36. Abenavoli, L.; Scarpellini, E.; Colica, C.; Boccuto, L.; Salehi, B.; Sharifi-Rad, J.; Aiello, V.; Romano, B.; De Lorenzo, A.; Izzo, A.A.; et al. Gut Microbiota and Obesity: A Role for Probiotics. *Nutrients* **2019**, *11*, 2690. [CrossRef]
37. Lee, C.J.; Sears, C.L.; Maruthur, N. Gut microbiome and its role in obesity and insulin resistance. *Ann. N. Y. Acad. Sci.* **2020**, *1461*, 37–52. [CrossRef] [PubMed]
38. Portincasa, P.; Bonfrate, L.; Vacca, M.; De Angelis, M.; Farella, I.; Lanza, E.; Khalil, M.; Wang, D.Q.-H.; Sperandio, M.; Di Ciaula, A. Gut Microbiota and Short Chain Fatty Acids: Implications in Glucose Homeostasis. *Int. J. Mol. Sci.* **2022**, *23*, 1105. [CrossRef]
39. Pérez-Reytor, D.; Puebla, C.; Karahanian, E.; García, K. Use of Short-Chain Fatty Acids for the Recovery of the Intestinal Epithelial Barrier Affected by Bacterial Toxins. *Front. Physiol.* **2021**, *12*, 650313. [CrossRef] [PubMed]
40. Silva, Y.P.; Bernardi, A.; Frozza, R.L. The Role of Short-Chain Fatty Acids From Gut Microbiota in Gut-Brain Communication. *Front. Endocrinol.* **2020**, *11*, 25. [CrossRef]
41. Feng, Y.; Wang, Y.; Wang, P.; Huang, Y.; Wang, F. Short-Chain Fatty Acids Manifest Stimulative and Protective Effects on Intestinal Barrier Function Through the Inhibition of NLRP3 Inflammasome and Autophagy. *Cell. Physiol. Biochem.* **2018**, *49*, 190–205. [CrossRef]

42. Bortolin, R.C.; Vargas, A.R.; Gasparotto, J.; Chaves, P.R.; Schnorr, C.E.; Martinello, K.B.; Silveira, A.K.; Rabelo, T.K.; Gelain, D.P.; Moreira, J.C.F. A new animal diet based on human Western diet is a robust diet-induced obesity model: Comparison to high-fat and cafeteria diets in term of metabolic and gut microbiota disruption. *Int. J. Obes.* **2017**, *42*, 525–534. [CrossRef]
43. Zinöcker, M.K.; Lindseth, I.A. The Western Diet–Microbiome-Host Interaction and Its Role in Metabolic Disease. *Nutrients* **2018**, *10*, 365. [CrossRef]
44. Gual-Grau, A.; Guirro, M.; Mayneris-Perxachs, J.; Arola, L.; Boqué, N. Impact of different hypercaloric diets on obesity features in rats: A metagenomics and metabolomics integrative approach. *J. Nutr. Biochem.* **2019**, *71*, 122–131. [CrossRef]
45. Seyyedin, S.; Nazem, M.N. Histomorphometric study of the effect of methionine on small intestine parameters in rat: An applied histologic study. *Folia Morphol.* **2017**, *76*, 620–629. [CrossRef]
46. Schmieder, R.; Edwards, R. Quality control and preprocessing of metagenomic datasets. *Bioinformatics* **2011**, *27*, 863–864. [CrossRef]
47. Cole, J.R.; Wang, Q.; Fish, J.A.; Chai, B.; McGarrell, D.M.; Sun, Y.; Brown, C.T.; Porras-Alfaro, A.; Kuske, C.R.; Tiedje, J.M. Ribosomal Database Project: Data and tools for high throughput rRNA analysis. *Nucleic Acids Res.* **2014**, *42*, D633–D642. [CrossRef] [PubMed]
48. Caporaso, J.G.; Kuczynski, J.; Stombaugh, J.; Bittinger, K.; Bushman, F.D.; Costello, E.K.; Fierer, N.; Peña, A.G.; Goodrich, J.K.; Gordon, J.I.; et al. QIIME allows analysis of high-throughput community sequencing data. *Nat. Methods* **2010**, *7*, 335–336. [CrossRef]
49. DeSantis, T.Z.; Hugenholtz, P.; Larsen, N.; Rojas, M.; Brodie, E.L.; Keller, K.; Huber, T.; Dalevi, D.; Hu, P.; Andersen, G.L. Greengenes, a Chimera-Checked 16S rRNA Gene Database and Workbench Compatible with ARB. *Appl. Environ. Microbiol.* **2006**, *72*, 5069–5072. [CrossRef]
50. Demers, G.; Roy, J.; Machuca-Parra, A.I.; pour, Z.D.; Bairamian, D.; Daneault, C.; Rosiers, C.D.; Ferreira, G.; Alquier, T.; Fulton, S.; et al. Fish oil supplementation alleviates metabolic and anxiodepressive effects of diet-induced obesity and associated changes in brain lipid composition in mice. *Int. J. Obes.* **2020**, *44*, 1936–1945. [CrossRef] [PubMed]
51. König, H.-G.; Kögel, D.; Rami, A.; Prehn, J.H. TGF-β1 activates two distinct type I receptors in neurons. *J. Cell Biol.* **2005**, *168*, 1077–1086. [CrossRef]
52. Segú, H.; Jalševac, F.; Pinent, M.; Ardévol, A.; Terra, X.; Blay, M.T. Intestinal Morphometric Changes Induced by a Western-Style Diet in Wistar Rats and GSPE Counter-Regulatory Effect. *Nutrients* **2022**, *14*, 2608. [CrossRef] [PubMed]
53. Dailey, M.J. Nutrient-induced intestinal adaption and its effect in obesity. *Physiol. Behav.* **2014**, *136*, 74–78. [CrossRef] [PubMed]
54. Mah, A.T.; Van Landeghem, L.; Gavin, H.E.; Magness, S.T.; Lund, P.K. Impact of Diet-Induced Obesity on Intestinal Stem Cells: Hyperproliferation but Impaired Intrinsic Function That Requires Insulin/IGF1. *Endocrinology* **2014**, *155*, 3302–3314. [CrossRef] [PubMed]
55. Zhou, W.; Davis, E.A.; Dailey, M.J. Obesity, independent of diet, drives lasting effects on intestinal epithelial stem cell proliferation in mice. *Exp. Biol. Med.* **2018**, *243*, 826–835. [CrossRef]
56. Guo, Z.; Martucci, N.J.; Moreno-Olivas, F.; Tako, E.; Mahler, G.J. Titanium dioxide nanoparticle ingestion alters nutrient absorption in an in vitro model of the small intestine. *NanoImpact* **2017**, *5*, 70–82. [CrossRef] [PubMed]
57. Hooper, L.V.; MacPherson, A.J. Immune adaptations that maintain homeostasis with the intestinal microbiota. *Nat. Rev. Immunol.* **2010**, *10*, 159–169. [CrossRef]
58. Fan, Y.; Pedersen, O. Gut microbiota in human metabolic health and disease. *Nat. Rev. Microbiol.* **2021**, *19*, 55–71. [CrossRef]
59. Cho, I.; Blaser, M.J. The human microbiome: At the interface of health and disease. *Nat. Rev. Genet.* **2012**, *13*, 260–270. [CrossRef]
60. Zhang, C.; Zhang, M.; Pang, X.; Zhao, Y.; Wang, L.; Zhao, L. Structural resilience of the gut microbiota in adult mice under high-fat dietary perturbations. *ISME J.* **2012**, *6*, 1848–1857. [CrossRef]
61. Simpson, H.L.; Campbell, B.J. Review article: Dietary fibre-microbiota interactions. *Aliment. Pharmacol. Ther.* **2015**, *42*, 158–179. [CrossRef]
62. Requena, T.; Martínez-Cuesta, M.C.; Peláez, C. Diet and microbiota linked in health and disease. *Food Funct.* **2018**, *9*, 688–704. [CrossRef] [PubMed]
63. Gentile, C.L.; Weir, T.L. The gut microbiota at the intersection of diet and human health. *Science* **2018**, *362*, 776–780. [CrossRef]
64. Wang, Q.-P.; Browman, D.; Herzog, H.; Neely, G.G. Non-nutritive sweeteners possess a bacteriostatic effect and alter gut microbiota in mice. *PLoS ONE* **2018**, *13*, e0199080. [CrossRef]
65. Woting, A.; Pfeiffer, N.; Loh, G.; Klaus, S.; Blaut, M. Clostridium ramosum Promotes High-Fat Diet-Induced Obesity in Gnotobiotic Mouse Models. *mBio* **2014**, *5*, e01530-14. [CrossRef] [PubMed]
66. Shin, N.-R.; Whon, T.W.; Bae, J.-W. Proteobacteria: Microbial signature of dysbiosis in gut microbiota. *Trends Biotechnol.* **2015**, *33*, 496–503. [CrossRef] [PubMed]
67. De Filippo, C.; Cavalieri, D.; Di Paola, M.; Ramazzotti, M.; Poullet, J.B.; Massart, S.; Collini, S.; Pieraccini, G.; Lionetti, P. Impact of diet in shaping gut microbiota revealed by a comparative study in children from Europe and rural Africa. *Proc. Natl. Acad. Sci. USA* **2010**, *107*, 14691–14696. [CrossRef] [PubMed]
68. Wang, J.; Tang, H.; Zhang, C.; Zhao, Y.; Derrien, M.; Rocher, E.; van-Hylckama Vlieg, J.E.T.; Strissel, K.; Zhao, L.; Obin, M.; et al. Modulation of gut microbiota during probiotic-mediated attenuation of metabolic syndrome in high fat diet-fed mice. *ISME J.* **2015**, *9*, 1–15. [CrossRef] [PubMed]

69. Kang, Y.; Kang, X.; Yang, H.; Liu, H.; Yang, X.; Liu, Q.; Tian, H.; Xue, Y.; Ren, P.; Kuang, X.; et al. Lactobacillus acidophilus ameliorates obesity in mice through modulation of gut microbiota dysbiosis and intestinal permeability. *Pharmacol. Res.* **2022**, *175*, 106020. [CrossRef] [PubMed]
70. Macchione, I.G.; Lopetuso, L.R.; Ianiro, G.; Napoli, M.; Gibiino, G.; Rizzatti, G.; Petito, V.; Gasbarrini, A.; Scaldaferri, F. Akkermansia muciniphila: Key player in metabolic and gastrointestinal disorders. *Eur. Rev. Med. Pharmacol. Sci.* **2019**, *23*, 8075–8083.
71. de Vos, W.M. Microbe Profile: Akkermansia muciniphila: A conserved intestinal symbiont that acts as the gatekeeper of our mucosa. *Microbiology* **2017**, *163*, 646–648. [CrossRef]
72. Ou, Z.; Deng, L.; Lu, Z.; Wu, F.; Liu, W.; Huang, D.; Peng, Y. Protective effects of Akkermansia muciniphila on cognitive deficits and amyloid pathology in a mouse model of Alzheimer's disease. *Nutr. Diabetes* **2020**, *10*, 1–10. [CrossRef] [PubMed]
73. Schneeberger, M.; Everard, A.; Gómez-Valadés, A.G.; Matamoros, S.; Ramírez, S.; Delzenne, N.; Gomis, R.; Claret, M.; Cani, P.D. Akkermansia muciniphila inversely correlates with the onset of inflammation, altered adipose tissue metabolism and metabolic disorders during obesity in mice. *Sci. Rep.* **2015**, *5*, 16643. [CrossRef] [PubMed]
74. Huwart, S.J.P.; D'Oplinter, A.D.W.; Rastelli, M.; Van Hul, M.; de Vos, W.M.; Luquet, S.; Cani, P.D.; Everard, A. An increase in the *Akkermansia* spp. population induced by metformin treatment improves glucose homeostasis in diet-induced obese mice. *Gut* **2015**, *33*, 496–503. [CrossRef]
75. Barra, N.G.; Anhê, F.F.; Cavallari, J.F.; Singh, A.M.; Chan, D.Y.; Schertzer, J.D. Micronutrients impact the gut microbiota and blood glucose. *J. Endocrinol.* **2021**, *250*, R1–R21. [CrossRef]
76. Hamilton, M.K.; Boudry, G.; Lemay, D.G.; Raybould, H.E. Changes in intestinal barrier function and gut microbiota in high-fat diet-fed rats are dynamic and region dependent. *Am. J. Physiol. Liver Physiol.* **2015**, *308*, G840–G851. [CrossRef]
77. Everard, A.; Belzer, C.; Geurts, L.; Ouwerkerk, J.P.; Druart, C.; Bindels, L.B.; Guiot, Y.; Derrien, M.; Muccioli, G.G.; Delzenne, N.M.; et al. Cross-talk between *Akkermansia muciniphila* and intestinal epithelium controls diet-induced obesity. *Proc. Natl. Acad. Sci. USA* **2013**, *110*, 9066–9071. [CrossRef] [PubMed]
78. Bailén, M.; Bressa, C.; Martínez-López, S.; González-Soltero, R.; Lominchar, M.G.M.; Juan, C.S.; Larrosa, M. Microbiota Features Associated With a High-Fat/Low-Fiber Diet in Healthy Adults. *Front. Nutr.* **2020**, *7*, 583608. [CrossRef]
79. Davis, S.C.; Yadav, J.; Barrow, S.D.; Robertson, B.K. Gut microbiome diversity influenced more by the Westernized dietary regime than the body mass index as assessed using effect size statistic. *MicrobiologyOpen* **2017**, *6*, e00476. [CrossRef] [PubMed]
80. Ozato, N.; Saito, S.; Yamaguchi, T.; Katashima, M.; Tokuda, I.; Sawada, K.; Katsuragi, Y.; Kakuta, M.; Imoto, S.; Ihara, K.; et al. Blautia genus associated with visceral fat accumulation in adults 20–76 years of age. *NPJ Biofilms Microbiomes* **2019**, *5*, 1–9. [CrossRef]
81. Ozato, N.; Yamaguchi, T.; Mori, K.; Katashima, M.; Kumagai, M.; Murashita, K.; Katsuragi, Y.; Tamada, Y.; Kakuta, M.; Imoto, S.; et al. Two *Blautia* Species Associated with Visceral Fat Accumulation: A One-Year Longitudinal Study. *Biology* **2022**, *11*, 318. [CrossRef] [PubMed]
82. Liu, X.; Mao, B.; Gu, J.; Wu, J.; Cui, S.; Wang, G.; Zhao, J.; Zhang, H.; Chen, W. *Blautia*—a new functional genus with potential probiotic properties? *Gut Microbes* **2021**, *13*, 1875796. [CrossRef]
83. Le Chatelier, E.; Nielsen, T.; Qin, J.; Prifti, E.; Hildebrand, F.; Falony, G.; Almeida, M.; Arumugam, M.; Batto, J.-M.; Kennedy, S.; et al. Richness of human gut microbiome correlates with metabolic markers. *Nature* **2013**, *500*, 541–546. [CrossRef]
84. Klingbeil, E.A.; de La Serre, C.B. Microbiota modulation by eating patterns and diet composition: Impact on food intake. *Am. J. Physiol. Integr. Comp. Physiol.* **2018**, *315*, R1254–R1260. [CrossRef]
85. Popovic, A.; Bourdon, C.; Wang, P.W.; Guttman, D.S.; Soofi, S.; Bhutta, Z.A.; Bandsma, R.H.J.; Parkinson, J.; Pell, L.G. Micronutrient supplements can promote disruptive protozoan and fungal communities in the developing infant gut. *Nat. Commun.* **2021**, *12*, 6729. [CrossRef]
86. Zackular, J.; Moore, J.; Jordan, A.T.; Juttukonda, L.J.; Noto, M.J.; Nicholson, M.; Crews, J.D.; Semler, M.W.; Zhang, Y.; Ware, L.B.; et al. Dietary zinc alters the microbiota and decreases resistance to Clostridium difficile infection. *Nat. Med.* **2016**, *22*, 1330–1334. [CrossRef]
87. Lin, H.V.; Frassetto, A.; Kowalik, E.J., Jr.; Nawrocki, A.R.; Lu, M.M.; Kosinski, J.R.; Hubert, J.A.; Szeto, D.; Yao, X.; Forrest, G.; et al. Butyrate and Propionate Protect against Diet-Induced Obesity and Regulate Gut Hormones via Free Fatty Acid Receptor 3-Independent Mechanisms. *PLoS ONE* **2012**, *7*, e35240. [CrossRef]
88. Valdes, A.; Walter, J.; Segal, E.; Spector, T.D. Role of the gut microbiota in nutrition and health. *BMJ* **2018**, *361*, k2179. [CrossRef]
89. Salvi, P.S.; Cowles, R.A. Butyrate and the Intestinal Epithelium: Modulation of Proliferation and Inflammation in Homeostasis and Disease. *Cells* **2021**, *10*, 1775. [CrossRef]
90. Louis, P.; Flint, H.J. Formation of propionate and butyrate by the human colonic microbiota. *Environ. Microbiol.* **2017**, *19*, 29–41. [CrossRef] [PubMed]
91. Morrison, D.J.; Preston, T. Formation of short chain fatty acids by the gut microbiota and their impact on human metabolism. *Gut Microbes* **2016**, *7*, 189–200. [CrossRef] [PubMed]
92. Frost, G.; Sleeth, M.L.; Sahuri-Arisoylu, M.; Lizarbe, B.; Cerdan, S.; Brody, L.; Anastasovska, J.; Ghourab, S.; Hankir, M.; Zhang, S.; et al. The short-chain fatty acid acetate reduces appetite via a central homeostatic mechanism. *Nat. Commun.* **2014**, *5*, 3611. [CrossRef] [PubMed]

93. Biesalski, H.K. Nutrition meets the microbiome: Micronutrients and the microbiota. *Ann. N. Y Acad. Sci.* **2016**, *1372*, 53–64. [CrossRef] [PubMed]
94. Pieper, R.; Vahjen, W.; Neumann, K.; Van Kessel, A.G.; Zentek, J. Dose-dependent effects of dietary zinc oxide on bacterial communities and metabolic profiles in the ileum of weaned pigs. *J. Anim. Physiol. Anim. Nutr.* **2011**, *96*, 825–833. [CrossRef]
95. Li, Y.; Dong, J.; Xiao, H.; Zhang, S.; Wang, B.; Cui, M.; Fan, S. Gut commensal derived-valeric acid protects against radiation injuries. *Gut Microbes* **2020**, *11*, 789–806. [CrossRef]
96. Stawarska, A.; Czerwonka, M.; Wyrębiak, R.; Wrzesień, R.; Bobrowska-Korczak, B. Zinc Affects Cholesterol Oxidation Products and Fatty Acids Composition in Rats' Serum. *Nutrients* **2021**, *13*, 1563. [CrossRef]
97. Wolters, M.; Ahrens, J.; Romaní-Pérez, M.; Watkins, C.; Sanz, Y.; Benítez-Páez, A.; Stanton, C.; Günther, K. Dietary fat, the gut microbiota, and metabolic health—A systematic review conducted within the MyNewGut project. *Clin. Nutr.* **2018**, *38*, 2504–2520. [CrossRef]
98. Fritsche, K.L. The Science of Fatty Acids and Inflammation. *Adv. Nutr. Int. Rev. J.* **2015**, *6*, 293S–301S. [CrossRef]
99. Hsu, T.M.; Kanoski, S.E. Blood-brain barrier disruption: Mechanistic links between Western diet consumption and dementia. *Front. Aging Neurosci.* **2014**, *6*, 88. [CrossRef]
100. Netto Candido, T.L.; Bressan, J.; de Alfenas, R.C.G. Dysbiosis and metabolic endotoxemia induced by high-fat diet. *Nutr. Hosp.* **2018**, *35*, 1432–1440. [CrossRef]
101. Wong, S.W.; Kwon, M.-J.; Choi, A.M.; Kim, H.-P.; Nakahira, K.; Hwang, D.H. Fatty Acids Modulate Toll-like Receptor 4 Activation through Regulation of Receptor Dimerization and Recruitment into Lipid Rafts in a Reactive Oxygen Species-dependent Manner. *J. Biol. Chem.* **2009**, *284*, 27384–27392. [CrossRef]
102. Pepping, J.K.; Freeman, L.R.; Gupta, S.; Keller, J.; Bruce-Keller, A.J. NOX2 deficiency attenuates markers of adiposopathy and brain injury induced by high-fat diet. *Am. J. Physiol. Metab.* **2013**, *304*, E392–E404. [CrossRef] [PubMed]
103. Suzuki, T. Regulation of the intestinal barrier by nutrients: The role of tight junctions. *Anim. Sci. J.* **2020**, *91*, e13357. [CrossRef] [PubMed]
104. Mercado, J.; Valenzano, M.C.; Jeffers, C.; Sedlak, J.; Cugliari, M.K.; Papanikolaou, E.; Clouse, J.; Miao, J.; Wertan, N.E.; Mullin, J.M. Enhancement of Tight Junctional Barrier Function by Micronutrients: Compound-Specific Effects on Permeability and Claudin Composition. *PLoS ONE* **2013**, *8*, e78775. [CrossRef] [PubMed]
105. Mayer, E.A.; Tillisch, K.; Gupta, A. Gut/brain axis and the microbiota. *J. Clin. Investig.* **2015**, *125*, 926–938. [CrossRef]
106. Agustí, A.; García-Pardo, M.P.; López-Almela, I.; Campillo, I.; Maes, M.; Romani-Pérez, M.; Sanz, Y. Interplay Between the Gut-Brain Axis, Obesity and Cognitive Function. *Front. Neurosci.* **2018**, *12*, 155. [CrossRef] [PubMed]
107. Margolis, K.G.; Cryan, J.F.; Mayer, E.A. The Microbiota-Gut-Brain Axis: From Motility to Mood. *Gastroenterology* **2021**, *160*, 1486–1501. [CrossRef]
108. Hill-Burns, E.M.; Debelius, J.W.; Morton, J.T.; Wissemann, W.T.; Lewis, M.R.; Wallen, Z.D.; Peddada, S.D.; Factor, S.A.; Molho, E.; Zabetian, C.P.; et al. Parkinson's disease and Parkinson's disease medications have distinct signatures of the gut microbiome. *Mov. Disord.* **2017**, *32*, 739–749. [CrossRef]
109. Li, Q.; Han, Y.; Dy, A.B.C.; Hagerman, R.J. The Gut Microbiota and Autism Spectrum Disorders. *Front. Cell. Neurosci.* **2017**, *11*, 120. [CrossRef]
110. Wu, W.-L.; Adame, M.D.; Liou, C.-W.; Barlow, J.T.; Lai, T.-T.; Sharon, G.; Schretter, C.E.; Needham, B.D.; Wang, M.I.; Tang, W.; et al. Microbiota regulate social behaviour via stress response neurons in the brain. *Nature* **2021**, *595*, 409–414. [CrossRef] [PubMed]
111. Sensi, S.L.; Granzotto, A.; Siotto, M.; Squitti, R. Copper and Zinc Dysregulation in Alzheimer's Disease. *Trends Pharmacol. Sci.* **2018**, *39*, 1049–1063. [CrossRef]
112. Granzotto, A.; Canzoniero, L.M.T.; Sensi, S.L. A Neurotoxic Ménage-à-trois: Glutamate, Calcium, and Zinc in the Excitotoxic Cascade. *Front. Mol. Neurosci.* **2020**, *13*, 600089. [CrossRef] [PubMed]
113. Jung, J.H.; Park, M.H.; Choi, S.Y.; Koh, J.Y. Activation of the Trk signaling pathway by extracellular zinc: Role of metalloproteinases. *J. Biol. Chem.* **2005**, *280*, 11995–12001. [CrossRef]
114. Corona, C.; Masciopinto, F.; Silvestri, E.; Del Viscovo, A.; Lattanzio, R.; La Sorda, R.; Ciavardelli, D.; Goglia, F.; Piantelli, M.; Canzoniero, L.M.T.; et al. Dietary zinc supplementation of 3xTg-AD mice increases BDNF levels and prevents cognitive deficits as well as mitochondrial dysfunction. *Cell Death Dis.* **2010**, *1*, e91. [CrossRef]
115. Nam, S.M.; Kim, J.W.; Kwon, H.J.; Yoo, D.Y.; Jung, H.Y.; Kim, D.W.; Hwang, I.K.; Seong, J.K.; Yoon, Y.S. Differential Effects of Low- and High-dose Zinc Supplementation on Synaptic Plasticity and Neurogenesis in the Hippocampus of Control and High-fat Diet-fed Mice. *Neurochem. Res.* **2017**, *42*, 3149–3159. [CrossRef]
116. Solati, Z.; Jazayeri, S.; Tehrani-Doost, M.; Mahmoodianfard, S.; Gohari, M.R. Zinc monotherapy increases serum brain-derived neurotrophic factor (BDNF) levels and decreases depressive symptoms in overweight or obese subjects: A double-blind, randomized, placebo-controlled trial. *Nutr. Neurosci.* **2014**, *18*, 162–168. [CrossRef]
117. Jafari, F.; Mohammadi, H.; Amani, R. The effect of zinc supplementation on brain derived neurotrophic factor: A meta-analysis. *J. Trace Elem. Med. Biol.* **2021**, *66*, 126753. [CrossRef]
118. Sandrini, L.; Di Minno, A.; Amadio, P.; Ieraci, A.; Tremoli, E.; Barbieri, S.S. Association between Obesity and Circulating Brain-Derived Neurotrophic Factor (BDNF) Levels: Systematic Review of Literature and Meta-Analysis. *Int. J. Mol. Sci.* **2018**, *19*, 2281. [CrossRef]

119. Martinelli, I.; Tomassoni, D.; Roy, P.; Amenta, F.; Tayebati, S.K. Altered Brain Cholinergic and Synaptic Markers in Obese Zucker Rats. *Cells* **2021**, *10*, 2528. [CrossRef]
120. Cai, M.; Wang, H.; Li, J.-J.; Zhang, Y.-L.; Xin, L.; Li, F.; Lou, S.-J. The signaling mechanisms of hippocampal endoplasmic reticulum stress affecting neuronal plasticity-related protein levels in high fat diet-induced obese rats and the regulation of aerobic exercise. *Brain Behav. Immun.* **2016**, *57*, 347–359. [CrossRef]

nutrients

MDPI

Article

Greater Ultra-Processed Food Intake during Pregnancy and Postpartum Is Associated with Multiple Aspects of Lower Diet Quality

Tonja R. Nansel [1,*], Jenna R. Cummings [1], Kyle Burger [2], Anna Maria Siega-Riz [3,†] and Leah M. Lipsky [1]

1 Social and Behavioral Sciences Branch, Division of Population Health Research, Eunice Kennedy Shriver National Institute of Child Health and Human Development, 6710B Rockledge Dr., MSC 7004, Bethesda, MD 20892, USA

2 Gillings School of Global Public Health, University of North Carolina Chapel Hill, 2204 McGavran-Greenberg Hall, CB# 7461, Chapel Hill, NC 27599, USA

3 Departments of Nutrition and Biostatistics and Epidemiology, School of Public Health and Health Sciences, University of Massachusetts, 109 Arnold House, 715 Pleasant St., Amherst, MA 01003, USA

* Correspondence: tonja.nansel@nih.gov; Tel.: +1-301-435-6937

† Any changed affiliations: Anna Maria Siega-Riz was affiliated with the Gillings School of Global Public Health, University of North Carolina Chapel Hill, Chapel Hill, NC 27599,USA, during the initiation of this study.

Abstract: Low diet quality during pregnancy and postpartum is associated with numerous adverse maternal and infant health outcomes. This study examined relations of ultra-processed food intake with diet quality during pregnancy and postpartum. Using data from 24-h recalls, ultra-processed food intake was operationalized as percent energy intake from NOVA-classified ultra-processed foods; diet quality was measured using Healthy Eating Index 2015 (HEI) total and component scores. Pearson correlations examined associations of ultra-processed food intake with HEI total and component scores, and food group intake was compared across four levels of ultra-processed food intake. On average, ultra-processed food comprised 52.6 ± 15.1% (mean ± SD) of energy intake in pregnancy and 50.6 ± 16.6% in postpartum. Ultra-processed food intake was inversely correlated with HEI total and 8 of 13 component scores. Compared to participants with the highest ultra-processed food intake (≥60% energy), those with the lowest ultra-processed food intake (<40% energy) had a 17.6-point higher HEI total score and consumed 2–3 times more fruit, vegetables, and seafood and plant proteins, and 1½ times more total protein. Additionally, they consumed 2/3 as much refined grains and 1/2 as much added sugar. Greater ultra-processed food intake was associated with lower diet quality across most HEI components. Reducing ultra-processed food intake may broadly improve adherence to dietary guidelines in pregnant and postpartum populations.

Keywords: ultra-processed food; diet quality; pregnancy; postpartum

Citation: Nansel, T.R.; Cummings, J.R.; Burger, K.; Siega-Riz, A.M.; Lipsky, L.M. Greater Ultra-Processed Food Intake during Pregnancy and Postpartum Is Associated with Multiple Aspects of Lower Diet Quality. *Nutrients* **2022**, *14*, 3933. https://doi.org/10.3390/nu14193933

Academic Editors: Monica Dinu and Daniela Martini

Received: 31 August 2022
Accepted: 19 September 2022
Published: 22 September 2022

1. Introduction

The importance of nutrition during pregnancy for supporting optimal maternal and child health is well-established and recognized by healthcare professionals [1–3]. However, dietary intake during pregnancy is typified by inadequate consumption of vegetables, fruit, and whole grains, as well as excessive intake of empty calories, fat, and sodium [4–6]. Limited evidence suggests diet quality is similarly low in the postpartum period [7,8]. Given associations of low maternal diet quality with a wide range of adverse outcomes for both mother and infant [9–13], there is a critical need to identify modifiable intervention targets.

Ultra-processed foods, that is, food products that are created from substances extracted from foods or derived from food constituents [14], contribute on average greater than half of energy intake in the general U.S. population [15]. Consistent evidence across countries indicates that ultra-processed food intake is associated with select nutrients, including lower intake of fiber [16–23]; higher total and saturated fat [16–20,23]; and higher free, added, or total sugar [16,17,20–25] in general population samples. Additionally, four studies have demonstrated an inverse relation of ultra-processed foods with overall diet quality [18,26–28]. Consumption of ultra-processed foods may adversely affect overall diet quality via excess intake of highly processed discretionary foods and also by displacing intake of unprocessed, whole plant foods; one study found that greater consumption of ultra-processed food was related to greater intake of sugar-sweetened beverages and processed meat, as well as lower intake of fruits/vegetables, nuts/seeds/legumes, and fish in a general population sample [26]. Understanding the impact of ultra-processed food consumption on intake across healthful and discretionary food groups is important for determining effective approaches to improve diet quality given the pervasiveness of ultra-processed foods in the food environment.

The importance of pregnancy and postpartum diet quality for fostering maternal and child health and the ubiquitous presence of ultra-processed foods in the environment are well-established; yet, the association of ultra-processed food intake with pregnancy and postpartum diet quality is largely unexamined. One study investigated relations of ultra-processed food intake with nutrient intake in pregnant women in Brazil, finding inverse associations with intake adequacy of several individual nutrients and with consumption of traditional foods including rice, beans, fruits, and vegetables [29]. However, this study did not examine associations with overall diet quality and may have limited generalizability to countries with high ultra-processed food intake given the low average intake of ultra-processed foods (22% of energy intake) in the sample. Additionally, no studies have investigated ultra-processed food intake, or the relations of ultra-processed food intake with diet quality, during postpartum. To address these knowledge gaps, this study examined associations of ultra-processed food intake with overall diet quality and diet quality components during pregnancy and postpartum in a U.S. sample.

2. Methods

2.1. Design, Participants, and Procedures

The Pregnancy Eating Attributes Study (PEAS) was a prospective observational study examining eating-related behaviors from the first trimester of pregnancy through one-year postpartum. Participants were enrolled at ≤12 weeks gestation from two university-based obstetrics clinics in Chapel Hill, North Carolina from November 2014 through October 2016; data collection was completed in June 2018 [30]. Inclusion criteria included the following: ≤12 weeks gestation at enrollment; body mass index ≥18.5 kg/m^2; age ≥18 and <45 at screening; uncomplicated singleton pregnancy anticipated; access to internet with email; able to complete self-report assessments in English; intention to deliver at the University of North Carolina Women's Hospital; plan to remain in the geographical vicinity of the clinical site for one year following delivery; and willing to provide informed consent for participation and assent for the baby's participation. Exclusion criteria included the following: multiple gestations; participant-reported eating disorder; pre-existing diabetes; any medical condition contraindicating participation in the study, such as chronic illnesses or use of medication that could affect diet or weight; and psychosocial condition contraindicating participation in the study. The primary study aims were to examine the roles of reward-related eating, self-control, and home food availability on dietary intake and weight change during pregnancy and postpartum. Power analyses to determine the sample size are described in the report on primary study aims [30].

Research staff screened clinical appointment data to identify potentially eligible patients, then verified eligibility at the time of the clinical appointment and obtained signed informed consent from persons choosing to participate in the study. Study visits were conducted once each trimester, and postpartum at 4–6 weeks, 6 months, and 12 months. Self-report measures including dietary recalls were completed online within designated study visit windows. The study was conducted in accordance with the Declaration of

Helsinki, and the protocol was approved by the University of North Carolina Institutional Review Board.

2.2. Measures

Dietary intake was assessed using a 24 h recall at each study visit window obtained through the National Cancer Institute's Automated Self-Administered 24-h Recall (ASA24). The ASA24 uses an online interface in which participants delineate all foods consumed for the specified time period. They are prompted to indicate information on food preparation, brands, portion size, and additions. Study staff provided participants with written information on how to access and use the program and also assisted them if they experienced difficulty using the program. Research staff at the University of North Carolina Nutrition and Obesity Research Core then reviewed the data to identify and corrected implausible entries (e.g., food items with implausible energy, fat or weight) and missing food or nutrient values and quantities. The ASA24 assigns food codes from the U.S. Department of Agriculture (USDA) Food and Nutrient Database for Dietary Surveys (FNDDS) to the participant-reported food items and outputs estimates of macronutrient, micronutrient, food categories, and USDA Food Patterns Equivalents Database food groups. The ASA24 has shown strong validity relative to the interviewer-administered automated multiple pass method [31,32]. Dietary records with daily energy intakes of <600 kcal (36 of 1883 records, 1.9%) or >4500 kcal (21 of 1883 records, 1.1%) were reviewed by the investigators for plausibility. All records with intakes <600 kcal were deemed to be likely incomplete and excluded from analysis. Those with intakes of >4500 kcal were determined to reflect plausible intake and were retained.

Ultra-processed food intake was estimated using the NOVA system (not an acronym). NOVA is a classification system developed by researchers at the University of São Paulo that categorizes foods and beverages into four groups based on the degree of processing [14]. Unprocessed or minimally processed foods are those which remain in their original natural form or which have been altered only by removal of inedible or unwanted parts, crushing, grinding, fractioning, refrigerating, freezing, drying, roasting, boiling, pasteurizing, placing in containers, or vacuum packaging (e.g., fresh and frozen fruit, vegetables, and meat; fresh or pasteurized milk). Processed culinary ingredients are substances that are obtained directly from unprocessed/minimally processed foods by pressing, centrifuging, refining, extracting, or mining (e.g., vegetable oils; butter and lard; sugar and molasses). Processed foods are food products made by adding processed culinary ingredients to unprocessed/minimally processed foods, using canning, bottling, or non-alcoholic fermentation (e.g., salted canned vegetables; fruit preserves; salted nuts; sweetened dried fruit). Ultra-processed foods are food products that are created from a series of industrial processes including chemical modification (e.g., hydrolysis), whole food fractioning, additions for palatability (e.g., colors, flavors, emulsifiers), assembly (e.g., pre-frying), and packaging with synthetic materials (e.g., 'instant' foods; ready-to-heat pre-prepared pies, pasta, and pizza; mass-produced packaged breads; reconstituted meats; sweet or savory packaged snacks; confectionery desserts; sweetened drinks). Standardized Stata (College Station, TX, USA) code available upon request from the University of São Paulo group was used to assign NOVA categories to ASA24 food codes. Food codes indicating a homemade recipe were classified using the underlying ingredient codes and correspondent nutrition information from the FNDDS [33]. Percent of daily energy intake from ultra-processed foods was calculated separately for pregnancy (n = 365) and postpartum (n = 266) by dividing the average total daily energy from ultra-processed foods by the average total daily energy intake across all dietary recalls for each time period.

The Healthy Eating Index 2015 (HEI) total and component scores [34] were used to indicate overall and specific aspects of diet quality. The HEI indicates degree of conformance to the 2015 US Dietary Guidelines for Americans [35] and is calculated from the ASA24 dietary intake data not including supplements. HEI scores include 9 adequacy components (total fruit, whole fruit, total vegetables, greens and beans, whole grains, dairy, total protein, seafood and plant proteins, fatty acids) and 4 moderation components (refined grains, sodium, added sugars, and saturated fats). All 13 component scores are summed to calculate the HEI total score, which ranges from 0–100. Scores are calculated on a per-1000 kcal or percent of kcal basis for comparability across persons having varying energy requirements. HEI total and component scores were calculated across pregnancy and across postpartum using all dietary recalls for each time period.

Demographic information including education, household income, household composition, race, ethnicity, and marital status were reported by participants at the initial assessment. Family income and household size were used to calculate income-to-poverty ratio [36]. Values were categorized as ≤1.85 (the threshold for eligibility for Special Supplemental Nutrition program for Women, Infants, and Children), 1.86–4.0, and ≥4.0. Participant age and parity were obtained from the electronic medical record; age was categorized by tertiles. Measured height and weight obtained at the initial clinic visit were used to calculate body mass index as kg/m^2 and categorized as 18.0–24.9, 25.0–29.9, and ≥30.0.

2.3. Analysis

Differences in ultra-processed food intake by participant characteristics were examined using analysis of variance or Kruskal–Wallis test when the variance homogeneity assumption was not met. Given expected correlations among sociodemographic characteristics, regression analysis was used to examine associations of ultra-processed food intake with the set of participant characteristics demonstrating significant bivariate associations. Pearson correlations were used to examine associations of ultra-processed food intake with HEI total and component scores. HEI scores and food group intake density values were then compared between participants consuming <40.0, 40.0 to 49.9, 50.0 to 59.9, and ≥60% energy intake from ultra-processed foods; these categories were selected to create consistent cut-points across pregnancy and postpartum based on the sample distribution (roughly quartiles). Food group intakes are provided as density values (e.g., cup or ounce equivalents per 1000 kcal) for ease of interpretation of the resulting mean values.

3. Results

Of 458 participants enrolled, 91 (20%) withdrew prior to delivery and 46 (10%) withdrew during the one-year postpartum period; 383 provided dietary intake data during pregnancy and/or postpartum. Mean ± SD baseline age was 30.8 ± 4.6, and 92% were married or living with their partner. Approximately half of the sample were nulliparous, three-quarters had a bachelor's degree or above, 22% had income ≤185% of the poverty level, and 31% identified as racial or ethnic minority (Table 1). Approximately half had normal weight, while one-quarter each had overweight and obesity. In bivariate analyses, ultra-processed food intake was higher among participants <29 years of age versus older age groups, those with less than a bachelor's degree versus participants with higher education attainment, those with an income-to-poverty ratio ≤1.85 versus above, those married versus those divorced/separated/widowed/single, and those with overweight or obesity compared to those having normal weight. Ultra-processed food intake was lower among Asian participants than White, Black, and Hispanic participants. In the regression model entering age, education, poverty-to-income ratio, marital status, and body mass index simultaneously, only age (standardized B = −0.13, p = 0.03) and education (standardized B = −0.25, p < 0.001) were significantly associated with ultra-processed food intake. (Race/ethnicity was not included in the model given the small cell size of the one race/ethnicity that demonstrated significant differences from others.)

Mean ± SD ultra-processed food was 52.6 ± 15.1% of energy intake in pregnancy and 50.6 ± 16.6% in postpartum. In both pregnancy and postpartum, ultra-processed food intake was inversely correlated with HEI total score and with 8 of the 13 component scores, including 6 of the 9 adequacy components and 2 of the 4 moderation components (Table 2).

Table 1. Sample characteristics and association with ultra-processed food intake.

	Sample Distribution *N* (%)	Percent of Energy Intake from Ultra-Processed Foods in Pregnancy Mean ± SD *
Age		
<29.00	110 (28.7)	58.0 ± 14.9 [a]
29.00–32.99	132 (34.5)	50.8 ± 15.2 [b]
≥33.00	141 (36.8)	50.2 ± 14.3 [b]
Education		
High school or less	29 (8.3)	59.3 ± 16.2 [a]
Associates/some college	65 (18.5)	59.5 ± 16.9 [a]
Bachelors	106 (30.2)	52.4 ± 12.9 [b]
Advanced degree	151 (43.0)	48.6 ± 13.9 [b]
Income poverty ratio		
≤1.85	77 (22.2)	57.4 ± 17.6 [a]
1.86–4.00	84 (24.2)	52.7 ± 15.6 [b]
≥4.00	186 (53.6)	51.0 ± 13.2 [b]
Race/ethnicity		
White, non-Hispanic	251 (68.8)	52.5 ± 13.8 [a]
Black, non-Hispanic	53 (14.5)	56.5 ± 17.7 [a]
Asian	17 (4.7)	39.9 ± 16.2 [b]
Hispanic	29 (7.9)	53.9 ± 16.3 [a]
Multi-race & other	15 (4.1)	50.7 ± 17.7 [ab]
Marital status		
Married/living with partner	322 (91.7)	51.9 ± 14.8 [a]
Divorced/separated/widowed/single	29 (8.3)	61.2 ± 14.8 [b]
Parity		
Nulliparous	188 (49.1)	52.3 ± 15.1
Parous	195 (50.9)	52.8 ± 15.2
Body mass index		
18.0–24.9	191 (49.9)	50.5 ± 14.2 [a]
25.0–29.9	99 (25.8)	54.0 ± 14.5 [b]
≥30.0	93 (24.3)	55.6 ± 17.2 [b]

N = 383 participants with dietary intake data during pregnancy and/or postpartum. Demographic data missing for 32 participants on education, 36 participants on income, 18 participants on race/ethnicity, and 32 participants on marital status. * Group differences tested using analysis of variance (age, education, marital status, parity, and body mass index) or Kruskal–Wallis test (income poverty ratio and race/ethnicity). Different superscript letters indicate statistically significant differences between groups at $p < 0.05$.

Table 2. Pearson correlations of ultra-processed food intake with Healthy Eating Index 2015 (HEI) total and component scores.

	Percent of Energy Intake from Ultra-Processed Foods	
	Pregnancy	Postpartum
HEI total	−0.55 **	−0.52 **
Total vegetables	−0.34 **	−0.40 **
Greens & beans	−0.36 **	−0.28 **
Total fruit	−0.41 **	−0.26 **
Whole fruit	−0.35 **	−0.28 **
Whole grains	−0.16	−0.16
Dairy	−0.09	0.06
Total protein foods	−0.24 **	−0.37 **
Seafood & plant protein	−0.43 **	−0.40 **
Fatty acid ratio	−0.09	−0.18
Sodium	0.02	−0.05
Refined grains	−0.35 **	−0.38 **
Saturated fat	−0.10	−0.16
Added sugar	−0.50 **	−0.49 **

** $p < 0.001$; Sidak-adjusted p-values.

HEI total scores and food group intake density by categories of ultra-processed food intake are detailed in Table 3. Compared to those consuming ≥60% of energy intake from ultra-processed food, participants consuming <40% energy from ultra-processed food had a 17.6-point higher total HEI score and consumed nearly 2 times more total vegetables, 3 times more greens and beans, 2–3 times more total and whole fruit, about 1½ times more total protein, and about 3 times more seafood and plant protein. Conversely, those consuming ≥60% of energy intake from ultra-processed food consumed about 1½ times more refined grains and more than twice as much added sugar as those consuming ≥60% of energy intake from ultra-processed food.

Table 3. Healthy Eating Index 2015 (HEI) total score and food group intake density by category of ultra-processed food intake.

	Percent of Energy Intake from Ultra-Processed Foods Mean (95% Confidence Interval)			
	<40.0 [a]	40.0–49.9 [a]	50.0–59.9 [a]	≥60 [a]
HEI total score				
Pregnancy	66.3 (63.6, 69.0)	62.6 (60.6, 64.6)	57.2 (55.1, 59.4)	48.8 (46.7, 50.9)
Postpartum	66.0 (63.1, 68.9)	62.6 (59.5, 65.8)	57.2 (54.1, 60.2)	48.5 (46.0, 51.0)
Total vegetables [b]				
Pregnancy	1.15 (1.00, 1.30)	1.00 (0.91, 1.09)	0.99 (0.90, 1.08)	0.67 (0.60, 0.73)
Postpartum	1.38 (1.20, 1.57)	1.13 (1.00, 1.26)	0.93 (0.82, 1.04)	0.73 (0.64, 0.82)
Greens & beans [b]				
Pregnancy	0.33 (0.26, 0.40)	0.27 (0.23, 0.31)	0.24 (0.18, 0.29)	0.10 (0.08, 0.12)
Postpartum	0.39 (0.30, 0.48)	0.32 (0.23, 0.41)	0.24 (0.18, 0.30)	0.13 (0.09, 0.17)
Total fruit [b]				
Pregnancy	1.23 (1.02, 1.45)	0.84 (0.73, 0.94)	0.67 (0.57, 0.76)	0.45 (0.38, 0.52)
Postpartum	0.71 (0.57, 0.86)	0.62 (0.48, 0.76)	0.53 (0.43, 0.64)	0.35 (0.26, 0.44)
Whole fruit [b]				
Pregnancy	0.97 (0.77, 1.16)	0.71 (0.60, 0.81)	0.56 (0.46, 0.65)	0.31 (0.25, 0.37)
Postpartum	0.62 (0.48, 0.75)	0.50 (0.38, 0.62)	0.45 (0.35, 0.55)	0.24 (0.17, 0.31)
Whole grains [c]				
Pregnancy	0.64 (0.47, 0.81)	0.56 (0.47, 0.66)	0.55 (0.45, 0.64)	0.39 (0.32, 0.47)
Postpartum	0.68 (0.52, 0.85)	0.77 (0.56, 0.99)	0.48 (0.37, 0.59)	0.45 (0.35, 0.55)
Dairy [b]				
Pregnancy	1.06 (0.92, 1.20)	1.03 (0.95, 1.12)	0.99 (0.91, 1.08)	0.90 (0.82, 0.97)
Postpartum	0.82 (0.67, 0.97)	0.96 (0.81, 1.11)	0.97 (0.86, 1.08)	0.83 (0.71, 0.95)
Total protein foods [c]				
Pregnancy	3.78 (3.38, 4.18)	3.26 (3.04, 3.47)	2.89 (2.63, 3.15)	2.43 (2.25, 2.61)
Postpartum	4.54 (4.17, 4.90)	3.68 (3.33, 4.02)	3.21 (2.90, 3.53)	2.88 (2.56, 3.21)
Seafood & plant protein [c]				
Pregnancy	1.56 (1.25, 1.85)	1.19 (1.04, 1.35)	0.96 (0.72, 1.20)	0.50 (0.37, 0.62)
Postpartum	1.92 (1.62, 2.22)	1.47 (1.19, 1.75)	1.13 (0.87, 1.39)	0.66 (0.46, 0.87)
Fatty acid ratio [d]				
Pregnancy	1.86 (1.73, 1.99)	1.70 (1.62, 1.79)	1.69 (1.61, 1.78)	1.75 (1.67, 1.83)
Postpartum	2.09 (1.92, 2.27)	1.93 (1.74, 2.11)	1.76 (1.64, 1.87)	1.80 (1.69, 1.91)
Sodium [e]				
Pregnancy	1.78 (1.68, 1.87)	1.77 (1.70, 1.84)	1.79 (1.72, 1.87)	1.76 (1.70, 1.81)
Postpartum	1.86 (1.74, 1.98)	1.83 (1.74, 1.91)	1.78 (1.71, 1.85)	1.88 (1.78, 1.98)
Refined grains [c]				
Pregnancy	2.18 (1.94, 2.41)	2.47 (2.28, 2.66)	2.75 (2.57, 2.93)	3.18 (3.01, 3.36)
Postpartum	1.93 (1.62, 2.24)	2.27 (2.04, 2.50)	2.78 (2.52, 3.05)	2.97 (2.72, 3.22)
Saturated fat [f]				
Pregnancy	11.5 (10.7, 12.4)	12.7 (12.2, 13.3)	12.9 (12.4, 13.4)	12.6 (12.0, 13.1)
Postpartum	12.2 (11.0, 13.4)	11.7 (10.8, 12.5)	12.7 (12.0, 13.4)	12.6 (11.9, 13.2)

Table 3. *Cont.*

	Percent of Energy Intake from Ultra-Processed Foods Mean (95% Confidence Interval)			
	<40.0 [a]	40.0–49.9 [a]	50.0–59.9 [a]	≥60 [a]
Added sugar [f]				
Pregnancy	6.5 (5.5, 7.6)	9.2 (8.5, 9.9)	10.7 (10.0, 11.5)	13.4 (12.3, 14.6)
Postpartum	5.9 (5.0, 6.8)	8.4 (7.6, 9.2)	10.3 (9.4, 11.2)	12.8 (11.3, 14.3)

[a] Sample distribution in pregnancy: <40.0 n = 65; 40.0–49.9 n = 95; 50.0–59.9 n = 101; ≥60 n = 104. Sample distribution in postpartum: <40.0 n = 68; 40.0–49.9 n = 61; 50.0–59.9 n = 62; ≥60 n = 75. [b] cup equivalents/1000 kcal. [c] oz equivalents/1000 kcal. [d] (PUFAs + MUFAs)/SFAs. [e] g/1000 kcal. [f] % of energy intake.

4. Discussion

Ultra-processed food comprised over half of energy intake during both pregnancy and postpartum in this North Carolina sample, similar to the general U.S. population, in which ultra-processed food contributes 57% of total energy intake [15]. Ultra-processed food intake was greater in younger participants and those having less than a bachelor's degree; associations of ultra-processed food intake with income-poverty ratio, marital status, and body mass index were not significant after adjusting for other participant characteristics. Consistent with previous research in general population samples demonstrating an inverse association of ultra-processed food intake with several indicators of overall diet quality [18,26–28], intake of ultra-processed food in pregnancy and postpartum was inversely associated with the HEI total score and with 8 of 13 HEI component scores. Differences in diet quality between those consuming <40% energy intake from ultra-processed food and those consuming ≥60% energy intake from ultra-processed food were of considerable magnitude, with more than a 17-point difference in the HEI total score and a 1.5- to 3-fold difference in intake of eight component food groups.

Ultra-processed food intake was inversely associated with intake of both adequacy and moderation HEI component scores. Greater intake of ultra-processed food was associated with lower intake of total vegetables, greens and beans, total fruit, whole fruit, total protein, and seafood and plant protein. Several of these food groups provide nutrients that are critically important during pregnancy, including protein (total protein, seafood and plant protein); iron, vitamin B6, and zinc (protein food groups, greens and beans, fruit), folate (greens and beans, fruit, seafood and plant protein), vitamin E (plant protein, fruit, vegetables), and magnesium (greens and beans, plant protein) [2]. Greater intake of ultra-processed food was also associated with greater intake of refined grains and added sugar, which provide minimal nutritional value and contribute to excess energy intake. Insufficient adequacy components and excess moderation components both represent risk to the developing fetus. As described by the developmental origins of health and disease hypothesis, inadequate intake of key nutrients, as well as excess energy intake during fetal development, may result in reprogramming within fetal tissues that predisposes the offspring to the development of future chronic disease [37]. The absence of an association of ultra-processed food intake with saturated fat intake is contrary to previous research in general population samples [16–21,23], and may be attributable to widespread reformulation of processed food products to avoid hydrogenated fats.

The relations of greater intake of ultra-processed food with lower intake of fruits and vegetables observed herein suggests that ultra-processed foods, which may be perceived as more palatable or convenient, displace intake of whole plant foods. Findings from rodent models indicate that exposure to highly palatable foods leads to rapid suppression of motivation and preference for and intake of standard chow [38], suggesting a mechanism by which widespread exposure to highly palatable, ultra-processed foods may displace consumption of unprocessed foods in free-living humans. In response to systematic findings that ultra-processed food intake is associated with obesity and chronic disease [39], previous authors have discussed the merits of public health policies targeting reformulation versus reduction of ultra-processed food [40]. Reformulation of ultra-processed food could improve diet quality through reduction of added sugar and refined grains, and by addition of fruit, vegetables, and plant protein where feasible (e.g., to packaged meals). However, to the extent that highly palatable ultra-processed foods displace intake of whole plant and protein foods, reformulating ultra-processed food to reduce undesirable nutrients is unlikely to be sufficient to negate the adverse impact of ultra-processed food on overall diet quality.

Given the critical role of nutrition during pregnancy and postpartum for maternal and child health [3] and emerging evidence that ultra-processed food is associated with adverse pregnancy and infant outcomes [41–45], findings from this study suggest that dietary guidance to reduce ultra-processed food intake may represent a singular intervention target with broad impact on adherence to dietary guidelines. Considering health care provider concerns that time constraints limit their ability to provide nutritional guidance during routine pregnancy visits [46], brief messages with broad-ranging impact may be especially valuable. Interventions promoting intake of minimally processed foods in Brazilian pregnant persons have resulted in reduced intake of ultra-processed foods [47] and reduced risk of gestational weight gain [48], providing evidence for the feasibility of such an approach Additionally, nutrition educators may assist pregnant persons in selecting the most nutritious options when they do choose to consume ultra-processed food, taking into consideration the resources and foods available to their clients.

Findings should be interpreted with consideration of the study's strengths and limitations. While all methods of self-reported dietary intake assessment have some measurement error due to faulty recall, over-reporting, and under-reporting, 24-h dietary recalls are the least biased self-report measure of dietary intake [49]. NOVA classifications were determined using standardized statistical code and disaggregating foods to their underlying ingredients [33] to ensure accurate application of the classification system. The sample size was relatively large with a wide range of diet quality represented and had similar sociodemographic characteristics to the geographic area of the study site. This county in North Carolina has somewhat higher median household income ($74,803) than the overall U.S. population ($64,994), but with a similar percent of persons in poverty (10% vs. 11% nationally). The area is also more highly educated (61% vs. 33% having a bachelor's degree or higher) and has a higher proportion of non-Hispanic White race/ethnicity (69% vs. 59% White) [50,51]. Future research in samples disproportionally affected by low resources or low access to grocery outlets would be informative, as the adverse impact of ultra-processed foods on diet quality could be even more pronounced in populations lacking access to a wide range of foods.

In conclusion, in this U.S. sample followed from early pregnancy through one-year postpartum, ultra-processed food accounted for more than half of energy intake and was associated with lower diet quality across most HEI components. Intervention studies testing the impact of approaches targeting reduction in ultra-processed food intake on diet quality are needed to determine their efficacy in improving adherence to dietary guidelines in pregnant populations.

Author Contributions: Conceptualization, T.R.N., J.R.C., K.B., A.M.S.-R. and L.M.L.; Formal analysis, T.R.N.; Data curation, L.M.L. and T.R.N.; Writing—Original draft preparation, T.R.N.; Writing—Review and editing, T.R.N., J.R.C., K.B., A.M.S.-R. and L.M.L.; Supervision, T.R.N. and A.M.S.-R.; Project administration, T.R.N., A.M.S.-R. and L.M.L.; Funding acquisition, T.R.N., L.M.L. and A.M.S.-R. All authors have read and agreed to the published version of the manuscript.

Funding: This research was supported by the Eunice Kennedy Shriver National Institute of Child Health and Human Development Intramural Research Program (contract # HHSN275201300012C and #HHSN275201300026I/HHSN27500002).

Institutional Review Board Statement: The PEAS study was approved by the University of North Carolina Institutional Review Board (IRB number 13-3848).

Informed Consent Statement: Signed informed consent was obtained from all study participants.

Data Availability Statement: Data described in the manuscript, code book, and analytic code will be made available upon request pending application and approval.

Acknowledgments: The authors acknowledge the contributions of the PEAS research team and study staff at University of North Carolina, EMMES Corporation who served as the data coordinating center, and the study participants.

Conflicts of Interest: The authors have no competing interest to declare.

References

1. Hanson, M.A.; McAuliffe, F.M.; Louise Killeen, S.; Maria Jacob, C.; Hod, M. New guidelines, position paper, and insights from the FIGO Pregnancy Obesity and Nutrition Initiative (PONI). *Int. J. Gynaecol. Obstet.* **2020**, *151*, 1–3. [CrossRef] [PubMed]
2. Kominiarek, M.A.; Rajan, P. Nutrition Recommendations in Pregnancy and Lactation. *Med. Clin. N. Am.* **2016**, *100*, 1199–1215. [CrossRef] [PubMed]
3. Procter, S.B.; Campbell, C.G. Position of the Academy of Nutrition and Dietetics: Nutrition and lifestyle for a healthy pregnancy outcome. *J. Acad. Nutr. Diet.* **2014**, *114*, 1099–1103. [CrossRef]
4. Bodnar, L.M.; Simhan, H.N.; Parker, C.B.; Meier, H.; Mercer, B.M.; Grobman, W.A.; Haas, D.M.; Wing, D.A.; Hoffman, M.K.; Parry, S.; et al. Racial or Ethnic and Socioeconomic Inequalities in Adherence to National Dietary Guidance in a Large Cohort of US Pregnant Women. *J. Acad. Nutr. Diet.* **2017**, *117*, 867–877.e863. [CrossRef] [PubMed]
5. Rifas-Shiman, S.L.; Rich-Edwards, J.W.; Kleinman, K.P.; Oken, E.; Gillman, M.W. Dietary quality during pregnancy varies by maternal characteristics in Project Viva: A US cohort. *J. Am. Diet. Assoc.* **2009**, *109*, 1004–1011. [CrossRef]
6. Siega-Riz, A.M.; Bodnar, L.M.; Savitz, D.A. What are pregnant women eating? Nutrient and food group differences by race. *Am. J. Obstet. Gynecol.* **2002**, *186*, 480–486. [CrossRef] [PubMed]
7. Shah, B.S.; Freeland-Graves, J.H.; Cahill, J.M.; Lu, H.; Graves, G.R. Diet quality as measured by the healthy eating index and the association with lipid profile in low-income women in early postpartum. *J. Am. Diet. Assoc.* **2010**, *110*, 274–279. [CrossRef] [PubMed]
8. van der Pligt, P.; Olander, E.K.; Ball, K.; Crawford, D.; Hesketh, K.D.; Teychenne, M.; Campbell, K. Maternal dietary intake and physical activity habits during the postpartum period: Associations with clinician advice in a sample of Australian first time mothers. *BMC Pregnancy Childbirth* **2016**, *16*, 27. [CrossRef]
9. Abdollahi, S.; Soltani, S.; de Souza, R.J.; Forbes, S.C.; Toupchian, O.; Salehi-Abargouei, A. Associations between Maternal Dietary Patterns and Perinatal Outcomes: A Systematic Review and Meta-Analysis of Cohort Studies. *Adv. Nutr.* **2021**, *12*, 1332–1352. [CrossRef] [PubMed]
10. Borge, T.C.; Aase, H.; Brantsæter, A.L.; Biele, G. The importance of maternal diet quality during pregnancy on cognitive and behavioural outcomes in children: A systematic review and meta-analysis. *BMJ Open* **2017**, *7*, e016777. [CrossRef] [PubMed]
11. Chen, L.W.; Aris, I.M.; Bernard, J.Y.; Tint, M.T.; Chia, A.; Colega, M.; Gluckman, P.D.; Shek, L.P.; Saw, S.M.; Chong, Y.S.; et al. Associations of Maternal Dietary Patterns during Pregnancy with Offspring Adiposity from Birth Until 54 Months of Age. *Nutrients* **2016**, *9*, 2. [CrossRef] [PubMed]
12. Zhang, Y.; Lin, J.; Fu, W.; Liu, S.; Gong, C.; Dai, J. Mediterranean diet during pregnancy and childhood for asthma in children: A systematic review and meta-analysis of observational studies. *Pediatr Pulmonol.* **2019**, *54*, 949–961. [CrossRef] [PubMed]
13. Schoenaker, D.A.; Mishra, G.D.; Callaway, L.K.; Soedamah-Muthu, S.S. The Role of Energy, Nutrients, Foods, and Dietary Patterns in the Development of Gestational Diabetes Mellitus: A Systematic Review of Observational Studies. *Diabetes Care* **2016**, *39*, 16–23. [CrossRef] [PubMed]
14. Monteiro, C.A.; Cannon, G.; Levy, R.B.; Moubarac, J.C.; Louzada, M.L.; Rauber, F.; Khandpur, N.; Cediel, G.; Neri, D.; Martinez-Steele, E.; et al. Ultra-processed foods: What they are and how to identify them. *Public Health Nutr.* **2019**, *22*, 936–941. [CrossRef] [PubMed]
15. Juul, F.; Parekh, N.; Martinez-Steele, E.; Monteiro, C.A.; Chang, V.W. Ultra-processed food consumption among US adults from 2001 to 2018. *Am. J. Clin. Nutr.* **2022**, *115*, 211–221. [CrossRef]
16. Calixto Andrade, G.; Julia, C.; Deschamps, V.; Srour, B.; Hercberg, S.; Kesse-Guyot, E.; Allès, B.; Chazelas, E.; Deschasaux, M.; Touvier, M.; et al. Consumption of Ultra-Processed Food and Its Association with Sociodemographic Characteristics and Diet Quality in a Representative Sample of French Adults. *Nutrients* **2021**, *13*, 682. [CrossRef] [PubMed]
17. Cediel, G.; Reyes, M.; Corvalán, C.; Levy, R.B.; Uauy, R.; Monteiro, C.A. Ultra-processed foods drive to unhealthy diets: Evidence from Chile. *Public Health Nutr.* **2021**, *24*, 1698–1707. [CrossRef]
18. Griffin, J.; Albaloul, A.; Kopytek, A.; Elliott, P.; Frost, G. Effect of ultraprocessed food intake on cardiometabolic risk is mediated by diet quality: A cross-sectional study. *BMJ Nutr. Prev. Health* **2021**, *4*, 174–180. [CrossRef]
19. Koiwai, K.; Takemi, Y.; Hayashi, F.; Ogata, H.; Matsumoto, S.; Ozawa, K.; Machado, P.P.; Monteiro, C.A. Consumption of ultra-processed foods decreases the quality of the overall diet of middle-aged Japanese adults. *Public Health Nutr.* **2019**, *22*, 2999–3008. [CrossRef]
20. Marrón-Ponce, J.A.; Flores, M.; Cediel, G.; Monteiro, C.A.; Batis, C. Associations between Consumption of Ultra-Processed Foods and Intake of Nutrients Related to Chronic Non-Communicable Diseases in Mexico. *J. Acad. Nutr. Diet.* **2019**, *119*, 1852–1865. [CrossRef]

21. Martini, D.; Godos, J.; Bonaccio, M.; Vitaglione, P.; Grosso, G. Ultra-Processed Foods and Nutritional Dietary Profile: A Meta-Analysis of Nationally Representative Samples. *Nutrients* **2021**, *13*, 3390. [CrossRef] [PubMed]
22. Neri, D.; Steele, E.M.; Khandpur, N.; Cediel, G.; Zapata, M.E.; Rauber, F.; Marrón-Ponce, J.A.; Machado, P.; da Costa Louzada, M.L.; Andrade, G.C.; et al. Ultraprocessed food consumption and dietary nutrient profiles associated with obesity: A multicountry study of children and adolescents. *Obes. Rev.* **2022**, *23*, e13387. [CrossRef] [PubMed]
23. Shim, J.S.; Shim, S.Y.; Cha, H.J.; Kim, J.; Kim, H.C. Association between Ultra-processed Food Consumption and Dietary Intake and Diet Quality in Korean Adults. *J. Acad. Nutr. Diet.* **2022**, *122*, 583–594. [CrossRef]
24. Machado, P.P.; Steele, E.M.; Louzada, M.; Levy, R.B.; Rangan, A.; Woods, J.; Gill, T.; Scrinis, G.; Monteiro, C.A. Ultra-processed food consumption drives excessive free sugar intake among all age groups in Australia. *Eur. J. Nutr.* **2020**, *59*, 2783–2792. [CrossRef]
25. Neri, D.; Martinez-Steele, E.; Monteiro, C.A.; Levy, R.B. Consumption of ultra-processed foods and its association with added sugar content in the diets of US children, NHANES 2009-2014. *Pediatr. Obes.* **2019**, *14*, e12563. [CrossRef] [PubMed]
26. Liu, J.; Steele, E.M.; Li, Y.; Karageorgou, D.; Micha, R.; Monteiro, C.A.; Mozaffarian, D. Consumption of Ultraprocessed Foods and Diet Quality Among U.S. Children and Adults. *Am. J. Prev. Med.* **2022**, *62*, 252–264. [CrossRef]
27. Gupta, S.; Rose, C.M.; Buszkiewicz, J.; Ko, L.K.; Mou, J.; Cook, A.; Aggarwal, A.; Drewnowski, A. Characterising percentage energy from ultra-processed foods by participant demographics, diet quality and diet cost: Findings from the Seattle Obesity Study (SOS) III. *Br. J. Nutr.* **2021**, *126*, 773–781. [CrossRef]
28. Lauria, F.; Dello Russo, M.; Formisano, A.; De Henauw, S.; Hebestreit, A.; Hunsberger, M.; Krogh, V.; Intemann, T.; Lissner, L.; Molnar, D.; et al. Ultra-processed foods consumption and diet quality of European children, adolescents and adults: Results from the I.Family study. *Nutr. Metab. Cardiovasc. Dis.* **2021**, *31*, 3031–3043. [CrossRef]
29. Graciliano, N.G.; Silveira, J.; Oliveira, A.C.M. The consumption of ultra-processed foods reduces overall quality of diet in pregnant women. *Cad. Saude Publica* **2021**, *37*, e00030120. [CrossRef]
30. Nansel, T.R.; Lipsky, L.M.; Siega-Riz, A.M.; Burger, K.; Faith, M.; Liu, A. Pregnancy eating attributes study (PEAS): A cohort study examining behavioral and environmental influences on diet and weight change in pregnancy and postpartum. *BMC Nutr.* **2016**, *2*, 45. [CrossRef]
31. Kirkpatrick, S.I.; Subar, A.F.; Douglass, D.; Zimmerman, T.P.; Thompson, F.E.; Kahle, L.L.; George, S.M.; Dodd, K.W.; Potischman, N. Performance of the Automated Self-Administered 24-hour Recall relative to a measure of true intakes and to an interviewer-administered 24-h recall. *Am. J. Clin. Nutr.* **2014**, *100*, 233–240. [CrossRef] [PubMed]
32. Thompson, F.E.; Dixit-Joshi, S.; Potischman, N.; Dodd, K.W.; Kirkpatrick, S.I.; Kushi, L.H.; Alexander, G.L.; Coleman, L.A.; Zimmerman, T.P.; Sundaram, M.E.; et al. Comparison of Interviewer-Administered and Automated Self-Administered 24-Hour Dietary Recalls in 3 Diverse Integrated Health Systems. *Am. J. Epidemiol.* **2015**, *181*, 970–978. [CrossRef] [PubMed]
33. Martínez Steele, E.; Baraldi, L.G.; Louzada, M.L.; Moubarac, J.C.; Mozaffarian, D.; Monteiro, C.A. Ultra-processed foods and added sugars in the US diet: Evidence from a nationally representative cross-sectional study. *BMJ Open* **2016**, *6*, e009892. [CrossRef] [PubMed]
34. Krebs-Smith, S.M.; Pannucci, T.E.; Subar, A.F.; Kirkpatrick, S.I.; Lerman, J.L.; Tooze, J.A.; Wilson, M.M.; Reedy, J. Update of the Healthy Eating Index: HEI-2015. *J. Acad. Nutr. Diet.* **2018**, *118*, 1591–1602. [CrossRef] [PubMed]
35. U.S. Department of Health and Human Services and U.S. Department of Agriculture. 2015–2020 Dietary Guidelines for Americans. 8th ed. December 2015. Available online: http://health.gov/dietaryguidelines/2015/guidelines/ (accessed on 30 August 2022).
36. U.S. Census Bureau. Fertility & Family Statistics Branch. Current Population Survey: Definitions and Explanations. Available online: http://www.census.gov/population/www/cps/cpsdef.html (accessed on 10 January 2019).
37. McMillen, I.C.; MacLaughlin, S.M.; Muhlhausler, B.S.; Gentili, S.; Duffield, J.L.; Morrison, J.L. Developmental origins of adult health and disease: The role of periconceptional and foetal nutrition. *Basic Clin. Pharmacol. Toxicol.* **2008**, *102*, 82–89. [CrossRef] [PubMed]
38. Lalanza, J.F.; Snoeren, E.M.S. The cafeteria diet: A standardized protocol and its effects on behavior. *Neurosci. Biobehav. Rev.* **2021**, *122*, 92–119. [CrossRef]
39. Chen, X.; Zhang, Z.; Yang, H.; Qiu, P.; Wang, H.; Wang, F.; Zhao, Q.; Fang, J.; Nie, J. Consumption of ultra-processed foods and health outcomes: A systematic review of epidemiological studies. *Nutr. J.* **2020**, *19*, 86. [CrossRef]
40. Tobias, D.K.; Hall, K.D. Eliminate or reformulate ultra-processed foods? Biological mechanisms matter. *Cell Metab.* **2021**, *33*, 2314–2315. [CrossRef]
41. Cummings, J.R.; Lipsky, L.M.; Schwedhelm, C.; Liu, A.; Nansel, T.R. Associations of ultra-processed food intake with maternal weight change and cardiometabolic health and infant growth. *Int. J. Behav. Nutr. Phys. Act.* **2022**, *19*, 61. [CrossRef]
42. Gomes, C.B.; Malta, M.B.; Benício, M.H.D.; Carvalhaes, M. Consumption of ultra-processed foods in the third gestational trimester and increased weight gain: A Brazilian cohort study. *Public Health Nutr.* **2021**, *24*, 3304–3312. [CrossRef]
43. Rodríguez-Cano, A.M.; González-Ludlow, I.; Suárez-Rico, B.V.; Montoya-Estrada, A.; Piña-Ramírez, O.; Parra-Hernández, S.B.; Reyes-Muñoz, E.; Estrada-Gutierrez, G.; Calzada-Mendoza, C.C.; Perichart-Perera, O. Ultra-Processed Food Consumption during Pregnancy and Its Association with Maternal Oxidative Stress Markers. *Antioxidants* **2022**, *11*, 1415. [CrossRef] [PubMed]
44. Silva, C.F.M.; Saunders, C.; Peres, W.; Folino, B.; Kamel, T.; Dos Santos, M.S.; Padilha, P. Effect of ultra-processed foods consumption on glycemic control and gestational weight gain in pregnant with pregestational diabetes mellitus using carbohydrate counting. *PeerJ* **2021**, *9*, e10514. [CrossRef] [PubMed]

45. Smit, A.J.P.; Hojeij, B.; Rousian, M.; Schoenmakers, S.; Willemsen, S.P.; Steegers-Theunissen, R.P.M.; van Rossem, L. A high periconceptional maternal ultra-processed food consumption impairs embryonic growth: The Rotterdam periconceptional cohort. *Clin. Nutr.* **2022**, *41*, 1667–1675. [CrossRef] [PubMed]
46. Killeen, S.L.; Callaghan, S.L.; Jacob, C.M.; Hanson, M.A.; McAuliffe, F.M. Examining the use of the FIGO Nutrition Checklist in routine antenatal practice: Multistakeholder feedback to implementation. *Int. J. Gynecol. Obstet.* **2020**, *151*, 51–56. [CrossRef] [PubMed]
47. Gomes, C.B.; Malta, M.B.; Louzada, M.; Benício, M.H.D.; Barros, A.J.D.; Carvalhaes, M. Ultra-processed Food Consumption by Pregnant Women: The Effect of an Educational Intervention with Health Professionals. *Matern Child Health J.* **2019**, *23*, 692–703. [CrossRef] [PubMed]
48. Sartorelli, D.S.; Crivellenti, L.C.; Baroni, N.F.; de Andrade Miranda, D.E.G.; da Silva Santos, I.; Carvalho, M.R.; de Lima, M.C.; Carreira, N.P.; Chaves, A.V.L.; Manochio-Pina, M.G.; et al. Effectiveness of a minimally processed food-based nutritional counselling intervention on weight gain in overweight pregnant women: A randomized controlled trial. *Eur. J. Nutr.* **2022**. [CrossRef] [PubMed]
49. Prentice, R.L.; Mossavar-Rahmani, Y.; Huang, Y.; Van Horn, L.; Beresford, S.A.; Caan, B.; Tinker, L.; Schoeller, D.; Bingham, S.; Eaton, C.B.; et al. Evaluation and comparison of food records, recalls, and frequencies for energy and protein assessment by using recovery biomarkers. *Am. J. Epidemiol.* **2011**, *174*, 591–603. [CrossRef]
50. U.S. Census Bureau. U.S.C. Quick Facts: Orange County, North Carolina. Available online: https://www.census.gov/quickfacts/fact/table/orangecountynorthcarolina/PST045221 (accessed on 30 August 2022).
51. U.S. Census Bureau. U.S.C. Quick Facts United States. Available online: https://www.census.gov/quickfacts/fact/table/US/PST045221 (accessed on 30 August 2022).

nutrients

Article

A Comparison of the Australian Dietary Guidelines to the NOVA Classification System in Classifying Foods to Predict Energy Intakes and Body Mass Index

Amanda Grech [1,2,*], Anna Rangan [1,3], Margaret Allman-Farinelli [1,3], Stephen J. Simpson [1,2], Tim Gill [1,3] and David Raubenheimer [1,2,*]

[1] Charles Perkins Centre, University of Sydney, Sydney, NSW 2006, Australia
[2] School of Life and Environmental Science, Faculty of Science, University of Sydney, Sydney, NSW 2006, Australia
[3] School of Nursing, Faculty of Medicine and Health, University of Sydney, Sydney, NSW 2006, Australia
* Correspondence: amanda.grech@sydney.edu.au (A.G.); david.raubenheimer@sydney.edu.au (D.R.)

Abstract: NOVA classification distinguishes foods by level of processing, with evidence suggesting that a high intake of ultra-processed foods (UPFs, NOVA category 4) leads to obesity. The Australian Dietary Guidelines, in contrast, discourage excess consumption of "discretionary foods" (DFs), defined according to their composition. Here, we (i) compare the classification of Australian foods under the two systems, (ii) evaluate their performance in predicting energy intakes and body mass index (BMI) in free-living Australians, and (iii) relate these outcomes to the protein leverage hypothesis of obesity. Secondary analysis of the Australian National Nutrition and Physical Activity Survey was conducted. Non-protein energy intake increased by 2.1 MJ ($p < 0.001$) between lowest and highest tertiles of DF intake, which was significantly higher than UPF (0.6 MJ, $p < 0.001$). This demonstrates that, for Australia, the DF classification better distinguishes foods associated with high energy intakes than does the NOVA system. BMI was positively associated with both DFs (−1. 0, $p = 0.0001$) and UPFs (−1.1, $p = 0.0001$) consumption, with no difference in strength of association. For both classifications, macronutrient and energy intakes conformed closely to the predictions of protein leverage. We account for the similarities and differences in performance of the two systems in an analysis of Australian foods.

Keywords: obesity; ultra-processed foods; dietary guidelines; non communicable disease; protein leverage hypothesis; macronutrient intake

Citation: Grech, A.; Rangan, A.; Allman-Farinelli, M.; Simpson, S.J.; Gill, T.; Raubenheimer, D. A Comparison of the Australian Dietary Guidelines to the NOVA Classification System in Classifying Foods to Predict Energy Intakes and Body Mass Index. *Nutrients* **2022**, *14*, 3942. https://doi.org/10.3390/nu14193942

Academic Editors: Monica Dinu and Daniela Martini

Received: 29 August 2022
Accepted: 19 September 2022
Published: 23 September 2022

1. Introduction

The proliferation of industrially formulated, affordable, palatable, aggressively marketed, and convenient foods has been linked to the rise in obesity and certain noncommunicable diseases [1]. Classification systems have thus been developed which distinguish foods according to the degree of industrial processing [2–4]. The NOVA system is the most applied processed food classification system in the academic literature [1]. It uses four categories of processing to classify foods. Category 1 features unprocessed and minimally processed foods, category 2 features culinary ingredients, category 3 features processed foods (combining ingredients from categories 1 and 2), and category 4 features ultra-processed foods (UPFs) [5]. UPFs are industrially produced foods that have been formulated with cosmetic food additives such as flavours, colours, and emulsifiers, with or without the addition of ingredients such as cheap oils, refined sugars and starches, and added salt [5]. Although the NOVA system is the most prominent processing-based classification system, despite accumulating evidence for the health risks associated with increased consumption of UPFs, it has not been adopted widely as the basis for dietary guidance [1]. Surprisingly, there are few direct comparisons of processing-based and alternative classification systems that are already widely used.

Most diet classification systems have attempted to identify unhealthy foods based on nutrient composition [6–11]. One example is the "discretionary foods" category in the Australian Dietary Guidelines (ADG). The ADG identifies healthy foods but contrasts with the USDA guidelines because it has a discrete category for less healthy foods [6,11]. These are defined as energy-dense, nutrient-poor foods that do not fit within one of five food group categories and are characterised by their high content of added sugars, saturated fats, added salt, and alcoholic beverages [6]. Discretionary foods are advised to be avoided to minimise the risk of noncommunicable disease and for weight maintenance, due to their high energy density and low nutrient content [6]. The systematic process adopted in the development of the ADG has been appraised to be amongst the most rigorous of 32 food-based dietary guidelines considered [12].

Fundamental to determining whether classification systems based on degree of industrial processing or nutrient content are more accurate at identifying foods that should be limited in the diet is understanding how these foods interact with appetite and metabolic biology to increase energy intakes and predispose populations to weight gain and poor health [13]. One possible mechanism is the protein leverage model, which proposes that, when faced with food environments where protein is diluted by fat and carbohydrate, the dominant human appetite for protein drives energy over consumption as an inadvertent outcome of the strong drive to meet the protein target [14–16]. There is a growing body of evidence to support the protein leverage model as a significant determinant of excess energy intake and obesity [16–18].

Two studies have directly linked UPF to the protein leverage model. An RCT conducted in a controlled environment demonstrated that UPF intake contributes to higher energy intakes and weight gain compared to a diet composed of minimally processed foods. As predicted by the protein leverage model, participants consumed foods that amounted to the same intake of protein (490 kcal) on both the minimally processed and the UPF diets, but non-protein energy increased by 509 kcal on the UPF diet [19]. Similarly, in a population study, increased UPF intake was associated with protein dilution and higher energy intakes in NHANES, whereas protein intake remained near constant [20]. However, no study has examined discretionary foods in the context of the protein leverage mechanism, nor compared the performance in predicting excess energy intake and BMI of this nutrient-based classification with processing-based classification systems.

The primary aim of the present analysis was to compare NOVA classification system with the ADG in classifying foods as healthy and unhealthy through their efficacy in predicting energy over-consumption and BMI within the Australian food system. We do so in an analysis of the National Nutrition and Physical Activity Survey, a large representative cross-sectional study of the Australian population, through (i) a comparison of the foods classified as healthy and unhealthy between the two systems, (ii) a comparison of the performance of the two systems in predicting energy intakes and body mass index, and (iii) a comparison of how discretionary food vs. UPF relate to the mechanistic model of protein leverage for predicting the relationship between dietary composition and energy intake.

2. Materials and Methods

2.1. Study Design

The data for the present analysis came from the National Nutrition and Physical Activity Survey, 2011–2012 [21]. The survey was conducted by the Australian Bureau of Statistics (ABS) using a stratified, multistage area sample of private dwellings across Australia to provide a representative sample of the population of the country. A random subsample of residents from each household were selected to participate in the survey. Eligible members of the household included one adult (aged 19 years and older) and (where applicable) one child aged 2–18 years (NNPAS). The survey was conducted between 29 May 2011 to 9 June 2012. Collection days included Monday through to Sunday. Data were collected by trained ABS interviewers, through computer-assisted personal interview

(CAPI) with eligible members of the household. For the present analysis, adults aged 19 years and over were included. Full details of the survey are published elsewhere [21]. Ethics approval was granted under the Census Act 1905 by the Australian Government in accordance with relevant guidelines and regulations, and written and informed consent was obtained from all subjects and/or their legal guardian(s) [21].

2.2. Variables

The exposure of interest was the percentage of total energy (%E) consumed from UPF and from discretionary foods. The outcomes included energy intake from protein and non-protein sources and body mass index.

Foods were categorised by the level of processing using the NOVA food classification system [5,22]. NOVA categories distinguish four groups of foods: (1) unprocessed or minimally processed foods (including fresh, frozen, squeezed, and dried foods); (2) culinary ingredients (e.g., flour, sugar, salt and vegetable oils); (3) processed foods, which are composite foods including ingredients from both group 1 and group 2 (e.g., canned or bottled group 1 foods); (4) UPF, which are often highly concentrated in fats and sugars from group 2 and include ingredients that have been highly processed or are industrially manufactured to aid processing and not normally found in culinary preparations. Examples of such ingredients include dyes, colours, flavour enhancers, flavourings, anti-caking agents, and emulsifiers or can include extracts derived from processing group 1 ingredients such as hydrolysed proteins, hydrogenated oils, high-fructose corn syrup, and maltodextrin. Detailed information on the classification system and the application of the NOVA classification system to the NNPAS is published in detail elsewhere [5,22].

Homemade mixed dishes and composite foods and recipes that were not deemed UPF, disaggregated into their individual components, and classified into their corresponding NOVA group. The recipes for 2585 mixed dishes and composite foods have been compiled by Food Standards Australia New Zealand (FSANZ) into the AUSNUT 2011–2013 food recipe file [23]. The recipe file was used to derive the ingredients for each of the mixed dishes/composite foods reported in the survey. An example is homemade banana bread (group 1: flour, egg, cinnamon, banana; group 2: butter, sugar). A small number of dishes (n = 38) did not have a recipe in the database. In this situation, a recipe with comparable ingredients and nutrient composition was selected from the AUSNUT 2011–2013 database [23,24]. Ingredients in recipes were matched to the AUSNUT 2011–2012 nutrient composition database. Recipe weight change factors were used to calculate the nutrient composition of the recipe accounting for differences in macronutrients of the raw and cooked weight of the recipe.

Discretionary foods are defined by the ADG as food and beverages that do not fit into the five food group foods (i.e., vegetables; fruits; milk, cheese, yoghurts, and non-dairy alternatives; lean meat, poultry, fish, seafood, nuts, seeds, and legumes; grains and cereals) and are high in saturated fat, added sugars, added salt, and/or alcohol [6]. Discretionary foods were identified using the defined ABS discretionary food list [21]. Examples of foods classified under the two systems are shown in Table 1.

Table 1. Contribution daily energy intake (%) of selected foods by NOVA classification system and Australian Dietary Guidelines.

Classification System	Five Food Groups Foods	Discretionary Foods	Disaggregated Discretionary Foods
Minimally processed foods	Tea, coffee, home squeezed juice, water, barley, cornmeal, millet, oats, quinoa, sago, rice bran, rice, wheat germ, wheat bran, couscous, flour, semolina, tapioca, noodles, pasta, natural muesli, fish, seafood, apple, pear and all frozen, fresh, and dried fruit, nuts, eggs, beef, lamb, pork, veal, goat, chicken, turkey, milk, plain yoghurt, seeds, psyllium, potato, carrot and all fresh, frozen and dried vegetables, herbs, lentils, beans		Homemade and takeaway foods † including sweet and savoury pastry, cakes, pies, French toast, cakes, muffins, slices, puddings, tarts, spring rolls, pizzas, waffles, deep fried fish and vegetables, cream-based desserts, sauces, jams and icings, pizzas (pepperoni, ham and cheese, meat lovers), quiche
Culinary ingredients	Olive oil, vinegar, flaxseed oil, rice bran oil, yeast, gelatine, canola oil, soybean oil, peanut oil, sunflower oil, vegetable oil, canola oil, gelatine, baking powder	Cream, butter, lard, ghee, sour cream, sugar, honey, and salt	
Processed foods	Homemade and artisan breads, salted nuts, nut spreads, cheese, tinned fruit, tinned meat and seafood, peanut butter, tinned vegetables, and legumes	Bacon, wine, beer, cider, chutneys, pickles, condensed milk, jam	
Ultra-processed foods	Commercial fruit juice, beverage (milo), commercial breads, commercial English muffins, instant noodles, breakfast cereals with low/no added sugar, savoury biscuits, commercial scones, fast food pizzas (<5 g saturated fat), fast food burgers (<5 g saturated fat), frozen meals, tinned spaghetti, commercial crumpets, margarine, sausages (<5 g saturated fats), breaded chicken, flavoured yoghurts, processed cheese, flavoured milks, soymilk, oat milk, tofu, tempeh, canned and packet soups (lower sodium), baked beans, intense sweeteners, oral supplements	Fruit drinks, sweetened drinks, cordial, soft drinks, flavoured beverage bases, sweet buns, breakfast cereals, commercial sweet biscuits, commercial garlic bread, ice cream cones, wafer commercial cakes, muffins, slices, pastries, commercial savoury pastries, fast food burgers, pizzas, frozen meals including pizzas, donuts, butter blends, Copha, frozen fish, sausages, ham, salami, other processed meats, chicken nuggets, ice cream, dairy desserts, packet soups, gravies, marinades, sauces, dressings and dips, fast foods and frozen potato fries, savoury snack foods, chocolates, confectionary, muesli bars, spirits, protein powder, yeast spreads	

† Discretionary mixed dishes that were disaggregated into their component ingredients.

2.3. Dietary Assessment

Data on the foods and beverages consumed from midnight to midnight on the day preceding the interview were collected. The 24 h dietary recall used was the five-pass, Automated Multiple-Pass Method (AMPM) originally developed by the Agricultural Research Service of the United States Department of Agriculture (USDA). The tool was modified by FSANZ and the ABS to reflect the Australian food supply and provided the details for over 10,000 individual and combined foods. Interviewers used the Food Model Booklet developed to aid participants in estimating portions and included life-sized photographs of food and beverages and food and beverage containers to reflect those used within Australia. A computer-assisted telephone interview (CATI) was used to conduct a second 24 h recall for a proportion (63.6%) of the survey participants, but only the first day of the survey was used, as this is an accurate representation of the mean population nutrient intake. The AUSNUT 2011–2013 food nutrient database, compiled by FSANZ specifically for NNPAS, was used to derive food and beverages nutrient composition. The database contains the nutrient composition for 5740 foods and beverages reported in the survey and reflects the nutrient composition of the Australian food supply [24]. The Australian food supply and food preparation practices from 2011 to 2012 were, therefore, captured in AUSNUT 2011–2013 and the accompanying measures database.

2.4. Implausible Energy Reporting

To identify participants with implausible energy intake, the Goldberg equation was used. Participants with an energy intake to basal metabolic rate ratio (EI:BMR) of <0.87 were classified as low energy reporters (LER) and high energy reporters were defined as those with an EI:BMR > 2.75. An EI:BMR ratio ≥0.87 and ≤2.75 is within the 95% CI of plausible energy intake assuming a sedentary physical activity level of 1.55 [25]. Sensitivity analysis was conducted by repeating all analyses with and without LER to determine the effect of implausible energy intakes on the outcomes. As there were no differences in the direction of the point estimates, all analysis presented include the full adult sample.

2.5. Quantitative Variables

Energy intake was calculated per day with Atwater factors as total protein energy (17 kJ/gram) and non-protein energy (kJ) (i.e., fat × 37 + total sugars × 16 + starch × 17 + other available carbohydrates (dextrin + maltodextrin + raffinose + stachyose + other undifferentiated oligosaccharides + glycogen) × 17 + alcohol × 29 + sorbitol/mannitol/glycerol × 16 + maltitol × 13 + citric/malic/quinic acids × 10 + lactic/acetic acids × 15 + dietary fibre × 8 + resistant starch × 8 + polydextrose × 5).

The percentage energy (%E) from discretionary foods and the %E from UPF was calculated for each person and participants were categorised into tertiles and quintiles of discretionary food intake and UPF (%E) using PROC RANK ($n = 3$ and $n = 5$, respectively).

Weight was measured using digital scales and height was measured using a stadiometer. All physical measurements were voluntary, and women who had identified they were pregnant were not measured. Respondents were encouraged to remove their shoes and any heavy clothing, but this was voluntary [21].

2.6. Statistical Methods

Univariate and multivariate linear regression was used to determine the relationship between discretionary food (%E) and UPF (%E) and the intake of protein energy and non-protein energy. Multivariate linear regression was used to determine the relationship between BMI kg/m^2 and discretionary food (%E) and UPF (%E) intake, classified into tertiles and quintiles. Multivariate analysis for energy intake was adjusted for the following factors: sex, age, smoking status. physical activity level, country of birth, and educational attainment. Complete case multivariate analysis for BMI was adjusted for sex, age, smoking status, and physical activity level. Estimates were weighted to reflect the Australian population distribution and probability of selection and replicate weights (the Jack-knife group delete one method) were used to compute standard errors. Analyses were conducted in SAS® version 9.4: SAS Institute Inc. Significant differences were considered for $p < 0{\cdot}05$.

3. Results

3.1. Participants

The demographics of participants by %E discretionary food intake and by UPF consumption is shown in Table 2. A total of 9341 adults reported dietary data on the first day of the survey and were included for analysis (Figure S1). Of these, 1486 adults were not included in the analysis of BMI due to missing data, as participants were only expected to record anthropometric data voluntarily, due to ethical considerations [21]. Participants who were older, female, higher socioeconomic index for area (SEIFA), university-educated, born in countries other than Australia or other English-speaking countries, and from major cities had lower mean intake from discretionary food (%E) and from UPF (%E) (Table 2).

Table 2. Percentage energy for discretionary food (DF) and ultra-processed foods (UPF) by demographics for Australian adults (n = 9341).

Demographics	%	(SE)	Mean DF %E		p-Trend	Mean UPF %E		p-Trend
Age								
19–30 years	23.1	0.3	34.0	(0.7)		43.9	(0.8)	
31–50 years	37.4	0.3	32.9	(0.4)		38.0	(0.4)	
51–70 years	28.7	0.2	31.2	(0.4)		34.5	(0.5)	
71+ years	10.8	0.1	31.6	(0.7)	0.0057	36.5	(0.7	<0.0001
Gender								
Female	49.4	(0.1)	30.7	(0.4)		37.5	(0.4)	
Male	50.6	(0.1)	34.3	(0.4)	<0.0001	38.8	(0.5)	0.0473
SEIFA								
Lowest (quintile 1)	18.1	(1.0)	33.8	(0.7)		40.1	(0.8)	
Middle (quintile 2–3)	59.7	(1.4)	32.6	(0.3)		38.4	(0.4)	
Highest (quintile 5)	22.2	(1.0)	31.2	(0.6)	0.0184	35.9	(0.7)	0.0013
Educational Attainment								
No tertiary education	38.8	(0.6)	33.5	(0.5)		40.1	(0.8)	
Vocational education	35.5	(0.7)	34.2	(0.4)		38.4	(0.4)	
University education	25.7	(0.7)	28.6	(0.5)	<0.0001	35.9	(0.7)	0.0013
Country of Birth								
Australia	68.8	(0.9)	34.6	(0.3)		40.3	(0.4)	
Other English-speaking countries	11.6	(0.4)	34.1	(0.8)		37.6	(0.8)	
Other	19.6	(0.8)	24.1	(0.6)	<0.0001	31.0	(0.7)	<0.0001
Geographic Area								
Major cities	71.5	(0.6)	31.3	(0.3)		37.3	(0.3)	
Inner regional	19.1	(0.8)	36.1	(0.6)		40.7	(0.7)	
Other	9.4	(0.8)	34.3	(0.9)	<0.0001	39.3	(1.1)	<0.0001
Energy Reporting Status								
Low (EI:BMR $\leq$ 0.87)	16.8	(0.5)	25.5	(0.7)		36.9	(0.7)	
Plausible (EI:BMR > 0.87)	69.2	(0.7)	34.6	(0.3)		38.7	(0.4)	
Missing	14.0	(0.4)	30.8	(0.7)	<0.0001	37.1	(0.7)	0.0427

SEIFA, socio-economic index for area.

3.2. Differences in UPFs and Discretionary Foods Classifications

In the AUSNUT 2011–2013 discretionary food list, 1631 foods of 5740 foods were classified as discretionary (28.4%). After disaggregation of the specified composite foods (e.g., homemade cakes) and mixed dishes (e.g., homemade pasta dishes) into individual ingredients, there were 2846 individual foods, of which 1016 (37.5%) were classed as discretionary foods and 1830 (64.3%) were five-food-group foods. Using the NOVA system to classify the disaggregated foods by degree of processing, 935 (32.9%) were classed as minimally processed foods, 60 (2.1%) were classed as culinary ingredients, 354 (12.4%) were classed as processed foods, and 1497 (52.6%) were classed as UPF.

All minimally processed foods, and most processed foods (83.9%) were classified as belonging to the five food groups under the ADG. A high proportion (63.3%) of culinary ingredients were classified as discretionary foods under the ADG. Similarly, a significant proportion of UPF (38.8%) were classified as belonging to the five food groups by the ADG (Figure 1a). The proportion of energy that came from discretionary foods and UPF, as classified by the ADG and NOVA, respectively, is shown in Figure 1b.

(**a**)

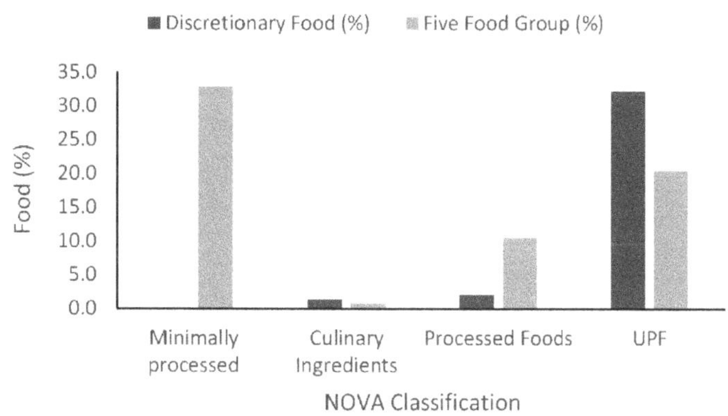

(**b**)

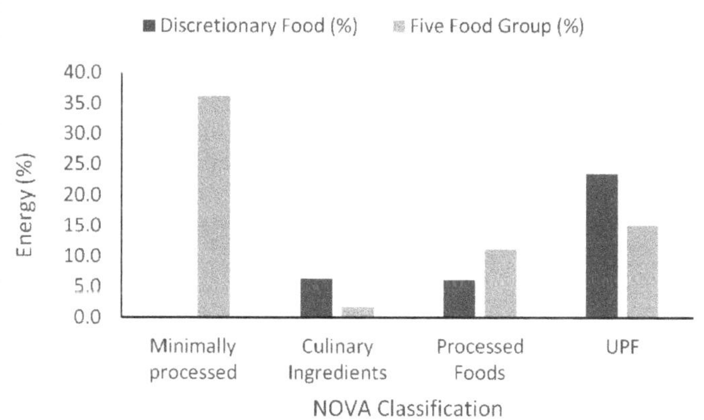

Figure 1. (**a**) Total number of foods reported in the survey (n = 2486); (**b**) proportion of daily energy from foods classified as discretionary foods or five-food-group foods (according to the Australian Dietary Guidelines) and by degree of processing (according to the NOVA classification system i.e., minimally processed, culinary ingredients, processed foods, or ultra-processed foods (UPF)) as reported by adults in the National Nutrition and Physical Activity Survey (n = 9431).

Across the surveyed population, discretionary foods accounted for 35.8 % and UPF accounted for 38.5% of total energy consumed. Almost 25% of daily energy came from foods classified as both discretionary food and UPF. However, there was a significant proportion of UPF classified as five-food-group foods, and 15.1% of daily energy came from five food group foods that were ultra-processed (Figure 1). Table S1 shows the proportion of energy from all foods reported in the National Nutrition and Physical Activity Survey classified by the NOVA classification system and the Australian Dietary Guidelines.

The observed agreement for participants to be classified in the same tertile of both discretionary food %E and UPF %E was 0.51, and the expected agreement was 0.3. A total of 38.4% of foods were classified in the adjacent tertile, and 10.4% were classified in the opposite tertile of discretionary food %E and UPF %E (i.e., the lowest discretionary food intake and the highest UPF intake and vice versa) (Table 3).

Table 3. Proportion of participants that were classified in the same, adjacent, and opposite tertile for percentage energy (%E) of discretionary food and ultra-processed food (UPF) (%E).

Discretionary Food (%E)	UPF (%E)	(*n*)	%
Tertile 1—lowest	Tertile 1—lowest	1721	18.4
Tertile 1—lowest	Tertile 2—middle	930	10.0
Tertile 1—lowest	Tertile 3—highest	462	4.9
Tertile 2—middle	Tertile 1—lowest	884	9.5
Tertile 2—middle	Tertile 2—middle	1319	14.1
Tertile 2—middle	Tertile 3—highest	911	9.8
Tertile 3—highest	Tertile 1—lowest	508	5.4
Tertile 3—highest	Tertile 2—middle	865	9.3
Tertile 3—highest	Tertile 3—highest	1741	18.6

Proportions in same tertile (51.2%), adjacent tertile (38.4%), and opposite tertile (10.4%). Kappa weights: observed agreement, 0.51; chance-expected agreement, 0.33.

3.3. Energy Intakes

The energy intake from protein (MJ) and non-protein energy (MJ) for each tertile of discretionary food and UPF intake is shown in Figure 2. Participants in all tertiles of discretionary food intake and UPF intake consumed similar amounts of protein, with ~1.5 MJ of energy from protein for all groups. In contrast, energy from non-protein macronutrients increased between tertiles 1 and 3 for both UPF %E intake and discretionary food %E. The mean adjusted non-protein energy was 2.0 MJ between lowest and highest tertile for discretionary food %E and 0.6 MJ for UPF %E (Figure 2). The unadjusted mean non-protein energy intake difference between tertiles 1 and 3 %E UPF was 0.8 MJ ($p < 0.001$), whereas that for discretionary food %E was threefold higher (+2.2 MJ) ($p < 0.001$). The total mean adjusted energy intake, protein energy intake, and energy intake from macronutrients excluding energy from alcohol is shown in Table 4 for participants classified as quintiles in UPF and discretionary food (%E).

Table 4. Total energy, protein energy and energy excluding alcohol for participants classified by proportion of energy from discretionary food (DF) and ultra-processed foods (UPF).

	DF (%E)	UPF (%E)	DF (%E)		UPF (%E)		DF (No Alcohol) (%E)	UPF
	Total Energy (MJ)	Total Energy (MJ)	Total Energy (MJ)	Protein (MJ)	Total Energy (MJ)	Protein (MJ)	P + C + F (MJ)	P + C + F (MJ)
Quintile	Mean	Mean	Adj. Mean †	Adj. Mean †	Adj. Mean †	Adj. Mean †	Adj. Mean †	Adj. Mean †
1	7.4	8.2	7.5	1.6	8.5	1.6	7.4	7.8
2	8.3	8.7	8.4	1.6	8.9	1.6	8.0	8.4
3	8.7	8.6	8.8	1.6	8.8	1.5	8.4	8.3
4	9.1	8.8	9.1	1.5	8.8	1.4	8.5	8.5
5	10.0 ***	9.1 ***	10.0 ***	1.4 ***	9.1 **	1.3 ***	9.0 ***	8.7 ***

† Adjusted (adj.) for age, sex, physical activity level, smoking status, educational attainment, and country of birth. P, protein; C, carbohydrate; F, fat. ** Significant linear trend across quintiles = 0.001; *** significant linear trend across quintiles < 0.0001.

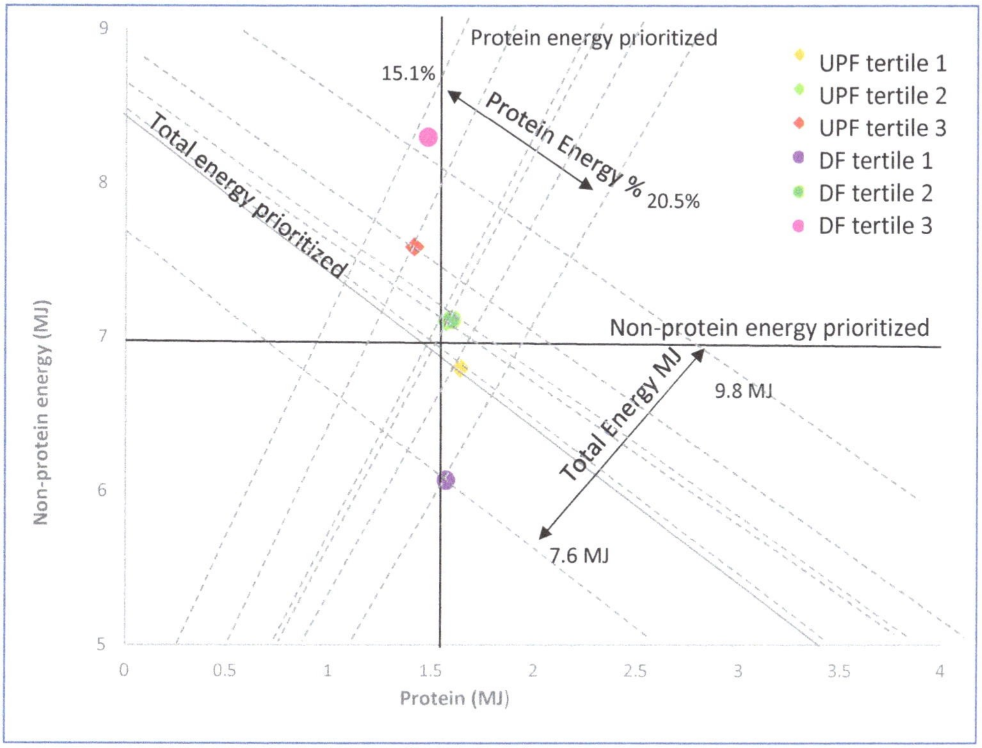

Figure 2. Mean protein and non-protein energy intake for participants categorised into tertiles of discretionary food (DF) and ultra-processed food (UPF). The positively sloped radials indicate the proportion of energy from protein from total energy intake and demonstrate protein dilution with increased intake of discretionary and UPF. The negatively sloped diagonals indicate total daily energy intake. The data points line up along the solid vertical line demonstrating that protein energy intake is prioritised. If total daily energy intake is prioritised, the values line up along the solid negative radial, while the horizontal line indicates the situation if non-protein energy is prioritised.

The difference in the unadjusted mean energy intake increased in a similar trend when participants were classified into quintiles, and the difference was 2.6 MJ between quintile 1 and quintile 5 between discretionary food, compared to 9.0 MJ between quintile 1 and quintile 5 for UPF. Excluding alcohol from the total energy intake reduced the difference in energy intake between quintile 1 and quintile 5 from 2.6 MJ to 1.6 MJ.

3.4. Body Mass Index and Intake of Discretionary Foods and UPF

In the unadjusted model, lower intakes of discretionary foods and UPF, categorised into tertiles by the proportion of energy from discretionary food and by the proportion of energy from UPF, were both significantly associated with a lower BMI (Table 5). In the multivariate model, participants classified into the lowest tertiles of discretionary food intake and UPF intake had the lowest BMIs by -1.0 and -1.1 kg/m^2, respectively, compared to the highest consumers (Table 5). The magnitude in the changes was similar for UPF and DF.

Table 5. Change in body mass index (BMI) with changes in intake of discretionary food as defined by the Australian Dietary Guidelines (ADG) and ultra-processed food (UPF) as a proportion of energy as defined by the NOVA classification system.

Food Intake (Range)		Model 1: Change in BMI (SE)		*p*-Value	Model 2: Change in BMI (SE)		*p*-Value
ADG classification							
DF—tertile 1	(0.0–≤21.8)	−0.8	(0.2)		−1.0	(0.2)	
DF—tertile 2	(>21.8–41.6)	−0.2	(0.2)		−0.4	(0.2)	
DF—tertile 3	(≥41.7–100)	Ref		0.0003	Ref		<0.0001
NOVA classification							
UPF—tertile 1	(≥0.0–<29.4)	−0.9	(0.2)		−1.1	(0.2)	
UPF—tertile 2	(≥29.4–<49.7)	−0.4	(0.2)		−0.5	(0.2)	
UPF—tertile 3	(≥49.7–100.0)	Ref		0.0003	Ref		<0.0001

%E, percentage energy. *p*-Values for trend were determined with linear and multiple liner regression. Model 1: univariate model. Model 2: adjusted for sex, age, smoking status (current smoker, daily; current smoker, <weekly; current smoker, at least once per week but not daily; ex-smoker, never smoked); physical activity level (sedentary (very low); sedentary (no exercise); not stated; low; moderate; high); energy intake: basal metabolic rate ratio. Survey weights were applied.

4. Discussion

The present analysis of the dietary intake of a nationally representative sample of the Australian population demonstrates considerable overlap in the NOVA and ADG classification systems. The two systems classified all minimally processed foods and many processed foods as foods to be encouraged, and international dietary advice converges in this regard. However, there was some discrepancy between which system best identifies foods to be avoided in Australia, with many culinary ingredients being classified as unhealthy in ADG and most international dietary guidelines but not in NOVA. Likewise, our findings show that some foods identified as healthier by the ADG are classified as ultra-processed in the NOVA system. From the perspective of energy balance, both NOVA and ADG identified dietary patterns that elevate energy intake and were associated with overall higher BMI. The discretionary classification system was associated with both higher (quintile 5 = 8.6 MJ, Table 4) and lower (quintile 1, 5.9 MJ, Table 4) acute non-protein energy intake than the UPF classification (7.8 MJ vs. 6.9) and, consequently, a wider gap between highest and lowest quintiles (2.5 MJ vs. 0.6 MJ, respectively). Higher intakes of both UPF and discretionary food were associated with a higher BMI overall.

The NOVA and the ADG systems performed differently in their association with energy intakes, with high consumption of discretionary food related to larger increases in energy intake. At a fundamental level, weight gain occurs when energy intake exceeds energy expenditure and excess energy is stored [26]. Discrepancies as small as an additional 30 kJ per day may trigger weight gain, but an increase of 9 MJ has been estimated as needed to sustain the weight increases of the USA population [27,28]. After adjusting for relevant confounding factors, both classification systems were able to identify dietary patterns associated with higher energy intake and positive energy balance sufficient to drive weight gain.

It is important to examine the food groups in which the NOVA and ADG differ in terms of dietary guidance. Our analysis showed one point of difference to be centred on culinary ingredients. For example, the ADG advises against the use of added sugar and limits saturated fats unless energy requirements allow for some additional energy after nutrient requirements have been met [6]. The NOVA system, in contrast, limits added sugar when it is incorporated into UPF, but not consumer-added sugar, which is considered a culinary ingredient associated with home cooking [29]. However, in our analysis of the Australian food system, most of the culinary ingredients were used to make discretionary foods rather than healthier homemade dishes that can also be high in saturated fats, such as homemade or cafe/restaurant made cakes, biscuits, sweet and savoury pastries, and deep-fried vegetable products that may be detrimental to health if consumed in excess.

A second difference between the systems is that a considerable proportion of five-food-group foods are classified as ultra-processed. Many of these that are low in saturated fat, e.g., ready-prepared-foods, may be harmful over time due to the impact of refined ingredients or exposure to harmful chemicals such as endocrine disruptors [2,30–33]. They may also have cumulative effects on energy storage over time if they contain highly refined, readily digestible sources of carbohydrates and fat relative to protein [34]. Additionally, alcohol, which is primarily classified as processed (rather than ultra-processed) in the NOVA system, also elevates acute energy intake, and sensitivity analysis demonstrated that some of the advantage held by the ADG system in predicting excess energy intake was due to the inclusion of alcohol in the discretionary category. The ADG and NOVA classification systems concur in identifying all minimally processed foods and many processed foods as healthy.

The mechanistic model of protein leverage for predicting the relationship between dietary composition and energy intake further explains the advantage of the ADG system compared to the UPF system in predicting high energy intakes in Australia. The protein leverage hypothesis posits that protein dilution of the food supply has been an important contributor to the obesity epidemic [16]. The comparatively smaller range of protein dilution in the NOVA system is due to the inclusion of relatively protein-dense foods as UPF. For example, flavoured yoghurt, classified as UPF but not discretionary food, has higher protein density. Conversely, we found that, in the Australian food system, many foods classified as discretionary food but not UPF have low protein density, e.g., homemade cakes. Additionally, low-protein diets that contain dietary fibre or resistant starch have been demonstrated to not lead to overconsumption [16,35]. Therefore, UPFs are unlikely to lead to higher energy intakes if they include unrefined sources of dietary fibre and resistant starch, such as commercial wholegrain bread or certain breakfast cereals with minimal added sugar.

It is important to note, however, that the close conformity between the patterns of macronutrient intakes observed in our analysis and predictions of the protein leverage hypothesis does not in itself provide definitive evidence of protein leverage. An alternative explanation could be that aggressive marketing, hyperpalatabilty, and other extrinsic factors drive excessive consumption of discretionary food and UPF, which both dilutes dietary protein and increases energy intake independent of protein appetite [32,36]. However, this hypothesis does not explain why absolute protein remained so constant; in contrast, constant protein intake is a central prediction of the protein leverage hypothesis. Furthermore, our interpretation is congruent with several other sources of evidence for protein leverage. These include recent advances in elucidating the biological mechanisms of protein appetites, demonstrating that fibroblast growth factor-21 (FGF-21) is the circulating metabolite when protein status is low in humans and rodents, acting on the brain to stimulate protein appetite [37,38]. Above all, protein leverage has been demonstrated in several RCTs in which diets were controlled for palatability, and aggressive marketing played no role [16,39,40]. It is, nonetheless, likely that, in ecological settings, protein leverage interacts with extrinsic factors. A two-stage model has been proposed for how this interaction might work [41], according to which hyper palatability, aggressive marketing, cheap price, and convenience associated with industrial foods attracts consumers to select and consume them. Due to their low protein content, this results in a reduction in the ratio of dietary protein to energy and, via protein leverage, energy over-consumption [20,34].

While our analysis implies that discretionary food consumption can explain the rising prevalence of obesity in Australia, these results may be country-specific or differ at an individual level. For example, an RCT examining the effect of UPF in the USA found similar increases in energy intake for discretionary food of 509 kcal (2118 kJ) greater compared to a control group consuming minimally processed foods [19]. Therefore, differences in the magnitude of effect could be due to difference in quality of the foods available in the Australian food supply or compared to those imposed on people in a laboratory setting, and the findings are likely to change dependent on overall diet quality and the relative fractions of UPF consumed. Analysis of NHANES dietary data derived a decline in %PE of 4.9% and an increase in energy intake of 0.9 MJ of non-protein energy between UPF quintiles, which is comparable to the increase in energy intake observed in the present study [20]. As discretionary foods led to greater protein dilution in Australia, this highlights the value of adopting an ecological model to examine the aetiology of the obesity epidemic.

A limitation of this study is that processing is not always adequately captured in the product description in the nutrient composition database, making the NOVA classification difficult to apply. Bread is one category where this is problematic as it is not possible to distinguish artisanal or homemade breads and mass-produced breads. To minimise differences, the NOVA coding system of Australian foods emulated that of previous research where foods were agreed upon by at least two researchers who applied expert knowledge in the Australian food supply, but there may still be some arbitrary misclassification of foods [22]. It should be noted that the Australian Bureau of Statistics discretionary food list was used to identify foods as discretionary according to the ADG. Although this list does not agree with the ADG in some aspects, as it was based on nutrient cut-points, it was used to more easily identify foods that were high in saturated fat, added sugars, and salt [42]. Misreporting dietary intake is a common limitation in dietary assessment [43], but sensitivity analysis revealed that this had a limited effect on the results presented here. Although the analysis was from a single day via a cross-sectional survey, the results are supported by other epidemiological studies and randomised controlled trials [3,44,45] and demonstrate the advantage of the discretionary food classification at a population level in the Australian food system. Our analysis used the most recent and comprehensive nutrition survey to have been conducted in Australia. Although this is 10 years old, we do not see reason to expect that factors driving the main conclusions would have changed over this period. Indeed, with the continued rise in industrial foods and obesity in Australia [46], we suspect that they might be even more relevant at present.

There is increased financial burden for individuals and communities amounting to 20.5 billion AUD annually in direct health costs in Australia due to poor diets [47]. The prevalent solution in Australia and other developed countries has been a focus on individual responsibility through setting-based approaches or public health information campaigns. Without further policy, legislative, and structural changes to the food environment, efforts to prevent noncommunicable disease will be frustrated [48]. There have been calls to level the playing field by matching the heavy marketing of UPF/discretionary food with government spending on advertising of healthful dietary patterns [49]. Even for some of the worst aspects of the food environment such as vending machines, there is good evidence that simply providing customers with healthier choices is enough to improve the purchases made [50]. Efforts to reduce the intake of foods that dilute protein energy have the potential to contribute to and strengthen obesity prevention.

5. Conclusions

UPFs have been associated with the rise in obesity in many countries, including Australia, but our study highlights the importance of monitoring dietary patterns with ecological studies to determine the efficacy of different food classification systems in different food environments. While there was considerable overlap in both the ADG and the NOVA classification systems, the apparent advantage of the discretionary classification system in Australia is that it detects a greater spread in the extent to which problem foods dilute dietary protein and, hence, their effect on energy intake via protein leverage. Dietary guidance targeted at reducing intake of discretionary foods in Australia may thus have greater potential for supporting obesity prevention than a focus on UPF. Excessive energy intake is not, however, the only aspect of health for which dietary guidelines are relevant. Further studies are needed to determine how different classification systems relate to other associations between diet and all aspects of health in Australia and other countries.

Supplementary Materials: The following supporting information can be downloaded at https://www.mdpi.com/article/10.3390/nu14193942/s1: Table S1. Proportion of energy from all foods reported in the National Nutrition and Physical Activity Survey classified by the NOVA classification system and the Australian Dietary Guidelines; Figure S1. Participant flow diagram.

Author Contributions: A.G. and D.R. designed the research; A.G. analysed the data; A.G. wrote the first draft of the paper; A.G., D.R., A.R., M.A.-F., S.J.S. and T.G. contributed to editing the manuscript and had primary responsibility for final content. All authors have read and agreed to the published version of the manuscript.

Funding: This research was funded by the National Health and Medical Research Council, Australia, Program Grant GNT1149976.

Institutional Review Board Statement: Ethical review and approval were waived for this study due to this being secondary analysis of a national survey conducted by the Australian Bureau of Statistics under the Census and Statistics Act, 1905.

Data Availability Statement: Data are available from the Australian Bureau of Statistics upon request https://www.abs.gov.au/ (accessed on 30 January 2021).

Conflicts of Interest: The authors declare no conflict of interest.

References

1. Monteiro, C.A.; Cannon, G.; Moubarac, J.-C.; Levy, R.B.; Louzada, M.L.C.; Jaime, P.C. The UN Decade of Nutrition, the NOVA food classification and the trouble with ultra-processing. *Public Health Nutr.* **2018**, *21*, 5–17. [CrossRef] [PubMed]
2. Rico-Campà, A.; Martínez-González, M.A.; Alvarez-Alvarez, I.; de Deus Mendonça, R.; De La Fuente-Arrillaga, C.; Gómez-Donoso, C.; Bes-Rastrollo, M. Association between consumption of ultra-processed foods and all cause mortality: SUN prospective cohort study. *BMJ* **2019**, *365*, l1949. [CrossRef] [PubMed]
3. Monteiro, C.A.; Cannon, G.; Lawrence, M.; Costa Louzada, M.D.; Pereira Machado, P. *Ultra-Processed Foods, Diet Quality, and Health Using the NOVA Classification System*; FAO: Rome, Italy, 2019.
4. Moubarac, J.-C.; Parra, D.C.; Cannon, G.; Monteiro, C.A. Food classification systems based on food processing: Significance and implications for policies and actions: A systematic literature review and assessment. *Curr. Obes. Rep.* **2014**, *3*, 256–272. [CrossRef] [PubMed]
5. Monteiro, C.A.; Cannon, G.; Levy, R.B.; Moubarac, J.-C.; Louzada, M.L.C.; Rauber, F.; Khandpur, N.; Cediel, G.; Neri, D.; Martinez-Steele, E.; et al. Ultra-processed foods: What they are and how to identify them. *Public Health Nutr.* **2019**, *22*, 936–941. [CrossRef] [PubMed]
6. National Health and Medical Research Council. *Australian Dietary Guidelines*; National Health and Medical Research Council: Canberra, Australia, 2013.
7. Public Health England. *The Eatwell Guide*; Public Health England: London, UK, 2016.
8. Ministry of Health. *Eating and Activity Guidelines for New Zealand Adults: Updated 2020*; Ministry of Health: Wellington, New Zealand, 2020.
9. Minister of Health Canada. *Evidence Review for Dietary Guidance: Summary of Results and Implications for Canada's Food Guide*; Minister of Health Canada: Ottawa, ON, Canada, 2015.
10. Nordic Council of Ministers. *Nordic Nutrition Recommendations 2012: Integrating Nutrition and Physical Activity*; Nordic Council of Ministers: Norden, Germany, 2014.
11. U.S. Department of Agriculture; U.S. Department of Health and Human Services. *Dietary Guidelines for Americans, 2020–2025*; USDA: Washington, DC, USA, 2020.
12. Blake, P.; Durão, S.; E Naude, C.; Bero, L. An analysis of methods used to synthesize evidence and grade recommendations in food-based dietary guidelines. *Nutr. Rev.* **2018**, *76*, 290–300. [CrossRef]
13. Raubenheimer, D.; Simpson, S.J. Nutritional Ecology and Human Health. *Annu. Rev. Nutr.* **2016**, *36*, 603–626. [CrossRef]
14. Simpson, S.J.; Raubenheimer, D. Obesity: The protein leverage hypothesis. *Obes. Rev.* **2005**, *6*, 133–142. [CrossRef]
15. Saner, C.; Tassoni, D.; Harcourt, B.E.; Kao, K.; Alexander, E.J.; McCallum, Z.; Olds, T.; Rowlands, A.V.; Burgner, D.P.; Simpson, S.J.; et al. Evidence for protein leverage in children and adolescents with obesity. *Obesity* **2020**, *28*, 822–829. [CrossRef]
16. Raubenheimer, D.; Simpson, S.J. Protein leverage: Theoretical foundations and ten points of clarification. *Obesity* **2019**, *27*, 1225–1238. [CrossRef]
17. Gosby, A.K.; Conigrave, A.D.; Raubenheimer, D.; Simpson, S.J. Protein leverage and energy intake. *Obes. Rev.* **2014**, *15*, 183–191. [CrossRef]
18. Hall, K.D. The Potential Role of Protein Leverage in the US Obesity Epidemic. *Obesity* **2019**, *27*, 1222–1224. [CrossRef] [PubMed]
19. Hall, K.D.; Ayuketah, A.; Brychta, R.; Cai, H.; Cassimatis, T.; Chen, K.Y.; Chung, S.T.; Costa, E.; Courville, A.; Darcey, V.; et al. Ultra-processed diets cause excess calorie intake and weight gain: An inpatient randomized controlled trial of ad libitum food intake. *Cell Metab.* **2019**, *30*, 67–77.e3. [CrossRef] [PubMed]

20. Steele, E.M.; Raubenheimer, D.; Simpson, S.J.; Baraldi, L.G.; A Monteiro, C. Ultra-processed foods, protein leverage and energy intake in the USA. *Public Health Nutr.* **2018**, *21*, 114–124. [CrossRef] [PubMed]
21. Australian Bureau of Statistics. *Australian Health Survey: User Guide*; Australian Bureau of Statistics: Canberra, Australia, 2012.
22. Machado, P.P.; Steele, E.M.; Levy, R.B.; Sui, Z.; Rangan, A.; Woods, J.; Gill, T.; Scrinis, G.; Monteiro, C.A. Ultra-processed foods and recommended intake levels of nutrients linked to non-communicable diseases in Australia: Evidence from a nationally representative cross-sectional study. *BMJ Open* **2019**, *9*, e029544. [CrossRef]
23. Food Standards Australia New Zealand. *AUSNUT 2011–2013 Food Recipe File*; FSANZ: Canberra, Australia, 2014.
24. Food Standards Australia New Zealand. *AUSNUT 2011–2013 Food Nutrient Database*; FSANZ: Canberra, Australia, 2014.
25. Black, A.E. Critical evaluation of energy intake using the Goldberg cut-off for energy intake: Basal metabolic rate. A practical guide to its calculation, use and limitations. *Int. J. Obes.* **2000**, *24*, 1119–1130. [CrossRef]
26. Hill, J.O. Understanding and addressing the epidemic of obesity: An energy balance perspective. *Endocr. Rev.* **2006**, *27*, 750–761. [CrossRef]
27. Hall, K.D.; Heymsfield, S.B.; Kemnitz, J.W.; Klein, S.; Schoeller, D.A.; Speakman, J.R. Energy balance and its components: Implications for body weight regulation. *Am. J. Clin. Nutr.* **2012**, *95*, 989–994. [CrossRef]
28. Hall, K.D.; Sacks, G.; Chandramohan, D.; Chow, C.C.; Wang, Y.C.; Gortmaker, S.L.; A Swinburn, B. Quantification of the effect of energy imbalance on bodyweight. *Lancet* **2011**, *378*, 826–837. [CrossRef]
29. Adams, J.; Hofman, K.; Moubarac, J.-C.; Thow, A.M. Public health response to ultra-processed food and drinks. *BMJ* **2020**, *369*, m2391. [CrossRef]
30. Mendonça, R.D.D.; Pimenta, A.M.; Gea, A.; de la Fuente-Arrillaga, C.; Martinez-Gonzalez, M.A.; Lopes, A.C.S.; Bes-Rastrollo, M. Ultraprocessed food consumption and risk of overweight and obesity: The University of Navarra Follow-Up (SUN) cohort study. *Am. J. Clin. Nutr.* **2016**, *104*, 1433–1440. [CrossRef]
31. Martínez Steele, E.; Monteiro, C.A. Association between Dietary Share of Ultra-Processed Foods and Urinary Concentrations of Phytoestrogens in the US. *Nutrients* **2017**, *9*, 209. [CrossRef] [PubMed]
32. Fardet, A.; Rock, E. Ultra-processed foods: A new holistic paradigm? *Trends Food Sci. Technol.* **2019**, *93*, 174–184. [CrossRef]
33. Buckley, J.P.; Kim, H.; Wong, E.; Rebholz, C.M. Ultra-processed food consumption and exposure to phthalates and bisphenols in the US National Health and Nutrition Examination Survey, 2013–2014. *Environ. Int.* **2019**, *131*, 105057. [CrossRef] [PubMed]
34. Wali, J.A.; Raubenheimer, D.; Senior, A.M.; Le Couteur, D.G.; Simpson, S.J. Cardio-metabolic consequences of dietary carbohydrates: Reconciling contradictions using nutritional geometry. *Cardiovasc. Res.* **2020**, *117*, 386–401. [CrossRef] [PubMed]
35. Wali, J.A.; Milner, A.J.; Luk, A.W.S.; Pulpitel, T.J.; Dodgson, T.; Facey, H.J.W.; Wahl, D.; Kebede, M.A.; Senior, A.M.; Sullivan, M.A.; et al. Impact of dietary carbohydrate type and protein-carbohydrate interaction on metabolic health. *Nat. Metab.* **2021**, *3*, 810–828. [CrossRef] [PubMed]
36. Fardet, A.; Lakhssassi, S.; Briffaz, A. Beyond nutrient-based food indices: A data mining approach to search for a quantitative holistic index reflecting the degree of food processing and including physicochemical properties. *Food Funct.* **2018**, *9*, 561–572. [CrossRef]
37. Hill, C.M.; Laeger, T.; Dehner, M.; Albarado, D.C.; Clarke, B.; Wanders, D.; Burke, S.J.; Collier, J.J.; Qualls-Creekmore, E.; Solon-Biet, S.M.; et al. FGF21 Signals Protein Status to the Brain and Adaptively Regulates Food Choice and Metabolism. *Cell Rep.* **2019**, *27*, 2934–2947.e3. [CrossRef]
38. Hill, C.M.; Qualls-Creekmore, E.; Berthoud, H.-R.; Soto, P.; Yu, S.; McDougal, D.H.; Münzberg, H.; Morrison, C.D. FGF21 and the Physiological Regulation of Macronutrient Preference. *Endocrinology* **2020**, *161*, bqaa019. [CrossRef]
39. Gosby, A.K.; Conigrave, A.D.; Lau, N.; Iglesias, M.A.; Hall, R.M.; Jebb, S.A.; Brand-Miller, J.; Caterson, I.D.; Raubenheimer, D.; Simpson, S.J. Testing protein leverage in lean humans: A randomised controlled experimental study. *PLoS ONE* **2011**, *6*, e25929. [CrossRef]
40. Campbell, C.P.; Raubenheimer, D.; Badaloo, A.V.; Gluckman, P.D.; Martinez, C.; Gosby, A.; Simpson, S.J.; Osmond, C.; Boyne, M.S.; Forrester, T.E. Developmental contributions to macronutrient selection: A randomized controlled trial in adult survivors of malnutrition. *Evol. Med. Public Health* **2016**, *2016*, 158–169. [CrossRef]
41. Raubenheimer, D.; Machovsky-Capuska, G.E.; Gosby, A.K.; Simpson, S. Nutritional ecology of obesity: From humans to companion animals. *Br. J. Nutr.* **2015**, *113*, S26–S39. [CrossRef] [PubMed]
42. Lee, A.; Rangan, A.; Allman-Farinelli, M.; Chen, J.; Grech, A.; McDonald, S.; Wilson, A. *A Rapid Review of Evidence: Discretionary Food and Drinks*; The Australian Prevention Partnership Centre: Sydney, Australia, 2018.
43. Freedman, L.S.; Commins, J.M.; Moler, J.E.; Arab, L.; Baer, D.J.; Kipnis, V.; Midthune, D.; Moshfegh, A.J.; Neuhouser, M.L.; Prentice, R.L.; et al. Pooled results from 5 validation studies of dietary self-report instruments using recovery biomarkers for energy and protein intake. *Am. J. Epidemiol.* **2014**, *180*, 172–188. [CrossRef] [PubMed]
44. Te Morenga, L.; Mallard, S.; Mann, J. Dietary sugars and body weight: Systematic review and meta-analyses of randomised controlled trials and cohort studies. *BMJ* **2013**, *345*, e7492. [CrossRef] [PubMed]
45. Mozaffarian, D.; Hao, T.; Rimm, E.B.; Willett, W.C.; Hu, F.B. Changes in Diet and Lifestyle and Long-Term Weight Gain in Women and Men. *N. Engl. J. Med.* **2011**, *364*, 2392–2404. [CrossRef]
46. Baker, P.; Machado, P.; Santos, T.; Sievert, K.; Backholer, K.; Hadjikakou, M.; Russell, C.; Huse, O.; Bell, C.; Scrinis, G.; et al. Ultra-processed foods and the nutrition transition: Global, regional and national trends, food systems transformations and political economy drivers. *Obes. Rev.* **2020**, *21*, e13126. [CrossRef]

47. Colagiuri, S.; Lee, C.M.; Colagiuri, R.; Magliano, D.; Shaw, J.E.; Zimmet, P.Z.; Caterson, I.D. The cost of overweight and obesity in Australia. *Med. J. Aust.* **2010**, *192*, 260–264. [CrossRef]
48. Morland, K.B.; Evenson, K.R. Obesity prevalence and the local food environment. *Health Place* **2009**, *15*, 491–495. [CrossRef]
49. O'Dowd, A. Spending on junk food advertising is nearly 30 times what government spends on promoting healthy eating. *BMJ* **2017**, *359*, j4677. [CrossRef]
50. Grech, A.; Allman-Farinelli, M. A systematic literature review of nutrition interventions in vending machines that encourage consumers to make healthier choices. *Obes. Rev.* **2015**, *16*, 1030–1041. [CrossRef]

nutrients MDPI

Article

Does Consumption of Ultra-Processed Foods Matter for Liver Health? Prospective Analysis among Older Adults with Metabolic Syndrome

Jadwiga Konieczna [1,2,*], Miguel Fiol [1,2], Antoni Colom [1,2], Miguel Ángel Martínez-González [2,3,4], Jordi Salas-Salvadó [2,5,6,7], Dolores Corella [2,8], María Trinidad Soria-Florido [2,9], J. Alfredo Martínez [2,10,11], Ángel M. Alonso-Gómez [2,12], Julia Wärnberg [2,13], Jesús Vioque [14,15], José López-Miranda [2,16], Ramon Estruch [2,17], M. Rosa Bernal-López [2,18], José Lapetra [2,19], Lluís Serra-Majem [2,20], Aurora Bueno-Cavanillas [14,21], Josep A. Tur [2,22], Vicente Martín Sánchez [14,23], Xavier Pintó [2,24], José J. Gaforio [14,25], Pilar Matía-Martín [26], Josep Vidal [27,28], Clotilde Vázquez [2,29], Lidia Daimiel [30], Emilio Ros [2,31], Maira Bes-Rastrollo [2,3], María Pascual [2,5,6,7], Jose V. Sorlí [2,8], Albert Goday [2,9,32], María Ángeles Zulet [2,10], Anai Moreno-Rodriguez [2,12], Francisco Jesús Carmona González [33], Rafael Valls-Enguix [34], Juana M. Janer [1,35], Antonio Garcia-Rios [2,16], Rosa Casas [2,17], Ana M. Gomez-Perez [2,36], José Manuel Santos-Lozano [2,19], F. Javier Basterra-Gortari [3,37], María Ángeles Martínez [2,5,6,7], Carolina Ortega-Azorin [2,8], Joan Bayó [38], Itziar Abete [2,10], Itziar Salaverria-Lete [2,12], Miguel Ruiz-Canela [2,3], Nancy Babio [2,5,6,7], Lourdes Carres [39] and Dora Romaguera [1,2]

1 Research Group on Nutritional Epidemiology & Cardiovascular Physiopathology (NUTRECOR), Health Research Institute of the Balearic Islands (IdISBa), University Hospital Son Espases (HUSE), 07120 Palma de Mallorca, Spain
2 Centro de Investigación Biomédica en Red Fisiopatología de la Obesidad y la Nutrición (CIBEROBN), Institute of Health Carlos III, 28029 Madrid, Spain
3 Department of Preventive Medicine and Public Health, IDISNA, University of Navarra, 31008 Pamplona, Spain
4 Department of Nutrition, Harvard T.H. Chan School of Public Health, Boston, MA 02115, USA
5 Universitat Rovira i Virgili, Departament de Bioquímica i Biotecnologia, Unitat de Nutrició, 43201 Reus, Spain
6 University Hospital of Sant Joan de Reus, Nutrition Unit, 43204 Reus, Spain
7 Institut d'Investigació Sanitària Pere Virgili (IISPV), 43204 Reus, Spain
8 Department of Preventive Medicine, University of Valencia,46010 Valencia, Spain
9 Unit of Cardiovascular Risk and Nutrition, Institut Hospital del Mar de Investigaciones Médicas Municipal d'Investigació Médica (IMIM), 08003 Barcelona, Spain
10 Department of Nutrition, Food Sciences, and Physiology, University of Navarra, IDISNA, 31008 Pamplona, Spain
11 Precision Nutrition and Cardiometabolic Health Program, IMDEA Food, CEI UAM + CSIC, 28049 Madrid, Spain
12 Bioaraba Health Research Institute, Cardiovascular, Respiratory and Metabolic Area, Osakidetza Basque Health Service, Araba University Hospital, University of the Basque Country UPV/EHU, 01009 Vitoria-Gasteiz, Spain
13 Department of Nursing, University of Málaga, Institute of Biomedical Research in Malaga (IBIMA), 29016 Málaga, Spain
14 CIBER de Epidemiología y Salud Pública (CIBERESP), Instituto de Salud Carlos III, 28029 Madrid, Spain
15 Alicante Institute for Health and Biomedical Research (ISABIAL-UMH), 03010 Alicante, Spain
16 Department of Internal Medicine, Maimonides Biomedical Research Institute of Cordoba (IMIBIC), Reina Sofia University Hospital, University of Cordoba, 14004 Cordoba, Spain
17 Department of Internal Medicine, Institut d'Investigacions Biomèdiques August Pi Sunyer (IDIBAPS), Hospital Clinic, University of Barcelona, 08007 Barcelona, Spain
18 Department of Internal Medicine, Regional University Hospital of Malaga, Instituto de Investigación Biomédica de Málaga (IBIMA), University of Málaga, 29009 Málaga, Spain
19 Department of Family Medicine, Research Unit, Distrito Sanitario Atención Primaria Sevilla, 41013 Sevilla, Spain
20 Research Institute of Biomedical and Health Sciences (IUIBS), University of Las Palmas de Gran Canaria & Centro Hospitalario Universitario Insular Materno Infantil (CHUIMI), Canarian Health Service, 35001 Las Palmas de Gran Canaria, Spain
21 Department of Preventive Medicine and Public Health, University of Granada, 18011 Granada, Spain
22 Research Group on Community Nutrition & Oxidative Stress, University of Balearic Islands, 07122 Palma de Mallorca, Spain
23 Institute of Biomedicine (IBIOMED), University of León, 24071 León, Spain

Citation: Konieczna, J.; Fiol, M.; Colom, A.; Martínez-González, M.Á.; Salas-Salvadó, J.; Corella, D.; Soria-Florido, M.T.; Martínez, J.A.; Alonso-Gómez, Á.M.; Wärnberg, J.; et al. Does Consumption of Ultra-Processed Foods Matter for Liver Health? Prospective Analysis among Older Adults with Metabolic Syndrome. *Nutrients* **2022**, *14*, 4142. https://doi.org/10.3390/nu14194142

Academic Editors: Monica Dinu and Daniela Martini

Received: 7 September 2022
Accepted: 27 September 2022
Published: 5 October 2022

[24] Lipids and Vascular Risk Unit, Internal Medicine, Hospital Universitario de Bellvitge Idibell-UB, Hospitalet de Llobregat, 08907 Barcelona, Spain
[25] Departamento de Ciencias de la Salud, Centro de Estudios Avanzados en Olivar y Aceites de Oliva, Universidad de Jaén, 23071 Jaén, Spain
[26] Department of Endocrinology and Nutrition, Instituto de Investigación Sanitaria Hospital Clínico San Carlos (IdISSC), 28040 Madrid, Spain
[27] CIBER Diabetes y Enfermedades Metabólicas (CIBERDEM), Instituto de Salud Carlos III (ISCIII), 28029 Madrid, Spain
[28] Department of Endocrinology, Institut d' Investigacions Biomédiques August Pi Sunyer (IDIBAPS), Hospital Clinic, University of Barcelona, 08007 Barcelona, Spain
[29] Department of Endocrinology and Nutrition, Hospital Fundación Jimenez Díaz, Instituto de Investigaciones Biomédicas IISFJD, University Autonoma, 28040 Madrid, Spain
[30] Nutritional Control of the Epigenome Group, Precision Nutrition and Obesity Program, IMDEA Food, CEI UAM + CSIC, 28049 Madrid, Spain
[31] Lipid Clinic, Department of Endocrinology and Nutrition, IDIBAPS, Hospital Clínic, 08036 Barcelona, Spain
[32] Departament de Medicina, Universitat Autònoma de Barcelona, 08193 Barcelona, Spain
[33] Unidad de Gestión Clínica Torrequebrada, Distrito de Atención Primaria Costa del Sol, Servicio Andaluz de Salud, 29640 Benalmádena, Spain
[34] Raval Health Care Centre, 03203 Elche, Spain
[35] Centro de Salud Camp Redó (UBS Son Sardina) Gerència Atenció Primària de Mallorca, 07010 Palma de Mallorca, Spain
[36] Virgen de la Victoria Hospital, Department of Endocrinology, Instituto de Investigación Biomédica de Málaga (IBIMA), University of Málaga, 29071 Málaga, Spain
[37] Servicio de Endocrinología. Complejo Hospitalario de Navarra, Servicio Navarro de Salud, 31003 Pamplona, Spain
[38] CAP El Clot, Institut Català de la Salut, 08018 Barcelona, Spain
[39] Atención Primaria Sant Martí, Institut Català de la Salut, 08020 Barcelona, Spain
* Correspondence: jadwiga.konieczna@ssib.es; Tel.: +34-871-20-5050 (ext. 64527)

Abstract: Background: Non-alcoholic fatty liver disease (NAFLD) includes a spectrum of liver alterations that can result in severe disease and even death. Consumption of ultra-processed foods (UPF) has been associated with obesity and related comorbidities. However, the link between UPF and NAFLD has not been sufficiently assessed. We aimed to investigate the prospective association between UPF consumption and liver health biomarkers. **Methods:** We followed for 1 year 5867 older participants with overweight/obesity and metabolic syndrome (MetS) from the PREDIMED-Plus trial. A validated 143-item semi-quantitative food frequency questionnaire was used to evaluate consumption of UPF at baseline, 6, and 12 months. The degree of processing for foods and beverages (g/day) was established according to the NOVA classification system. The non-invasive fatty liver index (FLI) and hepatic steatosis index (HSI) were used to evaluate liver health at three points in time. The associations between changes in UPF consumption (percentage of total daily dietary intake (g)) and liver biomarkers were assessed using mixed-effects linear models with repeated measurements. **Results:** In this cohort, UPF consumption at baseline was 8.19% (SD 6.95%) of total daily dietary intake in grams. In multivariable models, each 10% daily increment in UPF consumption in 1 year was associated with significantly greater FLI (β 1.60 points, 95% CI 1.24;1.96 points) and HSI (0.43, 0.29; 0.57) scores (all p-values < 0.001). These associations persisted statistically significant after adjusting for potential dietary confounders and NAFLD risk factors. **Conclusions:** A higher UPF consumption was associated with higher levels of NAFLD-related biomarkers in older adults with overweight/obesity and MetS.

Keywords: ultra-processed foods; liver health markers; fatty liver index; hepatic steatosis index; metabolic syndrome

1. Background

Excessive liver fat accumulation in the absence of alcoholism, known as non-alcoholic fatty liver disease (NAFLD), includes a spectrum of fat-liver alterations affecting approximately 30% of adults worldwide, but its prevalence is increased in persons with type 2 diabetes (70%) and morbid obesity (90%) [1,2]. Although it remains asymptomatic during long time, excess liver fat is an important cause of morbimortality from cardiovascular disease (CVD) and malignancy, in addition to chronic liver disease. Unfortunately, at present, there are gaps in NAFLD diagnosis and limited treatment options [1,2].

Along with physical activity (PA), dietary modifications are considered the cornerstone for NAFLD management [1]. A growing body of evidence suggests that dietary fiber intake and the Mediterranean diet (MedDiet) [3] might be protective, whereas saturated and trans fatty acids (FA), simple sugars, red and processed meat, sugar-sweetened beverages (SSB), and the Western dietary pattern might act as risk factors [4]. However, with the exception of small trials assessing interventions with the MedDiet [3], most of these findings were based on cross-sectional or case-control study designs.

Ultra-processed foods (UPF) are industrial formulations manufactured using a series of processes with no domestic equivalents (i.e., soft drinks, packed snacks, processed meats, and pre-prepared dishes) [5,6]. In addition to their usual poor nutritional composition (i.e., excessive calories, simple sugars, salt, poor quality fat, as well as fiber and vitamin deprivation) [5], UPF typically contains cosmetic additives and substances formed because of extensive processing during manufacturing [7]. On the other hand, UPF are highly palatable, appealing, convenient, microbiologically pure, inexpensive, accessible, and aggressively advertised [5,8]. All these factors explain the steadily increase in UPF consumption despite the health risks associated with their regular intake [9,10].

A recent study conducted in the PREvención con DIeta MEDiterránea Plus (PREDIMED-Plus) cohort found that consumption of UPF, classified according to the NOVA system [11], was associated with greater visceral and total fat accumulation [12]. Other prospective studies with adult cohorts found associations between UPF consumption and higher risks of obesity [13–15], CVD [16], type 2 diabetes [17,18], renal dysfunction [19], cancer [20], biological aging [21], and all-cause mortality [22]. However, the link between UPF and NAFLD has not been sufficiently assessed [23]. Only very recently a prospective study by Zhang et al. has been published, reporting a positive link between UPF and NAFLD diagnosed using ultrasonography, within the Tianjin Chronic Low-grade Systemic Inflammation and Health (TCLSIH) Cohort Study [24] after a median of 4.2 years of follow-up; but they only measured UPF at baseline, without repeating the dietary assessment during follow-up. Therefore, our aim was to prospectively investigate how concurrent changes in consumption of UPF were associated with liver health in older individuals with overweight/obesity and the metabolic syndrome (MetS) from the Mediterranean area, using three repeated measurements of diet and biomarkers related to NAFLD throughout a 1-year follow-up of PREDIMED-Plus trial.

2. Methods

2.1. Study Overview and Population

This study corresponds to longitudinal analyses nested in the ongoing PREDIMED-Plus trial, with data collected at baseline and during the first year of follow-up. Details about the study protocol have been described elsewhere [25,26], and are available at www.predimedplus.com. Briefly, PREDIMED-Plus is a 6-year parallel-group, multicenter randomized clinical trial, aimed to assess the effectiveness of a lifestyle intervention—energy-restricted (er) MedDiet, PA promotion, and behavioral support—on the primary prevention of CVD and weight loss in older individuals with overweight/obesity harboring MetS. The control group received usual health care and recommendations to follow the MedDiet, without advice on energy restriction or PA promotion. The trial was launched in 2013 (the recruitment finished at the end of 2016) in 23 centers across Spain. Community-dwelling older men and women (55–75 years), with body mass index (BMI) $\geq$ 27 and <40 kg/m^2), fulfilling $\geq$ three criteria for the MetS [27], but free of CVD at baseline, were invited to participate in the trial. Exclusion criteria included self-declared liver disease (chronic hepatitis or cirrhosis), therapy with immunosuppressive drugs, cytotoxic agents or systemic corticosteroids, alcohol abuse or addiction (defined as total daily alcohol intake > 50 g within past 6 months), history of inflammatory bowel disease, and active malignant cancer or history of malignancy within the last 5 years. For a small subset of participants sharing the same household, the randomization was performed as clusters (the couple was used as the unit of randomization). All participants provided written informed consent to a protocol approved by the Research Ethic Committees of all recruiting centers according to the ethical standards of the Declaration of Helsinki. The trial was registered at http://www.isrctn.com/ (ISRCTN89898870).

Out of the 6874 participants randomized for the PREDIMED-Plus trial, participants with history of liver cancer within >5 years before inclusion (n = 2), with missing data on variables of interest at baseline (n = 811), and those who were outside predefined limits for total daily energy intake (<500 or >3500 kcal for women, <800 or >4000 kcal for men) [28] at baseline and during follow-up (n = 194) were excluded from the analysis. After exclusions, a total of 5867 participants were included in the final analyses (see the flowchart of study participants in online Supplementary Figure S1).

2.2. Dietary Habits and Nutrient Intake Assessments

Dietary intake was assessed by trained dietitians during face-to-face interviews at baseline, 6-, and 12-month follow-up visits, using a semi-quantitative food frequency questionnaire (FFQ), repeatedly validated for the Spanish population [29,30]. Intakes of 143 foods and beverages (except water) were calculated by multiplying the common portion size by average consumption frequency (9 possible responses, from never to >6 times/day) over the last year (at baseline visit and over 6-month period at each follow-up visit). Daily intake of beverages was collected in cubic centimeters and then converted into milliliters (1 cc = 1 mL), and further into grams, assuming that 1 mL = 1 g. Food composition tables developed specifically for Spain [31] were used to derive nutrient (sodium(native and added in the form of salt), and cholesterol (both in mg/day), saturated and trans FA, fiber, and alcohol (all in g/day)), as well as total energy intake (kcal/day). The glycemic load was also calculated for each item taking into account the quality (glycemic index) and the amount of carbohydrate as previously described [32]. The glycemic index was determined using average values from the International Tables [33].

2.3. Dietary Data Processing-Based Classification and Ultra-Processed Foods

Items in the FFQ were classified according to the NOVA system [5,11] developed by the Public Health Faculty of the University of São Paulo in Brazil. This system classifies foods and beverages according to the nature, extent, and purpose of their industrial processing into four groups: (1) unprocessed or minimally processed foods (i.e., fresh or frozen fruits and vegetables, eggs, pasteurized milk, meat, seeds, nuts, grains, or plain yogurt); (2) processed culinary ingredients (i.e., oils, fats, sugar, and salt); (3) processed foods (i.e., canned vegetables, canned fish, fruits in syrup, cheeses, fresh bread, beer, and wine); and (4) UPF (i.e., soft drinks, sweet, or savory packed snacks, processed meats, pre-prepared frozen dishes, and 'instant' products). Details about the allocation of FFQ items to processing groups with examples are provided in online Supplementary Table S1. Furthermore, items belonging to the UPF group (foods and beverages) were allocated into the following subgroups: dairy products; processed meat; pre-prepared dishes, snacks and fast-foods; sweets; non-alcoholic beverages; and alcoholic beverages (Table 1).

Table 1. Relative contribution of different food groups to the consumption of UPF in diet of participants at baseline.

Subgroup	Contribution (%)	Item
Sweets	28	chocolate cookies, breakfast cereals, muffins, donuts, croissants, pastries, and confectionery
Non-alcoholic beverages	26	soft drinks (sugar- and artificially-sweetened) and commercial fruit juices
Processed meats	22	ham, chorizo, mortadella, sausages, hamburgers and meat balls, and pate and foie-gras
Pre-prepared dishes, snacks and fast-foods	11	potato chips, croquettes, pizza, instant soups, margarine, mayonnaise, mustard, ketchup, packed fried tomato sauce, and savory packed snacks
Dairy products	11	milkshakes, ice cream, Petit suisse, custard, flan, pudding, and creamy cheese spreads
Alcoholic beverages	2	distillated liquors

Daily intake of beverages was collected in cubic centimeters and then converted into milliliters (1 cc = 1 mL), and further into grams, assuming that 1 mL = 1 g.

Repeated data on the percentage of UPF consumption (and other NOVA groups) was computed as the sum of grams per day consumed from items in the UPF group (determined at baseline, 6-, and 12-month follow-up visits), divided by the total grams of all items consumed per day, and multiplied by 100.

2.4. Socio-Demographic, Lifestyle, Anthropometric, and Clinical Variables Assessment

Information on socio-demographics and health-related issues was collected from participants using a general questionnaire at baseline. Educational level, indicated by the highest educational qualification (professional or academic) achieved, was categorized into three groups (higher education or technician, secondary education, or primary education or less), smoking habits into four groups (current, never, ex-smoker, or insufficient data), whereas history of overweight was categorized into five groups (since childhood, adolescence, adulthood, after childbirth, or since menopause), and prevalence of type 2 diabetes was dichotomized (yes or no).

At baseline, 6-, and 12-month follow-up visits, total leisure-time PA (METs min/week) was assessed using the validated Minnesota-REGICOR short physical activity questionnaire [34] and sedentary behavior (SB) (h/day) by the validated Spanish version of the Nurses' Health Study questionnaire [35]. Adherence to the erMedDiet was determined using a 17-item screener, a modified version of a validated 14-item questionnaire [36].

At each visit, trained staff measured in duplicate height and weight according to the study´s protocol using a wall-mounted stadiometer and calibrated scales, respectively. Blood pressure was measured in triplicate using a calibrated oscillometer. Averages of these repeated measurements were taken for analyses. BMI was calculated by dividing weight (kg) by squared height (m), and waist circumference (cm) was determined midway between the lowest rib and the iliac crest using an anthropometric tape.

Fasting blood samples were collected at baseline and thereafter to quantify levels of alanine aminotransferase (ALT, U/L), aspartate aminotransferase (AST, U/L), gamma-glutamyl transferase (GGT, U/L), glucose (mg/dL), high-density lipoprotein cholesterol (HDL-c, mg/dL), triglyceride (mg/dL), and glycated hemoglobin (HbA1c, %) using standard methods. MetS components were ascertained according to guidelines from the International Diabetes Federation/National Heart, Lung and Blood Institute/American Heart Association (2009), namely: waist circumference $\geq$ 102 cm for men and $\geq$88 cm for women, triglycerides $\geq$ 150 mg/dL, HDL-c < 40 mg/dL in men and <50 mg/dL in women, systolic blood pressure $\geq$ 130 and/or diastolic blood pressure $\geq$ 85 mmHg (or antihypertensive drug treatment), and fasting glucose $\geq$ 100 mg/dL (or antidiabetic drug treatment) [27].

2.5. Outcome Assessment: Liver Health

Repeated data of the non-invasive liver health biomarkers were computed using anthropometric and biochemical data collected at baseline, 6-, and 12-month follow-up visits. The fatty liver index (FLI) and hepatic steatosis index (HSI) were used as surrogate measures of NAFLD [37,38]. FLI [37] and HSI [38] are diagnostic algorithms built upon a cluster of liver health biomarkers, including BMI, waist circumference, blood levels of triglycerides, the liver enzyme GGT, the AST/ALT ratio, type 2 diabetes status, and sex (see below). These scores have been previously validated in large populations against imaging techniques, showing high specificity and sensitivity in predicting excess liver fat, with values < 30 ruling out and values $\geq$ 60 (for FLI) or $\geq$36 (for HSI) confirming NAFLD [37,38].

$$\text{FLI} = (e^{(0.953\times\ln(\text{triglyceride})+0.139\times\text{BMI}+0.718\times\ln(\text{GGT})+0.053\times\text{waist circumference}-15.745)})/ (1 + e^{(0.953\times\ln(\text{triglyceride})+0.139\times\text{BMI}+0.718\times\ln(\text{GGT})+0.053\times\text{waist circumference}-15.745)}) \times 100$$

HSI = 8 * ALT/AST + BMI + (2 if type 2 diabetes, 0 otherwise) + (2 if women, 0 otherwise)

2.6. Statistical Analyses

All analyses were performed using the entire analytical sample as an observational cohort (both study arms combined). For descriptive analyses of a participant´s characteristics means and standard deviations (SD), and numbers and percentages (%), were calculated for continuous and categorical variables, respectively. Statistical differences in baseline characteristics by baseline sex-specific quintiles of UPF consumption were assessed using one-way ANOVA or χ^2 test, wherever appropriate. The differences in these characteristics over follow-up time were assessed using linear mixed-effects models with random intercepts at recruiting center, cluster family, and patient level.

The same approach with linear mixed-effects modelling with random intercepts (recruiting center, cluster family, and patient) was used to evaluate associations between concurrent 6-month changes in UPF consumption with changes in indices of NAFLD over the first year of follow-up. Our exposure was modelled as repeatedly measured continuous variable (per 10% increment) and as sex-specific quintiles, with the first quintile set as the reference category. The *p* for linear trend across increasing quintiles was calculated with the use of the median value for each category. In the main analyses, two sets of covariates were used. Model 1 was minimally adjusted for age at inclusion, sex, study arm, and follow-up time (months). Model 2 was additionally adjusted for baseline educational level, smoking habits (all categorical, height, as well as repeatedly measured at baseline and every 6 months thereafter PA, SB, and alcohol intake (all continuous). Selection of covariates was performed using a causal directed acyclic graph approach implemented in the DAGitty free web application [39].

Several sensitivity analyses were performed based on model 2. The potential influence of nutritional factors (characteristics of UPF) was addressed by additional adjustment for repeatedly measured intake of total energy, saturated and trans FA, cholesterol, fiber, glycemic load, sodium (individually and including all factors simultaneously in the model), as well as adherence to erMedDiet (all continuous). Moreover, we controlled for several NAFLD-related risk factors by additional adjustment for repeatedly measured BMI, waist circumference, HbA1c, and number of MetS factors (continuous), as well as for history of overweight self-reported at baseline and type 2 diabetes prevalence at baseline (both categorical). Finally, models were rerun eliminating outliers (1st and 99th percentile) from FLI and HSI, and imputing missing follow-up data (UPF, NAFLD indices and covariables) using the last observation carried forward (LOCF) method.

Additionally, the proportion mediated by nutritional factors (characteristics of UPF, and adherence to erMedDiet, as an indicator of a healthy dietary pattern) and NAFLD-related biomarkers (known risk factors and components of FLI and HSI scores) in the studied association were quantified. For this, standard steps proposed by Baron and Kenny (1986) with adjustments introduced by Iacobucci et al. [40] were followed using multicovariate-adjusted linear mixed-effects models (Model 2). Details about the method and analyses are presented in online Supplementary Text S1.

Subgroup analyses were also performed by rerunning the model for different strata at baseline: sex (men or women), age (<65 or ≥65 y), type 2 diabetes status (non-diabetics or diabetics), alcohol intake (<20/30 g/day for men/women or ≥20/30 g/day for men/women, respectively), and adherence to erMedDiet (<8 or ≥8 points). Median values were used as a threshold to stratify age and erMedDiet, whereas safe limits accepted by guidelines of scientific associations were used for alcohol intake [41]. The *p* values for interaction were computed for each scenario rerunning model 2 with a multiplicative interaction term inserted between these variables and exposure in continuous form.

In secondary analyses, model 2 was rerun to explore the associations between concurrent changes in consumption of different food subgroups within UPF (continuous variable) and NAFLD indices.

Analyses were performed using Stata v15. and statistical significances was set at $p < 0.05$. The last actualized version of the PREDIMED-Plus longitudinal database generated on 22nd December 2020 (202012220958_PREDIMEDplus 2020) was used.

3. Results

Table 2 presents baseline characteristics of study participants according to quintiles of UPF consumption. The analytical sample (n = 5867) comprised 47.8% women with average age at inclusion of 65.0 years (SD 4.9 years). Overall obesity and abdominal obesity (73.1 and 93.0%, respectively), as well as NAFLD screened using FLI and HSI (84.1 and 95.2%, respectively), were highly prevalent. Mean UPF consumption accounted for 8.19% (SD 6.95%) of total daily intake (in grams). Compared to participants with the lowest UPF consumption (Q1, mean consumption 2.12% (SD 0.81%) of total daily intake (in grams)), participants in the highest quintile (Q5, mean UPF consumption 19.0% (SD 7.9%)) were younger and showed less healthy lifestyle habits in terms of physical inactivity; higher sedentariness; intake of energy, saturated and trans FA, cholesterol, sodium, and glycemic load; as well as lower intake of fiber, alcohol, and adherence to MedDiet ($p < 0.001$ for all comparisons). Moreover, participants in the highest quintile of UPF consumption presented higher values of BMI and WC than their counterparts in the lowest quintile, as well as higher levels of liver health biomarkers, such as blood ALT/AST ratio and triglycerides, as well as both NAFLD indices ($p \leq 0.001$). Sweets (28%), non-alcoholic beverages (26%), and processed meats (22%) were the main food subgroups consumed within the UPF category at baseline (Table 1).

Table 2. Baseline characteristics of study participants according to baseline sex-specific quintiles of UPF.

	Quintiles of UPF Consumption						
	Total	Q1	Q2	Q3	Q4	Q5	
	Mean (SD)	Mean (SD)	Mean (SD)	Mean (SD)	Mean (SD)	Mean (SD)	*p*-Value
n	5867	1174	1173	1173	1173	1173	
Sociodemographic factors							
Women, *n* (%)	2807 (47.8)	562 (47.9)	561 (47.8)	562 (47.9)	561 (47.8)	561 (47.8)	
Age (years)	65.0 (4.9)	66.1 (4.7)	65.4 (4.8)	64.8 (4.8)	64.6 (4.9)	64.3 (5.0)	<0.001
Higher education, *n* (%)	1233 (21.0)	234 (19.9)	232 (19.8)	255 (21.7)	254 (21.7)	258 (22.0)	0.087
Current smokers, *n* (%)	732 (12.5)	128 (10.9)	128 (10.9)	146 (12.4)	159 (13.6)	171 (14.6)	0.167
Lifestyle factors							
Physical activity (METs min/week)	2477 (2297)	2743 (2481)	2646 (2449)	2439 (2245)	2383 (2160)	2174 (2081)	<0.001
Sedentary behavior (h/day)	6.00 (1.96)	5.68 (1.98)	5.89 (1.91)	6.04 (1.94)	6.18 (1.92)	6.23 (1.97)	<0.001
FFQ:							
Total energy intake (kcal/day)	2360 (550)	2203 (521)	2318 (511)	2371 (536)	2434 (545)	2473 (592)	<0.001
Saturated FA (% of energy intake)	9.95 (1.99)	8.95 (1.80)	9.63 (1.72)	10.1 (1.9)	10.4 (2.0)	10.7 (2.1)	<0.001
Trans FA (% of energy intake)	0.22 (0.13)	0.16 (0.10)	0.20 (0.11)	0.23 (0.12)	0.24 (0.13)	0.27 (0.14)	<0.001
Cholesterol (mg/day)	380 (115)	343 (105)	367 (107)	389 (115)	397 (115)	406 (123)	<0.001
Sodium (mg/day)	3281 (1016)	3021 (983)	3239 (972)	3298 (1010)	3397 (971)	3452 (1088)	<0.001
Glycemic load	131 (46)	123 (45)	128 (42)	131 (45)	133 (47)	138 (49)	<0.001
Fiber intake (g/day)	25.9 (8.7)	27.8 (9.5)	27.0 (8.7)	26.0 (8.3)	25.2 (8.4)	23.6 (7.9)	<0.001
Alcohol intake (g/day)	11.1 (15.1)	12.5 (17.3)	11.6 (15.6)	11.0 (14.3)	10.6 (14.1)	9.70 (14.0)	0.0001
Adherence to erMedDiet (17p score)	8.45 (2.67)	9.61 (2.55)	8.90 (2.67)	8.37 (2.48)	7.98 (2.53)	7.38 (2.55)	<0.001
NOVA food groups:							
Unprocessed or minimally processed foods (% of g/day)	68.1 (12.5)	73.3 (13.4)	71.6 (11.6)	69.7 (10.5)	66.7 (10.3)	59.1 (11.2)	<0.001
Processed culinary ingredients (% of g/day)	2.79 (1.28)	2.69 (1.23)	2.80 (1.28)	2.82 (1.24)	2.88 (1.37)	2.78 (1.26)	0.006
Processed foods (% of g/day)	20.9 (10.8)	21.9 (13.2)	21.4 (11.4)	21.2 (10.1)	21.0 (9.8)	19.0 (8.9)	<0.001
UPF (% of g/day)	8.19 (6.95)	2.12 (0.81)	4.17 (0.63)	6.23 (0.86)	9.44 (1.44)	19.0 (7.9)	<0.001
Liver health risk factors							
BMI (kg/m^2)	32.5 (3.4)	32.0 (3.3)	32.5 (3.4)	32.6 (3.4)	32.7 (3.5)	32.8 (3.6)	<0.001
Overall obesity prevalence, *n* (%)	4289 (73.1)	819 (69.8)	848 (72.3)	870 (74.1)	864 (73.7)	888 (75.7)	0.018
History of overweight from childhood, *n* (%)	334 (5.69)	53 (4.51)	64 (5.46)	78 (6.64)	60 (5.12)	79 (6.73)	0.595
Waist circumference (cm)	107.5 (9.6)	106.3 (9.1)	107.1 (9.3)	107.7 (9.7)	108.2 (9.7)	108.4 (9.9)	<0.001
Abdominal obesity prevalence, *n* (%)	5454 (93.0)	1075 (91.6)	1093 (93.2)	1103 (94.0)	1095 (93.4)	1088 (92.8)	0.224
HbA1c (%)	6.12 (0.87)	6.12 (0.82)	6.14 (0.86)	6.08 (0.82)	6.14 (0.93)	6.11 (0.90)	0.570
Type 2 diabetes prevalence at baseline, *n* (%)	1828 (31.2)	385 (32.8)	363 (31.0)	356 (30.3)	384 (32.7)	340 (29.0)	0.213
Number of MetS factors at baseline	3.38 (0.98)	3.31 (0.99)	3.37 (0.99)	3.40 (0.96)	3.38 (0.98)	3.42 (0.97)	0.069
Liver health biomarkers							
FLI (arbitrary units)	77.9 (17.1)	75.8 (17.3)	77.5 (17.4)	78.4 (17.1)	78.7 (16.7)	79.4 (16.8)	<0.001
NAFLD prevalence (FLI $\geq$ 60), *n* (%)	4934 (84.1)	962 (81.9)	970 (82.7)	990 (84.3)	1000 (85.3)	1012 (86.3)	0.132
HSI (arbitrary units)	43.4 (5.9)	42.7 (4.6)	43.5 (6.5)	43.3 (5.5)	43.4 (4.8)	44.0 (7.4)	<0.001

Table 2. *Cont.*

	Quintiles of UPF Consumption						
	Total	Q1	Q2	Q3	Q4	Q5	
	Mean (SD)	Mean (SD)	Mean (SD)	Mean (SD)	Mean (SD)	Mean (SD)	*p*-Value
NAFLD prevalence (HSI ≥ 36), *n* (%)	5585 (95.2)	1117 (95.1)	1124 (95.8)	1119 (95.3)	1115 (95.1)	1110 (94.6)	0.750
ALT (U/L)	27.0 (15.4)	26.4 (15.3)	27.7 (16.2)	26.7 (15.0)	26.5 (14.3)	27.8 (16.0)	0.065
AST (U/L)	23.3 (9.9)	23.1 (9.7)	23.6 (10.4)	23.3 (10.1)	23.2 (9.6)	23.4 (9.8)	0.771
ALT/AST ratio	1.16 (0.53)	1.13 (0.34)	1.18 (0.64)	1.15 (0.46)	1.14 (0.34)	1.21 (0.76)	0.001
AST/ALT ratio	0.95 (0.30)	0.96 (0.28)	0.94 (0.29)	0.96 (0.30)	0.96 (0.30)	0.94 (0.33)	0.204
GGT (U/L)	37.6 (37.2)	37.3 (34.7)	38.5 (40.8)	37.9 (39.7)	37.1 (35.8)	37.1 (34.7)	0.889
Triglycerides (mg/dL)	151 (77)	145 (73)	149 (79)	151 (70)	151 (73)	158 (88)	0.001

Abbreviations: ALT—alanine aminotransferase; AST—aspartate aminotransferase; BMI—body mass index; erMedDiet—energy-restricted Mediterranean Diet; GGT—gamma-glutamyl transferase; FA—fatty acids; FFQ—Food frequency questionnaire; FLI—fatty liver index; HbA1c—glycated hemoglobin; HSI—hepatic steatosis index; MetS—metabolic syndrome; NAFLD—non-alcoholic fatty liver disease; UPF—ultra-processed foods. Sex-specific ranges for quintiles of UPF (%): men—Q1 (lowest): 0.00–3.53, Q2: 3.53–5.44, Q3: 5.45–8.17, Q4: 8.18–12.60, Q5 (highest): 12.62–57.67; women—Q1 (lowest): 0.10–2.97, Q2: 2.97–4.69, Q3: 4.70–6.91, Q4: 6.92–10.64, and Q5 (highest): 10.65–59.48. Values shown are mean (SD) unless otherwise specified. Overall obesity was defined as body mass index ≥ 30.0 kg/m^2, and abdominal obesity as waist circumference ≥ 88 cm in women or ≥102 cm in men. The consumption of NOVA food groups was expressed as a percentage of total food and beverage intake in g/day. Daily intake of beverages was collected in cubic centimeters and then converted into milliliters (1 cc = 1 mL), and further into grams, assuming that 1 mL = 1 g. *p*-values for comparisons between baseline quintiles of UPF consumption were calculated by one-way ANOVA test for continuous variables and $\chi 2$ test for categorical variables.

As shown in online Supplementary Table S2, lifestyle factors and liver health markers improved over time compared to the baseline ($p < 0.05$ for all comparisons). An increase in the consumption of unprocessed or minimally processed foods and decrease in products with higher degree of processing was observed during the 1-year follow-up period, probably as consequence of MedDiet recommendations, which were given to participants in both study arms ($p < 0.001$).

Results from the main analysis evaluating the association between concurrent changes in UPF consumption and changes in indices of NAFLD are summarized in Figure 1 and presented in detail in online Supplementary Table S3. In model 2, we observed significant ($p < 0.001$) associations between each daily 10% increment in UPF consumption and greater FLI (β 1.60, 95% CI 1.24; 1.96) and HSI (0.43, 0.29; 0.57) over the 1-year follow-up. Comparison across increasing quintiles of UPF consumption revealed a significant dose–response relationship with both NAFLD indices (*p for trend < 0.001*): FLI (β estimates for Q5 3.73, 95% CI 3.10; 4.35) and HSI (0.93, 0.67; 1.18).

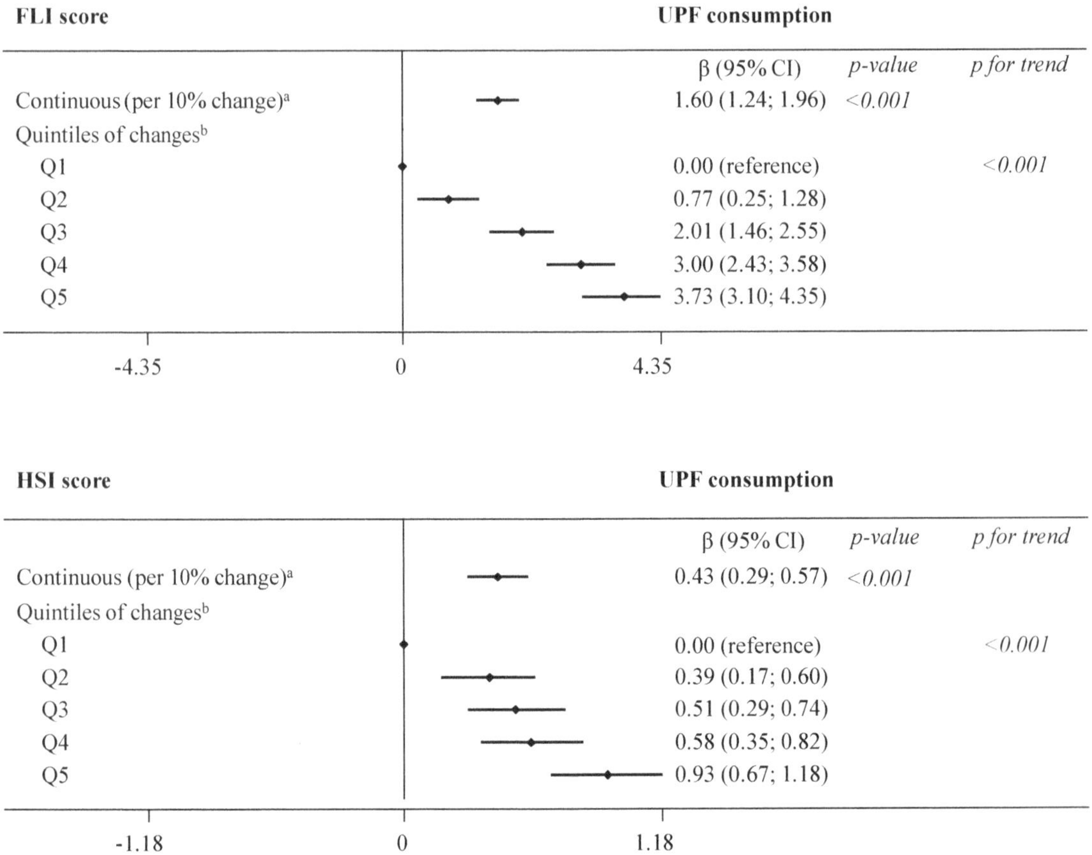

Figure 1. Dose–response relationship in the association between concurrent changes in UPF consumption (% of g/day) and changes in NAFLD indices during 1 year of follow-up (fully adjusted model 2). The consumption of UPF was expressed as a percentage of total food and beverage intake in g/day. Daily intake of beverages was collected in cubic centimeters and then converted into milliliters (1 cc = 1 mL), and further into grams, assuming that 1 mL = 1 g. Mixed-effects linear modelling for repeated measures with random intercepts at recruiting center, cluster family, and patient level were used after controlling in fully adjusted model 2 for baseline variables, such as age, sex, study arm, educational level, smoking habits, and height, as well as repeatedly measured physical activity, sedentary behavior, alcohol intake, and follow-up time. [a] Estimates β are interpreted as changes in NAFLD associated with increments of 10% in UPF consumption. [b] Estimates β are interpreted as changes in NAFLD indices in each sex-specific quintile of UPF consumption, compared to quintile 1, the reference category. Abbreviations: FLI—fatty liver index; HSI—hepatic steatosis index; NAFLD—non-alcoholic fatty liver disease; UPF—ultra-processed foods.

As highlighted in Figure 2 (UPF coded as continuous variable) and shown in detail in online Supplementary Table S4 (UPF coded as continuous and sex-specific quintiles), results remained statistically significant after further adjustments in sensitivity analysis. Only simultaneous adjustment for several factors related to nutritional quality of the diet (saturated and trans FA, cholesterol, fiber, glycemic load and sodium) and an adherence to erMedDiet, as well as BMI and waist circumference, decreased point estimates, yet the associations of UPF consumption with both NAFLD indices remained statistically significant. A similar pattern was observed when the exposure was coded in quintiles (online Supplementary Table S4).

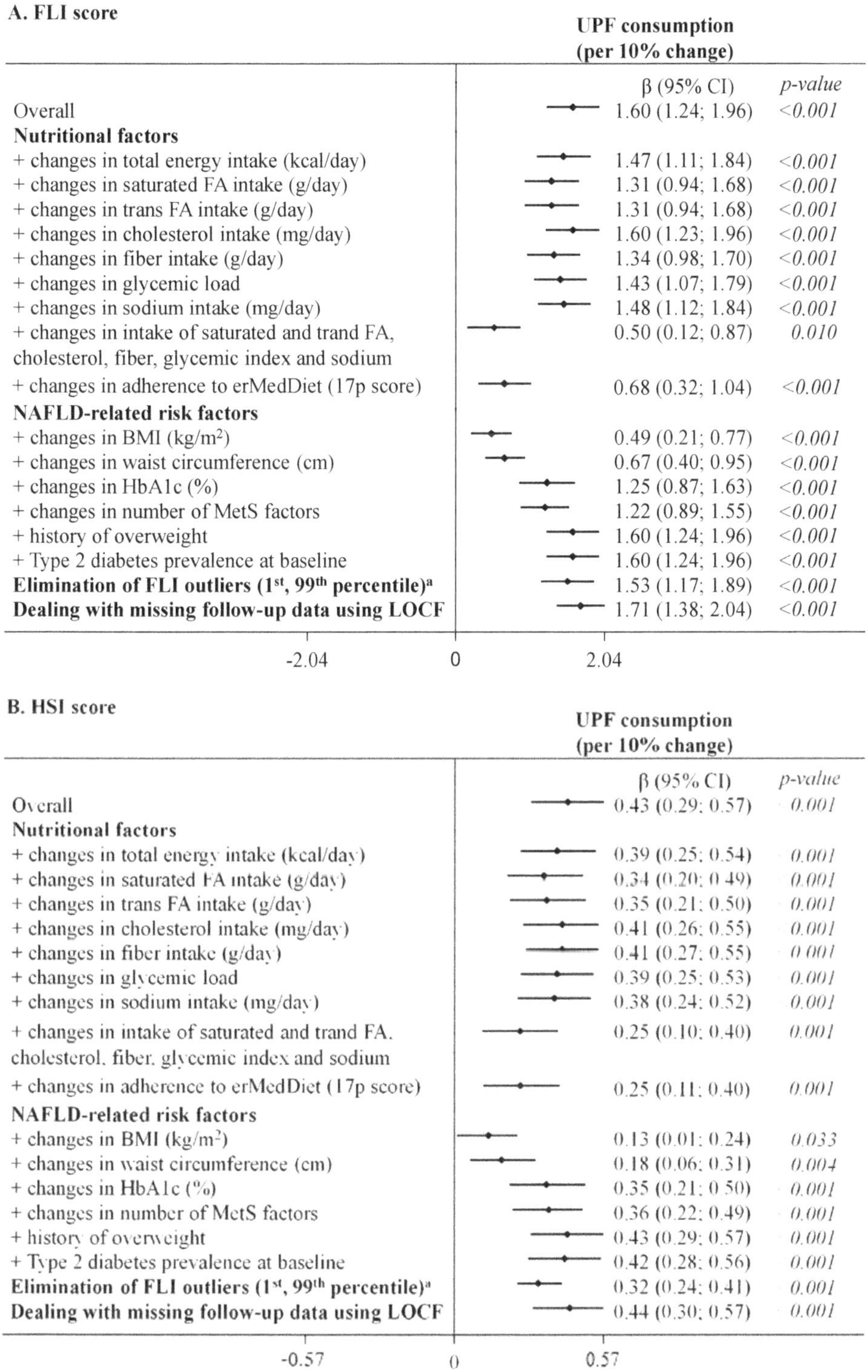

Figure 2. Summary of the sensitivity analysis for the association between concurrent changes in UPF consumption (% of g/day, continuous variable) and changes in NAFLD indices during 1 year of

follow-up (fully adjusted model 2). The consumption of UPF was expressed as a percentage of total food and beverage intake in g/day. Daily intake of beverages was collected in cubic centimeters and then converted into milliliters (1 cc = 1 mL), and further into grams, assuming that 1 mL = 1 g. Mixed-effects linear modelling for repeated measures with random intercepts at recruiting center, cluster family and patient level were used after controlling in fully adjusted model 2 for baseline variables, such as age, sex, study arm, educational level, smoking habits, and height, as well as repeatedly measured physical activity, sedentary behavior, alcohol intake, follow-up time, and use of antidiabetic medications (for models with HbA1c). Estimates β are interpreted as changes in NAFLD indices associated with increments of 10% in UPF consumption. [a] Outliers (1st, 99th percentile) in the outcome variables were eliminated at baseline and follow-up (for FLI total n = 318, for HSI total n = 316). Abbreviations: BMI—body mass index; erMedDiet—energy-restricted Mediterranean Diet; FA—fatty acids; FLI—Fatty liver index; HbA1c—glycated hemoglobin; HSI—Hepatic steatosis index; LOCF—last observation carried forward; MetS—metabolic syndrome; NAFLD—non-alcoholic fatty liver disease; UPF—ultra-processed foods.

In the mediation analysis, we found that changes in nutritional factors partly mediated the association with concurrent changes in NAFLD indices (Table 3 and online Supplementary Table S5). Among them, changes in nutritional characteristics of UPF, such as saturated and trans FA, mediated 17–21% of the associations for both indices, and fiber and glycemic load explained 15% and 11% of the association for FLI, respectively; whereas changes in intake of total energy, sodium, and cholesterol did not mediate any of the associations. Moreover, changes in adherence to erMedDiet acted as mediator in 58% and 43% for FLI and HSI, respectively. As far as NAFLD-related biomarkers (known risk factors and components of both scores) are concerned, changes in BMI were responsible in 69% for the association between concurrent changes in UPF consumption and both FLI and HSI; whereas, waist circumference was responsible in 56% of the association for FLI and in 82% for HSI. Furthermore, in the case of FLI, the association with UPF was driven by changes in triglycerides and MetS factors (both in 26%), followed by changes in Hba1c (14%). In turn, in case of HSI, the association was driven by the ALT/AST ratio (39%), ALT and MetS factors (both in 16%), and changes in HbA1c (15%). Changes in GGT and AST did not mediate the respective associations for FLI and HSI.

Table 3. Proportion of the association between concurrent changes in UPF consumption (% of g/day, continuous variable) and changes in NAFLD indices during 1 year of follow-up mediated through nutritional factors and NAFLD-related biomarkers (fully adjusted model 2).

(A) FLI Score	
Mediator	**% Mediated**
Nutritional factors	
+changes in total energy intake (kcal/day)	0%
+changes in saturated FA intake (g/day)	19%
+changes in trans FA intake (g/day)	18%
+changes in cholesterol intake (mg/day)	0%
+changes in fiber intake (g/day)	15%
+changes in glycemic load	11%
+changes in sodium intake (mg/day)	0%
+changes in adherence to erMedDiet (17p score)	58%
NAFLD-related biomarkers	
+changes in BMI (kg/m^2)	69%
+changes in waist circumference (cm)	56%
+changes in HbA1c (%)	14%
+changes in number of MetS factors	26%
+changes in GGT (U/L)	0%
+changes in triglycerides (mg/dL)	26%

Table 3. *Cont.*

(B) HSI Score	
Mediator	**% Mediated**
Nutritional factors	
+changes in total energy intake (kcal/day)	0%
+changes in saturated FA intake (g/day)	21%
+changes in trans FA intake (g/day)	17%
+changes in cholesterol intake (mg/day)	0%
+changes in fiber intake (g/day)	0%
+changes in glycemic load	0%
+changes in sodium intake (mg/day)	0%
+changes in adherence to erMedDiet (17p score)	43%
NAFLD-related biomarkers	
+changes in BMI (kg/m^2)	69%
+changes in waist circumference (cm)	82%
+changes in HbA1c (%)	15%
+changes in number of MetS factors	16%
+changes in ALT (U/L)	16%
+changes in AST (U/L)	0%
+changes in ALT/AST	39%

Abbreviations: ALT—alanine aminotransferase; AST—aspartate aminotransferase; BMI—body mass index; erMedDiet—energy-restricted Mediterranean Diet; FA—fatty acids; FLI—Fatty liver index; GGT—gamma-glutamyl transferase; HbA1c—glycated hemoglobin; HSI—Hepatic steatosis index; MetS—metabolic syndrome; NAFLD—non-alcoholic fatty liver disease; UPF—ultra-processed foods. Summary of the mediation analyses was performed to determine the extent to which the association between independent variable (UPF consumption, continuous variable) and each dependent variable (FLI and HSI scores) was mediated through individual nutritional factors (characteristics of UPF and adherence to erMedDiet, as an indicator of healthy dietary pattern) and NAFLD-related biomarkers (known risk factors and components of NAFLD indices). Mediation analyses were performed following standard steps proposed by Baron and Kenny (1986) with adjustments introduced by Iacobucci et al. [39] to evaluate direct and indirect effect and the proportion mediated by each of these variables. More details of this analysis are presented in Supplementary Text S1 and Supplementary Table S5).

In subgroup analyses (online Supplementary Table S6), we found that the direct association between UPF consumption and FLI was slightly more pronounced in non-diabetics (β 1.73, 95% CI 1.29; 2.18, $p < 0.001$) than in diabetics (1.29, 0.68; 1.90, <0.001) (*p for interaction = 0.027*).

In additional analyses (online Supplementary Table S7), we found that all UPF subgroups contributed to observed associations with NAFLD indices. In particular, pre-prepared dishes, snacks, and fast-foods, as well as processed meats and sweets, showed the strongest statistically significant associations (all p-values < 0.001) with both liver scores. In turn, the subgroup of alcoholic beverages was only strongly associated with FLI.

4. Discussion

In this large prospective cohort study of older adults with overweight/obesity and MetS from Spain, a Mediterranean country, we found that UPF consumption was associated with worse liver health, assessed using biomarkers related to NAFLD. This direct association was ascertained using sophisticated analyses and validated tools, and was robust after accounting for a wide range of indicators related to nutritional quantity and quality of the diet and NAFLD risk.

Given the mounting body of evidence showing associations of UPF consumption with well-known risk factors for NAFLD, such as obesity [12–15], type 2 diabetes [17], or hypertension [42,43], our findings were not unexpected. Moreover, they are in line with recent evidence on the role of diet in NAFLD [4], pointing to several UPF, such as processed meat [44] and SSB [45], as culprits in objectively-determined NAFLD development. However, studies on diet-NAFLD risk have rarely been prospective, and few of them aggregated foods according to the nature, extent, and purpose of processing [23,24]. In this sense, our findings corroborate recent results from TCLSIH prospective cohort in China, showing the link between UPF consumption and risk of developing NAFLD, diagnosed using ultrasonography [24]. Here, we support this link using repeatedly measured dietary habits in a different population from a Mediterranean country. It needs to be underlined that processing not always has to be negative, as it also can increase the safety and shelf-life of foods and beverages. However, ultra-processing combines several ingredients with little, if any, intact whole foods, which results in the creation of new products with nutritionally imbalanced properties [5,6].

Several putative mechanisms of action could be responsible for the link between UPF and liver fat accumulation. The first is the nutritional characteristics of UPF, which is poor due to the industrial manipulations that they undergo. For instance, the incorporation of sizable amounts of saturated and trans FA may increase product stability and palatability [46], and these nutrients have been associated with increased liver fat in humans and rodents [4,47]. It should be mentioned that there has been significant progress towards 2023, the target for elimination of industrially produced trans FA around the world [48]. In the European Union countries, the regulation limiting the use of artificial trans FA came into force just recently (April 2021) [49]. However, given that dietary data used in this study were collected before that (2013–2017), we could still estimate the intake of trans FA in this population. Furthermore, low fiber content is a common attribute of UPF, and the breakdown of natural food matrix during ultra-transformation might also reduce its quantity. Recent findings from a large cross-sectional study showed an inverse association of dietary fiber with NAFLD [50], and this could be explained through the effects of fiber on microbiota as well as on satiation and satiety. UPF, including beverages, usually lead to postprandial hyperglycemia due to a high content of refined carbohydrates, such as white flour and sugar, and simultaneous fiber, water, and protein deprivation [51]. Food glycemic responses have been implicated in liver fat mass accretion through alterations in glucose, insulin, and lipid metabolism [52,53]. In our multivariate analyses, we found that the associations between UPF and liver scores were attenuated but remained statistically significant after further adjustment for these nutritional factors (i.e., saturated and trans FA, fiber, and glycemic load). Mediation analyses revealed that the quality of fat (saturated and trans FA) and carbohydrate (fiber and glycemic load) contributed (11 to 21%) to this pooled effect. This could indicate that these nutritional attributes of UPF might explain part of the observed associations, but clearly not the overall effect. This suggests that mechanisms beyond the nutritional dimension of UPF might also be responsible for the observed associations.

Another potential mechanism responsible for the association between UPF and liver health could be related to additives used during the ultra-processing of these products. In this regard, although the health properties of non-nutritional additives are relatively underexplored in humans, current research performed in rodents and cell lines suggests that they can be harmful for the liver, and the effect could be partly mediated through imbalances of gut microbiota [54]. Some artificial sweeteners (i.e., saccharin, aspartame), emulsifiers (i.e., polysorbate 80), preservatives (i.e., benzoic acid), and flavor enhancers (i.e., monosodium glutamate) could lead to transaminitis, steatosis, degradation, and toxicity in the liver of rodents [54–57]. We could not explore the potential mediating role of additive content because this information is not yet available in most food composition tables.

In addition, UPF composition may indirectly lead to excessive hepatic lipid accumulation, given their ability to displace healthy foods affecting overall diet quality. In mediation analyses we confirmed that a low adherence to MedDiet explained approximately half of the studied association. Overall diet quality and consumption of UPF are two different but complementary nutritional dimensions to consider in relation to health, given that they could offset one another. A high-quality dietary pattern such as the MedDiet is presumed to be beneficial for NAFLD management [3,4]. An interesting result obtained in our sensitivity analyses suggests that the associations between consumption of UPF and NAFLD indices were attenuated after adjusting for MedDiet adherence, albeit it remained statistically significant. All in all, the potential mechanisms by which UPF consumption may be related to NAFLD is speculative and warrants future studies.

Although not the sole determinant, obesity is recognized as a major risk factor for NAFLD [1,2], whereas type 2 diabetes and MetS are considered as clinical factors that coexist with NAFLD in a bi-directional relationship [58,59]. In this sense, we found that the direct associations between UPF consumption and liver health scores remained significant after adjustment for BMI or waist circumference, albeit their strength markedly diminished. Mediation analyses confirmed that a substantial proportion of the association was driven by obesity, either overall (69%) or abdominal (56% for FLI and 82% for his). In turn, the proportion mediated by type 2 diabetes, MetS, or triglycerides was lower. We have previously reported in the PREDIMED-Plus cohort by using imaging technique that consumption of UPF affected to a similar extent visceral and total fat [12], potentially leading to NAFLD and other diseases. Other large cross-sectional and prospective studies in adults have also shown strong and direct association between UPF and type 2 diabetes [17], MetS [60], and hypertension [42,43]. Regarding markers of liver function, only in the case of the HSI score was the association driven in part by its enzymatic component ALT, and particularly the ALT/AST ratio. The latter has been considered as more accurate than each of these enzymes alone [61]. There is evidence from cross-sectional and cohort studies with healthy adults on a relationship of SSB and fast foods with greater levels of ALT and the ALT/AST ratio [45,62,63]. It needs to be underlined that all PREDIMED-Plus participants were overweight/obese with MetS, and some also had type 2 diabetes (≈27%); hence, future longitudinal studies with healthier individuals are warranted to ascertain what mechanisms underly the association between UPF and liver health.

Beyond the prospective design, control for a wide set of confounders, and performance of a series of sensitivity and stratified analyses, a marked strength of the present study was the use of a large and homogenous sample of men and women within a narrow range of age, BMI, and health conditions. Of note, a unique feature of the present study is that both exposure and outcome were repeatedly measured at the same points in time, potentially decreasing the risk of reverse causality. This is relevant as eating behaviors change over time, given the nutritional intervention given to participants in the PREDIMED-Plus trial [64].

Our study also has limitations. First, its observational nature enabled the identifications of associations only. Second, the participants were older individuals with overweight/obesity and MetS from a Mediterranean area, which limits the generalizability of our findings. However, the described health profile is quite common in modern societies. Third, measurement error is unavoidable when using self-reported dietary data, even though we undertook some actions to improve measurement precision; namely, the FFQ was previously validated in a Spanish population and administrated repeatedly (each 6 months) to the participants by trained dietitians during face-to-face interviews. Moreover, participants with implausible total energy intake values were a priori excluded from anal yses, and models were adjusted for total energy intake changes in sensitivity analyses. Fourth, some misclassification in NOVA groups cannot be ruled out, as the FFQ used was not designed to capture details on food processing, and the definition of UPF in the NOVA system is rather broad, allowing multiple interpretations. However, the FFQ items were classified into processing groups with caution and consensus was reached between experts in nutrition and epidemiology. Last but not least, liver fat was estimated based on two surrogate indices, but was not directly measured using imaging techniques. However, both NAFLD algorithms have been validated and have shown good agreement with ultrasonography [37,38].

5. Conclusions

In this prospective study we revealed that in older adults with chronic health conditions, consumption of UPF was directly and robustly associated with FLI and HSI scores. Furthermore, these associations were only to a lesser extent explained by the nutritional characteristics of UPF, pointing out the potential and uncovered role of factors related to the processing itself (i.e., non-nutritional chemicals and food matrix breakdown). Future prospective studies in different contexts and with more precise imaging techniques are warranted to confirm our findings on liver fat accumulation, as well as future toxicological, technological, and human experimental studies to clarify underlying mechanisms and develop detection methods for components generated through food processing. With this study we provide novel insights into the recently growing body of evidence on food processing and health risk. The accumulation of firm and high-quality evidence would help global health authorities to update dietary recommendations and food policies by considering criteria of food processing, imposing restrictions to marketing, use of additives, and types of packaging in food technology and trade. Discouragement of UPF consumption and favoring instead fresh or minimally processed foods should be considered by health care providers as a valid preventive and treatment strategy for NAFLD.

Supplementary Materials: The following supporting information can be downloaded at: https://www.mdpi.com/article/10.3390/nu14194142/s1, Figure S1: Flow chart for the selection of participants for analysis; Figure S2: Directed acyclic graph (DAG); Table S1: Examples of food and beverage items considered as NOVA processing groups; Table S2: Characteristics of the study participants at baseline, 6 months and 12 months of follow-up; Table S3: Association between concurrent changes in UPF consumption (% of g/day) and changes in NAFLD indices during 1-year of follow-up; Table S4: Sensitivity analysis for the association between concurrent changes in UPF consumption (% of g/day) and changes in NAFLD indices during 1-year of follow-up; Table S5: Mediation analysis for the association between concurrent changes in UPF consumption (% of g/day, continuous variable) and changes in NAFLD indices during 1-year of follow-up, through nutritional factors and NAFLD-related biomarkers; Table S6: Association between concurrent changes in UPF consumption (% of g/day) and changes in NAFLD indices during 1-year of follow-up by subgroups; Table S7: Association between concurrent changes in consumption of specific food subgroups within UPF (% of g/day) and changes in NAFLD indices during 1-year of follow-up; Text S1: Procedure for mediation analysis; File S1: Group information.

Author Contributions: Conceptualization, J.K. and D.R.; Methodology, J.K. and D.R.; Validation, J.K. and D.R.; Formal Analysis, J.K. and D.R.; Investigation, J.K., M.F., A.C., M.Á.M.-G., J.S.-S., D.C., M.T.S.-F., J.A.M., Á.M.A.-G., J.W., J.V. (Jesus Vioque), J.L.-M., R.E., M.R.B.-L., J.L., L.S.-M., A.B.-C., J.A.T., V.M.S., X.P., J.J.G., P.M.-M., J.V. (Josep Vidal), C.V., L.D., E.R., M.B.-R., M.P., J.V.S., A.G., M.Á.Z., A.M.-R., F.J.C.G., R.V.-E., J.M.J., A.G.-R., R.C., A.M.G.-P., J.M.S.-L., F.J.B.-G., M.Á.M., C.O.-A., J.B., I.A., I.S.-L., M.R.-C., N.B., L.C. and D.R.; Resources, J.K., M.F., A.C., M.Á.M.-G., J.S.-S., D.C., M.T.S.-F., J.A.M., Á.M.A.-G., J.W., J.V. (Jesus Vioque), J.L.-M., R.E., M.R.B.-L., J.L., L.S.-M., A.B.-C., J.A.T., V.M.S., X.P., J.J.G., P.M.-M., J.V. (Josep Vidal), C.V., L.D., E.R., M.B.-R., M.P., J.V.S., A.G., M.Á.Z., A.M.-R., F.J.C.G., R.V.-E., J.M.J., A.G.-R., R.C., A.M.G.-P., J.M.S.-L., F.J.B.-G., M.Á.M., C.O.-A., J.B., I.A., I.S.-L., M.R.-C., N.B., L.C. and D.R.; Data Curation, J.K., M.F., A.C., M.Á.M.-G., J.S.-S., D.C., M.T.S.-F., J.A.M., Á.M.A.-G., J.W., J.V. (Jesus Vioque), J.L.-M., R.E., M.R.B.-L., J.L., L.S.-M., A.B.-C., J.A.T., V.M.S., X.P., J.J.G., P.M.-M., J.V. (Josep Vidal), C.V., L.D., E.R., M.B.-R., M.P., J.V.S., A.G., M.Á.Z., A.M.-R., F.J.C.G., R.V.-E., J.M.J., A.G.-R., R.C., A.M.G.-P., J.M.S.-L., F.J.B.-G., M.Á.M., C.O.-A., J.B., I.A., I.S.-L., M.R.-C., N.B., L.C. and D.R.; Writing—Original Draft Preparation, J.K. and D.R.; Writing—Review & Editing, J.K., M.F., A.C., M.Á.M.-G., J.S.-S., D.C., M.T.S.-F., J.A.M., Á.M.A.-G., J.W., J.V. (Jesus Vioque), J.L.-M., R.E., M.R.B.-L., J.L., L.S.-M., A.B.-C., J.A.T., V.M.S., X.P., J.J.G., P.M.-M., J.V. (Josep Vidal), C.V., L.D., E.R., M.B.-R., M.P., J.V.S., A.G., M.Á.Z., A.M.-R., F.J.C.G., R.V.-E., J.M.J., A.G.-R., R.C., A.M.G.-P., J.M.S.-L., F.J.B.-G., M.Á.M., C.O.-A., J.B., I.A., I.S.-L., M.R.-C., N.B., L.C. and D.R.; Visualization, J.K., M.F., A.C., M.Á.M.-G., J.S.-S., D.C., M.T.S.-F., J.A.M., Á.M.A.-G., J.W., J.V.(Jesus Vioque), J.L.-M., R.E., M.R.B.-L., J.L., L.S.-M., A.B.-C., J.A.T., V.M.S., X.P., J.J.G., P.M.-M., J.V. (Josep Vidal), C.V., L.D., E.R., M.B.-R., M.P., J.V.S., A.G., M.Á.Z., A.M.-R., F.J.C.G., R.V.-E., J.M.J., A.G.-R., R.C., A.M.G.-P., J.M.S.-L., F.J.B.-G., M.Á.M., C.O.-A., J.B., I.A., I.S.-L., M.R.-C., N.B., L.C. and D.R.; Supervision, D.R.; Project Administration, J.S.-S., M.Á.M.-G., R.E., E.R., D.C.; Funding Acquisition, J.K., M.F., A.C., M.Á.M.-G., J.S.-S., D.C., M.T.S.-F., J.A.M., Á.M.A.-G., J.W., J.V. (Jesus Vioque), J.L.-M., R.E., M.R.B.-L., J.L., L.S.-M., A.B.-C., J.A.T., V.M.S., X.P., J.J.G., P.M.-M., J.V. (Josep Vidal), C.V., L.D., E.R., M.B.-R., M.P., J.V.S., A.G., M.Á.Z., A.M.-R., F.J.C.G., R.V.-E., J.M.J., A.G.-R., R.C., A.M.G.-P., J.M.S.-L., F.J.B.-G., M.Á.M., C.O.-A., J.B., I.A., I.S.-L., M.R.-C., N.B., L.C. and D.R. All authors have read and agreed to the published version of the manuscript.

Funding: This work was supported by the European Research Council (Advanced Research Grant 2014–2019; agreement #340918) granted to MÁM-G and the Spanish National Health Institute of

Health Carlos III (ISCIII), through CIBEROBN and "Fondo de Investigación para la Salud" (FIS), which is co-funded by the European Regional Development Fund (six coordinated FIS projects led by JS-S and JVi, including the following projects: PI13/00673, PI13/00492, PI13/00272, PI13/01123, PI13/00462, PI13/00233, PI13/02184, PI13/00728, PI13/01090, PI13/01056, PI14/01722, PI14/00636, PI14/00618, PI14/00696, PI14/01206, PI14/01919, PI14/00853, PI14/01374, PI14/00972, PI14/00728, PI14/01471, PI16/00473, PI16/00662, PI16/01873, PI16/01094, PI16/00501, PI16/00533, PI16/00381, PI16/00366, PI16/01522, PI16/01120, PI17/00764, PI17/01183, PI17/00855, PI17/01347, PI17/00525, PI17/01827, PI17/00532, PI17/00215, PI17/01441, PI17/00508, PI17/01732, PI17/00926, PI19/00957, PI19/00386, PI19/00309, PI19/01032, PI19/00576, PI19/00017, PI19/01226, PI19/00781, PI19/01560, PI19/01332, PI20/01802, PI20/00138, PI20/01532, PI20/00456, PI20/00339, PI20/00557, PI20/00886, PI20/01158, PI21/00465); the Especial Action Project entitled: Implementación y evaluación de una intervención intensiva sobre la actividad física Cohorte PREDIMED-Plus grant to JS-S; the Recercaixa (number 2013ACUP00194) grant to JS-S; grants from the Consejería de Salud de la Junta de Andalucía (PI0458/2013, PS0358/2016, PI0137/2018, RH-0024-2021); the PROMETEO/2017/017 grant from the Generalitat Valenciana; and the SEMERGEN grant; Juan de la Cierva-Incorporación research grant (IJC2019-042420-I) of the Spanish Ministry of Economy, Industry and Competitiveness and European Social Funds to JK; Miguel Servet Tipo 2 research grant (CPII20/00014) of the Spanish National Health Institute of Health Carlos III (ISCIII) to MRBL This work was also partially supported by ICREA under the ICREA Academia programme. None of the funding sources took part in the design, collection, analysis, interpretation of the data, or writing the report, or in the decision to submit the manuscript for publication.

Institutional Review Board Statement: The study protocol was approved by the Research Ethics Committees from all recruiting centers according to the ethical standards of the Declaration of Helsinki: CEI Provincial de Málaga-Servicio Andaluz de Salud (O01_feb_PR2); CEI de los Hospitales Universitarios Virgen Macarena y Virgen del Rocío-Servicio Andaluz de Salud (PI13/00673); CEIC Universidad de Navarra (053/2013); CEI de las Illes Balears—Conselleria de Salut Direcció General de Salut Publica i Consum (IB 2242/14 PI); CEIC del Hospital Clínic de Barcelona (HCB/2016/0287); CEIC Parc de Salut Mar y IDIAP Jordi Gol (PI13/120); CEIC del Hospital Universitari Sant Joan de Reus y IDIAB Jordi Gol (13-07-25/7proj2); CEI de la Provincia de Granada- Servicio Andaluz de Salud (MAB/BGP/pg); CEIC de la Fundacion Jiménez Díaz (EC 26-14/IIS-FJD); CEIC Universidad de Navarra (053/2013); CEIC Euskadi (PI2014044); CEIC Corporativo de Atención Primaria de la Comunitat Valenciana (2011-005398-22); CEI Humana de la Universidad de las Palmas de Gran Canaria (CEIH-2013-07); CEIC del Hospital de Bellvitge (PR240/13); CEI de Cordoba-Junta de Salud (3078); CEI de la Fundación IMDEA Alimentación (PI-012); CEIC Hospital Clínico San Carlos de Madrid-Piloto-CEIC Servicio Madrileño de salud-General (30/15); CEI Provincial de Málaga-Servicio Andaluz de Salud; CEI de las Illes Balears—Conselleria de Salut Direcció General de Salut Publica i Consum (IB 2251/14 PI); CEIC del Hospital Clínic de Barcelona (HCB/2017/0351); CEIC del Hospital General Universitario de Alicante (CEIC PI2017/02); CEIC de la Investigación Biomédica de Andalucía (CCEIBA); and CEI de la Universidad de Leon (ÉTICA-ULE-014-2015).

Informed Consent Statement: All participants provided written informed consent.

Data Availability Statement: There are restrictions on data availability for the PREDIMED-Plus trial due to the signed consent agreements around data sharing, which only allow access to external researchers for studies following the project purposes. Requestors wishing to access the PREDIMED-Plus trial data used in this study can make a request to the PREDIMED-Plus trial Steering Committee chair: jordi.salas@urv.cat. The request will then be passed to members of the PREDIMED-Plus Steering Committee for deliberation.

Trial Registration: The trial was registered at the International Standard Randomized Controlled Trial (ISRCTN: http://www.isrctn.com/ISRCTN89898870) with number 89898870 and registration date of 24 July 2014, retrospectively registered.

Acknowledgments: We are grateful to the PREDIMED-Plus volunteers for their participation and to the personnel, investigators, and primary care centers for their contribution. (Supplementary File S1).

Conflicts of Interest: E Ros reports grants, personal fees, non-financial support, and other support from the California Walnut Commission during the conduct of the study; grants, personal fees, non-financial support and other support from Alexion; personal fees, non-financial support and other support from Ferrer International and Danone; and personal fees and other from Amarin, outside the submitted work. R Estruch reports grants from Fundación Dieta Mediterránea, Spain and Cerveza y Salud, Spain. In addition, personal fees for lectures given from Brewers of Europe, Belgium, Fundación Cerveza y Salud, Spain, Pernaud-Ricard, Mexico, Instituto Cervantes, Alburquerque, USA, Instituto Cervantes, Milan, Italy, Instituto Cervantes, Tokyo, Japan, Lilly Laboratories, Spain, and Wine and Culinary International Forum, Spain, and non-financial support to organize a National Congress on Nutrition. In addition, feeding trials with product from Grand Fountain and Uriach Laboratories, Spain. J Salas-Salvadó reports receiving research support from the Instituto de Salud Carlos III, Ministerio de Educación y Ciencia, the European Commission, the USA National Institutes of Health; receiving consulting fees or travel expenses from Eroski Foundation, Instituto Danone, Nestle, and Abbott Laboratories; receiving nonfinancial support from Hojiblanca, Patrimonio Comunal Olivarero, the California Walnut Commission, Almond Board of California, La Morella Nuts, Pistachio Growers, and Borges S.A; serving on the board of and receiving grant support through his institution from the International Nut and Dried Foundation and the Eroski Foundation; personal fees from Instituto Danone; and serving in the Board of Danone Institute International. R Casas reports receiving fees of educational conferences from Fundación para la investigación del Vino y la Nutrición (FIVIN), Spain. The other authors declare no conflicts of interest.

Abbreviations

ALT	alanine aminotransferase
AST	aspartate aminotransferase
CVD	cardiovascular disease
er	energy-restricted
FA	fatty acids
FFQ	food frequency questionnaire
FLI	fatty liver Index
GGT	gamma-glutamyl transferase
HbA1c	glycated hemoglobin
HDL-c	high-density lipoprotein cholesterol
HSI	hepatic steatosis index
LOCF	last observation carried forward
MedDiet	Mediterranean diet
MetS	metabolic syndrome
NAFLD	non-alcoholic fatty liver disease
PA	physical activity
PREDIMED-Plus	PREvención con DIeta MEDiterránea Plus
SB	sedentary behavior
SSB	sugar-sweetened beverages
TCLSIH	Tianjin Chronic Low-grade Systemic Inflammation and Health Cohort Study
UPF	ultra-processed foods

References

1. Ofosu, A.; Ramai, D.; Reddy, M. Non-alcoholic fatty liver disease: Controlling an emerging epidemic, challenges, and future directions. *Ann. Gastroenterol.* **2018**, *31*, 288–295. [CrossRef] [PubMed]
2. Ahmed, M. Non-alcoholic fatty liver disease in 2015. *World J. Hepatol.* **2015**, *7*, 1450. [CrossRef]
3. Martínez-González, M.A.; Bastarrika, G. Mediterranean diet as the ideal model for preventing non-alcoholic fatty liver disease (NAFLD). *Hepatobiliary Surg. Nutr.* **2020**, *9*, 379–381. [CrossRef] [PubMed]
4. Berná, G.; Romero-Gomez, M. The role of nutrition in non-alcoholic fatty liver disease: Pathophysiology and management. *Liver Int.* **2020**, *40*, 102–108. [CrossRef] [PubMed]
5. Monteiro, C.A.; Cannon, G.; Moubarac, J.-C.; Levy, R.B.; Louzada, M.L.C.; Jaime, P.C. The UN Decade of Nutrition, the NOVA food classification and the trouble with ultra-processing. *Public Health Nutr.* **2018**, *21*, 5–17. [CrossRef] [PubMed]
6. Monteiro, C.A.; Cannon, G.; Levy, R.; Moubarac, J.-C.; Jaime, P.; Paula Martins, A.; Canella, D.; Louzada, M.; Parra, D. NOVA. The star shines bright. [Food classification. Public health]. *World Nutr.* **2016**, *7*, 28–38.
7. Morales, F.J.; Mesías, M.; Delgado-Andrade, C. Association between Heat-Induced Chemical Markers and Ultra-Processed Foods: A Case Study on Breakfast Cereals. *Nutrients* **2020**, *12*, 1418. [CrossRef]
8. Monteiro, C.A.; Moubarac, J.-C.; Cannon, G.; Ng, S.W.; Popkin, B. Ultra-processed products are becoming dominant in the global food system. *Obes. Rev.* **2013**, *14*, 21–28. [CrossRef]

9. Elizabeth, L.; Machado, P.; Zinöcker, M.; Baker, P.; Lawrence, M. Ultra-Processed Foods and Health Outcomes: A Narrative Review. *Nutrients* **2020**, *12*, 1955. [CrossRef]
10. Martínez-González, M.Á.; Martín-Calvo, N. Ultraprocessed Foods and Public Health: A Need for Education. *Mayo Clin. Proc.* **2019**, *94*, 2156–2157. [CrossRef]
11. Monteiro, C.A.; Cannon, G.; Levy, R.B.; Moubarac, J.-C.; Louzada, M.L.; Rauber, F.; Khandpur, N.; Cediel, G.; Neri, D.; Martinez-Steele, E.; et al. Ultra-processed foods: What they are and how to identify them. *Public Health Nutr.* **2019**, *22*, 936–941. [CrossRef]
12. Konieczna, J.; Morey, M.; Abete, I.; Bes-Rastrollo, M.; Ruiz-Canela, M.; Vioque, J.; Gonzalez-Palacios, S.; Daimiel, L.; Salas-Salvadó, J.; Fiol, M.; et al. Contribution of ultra-processed foods in visceral fat deposition and other adiposity indicators: Prospective analysis nested in the PREDIMED-Plus trial. *Clin. Nutr.* **2021**, *40*, 4290–4300. [CrossRef]
13. Mendonça, R.D.; Pimenta, A.M.; Gea, A.; de la Fuente-Arrillaga, C.; Martinez-Gonzalez, M.A.; Lopes, A.C.S.; Bes-Rastrollo, M. Ultraprocessed food consumption and risk of overweight and obesity: The University of Navarra Follow-Up (SUN) cohort study. *Am. J. Clin. Nutr.* **2016**, *104*, 1433–1440. [CrossRef]
14. Canhada, S.L.; Luft, V.C.; Giatti, L.; Duncan, B.B.; Chor, D.; da Fonseca, M.d.J.M.; Matos, S.M.A.; Molina, M.D.C.B.; Barreto, S.M.; Levy, R.B.; et al. Ultra-processed foods, incident overweight and obesity, and longitudinal changes in weight and waist circumference: The Brazilian Longitudinal Study of Adult Health (ELSA-Brasil). *Public Health Nutr.* **2020**, *23*, 1076–1086. [CrossRef]
15. Beslay, M.; Srour, B.; Méjean, C.; Allès, B.; Fiolet, T.; Debras, C.; Chazelas, E.; Deschasaux, M.; Wendeu-Foyet, M.G.; Hercberg, S.; et al. Ultra-processed food intake in association with BMI change and risk of overweight and obesity: A prospective analysis of the French NutriNet-Santé cohort. *PLoS Med.* **2020**, *17*, e1003256. [CrossRef]
16. Srour, B.; Fezeu, L.K.; Kesse-Guyot, E.; Allès, B.; Méjean, C.; Andrianasolo, R.M.; Chazelas, E.; Deschasaux, M.; Hercberg, S.; Galan, P.; et al. Ultra-processed food intake and risk of cardiovascular disease: Prospective cohort study (NutriNet-Santé). *BMJ* **2019**, *365*, l1451. [CrossRef]
17. Srour, B.; Fezeu, L.K.; Kesse-Guyot, E.; Allès, B.; Debras, C.; Druesne-Pecollo, N.; Chazelas, E.; Deschasaux, M.; Hercberg, S.; Galan, P.; et al. Ultraprocessed food consumption and risk of type 2 diabetes among participants of the NutriNet-Santé prospective cohort. *JAMA Intern. Med.* **2020**, *180*, 283. [CrossRef]
18. Llavero-Valero, M.; Escalada-San Martín, J.; Martínez-González, M.A.; Basterra-Gortari, F.J.; de la Fuente-Arrillaga, C.; Bes-Rastrollo, M. Ultra-processed foods and type-2 diabetes risk in the SUN project: A prospective cohort study. *Clin. Nutr.* **2021**, *40*, 2817–2824. [CrossRef]
19. Rey-García, J.; Donat-Vargas, C.; Sandoval-Insausti, H.; Bayan-Bravo, A.; Moreno-Franco, B.; Banegas, J.R.; Rodríguez-Artalejo, F.; Guallar-Castillón, P. Ultra-Processed food consumption is associated with renal function decline in older adults: A prospective cohort study. *Nutrients* **2021**, *13*, 428. [CrossRef]
20. Fiolet, T.; Srour, B.; Sellem, L.; Kesse-Guyot, E.; Allès, B.; Méjean, C.; Deschasaux, M.; Fassier, P.; Latino-Martel, P.; Beslay, M.; et al. Consumption of ultra-processed foods and cancer risk: Results from NutriNet-Santé prospective cohort. *BMJ* **2018**, *360*, k322. [CrossRef]
21. Alonso-Pedrero, L.; Ojeda-Rodríguez, A.; Martínez-González, M.A.; Zalba, G.; Bes-Rastrollo, M.; Marti, A. Ultra-processed food consumption and the risk of short telomeres in an elderly population of the Seguimiento Universidad de Navarra (SUN) Project. *Am. J. Clin. Nutr.* **2020**, *111*, 1259–1266. [CrossRef] [PubMed]
22. Kim, H.; Hu, E.A.; Rebholz, C.M. Ultra-processed food intake and mortality in the USA: Results from the Third National Health and Nutrition Examination Survey (NHANES III, 1988–1994). *Public Health Nutr.* **2019**, *22*, 1777–1785. [CrossRef] [PubMed]
23. Ivancovsky-Wajcman, D.; Fliss-Isakov, N.; Webb, M.; Bentov, I.I.; Shibolet, O.; Kariv, R.; Zelber-Sagi, S. Ultra processed food is associated with features of metabolic syndrome and non-alcoholic fatty liver disease. *Liver Int.* **2021**, *41*, 2635–2645. [CrossRef] [PubMed]
24. Zhang, S.; Gan, S.; Zhang, Q.; Liu, L.; Meng, G.; Yao, Z.; Wu, H.; Gu, Y.; Wang, Y.; Zhang, T.; et al. Ultra processed food consumption and the risk of non-alcoholic fatty liver disease in the Tianjin Chronic Low-grade Systemic Inflammation and Health Cohort Study. *Int. J. Epidemiol.* **2021**, *51*, 237–249. [CrossRef]
25. Martínez-González, M.A.; Buil-Cosiales, P.; Corella, D.; Bulló, M.; Fitó, M.; Vioque, J.; Romaguera, D.; Martínez, J.A.; Wärnberg, J.; López-Miranda, J.; et al. Cohort Profile: Design and methods of the PREDIMED-Plus randomized trial. *Int. J. Epidemiol.* **2019**, *48*, 387–388. [CrossRef]
26. Sayón-Orea, C.; Razquin, C.; Bulló, M.; Corella, D.; Fitó, M.; Romaguera, D.; Vioque, J.; Alonso-Gómez, Á.M.; Wärnberg, J.; Martínez, J.A.; et al. Effect of a Nutritional and Behavioral Intervention on Energy-Reduced Mediterranean Diet Adherence among Patients with Metabolic Syndrome. *JAMA* **2019**, *322*, 1486. [CrossRef]
27. Alberti, K.G.M.M.; Eckel, R.H.; Grundy, S.M.; Zimmet, P.Z.; Cleeman, J.I.; Donato, K.A.; James, W.P.; Loria, C.M.; Smith, S.C., Jr.; International Diabetes Federation Task Force on Epidemiology and Prevention; et al. Harmonizing the Metabolic Syndrome: A Joint Interim Statement of the International Diabetes Federation Task Force on Epidemiology and Prevention; National Heart, Lung, and Blood Institute; American Heart Association; World Heart Federation; International. *Circulation* **2009**, *120*, 1640–1645.
28. Willett, W. *Nutritional Epidemiology*, 3rd ed.; Oxford University Press: Oxford, UK, 2013; 529p.
29. Fernández-Ballart, J.D.; Piñol, J.L.; Zazpe, I.; Corella, D.; Carrasco, P.; Toledo, E.; Perez-Bauer, M.; Martínez-González, M.Á.; Salas-Salvadó, J.; Martín-Moreno, J.M. Relative validity of a semi-quantitative food-frequency questionnaire in an elderly Mediterranean population of Spain. *Br. J. Nutr.* **2010**, *103*, 1808–1816. [CrossRef]

30. de la Fuente-Arrillaga, C.; Vázquez Ruiz, Z.; Bes-Rastrollo, M.; Sampson, L.; Martinez-González, M.A. Reproducibility of an FFQ validated in Spain. *Public Health Nutr.* **2010**, *13*, 1364–1372. [CrossRef]
31. Moreiras, O.; Carbajal, A.; Cabrera, L.; Cuadrado, C. *Tablas de Composición de Alimentos "Food Composition Table"*, 16th ed.; Pirámide: Madrid, Spain, 2013.
32. Salmerón, J. Dietary Fiber, Glycemic Load, and Risk of Non—Insulin-dependent Diabetes Mellitus in Women. *JAMA J. Am. Med. Assoc.* **1997**, *277*, 472. [CrossRef]
33. Atkinson, F.S.; Foster-Powell, K.; Brand-Miller, J.C. International Tables of Glycemic Index and Glycemic Load Values: 2008. *Diabetes Care* **2008**, *31*, 2281–2283. [CrossRef]
34. Molina, L.; Sarmiento, M.; Peñafiel, J.; Donaire, D.; Garcia-Aymerich, J.; Gomez, M.; Ble, M.; Ruiz, S.; Frances, A.; Schroder, H.; et al. Validation of the Regicor Short Physical Activity Questionnaire for the Adult Population. *PLoS ONE* **2017**, *12*, e0168148. [CrossRef]
35. Martínez-González, M.A.; López-Fontana, C.; Varo, J.J.; Sánchez-Villegas, A.; Martinez, J.A. Validation of the Spanish version of the physical activity questionnaire used in the Nurses' Health Study and the Health Professionals' Follow-up Study. *Public Health Nutr.* **2005**, *8*, 920–927. [CrossRef]
36. Schröder, H.; Fitó, M.; Estruch, R.; Martínez-González, M.A.; Corella, D.; Salas-Salvadó, J.; Lamuela-Raventós, R.; Ros, E.; Salaverría, I.; Fiol, M.; et al. A Short Screener Is Valid for Assessing Mediterranean Diet Adherence among Older Spanish Men and Women. *J. Nutr.* **2011**, *141*, 1140–1145. [CrossRef]
37. Bedogni, G.; Bellentani, S.; Miglioli, L.; Masutti, F.; Passalacqua, M.; Castiglione, A.; Tiribelli, C. The Fatty Liver Index: A simple and accurate predictor of hepatic steatosis in the general population. *BMC Gastroenterol.* **2006**, *6*, 33. [CrossRef]
38. Lee, J.-H.; Kim, D.; Kim, H.J.; Lee, C.-H.; Yang, J.I.; Kim, W.; Kim, Y.; Yoon, J.; Cho, S.; Sung, M.; et al. Hepatic steatosis index: A simple screening tool reflecting nonalcoholic fatty liver disease. *Dig. Liver Dis.* **2010**, *42*, 503–508. [CrossRef]
39. Textor, J.; van der Zander, B.; Gilthorpe, M.S.; Liśkiewicz, M.; Ellison, G.T. Robust causal inference using directed acyclic graphs: The R package "dagitty". *Int. J. Epidemiol.* **2016**, *45*, 1887–1894. [CrossRef]
40. Iacobucci, D.; Saldanha, N.; Deng, X. A Meditation on Mediation: Evidence That Structural Equations Models Perform Better Than Regressions. *J. Consum. Psychol.* **2007**, *17*, 139–153. [CrossRef]
41. Petroni, M.L.; Brodosi, L.; Marchignoli, F.; Musio, A.; Marchesini, G. Moderate Alcohol Intake in Non-Alcoholic Fatty Liver Disease: To Drink or Not to Drink? *Nutrients* **2019**, *11*, 3048. [CrossRef]
42. Mendonça, R.d.D.; Lopes, A.C.S.; Pimenta, A.M.; Gea, A.; Martinez-Gonzalez, M.A.; Bes-Rastrollo, M. Ultra-Processed Food Consumption and the Incidence of Hypertension in a Mediterranean Cohort: The Seguimiento Universidad de Navarra Project. *Am. J. Hypertens.* **2016**, *30*, hpw137. [CrossRef]
43. De Oliveira da Silva Scaranni, P.; De Oliveira Cardoso, L.; Chor, D.; Prates Melo, E.C.; Alvim Matos, S.M.; Giatti, L.; Barreto, S.M.; da Fonseca, M.J.M. Ultra-processed foods, changes in blood pressure and incidence of hypertension: The Brazilian Longitudinal Study of Adult Health (ELSA-Brasil). *Public Health Nutr.* **2021**, *24*, 3352–3360. [CrossRef] [PubMed]
44. Zelber-Sagi, S.; Ivancovsky-Wajcman, D.; Fliss Isakov, N.; Webb, M.; Orenstein, D.; Shibolet, O.; Kariv, R. High red and processed meat consumption is associated with non-alcoholic fatty liver disease and insulin resistance. *J. Hepatol.* **2018**, *68*, 1239–1246. [CrossRef] [PubMed]
45. Ma, J.; Fox, C.S.; Jacques, P.F.; Speliotes, E.K.; Hoffmann, U.; Smith, C.E.; Saltzman, E.; McKeown, N.M. Sugar-sweetened beverage, diet soda, and fatty liver disease in the Framingham Heart Study cohorts. *J. Hepatol.* **2015**, *63*, 462–469. [CrossRef] [PubMed]
46. Ludwig, D.S. Technology, Diet, and the Burden of Chronic Disease. *JAMA* **2011**, *305*, 1352. [CrossRef]
47. Mazidi, M.; Katsiki, N.; Mikhailidis, D.P.; Banach, M. Link between plasma trans-fatty acid and fatty liver is moderated by adiposity. *Int. J. Cardiol.* **2018**, *272*, 316–322. [CrossRef]
48. World Health Organization. *Countdown to 2023: WHO Report on Global Trans-Fat Elimination 2021*; World Health Organization: Geneva, Switzerland, 2021.
49. Pipoyan, D.; Stepanyan, S.; Stepanyan, S.; Beglaryan, M.; Costantini, L.; Molinari, R.; Merendino, N. The Effect of Trans Fatty Acids on Human Health: Regulation and Consumption Patterns. *Foods* **2021**, *10*, 2452. [CrossRef]
50. Xia, Y.; Zhang, S.; Zhang, Q.; Liu, L.; Meng, G.; Wu, H.; Bao, X.; Gu, Y.; Sun, S.; Wang, X.; et al. Insoluble dietary fibre intake is associated with lower prevalence of newly-diagnosed non-alcoholic fatty liver disease in Chinese men: A large population-based cross-sectional study. *Nutr. Metab.* **2020**, *17*, 4. [CrossRef]
51. Schulte, E.M.; Avena, N.M.; Gearhardt, A.N. Which Foods May Be Addictive? The Roles of Processing, Fat Content, and Glycemic Load. Weir TL, editor. *PLoS ONE* **2015**, *10*, e0117959. [CrossRef]
52. Parker, A.; Kim, Y. The Effect of Low Glycemic Index and Glycemic Load Diets on Hepatic Fat Mass, Insulin Resistance, and Blood Lipid Panels in Individuals with Nonalcoholic Fatty Liver Disease. *Metab. Syndr. Relat. Disord.* **2019**, *17*, 389–396. [CrossRef]
53. Valtueña, S.; Pellegrini, N.; Ardigò, D.; Del Rio, D.; Numeroso, F.; Scazzina, F.; Monti, L.; Zavaroni, I.; Brighenti, F. Dietary glycemic index and liver steatosis. *Am. J. Clin. Nutr.* **2006**, *84*, 136–142. [CrossRef]
54. Golonka, R.M.; Yeoh, B.S.; Vijay-Kumar, M. Dietary Additives and Supplements Revisited: The Fewer, the Safer for Gut and Liver Health. *Curr. Pharmacol. Rep.* **2019**, *5*, 303–316. [CrossRef]
55. Hassan, M.M.; Elrrigieg, M.A.A.; Sabahelkhier, M.; Idris, O. Impacts of the food additive benzoic acid on liver function of Wistar rats. *Int. J. Adv. Res.* **2016**, *4*, 568–575. [CrossRef]

56. Elbassuoni, E.A.; Ragy, M.M.; Ahmed, S.M. Evidence of the protective effect of l-arginine and vitamin D against monosodium glutamate-induced liver and kidney dysfunction in rats. *Biomed. Pharmacother.* **2018**, *108*, 799–808. [CrossRef]
57. Singh, R.K. Food additive P-80 impacts mouse gut microbiota promoting intestinal inflammation, obesity and liver dysfunction. *SOJ Microbiol. Infect. Dis.* **2016**, *4*, 1–10. [CrossRef]
58. Xia, M.-F.; Bian, H.; Gao, X. NAFLD and Diabetes: Two Sides of the Same Coin? Rationale for Gene-Based Personalized NAFLD Treatment. *Front. Pharmacol.* **2019**, *10*, 877. [CrossRef]
59. Lonardo, A.; Leoni, S.; Alswat, K.A.; Fouad, Y. History of Nonalcoholic Fatty Liver Disease. *Int. J. Mol. Sci.* **2020**, *21*, 5888. [CrossRef]
60. Martínez Steele, E.; Juul, F.; Neri, D.; Rauber, F.; Monteiro, C.A. Dietary share of ultra-processed foods and metabolic syndrome in the US adult population. *Prev. Med.* **2019**, *125*, 40–48. [CrossRef]
61. Long, M.T.; Pedley, A.; Colantonio, L.D.; Massaro, J.M.; Hoffmann, U.; Muntner, P.; Fox, C.S. Development and Validation of the Framingham Steatosis Index to Identify Persons With Hepatic Steatosis. *Clin. Gastroenterol. Hepatol.* **2016**, *14*, 1172–1180.e2. [CrossRef]
62. Shimony, M.K.; Schliep, K.C.; Schisterman, E.F.; Ahrens, K.A.; Sjaarda, L.A.; Rotman, Y.; Perkins, N.J.; Pollack, A.Z.; Wactawski-Wende, J.; Mumford, S.L.; et al. The relationship between sugar-sweetened beverages and liver enzymes among healthy premenopausal women: A prospective cohort study. *Eur. J. Nutr.* **2016**, *55*, 569–576. [CrossRef]
63. Mirmiran, P.; Gaeini, Z.; Bahadoran, Z.; Azizi, F. Elevated serum levels of aminotransferases in relation to unhealthy foods intake: Tehran lipid and glucose study. *BMC Endocr. Disord.* **2019**, *19*, 100. [CrossRef]
64. Smith, J.D.; Hou, T.; Hu, F.B.; Rimm, E.B.; Spiegelman, D.; Willett, W.C.; Mozaffarian, D. A Comparison of different methods for evaluating diet, physical activity, and long-term weight gain in 3 prospective cohort studies. *J. Nutr.* **2015**, *145*, 2527–2534. [CrossRef] [PubMed]

nutrients

MDPI

Article

Association between Ultra-Processed Food Consumption and Diabetes in Chinese Adults—Results from the China Health and Nutrition Survey

Ming Li [1,*] and Zumin Shi [2]

1 Centre for Population Health Research, Division of Health Sciences, University of South Australia, Adelaide, SA 5000, Australia
2 Human Nutrition Department, College of Health Sciences, QU Health, Qatar University, Doha 2713, Qatar
* Correspondence: ming.li@unisa.edu.au; Tel.: +61-8-830-21051

Abstract: Aims: We aimed to assess the association between ultra-processed food (UPF) consumption with diabetes in Chinese adults. Methods: This study included 12,849 eligible adults aged 20 years and over attending at least two surveys in the China Nutrition and Health Survey during 1997–2011. Food intake at each survey was assessed by a 3-day 24-h dietary recall method. UPF was defined based on the NOVA classification. Diabetes was obtained from questionnaires and/or ascertained by fasting blood tests. The association of diabetes with UPF was examined using mix effect logistic regression adjusting for potential confounding factors. Results: The mean age of the participants was 43.3 (SD 14.8) years. The age and gender adjusted mean UPF intake increased four times and the prevalence of diabetes increased eight times in 1997–2011. Compared with non-consumers, the odds ratios (95% CI) of diabetes for those with mean UPF consumption of 1–19 g/day, 20–49 g/day, and ≥50 g/day were 1.21 (0.98, 1.48), 1.49 (1.19, 1.86), and 1.40 (1.08, 1.80), respectively (p trend < 0.001) after adjusted for the measured covariates including lifestyle factors (smoking, alcohol drinking, and physical activity), BMI and hypertension. Conclusions: both UPF consumption and prevalence of diabetes increased among adults in China during 1997–2011. Higher UPF consumption was positively associated with diabetes.

Keywords: ultra-processed food; long-term consumption; diabetes; China; adults

Citation: Li, M.; Shi, Z. Association between Ultra-Processed Food Consumption and Diabetes in Chinese Adults—Results from the China Health and Nutrition Survey. *Nutrients* **2022**, *14*, 4241. https://doi.org/10.3390/nu14204241

Academic Editors: Monica Dinu and Daniela Martini

Received: 14 September 2022
Accepted: 10 October 2022
Published: 12 October 2022

1. Introduction

Diabetes is a global health issue contributing to many severe complications and posing huge economic burden [1]. It affected 10.5% in 20–79 years old of the general population worldwide and China has the most people with diabetes with estimates of over 140 million in 2021 with projections of 174.4 million in 2045 [2]. In addition to the known risk factors including overweight/obesity, sedentary lifestyle, family history, hypertension, and elevated levels of triglycerides, diet attributed to 34.9% of disability-adjusted life years of diabetes [3], such that processed meat, refined grains, and fried products were positively associated with diabetes [4].

The classifications based solely on nutrient composition failed to explain the entire influence of food consumption on diabetes [5]. NOVA classifies foods and drinks into four groups based on food processing and brings a perspective insight into the diabetes epidemic [6]. Ultra-processed food (UPF) is the 4th Group in NOVA including products of entirely industrial formulations or made from substances extracted from foods, with minimal whole foods [6]. UPF is commonly high in energy density, sugars, salt, trans fats as well as additives, but low in protein, micronutrients, and fibers. UPF takes up more than half of total daily energy intake in high-income countries and its consumption is increasing rapidly in middle-income countries [7].

Accumulating evidence have indicated an adverse impact of high UPF intake on metabolic health, including cardiovascular diseases and mortality [8,9]. The evidence from the animal experiment indicates that UPF is a significant risk factor hyperinsulinemia and glucose intolerance [10], and certain types of UPF (e.g., soda and processed meats) were correlated with diabetes [11,12]. A recent meta-analysis of five observational studies from France, Netherland, Spain, UK, and Canada indicated each 10% increase UPF consumption was associated with the increased risk of diabetes by 15% in adults after adjusted for potential socioeconomic and lifestyle factors [13], while a cross-sectional study found among Brazil's pregnant women that UPF intake was not associated with gestational diabetes mellitus [14]. There is no investigation of UPF intake and diabetes yet in China.

Despite the emerging evidence of UPF and its association with health risks, the consumption of the poor-quality food has been increasing in line with the economic development and urbanization, especially in nutrition transition countries (e.g., India, Indonesia, and Brazil) [15]. Studies has shown that food choice is based not only on nutrients profile but also on the taste, convenience and cost [16] which may partly drive the trend.

China had experienced a remarkable nutrition transition in the past several decades. Diet changed from dominantly a traditional pattern of home-made food out of natural food sources towards a modern one of increased processed food and drink packs from supermarket [17] that is associated with cardiometabolic risks [18]. We recently reported using national representative data from China Nutrition and Health Survey (CNHS) that UPF consumption per capita was increased fourfold during 1997–2011 among Chinese adults aged over 20 years and higher UPF consumption was associated with overweight/obesity [19]. However, its long-term association with diabetes and the impact of overweight/obesity on the association have not been investigated in this population. We aimed to fill the knowledge gap among adults attending the CHNS.

2. Materials and Methods

2.1. Study Design and Sample

This is an association study between repeated measurements of dietary intake and diabetes status during 1997–2011 using public access CHNS data.

The CHNS study was a continuing open household-based cohort study conducted in nine provinces in China [20]. Samples in both urban and rural areas were drawn by a multistage random-cluster sampling method. So far, ten waves of dietary data collection have been completed (1989, 1991, 1993, 1997, 2000, 2004, 2006, 2009, 2011 and 2015). Blood samples were collected in the 2009 and 2015 surveys. However, blood glucose data in 2015 were not open to the public. The overall response rate was >60% based on the first survey in 1989 and >80% based on the previous year [20]. In this study, a total of 12,849 eligible adults were included based on the following criteria: aged $\geq$ 20 years; having self-reported diagnosis of diabetes and/or fasting blood tests; having attended at least two nutrition surveys during 1997–2011; having plausible energy intake (800–6000 kcal/d for men, and 600–4000 kcal/d for women) (Figure 1). Informed consent was obtained from all participants. The survey was approved by the institutional review committees [20].

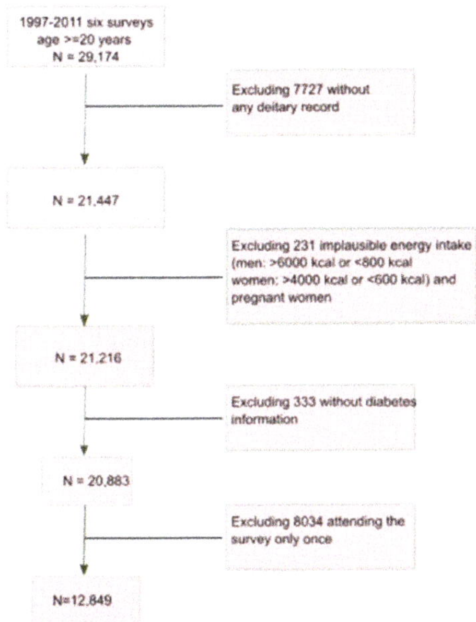

Figure 1. Sample flowchart of participants attending CHNS 1997−2011.

2.2. Outcome Variable

The primary outcome was diabetes. Diabetes was self-reported at each survey during 1997–2011. It was ascertained if a participant answered yes to either of the following questions: "Has the doctor ever told you that you suffer from diabetes?" "if yes, How old were you when the doctor told you about such a situation" "Did you use any of the treatment methods for diabetes (for example, on diet, weight control, oral medicine, Injection of insulin, Chinese, home remedies, Qigong)?". In addition, fasting plasma glucose was obtained in 2009 with diabetes defined as fasting plasma glucose $\geq$ 7.0 mmol/L, HbA1c $\geq$ 48 mmol/mol (equivalent to 6.5%). Diabetes in 2009 was ascertained if a participant self-reported being told having diabetes, or if self-reported not been told having diabetes but blood tests results met the diagnostic criteria. Fasting blood was taken in the morning and prepared for a further test in a national central lab in Beijing (medical laboratory accreditation certificate ISO 15189: 2007). Fasting plasma glucose was measured with the GOD-PAP method (Randox Laboratories Ltd., Crumlin, UK). All the measurements and tests were collected using standard protocol by trained staff. The detailed data collection protocol was described elsewhere [20].

2.3. UPF Assessment

At each survey, individual dietary intake was collected by a trained investigator conducting a 24-h dietary recall on each of 3 consecutive days [21]. Foods and condiments in the home inventory, foods purchased from markets or picked from gardens, and food waste were weighed and recorded by interviewers at the beginning and end of the three-day survey period. The types and amount of food, the type of meal and the place of consumption for a participant were from both dietary recall and the records kept by the individual. Cooking oil and condiments consumption for everyone in the household was estimated using individual energy-weighted intake. Detailed description of the dietary measurement has been published previously [22]. The food intake data in 1997–2011 was recoded and converted to nutrient intake using the corresponding updated food composition tables [23]. Around 3000 food items in the food composition tables since 1997 were categorized into four groups by the NOVA classification [6,19]. Long-term cumulative mean UPF intake at each survey was calculated from all the proceeding surveys to reduce within individual variation. For instance, if the UPF intake of a participant was a, b, c in 1997, 2004, and 2009, the corresponding mean UPF intake in 1997, 2004 and 2009 was a, (a + b)/2, and (a + b + c)/3.

2.4. Covariates

Sociodemographic and lifestyle factors were collected at each survey using a structured questionnaire. The socioeconomic status included: education (low: illiterate/primary school; medium: junior middle school; high: high middle school or higher), annual family income (recoded into tertiles as low, medium and high), urbanization levels (recoded into tertiles as low, medium and high).

Height, weight, and blood pressure were measured at each survey round. Overweight/obesity was defined as BMI $\geq$ 25 kg/m^2. Hypertension was defined as systolic blood pressure $\geq$ 140 mmHg and/or diastolic blood pressure $\geq$ 90 mmHg or having known hypertension.

Physical activity level (metabolic equivalent of task) was estimated based on self-reported activities and duration using a Compendium of Physical Activities. Smoking status was categorized as non-smokers, ex-smokers and current smokers. Alcohol consumption was recorded as yes or no. Two dietary patterns (traditional and modern) were identified in this population using principal components analysis from thirty-five food groups of similar nutrient profiles or culinary uses [18]. The traditional one was characterized by high intakes of rice, meat, and vegetables, while the modern pattern was highly correlated with fast food, milk, and deep-fried food [18].

2.5. Statistical Analysis

Mean UPF intake was grouped into: non-consumers, 1–19, 20–49, and $\geq$50 g/day based on that the serving size in the context of Chinese food is *Liang* (50 g). Sample characteristics were presented and compared by UPF intake levels using ANOVA for continuous measures or chi-square tests for categorical ones.

The association between UPF intake and diabetes were assessed using mixed effect logistic regression models. Unadjusted and adjusted odds ratios (95% CI) of the fixed part of the models were reported. Adjusted models were built by including age, sex, and energy intake initially in Model 1; further adding fat intake, socioeconomic status (income, urbanization, and education), and lifestyle factors (smoking, alcohol drinking, and physical activity) in Model 2, and next adjusted for overweight/obesity or hypertension in Model 3 or Model 4. Model 5 included BMI, hypertension, and dietary patterns; Sensitivity analysis was presented as Model 6 from Model 5 among participants attended at least four waves of the surveys (n = 7263).

A subgroup analysis was conducted among 8382 participants in 2009 with self-report diagnosis of diabetes and/or fasting blood records. The association between UPF intake in 1997–2009 or in 2009 and diabetes was assessed using logistic regression analysis.

To test the interaction between UPF intake and other covariates (sex, sociodemographic, lifestyle, diet, and health factors), a product term of these two variables was put in the regression model.

The analyses were performed using STATA 17.0 (Stata Corporation, College Station, TX, USA). Statistical significance was considered when $p < 0.05$ (two-sided).

3. Results

3.1. Population Profile

At entry, the mean age of the participants was 43.3 years old (SD 14.8). In total, 49.0% were men, one third had medium level of income or lived in high urbanized areas, 44.9% attained low level of education, more than 30% were current smokers or alcohol drinkers. The prevalence of hypertension and diabetes were 15.9% and 2.1%, respectively. The percentages of UPF energy over total energy intake for non-consumers, 1–19 g/d,

20–49 g/d, and ≥50 g/d were 0, 1.6%, 4.9%, 14.3%. And the corresponding weight percentages of UPF over total food intake in gram per day were 0, 1.2%, 3.2%, and 10.4%.

3.2. Consumption of UPF during 1997–2011

The mean UPF consumption (age- and sex-adjusted) increased continuously from 12.6 g/day in 1997 to 41.3 g/day in 2011 with sharp increase since 2004. The daily energy contribution of UPF increased from 1.4% in 1997 to 4.9% in 2011 and the daily food weight proportion of UPF from 1.1% to 3.6%. At entry, 11% (n = 1396) of the participants had UPF intake greater than 50 g/d.

Compared to those with no or lower UPF intake of 1–19 g/d, participants having UPF ≥ 50 g/d at entry were more likely: being males, or having higher level of education or income, or living in the higher urbanized areas, or being smokers or alcohol drinkers, or having higher BMI. Energy, fat, protein intakes, and modern dietary pattern score were higher, while intake of carbohydrate and traditional dietary pattern score were lower (Table 1).

Table 1. Sample characteristics by ultra-processed food intake among participants attending China Health and Nutrition Survey (n = 12,849).

	None	1–19 g/d	20–49 g/d	≥50 g/d	p-Value
N	n = 10,129	n = 616	n = 708	n = 1396	
Survey year at entry					<0.001
1997	58.7%	63.6%	48.2%	39.9%	
2000	16.4%	13.3%	15.3%	15.9%	
2004	12.4%	11.5%	12.9%	12.0%	
2006	5.0%	3.4%	9.9%	9.7%	
2009	7.5%	8.1%	13.8%	22.6%	
Age, mean (years)	43.2 (14.7)	43.7 (15.9)	43.2 (15.2)	44.2 (14.7)	0.091
Sex					<0.001
Men	46.8%	44.2%	50.8%	66.0%	
Women	53.2%	55.8%	49.2%	34.0%	
Income					<0.001
Low	31.7%	24.2%	20.3%	21.8%	
Medium	33.2%	34.5%	32.2%	31.6%	
High	35.0%	41.3%	47.5%	46.7%	
Education					<0.001
Low	47.5%	42.6%	30.9%	33.5%	
Medium	32.0%	33.0%	32.7%	30.8%	
High	20.4%	24.4%	36.4%	35.7%	
Urbanization					<0.001
Low	36.4%	28.2%	19.8%	20.5%	
Medium	30.1%	26.9%	25.3%	28.2%	
High	33.5%	44.8%	54.9%	51.4%	
Energy intake, mean (kcal/d)	2242.9 (633.1)	2153.6 (595.6)	2214.7 (600.3)	2480.9 (702.5)	<0.001
Percent (%) of UPF over total energy intake/d	0.0 (0.0)	1.6 (1.5)	4.9 (2.8)	14.3 (11.1)	<0.001
Percent (%) of UPF over total food intake/d	0.0 (0.0)	1.2 (0.7)	3.2 (1.2)	10.4 (6.8)	<0.001
Fat intake, mean (g/d)	65.3 (35.7)	65.3 (33.7)	75.6 (35.7)	82.1 (39.9)	<0.001
Protein intake, mean (g/d)	67.5 (22.1)	67.5 (21.7)	71.1 (22.9)	76.6 (25.0)	<0.001
Carbohydrate intake, mean (g/d)	345.7 (120.6)	322.2 (114.2)	308.7 (112.0)	323.1 (113.6)	<0.001
Traditional dietary pattern score, mean	−0.0 (1.0)	0.1 (0.9)	0.1 (1.0)	0.1 (1.0)	<0.001
Modern dietary pattern score, mean	−0.3 (0.7)	−0.2 (0.8)	0.2 (1.0)	0.7 (1.2)	<0.001

Table 1. *Cont.*

	None	1–19 g/d	20–49 g/d	≥50 g/d	*p*-Value
N	*n* = 10,129	*n* = 616	*n* = 708	*n* = 1396	
Smoking					<0.001
Non smoker	69.1%	69.8%	66.1%	53.5%	
Ex-smokers	1.3%	1.0%	2.1%	3.0%	
Current smokers	29.6%	29.3%	31.8%	43.5%	
Alcohol drinking	32.1%	34.5%	39.8%	57.8%	<0.001
Physical activity, mean (MET-hrs/week)	141.0 (117.0)	135.6 (117.2)	132.2 (112.5)	143.1 (118.9)	0.15
BMI (kg/m^2), mean (SD)	22.6 (3.2)	22.8 (3.3)	23.1 (3.3)	23.0 (3.3)	<0.001
Diabetes	2.0%	1.3%	2.5%	2.8%	0.087
Hypertension	15.1%	19.4%	16.5%	19.5%	<0.001

Data in table as *n* (%) or mean (SD). *p* values from ANOVA or chi square test.

3.3. *Diabetes and UPF Consumption Level*

The prevalence of diabetes increased eight times from 1.5% in 1997 to 11.2% in 2009 and to 12.1% in 2011. The unadjusted ORs (95% CI) of diabetes for UPF consumption levels of none, 1–19 g/d, 20–49 g/d, and >50 g/d were 1 (reference), 2.13 (1.76, 2.56), 2.79 (2.29, 3.40), and 2.60 (2.10, 3.23), respectively ($p < 0.001$). The odds ratios remained significant after adjusted for age, sex, and energy intake (aOR 2.21; 95% CI 1.76, 2.77 for ≥50 g/d Model 1) and after further adjusted for fat, behavioural and sociodemographic factors (aOR 1.96; 95% CI 1.53, 2.51 in Model 2). Adjusted for either BMI or hypertension did not change the relative odds substantially in Model 3 or Model 4. Nor did BMI and hypertension, and overall dietary patterns. Specifically, the aORs (95% CI) of diabetes for UPF level of 20–49 g/d and ≥50 g/d were 1.49 (1.19–1.86), 1.40 (1.08–1.80), respectively. Sensitivity analysis among participants attending four waves of the surveys showed the corresponding aORs (95% CI) of 1.55 (1.20–2.00) and 1.37 (1.00–1.88) (Table 2).

Table 2. Odds ratio (95% CI) for diabetes by cumulative ultra-processed food intake in 1997–2011 among participants attending China Health and Nutrition Survey.

		Cumulative UPF Intake (g/day)			
	None	1–19	20–49	≥50	*p* for Trend
Unadjusted	1.00	2.13 (1.76–2.56)	2.79 (2.29–3.40)	2.60 (2.10–3.23)	<0.001
Model 1	1.00	1.53 (1.27–1.85)	2.15 (1.76–2.64)	2.21 (1.76–2.77)	<0.001
Model 2	1.00	1.34 (1.09–1.65)	1.87 (1.50–2.34)	1.96 (1.53–2.51)	<0.001
Model 3	1.00	1.29 (1.05–1.58)	1.79 (1.43–2.23)	1.85 (1.45–2.36)	<0.001
Model 4	1.00	1.29 (1.05–1.58)	1.74 (1.40–2.17)	1.79 (1.40–2.29)	<0.001
Model 5	1.00	1.21 (0.98–1.48)	1.49 (1.19–1.86)	1.40 (1.08–1.80)	<0.001
Model 6	1.00	1.22 (0.97–1.53)	1.55 (1.20–2.00)	1.37 (1.00–1.88)	0.003

Odds ratios from mixed effect logistic regression. Model 1: adjusted for age, gender and energy intake. Model 2: model 1further adjusted for intake of fat, income, urbanicity, education, smoking, alcohol drinking, and physical activity. Model 3: model 2 further adjusted hypertension. Model 4: model 2 further adjusted BMI. Model 5: model 2 further adjusted hypertension, BMI and dietary patterns [18]. Model 6: model 5 among all participants who attended at least four waves of survey (*n* = 7263).

The cross-sectional analysis of 8382 participants in 2009 showed both UPF intake in 1997–2009 or in 2009 was positively associated with diabetes. After adjusted for sociodemographic and lifestyle factors, the ORs (95% CI) of diabetes for UPF intake of 1–19 g/d, 20–49 g/d, and ≥50 g/d were 1.05 (0.86–1.28), 1.21 (0.96–1.51), and 1.31 (1.04–1.65) (*p* for trend = 0.015), respectively, compared with no UPF intake. Similarly, BMI slightly attenuated the association. The cross-sectional association using UPF intake in 2009 showed the corresponding adjusted ORs (95% CI) were 1.16 (0.79–1.68), 0.85 (0.62–1.15), and 1.23 (1.01–1.50) (*p* for trend = 0.037) (Table 3).

Table 3. Odds ratio (95% CI) for diabetes by cumulative ultra-processed food intake among participants attending China Health and Nutrition Survey in 2009 (n = 8382).

	None	1–19 g/d	20–49 g/d	≥50 g/d	*p* for Trend
	n = 3764	*n* = 1947	*n* = 1323	*n* = 1348	
Diabetes cases	364	227	169	180	<0.001
Unadjusted	1.00	1.23 (1.03–1.66)	1.37 (1.13–1.66)	1.44 (1.19–1.74)	0.003
Model 1	1.00	1.10 (0.92–1.31)	1.27 (1.04–1.55)	1.46 (1.19–1.77)	<0.001
Model 2	1.00	1.05 (0.86–1.28)	1.21 (0.96–1.51)	1.31 (1.04–1.65)	0.015
Model 3	1.00	1.07 (0.87–1.31)	1.16 (0.92–1.46)	1.24 (0.97–1.57)	0.060
Sensitivity analysis	1.00	1.16 (0.79–1.68)	0.85 (0.62–1.15)	1.23 (1.01–1.50)	0.037

Odds ratio from logistic regression analysis using diabetes in 2009 as outcome and UPF intake in 1997–2009 as study factor. Model 1: adjusted for age, gender and energy intake. Model 2: Model 1 further adjusted for intake of fat, smoking, alcohol drinking, income, urbanicity, education, physical activity, intake of fruit and vegetable. Model 3 further adjusted for BMI. Sensitivity analysis: model 2 among those with UPF intake in 2009 (n = 8382).

The association was consistent across subgroups by sex, education, income, urbanization, smoking, overweight/obesity, and hypertension status (Table 4).

Table 4. Stratified analysis of the association between cumulative UPF consumption in 1997–2011 and diabetes by sample characteristics.

	None	Cumulative UPF Intake (g/day) 1–19	20–49	≥50	*p* Value	*p* Interaction
Sex						0.637
Men	1.00	1.51 (1.10–2.09)	2.00 (1.45–2.76)	2.22 (1.60–3.09)	<0.001	
Women	1.00	1.23 (0.94–1.61)	1.82 (1.34–2.49)	1.78 (1.21–2.61)	<0.001	
Education						0.146
Low	1.00	1.59 (1.21–2.10)	2.62 (1.90–3.61)	2.22 (1.50–3.30)	<0.001	
Medium	1.00	0.95 (0.63–1.45)	1.51 (0.98–2.32)	1.77 (1.11–2.80)	0.008	
High	1.00	1.03 (0.65–1.63)	1.12 (0.70–1.79)	1.39 (0.86–2.25)	0.195	
Income						0.475
Low	1.00	1.14 (0.80–1.64)	2.25 (1.50–3.37)	2.11 (1.32–3.38)	<0.001	
Medium	1.00	1.25 (0.89–1.76)	1.40 (0.94–2.10)	1.56 (1.00–2.45)	0.020	
High	1.00	1.34 (0.94–1.91)	1.99 (1.38–2.85)	2.02 (1.37–2.98)	<0.001	
Urbanization						0.469
Low	1.00	1.12 (0.71–1.77)	1.34 (0.73–2.45)	1.36 (0.69–2.68)	0.223	
Medium	1.00	1.45 (1.00–2.11)	1.88 (1.24–2.83)	2.49 (1.61–3.85)	<0.001	
High	1.00	1.24 (0.93–1.67)	1.99 (1.45–2.71)	1.81 (1.27–2.57)	<0.001	
Smoking						0.987
Non smoker	1.00	1.37 (1.08–1.73)	1.99 (1.52–2.61)	2.18 (1.61–2.97)	<0.001	
Current smokers	1.00	1.29 (0.86–1.93)	1.56 (1.04–2.35)	1.68 (1.11–2.52)	0.006	
Overweight/obesity						0.366
No	1.00	1.14 (0.85–1.53)	1.77 (1.29–2.43)	1.70 (1.19–2.41)	<0.001	
Yes	1.00	1.43 (1.06–1.92)	1.72 (1.24–2.38)	1.91 (1.35–2.71)	<0.001	
Hypertension						0.684
No	1.00	1.19 (0.90–1.56)	1.61 (1.19–2.16)	1.77 (1.28–2.43)	<0.001	
Yes	1.00	1.34 (1.00–1.79)	1.94 (1.41–2.66)	1.87 (1.32–2.66)	<0.001	

Odds ratio (95% CI) from mixed effect logistic regression. Model adjusted for age, sex, intake of energy and fat, education levels, income, urbanization, smoking, alcohol drinking, and physical activity.

4. Discussion

Among the 12,849 participants in the CHNS, the mean per capita UPF consumption increased from 12.6 g/day in 1997 to 41.3 g/day in 2011 and the UPF contribution to daily total energy or daily total foods rose from 1.4 to 4.9% or 1.1 to 3.6%. Meanwhile, the prevalence of diabetes increased eight times from 1.5 to 12.1% in 2011. UPF intake ≥ 50 g/d increased the risk of diabetes by 40% compared with non-consumers.

Although the per capita UPF consumption and proportion of diet weight in China was below the level observed in other countries [8] and it is impossible to compare directly due to different UPF items, methodology and study period among these studies, it is unquestionable that the increased trend in China was dramatic, especially among those who were younger, or had higher educational attainment, or resided in highly urbanized areas. The younger people were more likely to eat out compared to older adults in China as home-prepared food are of better quality [24] while eating out increased the consumption of UPF by 41% compared with preparing meals exclusively at home [25]. The subgroup had higher educational levels and lived in highly urbanized area facilitating UPF consumption for time saving, savory taste, attractive packaging, and affordability [26].

The association between UPF and diabetes among this Chines population was consistent with the synthesized result of observational studies among adults in France, Netherland, UK, Spain, and Canada [13]. All studies applied the NOVA classification and four of them had follow-ups of 3.4–12 years [27–30] with HR/OR ranging from 1.13 to 1.53. The Canadian cross-sectional survey data reported 37% increased odds of self-reported diabetes in 2015 [31] while our result using UPF consumption data in 2009 reported the increased odds of 23% for diabetes.

Potential mechanisms underlying the association should be noted. Studies had shown that UPF was rich in added/free sugars and saturated fats, which were positively associated with diabetes [32,33]. Grains, meat, vegetables, and fruits lost the physical and structural characteristics of the food matrix during processing, which would result in a high glycaemic index [34]. In addition, as satiety mechanisms showed, humans are more sensitive to volume than energetic content [35], therefore, UPF with higher energy density may facilitate excessive energy intakes, leading to obesity as showed in our previous study [36]. We found in this study that obesity attenuated partly the association between UPF and diabetes. This is supported by a follow up study indicating a 23% increased risk of incident diabetes with each kg/m^2 increase in BMI (95% CI 1.22 to 1.24) among 211,833 Chinese persons >20 years old across 32 sites and 11 cities in China [37].

Food additives in UPF should not be ignored since the association was independent of energy and fat intake. More than 2000 food additives in 23 different categories have been added during food processing in China [38]. Although a maximum dose limitation for each additive has been set, it's unknown whether the long-term intake of these safe-dose food additives, whether single or combined, has cumulative or synergetic adverse effects on health. Emerging evidence has suggested that very low concentrations of polysorbate 80, the common food emulsifier, might change the gut microbiota, increase bacterial translocation, cause intestinal inflammation and promote type 2 diabetes [39]. Exposure to Carrageenan would result in glucose intolerance and fasting hyperglycaemia [40]. Sucralose, as a non-caloric artificial sweetener, could alter the metabolic response to the glucose load and slow down insulin clearance from plasma [41]. Furthermore, heat treatment during food processing, in particularly, could pose exposure to contaminants such as acrylamide which was associated with insulin resistance [42]. Finally, UPF could be contaminated by the package material with endocrine-disrupting properties (e.g., bisphenols A) [43] in order to keep within the extended expiration date. The impact of food additives on health and food processing in China should be closely regulated and monitored.

To our best knowledge, this is the first association study between long-term UPF consumption and diabetes in Chinese adults. The use of mean UPF intake during 1997–2011 from 3-day dietary intake in combination with household food inventory provided a robust estimate of long-term habitual intake. An updated NOVA classification system was used to classify UPF in this population. The association was confirmed by sensitivity analysis. Potential confounding factors including sociodemographic, behavioural, health, and dietary factors were adjusted.

Limitations should not be ignored. First, misclassification was possible due to lack of completing information about food processing in the CHNS survey that was not specifically designed for NOVA classification. Second, we used the weight unit (gram) to estimate the consumption of UPF which might not be precise for the diverse UPF items. Third, due to the complexity of food processing and variabilities in additive composition between brands for a similar type of product, we could only roughly group some food items therefore the association could be biased. The NOVA classification has been criticized on its lack of specificity at an individual nutrient level or overall adequacy of dietary patterns [44]. We have incorporated both nutrients and dietary pattern in the study to overcome the pitfalls. Fourth, the ascertainment of diabetes was self-reported except for 2009 which might pose misclassification of the outcome. However, out subgroup analysis using diabetes identified in 2009 and UPF intake either in 1997–2009 or in 2009 showed consistent results. Also, the prevalence and the temporal trend of diabetes in the study period were consistent with other national estimates [45–47], especially the prevalence in 2009 when both self-report and blood tests were applied to ascertain diabetes. This study did not distinguish between type 1 and type 2 diabetes. The association was unlikely to change much, as the data showed among the 1219 participant self-reported having been told to have diabetes in 1997–2011, only 30 cases (2.5%) were identified at the age of under 20. In addition, population-based data indicates Type 1 diabetes onset peak is in the 10–14-year-old age group in Chinese population [48]. Finally, residual confounding was still possible, for example, there was no record on family history of diabetes and ethnicity which is closely related to culinary culture in China.

5. Conclusions

Both UPF consumption and the prevalence of diabetes increased during 1997–2011 in Chinese adults. Higher UPF consumers had a significantly higher risk of diabetes than non-consumers. The association between UPF consumption and diabetes was partly mediated by overweight/obesity. In facing with the diabetes epidemic in China, nutrition education should focus on in part the modification of the unhealthy dietary factor and the maintenance for healthy weight.

Author Contributions: M.L. and Z.S. conceived the study. M.L. drafted the manuscript, interpreted the results, and revised the manuscript. Z.S. analysed data and reviewed the manuscript. All authors have read and agreed to the published version of the manuscript.

Funding: This research received no specific grant from any funding agency in the public, commercial or not-for-profit sectors.

Institutional Review Board Statement: Not applicable.

Informed Consent Statement: Not applicable.

Data Availability Statement: The current research uses data from the China Health and Nutrition Survey (CHNS). Data described in the manuscript, code book, and analytic code are made publicly and freely available without restriction at https://www.cpc.unc.edu/projects/china accessed on 15 January 2019.

Acknowledgments: This research uses data from China Health and Nutrition Survey. The authors thank the National Institute of Nutrition and Food Safety, China Centre for Disease Control and Prevention, Carolina Population Centre, the University of North Carolina at Chapel Hill, the NIH (R01-HD30880, DK056350, and R01-HD38700) and the Fogarty International Centre for financial support for the CHNS data collection and analysis files from 1989 to 2006 and both parties plus the China-Japan Friendship Hospital, Ministry of Health for support for CHNS 2009 and future surveys.

Conflicts of Interest: The authors declare no conflict of interest.

References

1. Harding, J.L.; Pavkov, M.E.; Magliano, D.J.; Shaw, J.E.; Gregg, E.W. Global trends in diabetes complications: A review of current evidence. *Diabetologia* **2019**, *62*, 3–16. [CrossRef] [PubMed]
2. International Diabetes Federation. IDF Diabetes Atlas. 2019. Available online: www.diabetesatlas.org/en/ (accessed on 27 April 2020).
3. Dicker, D.; Nguyen, G.; Abate, D.; Abate, K.H.; Abay, S.M.; Abbafati, C.; Abbasi, N.; Abbastabar, H.; Abd-Allah, F.; Abdela, J.; et al. Global, regional, and national age-sex-specific mortality and life expectancy, 1950–2017: A systematic analysis for the Global Burden of Disease Study 2017. *Lancet* **2018**, *392*, 1684–1735. [CrossRef]
4. Jannasch, F.; Kroger, J.; Schulze, M.B. Dietary Patterns and Type 2 Diabetes: A Systematic Literature Review and Meta-Analysis of Prospective Studies. *J. Nutr.* **2017**, *147*, 1174–1182. [CrossRef] [PubMed]
5. Monteiro, C.; Cannon, G.; Levy, R.B.; Claro, R.; Moubarac, J.; Martins, A.P.B.; Louzada, M.L.; Baraldi, L.G.; Canella, D. The Food System. Ultra-processing: The big issue for nutrition, disease, health, well-being. *World Nutr.* **2012**, *3*, 527–569.
6. Monteiro, C.A.; Cannon, G.; Moubarac, J.C.; Levy, R.B.; Louzada, M.L.C.; Jaime, P.C. The UN Decade of Nutrition, the NOVA food classification and the trouble with ultra-processing. *Public Health Nutr.* **2018**, *21*, 5–17. [CrossRef]
7. Euromonitor International. Passport Global Market Information Database. 2015. Available online: http://go.euromonitor.com/Passport-Home (accessed on 9 April 2022).

8. Matos, R.A.; Adams, M.; Sabate, J. Review: The Consumption of Ultra-Processed Foods and Non-communicable Diseases in Latin America. *Front. Nutr.* **2021**, *8*, 622714. [CrossRef]
9. Costa de Miranda, R.; Rauber, F.; Levy, R.B. Impact of ultra-processed food consumption on metabolic health. *Curr. Opin. Lipidol.* **2021**, *32*, 24–37. [CrossRef]
10. Gomez-Smith, M.; Karthikeyan, S.; Jeffers, M.S.; Janik, R.; Thomason, L.A.; Stefanovic, B.; Corbett, D. A physiological characterization of the Cafeteria diet model of metabolic syndrome in the rat. *Physiol. Behav.* **2016**, *167*, 382–391. [CrossRef]
11. Jensen, P.N.; Howard, B.V.; Best, L.G.; O'Leary, M.; Devereux, R.B.; Cole, S.A.; Maccluer, J.W.; Ali, T.; Lee, E.T.; Yeh, F.L.; et al. Associations of diet soda and non-caloric artificial sweetener use with markers of glucose and insulin homeostasis and incident diabetes: The Strong Heart Family Study. *Eur. J. Clin. Nutr.* **2020**, *74*, 322–327. [CrossRef]
12. Micha, R.; Michas, G.; Mozaffarian, D. Unprocessed red and processed meats and risk of coronary artery disease and type 2 diabetes—An updated review of the evidence. *Curr. Atheroscler. Rep.* **2012**, *14*, 515–524. [CrossRef]
13. Moradi, S.; Hojjati Kermani, M.a.; Bagheri, R.; Mohammadi, H.; Jayedi, A.; Lane, M.M.; Asbaghi, O.; Mehrabani, S.; Suzuki, K. Ultra-Processed Food Consumption and Adult Diabetes Risk: A Systematic Review and Dose-Response Meta-Analysis. *Nutrients* **2021**, *13*, 4410. [CrossRef] [PubMed]
14. Sartorelli, D.S.; Crivellenti, L.C.; Zuccolotto, D.C.C.; Franco, L.J. Relationship between minimally and ultra-processed food intake during pregnancy with obesity and gestational diabetes mellitus. *Cad. Saude Publica* **2019**, *35*, e00049318. [CrossRef] [PubMed]
15. Popkin, B.M. Nutrition Transition and the Global Diabetes Epidemic. *Curr. Diab. Rep.* **2015**, *15*, 64. [CrossRef] [PubMed]
16. Drewnowski, A. The carbohydrate-fat problem: Can we construct a healthy diet based on dietary guidelines? *Adv. Nutr.* **2015**, *6*, 318s–325s. [CrossRef] [PubMed]
17. Zhai, F.; Wang, H.; Du, S.; He, Y.; Wang, Z.; Ge, K.; Popkin, B.M. Prospective study on nutrition transition in China. *Nutr. Rev.* **2009**, *67*, S56–S61. [CrossRef] [PubMed]
18. Li, M.; Shi, Z. Dietary Pattern during 1991–2011 and Its Association with Cardio Metabolic Risks in Chinese Adults: The China Health and Nutrition Survey. *Nutrients* **2017**, *9*, 1218. [CrossRef]
19. Li, M.; Shi, Z. Ultra-Processed Food Consumption Associated with Overweight/Obesity among Chinese Adults-Results from China Health and Nutrition Survey 1997–2011. *Nutrients* **2021**, *13*, 2796. [CrossRef]
20. Chinese Center for Disease Control and Prevention. China Health and Nutrition Survey. 2011. Available online: www.cpc.unc.edu/projects/china (accessed on 6 April 2022).
21. Popkin, B.M.; Du, S.; Zhai, F.; Zhang, B. Cohort Profile: The China Health and Nutrition Survey–monitoring and understanding socio-economic and health change in China, 1989–2011. *Int. J. Epidemiol.* **2010**, *39*, 1435–1440. [CrossRef]
22. Zhai, F.Y.; Du, S.F.; Wang, Z.H.; Zhang, J.G.; Du, W.W.; Popkin, B.M. Dynamics of the Chinese diet and the role of urbanicity, 1991–2011. *Obes. Rev.* **2014**, *15*, 16–26. [CrossRef]
23. Wang, G. *Chinese Food Composition Table 1991*; People's Medical Publishing House: Beijing, China, 1991.
24. Wolfson, J.A.; Bleich, S.N. Is cooking at home associated with better diet quality or weight-loss intention? *Public Health Nutr.* **2015**, *18*, 1397–1406. [CrossRef]
25. Andrade, G.C.; Gombi-Vaca, M.F.; Louzada, M.; Azeredo, C.M.; Levy, R.B. The consumption of ultra-processed foods according to eating out occasions. *Public Health Nutr.* **2020**, *23*, 1041–1048. [CrossRef] [PubMed]
26. Monteiro, C.A.; Moubarac, J.C.; Cannon, G.; Ng, S.W.; Popkin, B. Ultra-processed products are becoming dominant in the global food system. *Obes. Rev.* **2013**, *14*, 21–28. [CrossRef] [PubMed]
27. Duan, M.J.; Vinke, P.C.; Navis, G.; Corpeleijn, E.; Dekker, L.H. Ultra-processed food and incident type 2 diabetes: Studying the underlying consumption patterns to unravel the health effects of this heterogeneous food category in the prospective Lifelines cohort. *BMC Med.* **2022**, *20*, 7. [CrossRef] [PubMed]
28. Srour, B.; Fezeu, L.K.; Kesse-Guyot, E.; Allès, B.; Debras, C.; Druesne-Pecollo, N.; Chazelas, E.; Deschasaux, M.; Hercberg, S.; Galan, P.; et al. Ultraprocessed Food Consumption and Risk of Type 2 Diabetes Among Participants of the NutriNet-Sante Prospective Cohort. *JAMA Intern. Med.* **2020**, *180*, 283–291. [CrossRef] [PubMed]
29. Llavero-Valero, M.; Escalada-San Martin, J.; Martinez-Gonzalez, M.A.; Basterra-Gortari, F.J.; de la Fuente-Arrillaga, C.; Bes-Rastrollo, M. Ultra-processed foods and type-2 diabetes risk in the SUN project: A prospective cohort study. *Clin. Nutr.* **2021**, *40*, 2817–2824. [CrossRef] [PubMed]
30. Levy, R.B.; Rauber, F.; Chang, K.; da Costa Louzada, M.L.; Monteiro, C.A.; Millett, C.; Vamos, E.P. Ultra-processed food consumption and type 2 diabetes incidence: A prospective cohort study. *Clin. Nutr.* **2021**, *40*, 3608–3614. [CrossRef]
31. Nardocci, M.; Polsky, J.Y.; Moubarac, J.C. Consumption of ultra-processed foods is associated with obesity, diabetes and hypertension in Canadian adults. *Can. J. Public Health* **2021**, *112*, 421–429. [CrossRef] [PubMed]
32. Tappuni, A.R.; Al-Kaabi, R.; Joury, E. Effect of Free Sugars on Diabetes, Obesity, and Dental Caries. *J. Dent. Res.* **2017**, *96*, 116. [CrossRef] [PubMed]
33. Popov, D.; Simionescu, M.; Shepherd, P.R. Saturated-fat diet induces moderate diabetes and severe glomerulosclerosis in hamsters. *Diabetologia* **2003**, *46*, 1408–1418. [CrossRef] [PubMed]
34. Fardet, A. Minimally processed foods are more satiating and less hyperglycemic than ultra-processed foods: A preliminary study with 98 ready-to-eat foods. *Food Funct.* **2016**, *7*, 2338–2346. [CrossRef] [PubMed]
35. Rolls, B.J. The relationship between dietary energy density and energy intake. *Physiol. Behav.* **2009**, *97*, 609–615. [CrossRef] [PubMed]

36. Malone, J.I.; Hansen, B.C. Does obesity cause type 2 diabetes mellitus (T2DM)? Or is it the opposite? *Pediatr. Diabetes* **2019**, *20*, 5–9. [CrossRef] [PubMed]
37. Chen, Y.; Zhang, X.-P.; Yuan, J.; Cai, B.; Wang, X.-L.; Wu, X.-L.; Zhang, Y.-H.; Yin, T.; Zhu, X.-H.; Gu, Y.-J.; et al. Association of body mass index and age with incident diabetes in Chinese adults: A population-based cohort study. *BMJ Open* **2018**, *8*, e021768. [CrossRef]
38. Zhong, Y.; Wu, L.; Chen, X.; Huang, Z.; Hu, W. Effects of Food-Additive-Information on Consumers' Willingness to Accept Food with Additives. *Int. J. Environ. Res. Public Health* **2018**, *15*, 2394. [CrossRef]
39. Partridge, D.; Lloyd, K.A.; Rhodes, J.M.; Walker, A.W.; Johnstone, A.M.; Campbell, B.J. Food additives: Assessing the impact of exposure to permitted emulsifiers on bowel and metabolic health-introducing the FADiets study. *Nutr. Bull.* **2019**, *44*, 329–349. [CrossRef] [PubMed]
40. Bhattacharyya, S.; Feferman, L.; Unterman, T.; Tobacman, J.K. Exposure to common food additive carrageenan alone leads to fasting hyperglycemia and in combination with high fat diet exacerbates glucose intolerance and hyperlipidemia without effect on weight. *J. Diabetes Res.* **2015**, *2015*, 513429. [CrossRef] [PubMed]
41. Pepino, M.Y.; Tiemann, C.D.; Patterson, B.W.; Wice, B.M.; Klein, S. Sucralose affects glycemic and hormonal responses to an oral glucose load. *Diabetes Care* **2013**, *36*, 2530–2535. [CrossRef] [PubMed]
42. Lin, C.Y.; Lin, Y.C.; Kuo, H.K.; Hwang, J.J.; Lin, J.L.; Chen, P.C.; Lin, L.Y. Association among acrylamide, blood insulin, and insulin resistance in adults. *Diabetes Care* **2009**, *32*, 2206–2211. [CrossRef]
43. Gramec Skledar, D.; Peterlin Mašič, L. Bisphenol A and its analogs: Do their metabolites have endocrine activity? *Environ. Toxicol. Pharmacol.* **2016**, *47*, 182–199. [CrossRef] [PubMed]
44. Gibney, M.J.; Forde, C.G.; Mullally, D.; Gibney, E.R. Ultra-processed foods in human health: A critical appraisal. *Am. J. Clin. Nutr.* **2017**, *106*, 717–724. [CrossRef]
45. Yang, W.; Lu, J.; Weng, J.; Jia, W.; Ji, L.; Xiao, J.; Shan, Z.; Liu, J.; Tian, H.; Ji, Q.; et al. Prevalence of diabetes among men and women in China. *N. Engl. J. Med.* **2010**, *362*, 1090–1101. [CrossRef]
46. Zuo, H.; Shi, Z.; Hussain, A. Prevalence, trends and risk factors for the diabetes epidemic in China: A systematic review and meta-analysis. *Diabetes Res. Clin. Pract.* **2014**, *104*, 63–72. [CrossRef]
47. Shen, X.; Vaidya, A.; Wu, S.; Gao, X. The Diabetes Epidemic in China: An Integrated Review of National Surveys. *Endocr. Pract.* **2016**, *22*, 1119–1129. [CrossRef] [PubMed]
48. Liu, C.; Yuan, Y.C.; Guo, M.N.; Xin, Z.; Chen, G.J.; Bentley, A.R.; Hua, L.; Zheng, J.P.; Ekoru, K.; Yang, J.K. Incidence of Type 1 Diabetes May Be Underestimated in the Chinese Population: Evidence From 21.7 Million People Between 2007 and 2017. *Diabetes Care* **2021**, *44*, 2503–2509. [CrossRef] [PubMed]

nutrients MDPI

Article

Description of Ultra-Processed Food Intake in a Swiss Population-Based Sample of Adults Aged 18 to 75 Years

Valeria A. Bertoni Maluf [1], Sophie Bucher Della Torre [1], Corinne Jotterand Chaparro [1], Fabiën N. Belle [2,3], Saman Khalatbari-Soltani [4,5], Maaike Kruseman [6], Pedro Marques-Vidal [7] and Angeline Chatelan [1,2,*]

1 Department of Nutrition and Dietetics, Geneva School of Health Sciences, HES-SO University of Applied Sciences and Arts Western Switzerland, Rue des Caroubiers 25, 1227 Carouge, Switzerland
2 Institute of Social and Preventive Medicine (ISPM), University of Bern, Mittelstrasse 43, 3012 Bern, Switzerland
3 Center for Primary Care and Public Health (Unisanté), University of Lausanne, Route de la Corniche 10, 1010 Lausanne, Switzerland
4 The University of Sydney School of Public Health, Faculty of Medicine and Health, University of Sydney, Sydney, NSW 2006, Australia
5 ARC Centre of Excellence in Population Aging Research (CEPAR), University of Sydney, Sydney, NSW 2006, Australia
6 MK-Nutrition, Rue Grosselin 25, 1227 Carouge, Switzerland
7 Department of Medicine, Internal Medicine, Lausanne University Hospital, Rue de Bugnon 21, 1011 Lausanne, Switzerland
* Correspondence: angeline.chatelan@hesge.ch; Tel.: +41-22-558-51-16

Abstract: Ultra-processed foods (UPFs) are associated with lower diet quality and several non-communicable diseases. Their consumption varies between countries/regions of the world. We aimed to describe the consumption of UPFs in adults aged 18–75 years living in Switzerland. We analysed data from the national food consumption survey conducted among 2085 participants aged 18 to 75 years. Foods and beverages resulting from two 24-h recalls were classified as UPFs or non-UPFs according to the NOVA classification, categorized into 18 food groups, and linked to the Swiss Food Composition Database. Overall, the median energy intake [P25–P75] from UPFs was 587 kcal/day [364–885] or 28.7% [19.9–38.9] of the total energy intake (TEI). The median intake of UPFs relative to TEI was higher among young participants (<30 years, $p = 0.001$) and those living in the German-speaking part of Switzerland ($p = 0.002$). The food groups providing the most ultra-processed calories were confectionary, cakes & biscuits (39.5% of total UPF kcal); meat, fish & eggs (14.9%); cereal products, legumes & potatoes (12.5%), and juices & soft drinks (8.0%). UPFs provided a large proportion of sugars (39.3% of total sugar intake), saturated fatty acids (32.8%), and total fats (31.8%) while providing less than 20% of dietary fibre. Consumption of UPFs accounted for nearly a third of the total calories consumed in Switzerland. Public health strategies to reduce UPF consumption should target sugary foods/beverages and processed meat.

Keywords: food processing; ultra-processed; NOVA classification; food group; macronutrients; Switzerland; Swiss adults; menuCH

Citation: Bertoni Maluf, V.A.; Bucher Della Torre, S.; Jotterand Chaparro, C.; Belle, F.N.; Khalatbari-Soltani, S.; Kruseman, M.; Marques-Vidal, P.; Chatelan, A. Description of Ultra-Processed Food Intake in a Swiss Population-Based Sample of Adults Aged 18 to 75 Years. *Nutrients* **2022**, *14*, 4486. https://doi.org/10.3390/nu14214486

Academic Editors: Monica Dinu and Daniela Martini

Received: 27 September 2022
Accepted: 21 October 2022
Published: 25 October 2022

1. Introduction

Ultra-processed foods (UPFs) are defined as "formulations of ingredients that result from a series of industrial processes (hence 'ultra-processed'), many requiring sophisticated equipment and technology" [1]. UPFs include soft drinks, energy drinks, ready-to-eat salty snacks, chocolate, confectionery, ice cream, mass-produced packaged breads, margarines, pre-packaged biscuits, breakfast cereals, pre-prepared pies, pasta and pizza dishes, poultry and fish nuggets and sticks, sausages, burgers, hot dogs and other reconstituted meat products, industrial soups and sauces, and many other products [1]. In addition to added salt, sugars, oils, and fats, these industrial formulations include substances not used in homemade food preparations like colours, flavours, emulsifiers, and other additives, which are known as ultra-processing markers [1]. The NOVA classification designates four categories according to the extent of food processing: (group 1) unprocessed or minimally processed foods; (group 2) processed culinary ingredients; (group 3) processed foods; and (group 4) ultra-processed food and drink products (1). NOVA has been used to study the consumption of UPFs in different countries and regions of the world, their nutritional quality, and their association with various non-communicable diseases. These studies have shown that UPFs have unbalanced nutrient profiles, with high contribution of energy, saturated fatty acids (SFAs), added sugars, and sodium and low contribution of proteins, fibre, and most micronutrients [2–4]. In addition, their food matrix is modified so that the complex physical and nutritional structures of whole foods are lost during the food ultra-processing [5,6]. High consumption of UPFs has been associated with overweight/obesity [7–11], high waist circumference, metabolic syndrome, reduced high-density lipoprotein (HDL) cholesterol [7], as well as an increased risk of cardiovascular disease, cerebrovascular disease [7,8], cancers [8], and death [7].

The level of UPF consumption was reviewed in 21 countries with widely varying results [12], including a total of 1,378,454 subjects living in America, Europe, Asia, and Australia (no study in Switzerland). The United States (US) and the United Kingdom (UK) had the highest levels of consumption, reaching more than 50% of total energy intake (TEI); conversely, Italy had the lowest consumption (10–11%) [12]. Because Switzerland is a multilingual country (speaking mainly German, French, and Italian) and surrounded by three countries with differing dietary habits (Germany, France, and Italy), language-regional differences in UPF consumption are expected [13]. Furthermore, associations between consumption of UPFs and sociodemographic characteristics (e.g., sex, age, educational level, household income) as well as weight status have been found in several countries [14–16]. Considering sex, the levels of UPF intake appeared comparable, with men having often an overall slightly higher intake compared to women [12]. Regarding age, the highest levels of consumption were observed in children and adolescents and the lowest in older participants [12]. The association between education and consumption of UPFs is not consistent. In France, UPFs are consumed less by individuals with incomplete high school [15]. Conversely, in countries like Australia [17], Canada [18], and the US [14], the percentage of energy from UPFs was higher in lower educated participants. In Belgium, on the other hand, there were no differences in the consumption of UPFs between different levels of education [19]. When investigating the level of consumption of UPFs according to BMI, it was found that generally, the UPF intake was slightly higher in people with higher BMI [12]. In Switzerland, UPF consumption has been associated with excess body weight in women but not in men [11], but there is no information regarding the differential intake of UPFs by sociodemographic characteristics nor the contribution of UPFs to total nutrient intake.

Nutritional surveillance of population-level dietary intake according to the level of processing by food group is necessary for setting goals, orienting policies, and monitoring the changes in diet quality and diet-related chronic diseases. Similarly, knowing how much of healthy or unhealthy nutrients is provided by UPFs in a standard diet is important for tailoring specific recommendations. Finally, determining whether the consumption of UPFs varies by sociodemographic subgroups makes it possible to tackle health disparities. These data are currently lacking in Switzerland. Therefore, the aims of this analysis of the first Swiss national food consumption survey, menuCH, were to (i) describe the consumption of UPFs according to sociodemographic characteristics; (ii) determine food groups that provide the most ultra-processed energy, and (iii) define the percentage of nutrients provided by UPFs in the Swiss diet.

2. Materials and Methods

2.1. Study Design and Population

We analysed the data from the Swiss National Nutrition Survey (menuCH; https://menuch.iumsp.ch, accessed on 21 April 2020), a cross-sectional survey conducted among non-institutionalised residents aged 18–75 years old (N = 2085) [13]. The stratified random sample was provided by the Federal Statistical Office. The participants were representative of the seven main regions of Switzerland and lived in the cantons of Aargau, Basel–Land, Basel–Stadt, Bern, Lucerne, St. Gallen, and Zurich (German-speaking region); Geneva, Jura, Neuchatel, and Vaud (French-speaking region); and Ticino (Italian-speaking region). The survey was conducted between January 2014 and February 2015. Pregnant and breastfeeding women were included. Institutionalised people or those with insufficient mobility to access a study centre were excluded, as well as people with insufficient oral and written language skills. The study was registered in the trial registry (identification number: ISRCTN16778734). Detailed information on the menuCH study design can be found in these references [13,20,21].

2.2. Dietary Assessment in the Swiss National Nutrition Survey

Fifteen trained dieticians assessed dietary intake via two non-consecutive 24-h recalls (24HDR), the first being conducted face-to-face and the second by phone 2–6 weeks later. 24HDR were spread over all weekdays and seasons. To conduct 24HDR, dieticians used the computer-directed interview program GloboDiet® (GD, formerly EPIC-Soft®, version CH-2016.4.10, International Agency for Research on Cancer (IARC), Lyon, France). The procedure was standardized and followed 3 steps: (i) general information about the participant (e.g., special diet, special day); (ii) quick list of food consumption occasions and items; and (iii) detailed description and quantification of all the consumed foods and beverages, including cooking and preservation methods, brand name, and portion size [22,23]. A book containing photos of standardised portions and a set of 60 household utensils (e.g., glasses, cups, plates) was used to estimate the consumed quantities [24]. The FoodCASE tool (Premotec GmbH, Winterthur, Switzerland) linked all consumed foods with the best match item of the Swiss Food Composition Database (2015 version) [25]. We included in our analysis energy and 7 nutrients: proteins; total carbohydrates; sugars (including all the mono and disaccharides, e.g., glucose, fructose, lactose, saccharose); dietary fibre; total fats; SFAs; and sodium. Other nutrients were excluded because more than 5% of the reported foods had missing data for these nutrients (e.g., calcium, vitamin D).

2.3. Food Classification according to Processing

A registered dietician (VBM) coded each food item as belonging (1) or not (0) to group 4 of the NOVA classification (foods and drinks). For foods considered as recipes by the GD software (e.g., sandwiches, salads, pizzas, lasagne), we classified each underlying ingredient independently. Alcoholic beverages were also classified according to their degree of processing. As previously described [26,27], we used "food descriptors" and "brand name" to ensure more accurate classification. For instance, the words "fresh", "raw", and "homemade" were characteristic of foods classified as not ultra-processed. Conversely, we considered descriptors such as "with flavour", "industrial", "pre-fried", and "with artificial sweetener" as markers of ultra-processing. The online database Open Food Facts [28] and the websites of Swiss supermarkets were used to check the ingredient list of products and to facilitate decision-making, when relevant. When the level of processing was unclear for a food/beverage, the dietitian referred to a senior dietician (AC). In the absence of clear evidence of ultra-processing markers, a conservative attitude was adopted to avoid an overestimation of UPF consumption.

2.4. Food Grouping

The GD software contains 18 main food groups. For this study, we reclassified foods into slightly modified groups according to their nutritional characteristics when there were discrepancies between GD and the Swiss Food Pyramid [29]. We (i) gather legumes, tubers, and cereal products; (ii) gather fruits and vegetables; (iii) separate nuts and seeds from fruits; (iv) separate ice-creams and milk-based desserts from dairy products; (v) gather meat with fish and eggs; (vi) separate breakfast cereals from cereal products; (vii) put avocado and olives with nuts and seeds. After reclassification, our 18 food groups were: cereal products, legumes & potatoes; fruit & vegetables; dairy products; meat, fish & eggs; added fats; nuts & seeds; industrial dishes; soups & broth; juices & soft drinks; other non-alcoholic beverages; alcoholic beverages & substitutes; sugar, honey, jam, sweet sauces & syrups; ice-creams & milk-based desserts; breakfast cereals; confectionary, cakes & biscuits; salty snacks; seasoning, spices, yeast & herbs; and other foods. Supplemental Table S1 provides examples of foods from each food group.

2.5. Sociodemographic Characteristics

The participants completed a 49-item questionnaire at home, which was checked for completeness by the dieticians at the first interview [13]. The linguistic region was defined according to the home address of participants. An open question assessed the nationality (up to two countries) and participants were classified as Swiss or non-Swiss (foreigners). The number of people in the household was categorized into four categories: one, two, three, and four or more people. Education was dichotomized into (i) primary/secondary education (from no compulsory school to high school or specialized professional or vocational school) and (ii) tertiary education (university and higher vocational training, at least 5–7 years after compulsory school).

2.6. Statistical Analyses

Descriptive statistics were used. Daily nutrient intake per survey participant was calculated as the mean intake of the two 24HDR. If the second 24HDR was missing (N = 28, 1.3% of the sample), data from the first 24HDR were used.

Medians and 25th and 75th percentiles (P25–P75) of TEI and energy intake from UPFs were calculated for the whole sample and by subgroups of participants. Medians were preferred over means because of the skewed distribution. Two-sample Wilcoxon rank-sum (Mann–Whitney) tests and Kruskal–Wallis equality-of-populations rank tests were used to determine if there were significant differences in the consumption of UPFs between groups, i.e., sex, age, linguistic region of residency, nationality, household size, and education (bivariate analyses). We also used multiple quantile regressions to test whether the potential differences between groups were still observed after adjustment for all the other parameters and monthly net household income (4499 CHF; 4500–8999 CHF; $\geq$9000 CHF; no answer) (1.00 CHF = 1.05 USD = 1.04 EUR, values as of 14 September 2022) (multivariable analyses).

To assess the energy from UPFs (in kcal/day) for each of the 18 groups, means $\pm$ SD were computed because some medians were 0 and therefore not very informative. Weight of UPFs (in grams/day) in the total diet and by food group was also considered to better take heavy foods (e.g., beverages) and low-calorie foods (e.g., foods with artificial sweeteners) into account and to test whether the contribution of the food groups changed while taking weight or energy (kcal) into account.

We also calculated the medians and P25–P75 intake for 7 nutrients to understand how much UPFs contribute to total nutrient intake and therefore the nutritional benefits (and potential risks) of reducing UPF consumption. For these calculations, alcoholic beverages were excluded, as they are not part of an ideal diet [30]. The relative nutrient intakes of UPFs compared to total nutrient intakes were based on median intakes.

All statistical analyses were performed using STATA software, version 15 (Stata Corporation, College Station, TX, USA). A p-value of < 0.05 was considered statistically significant.

3. Results

3.1. Characteristics of the Participants

The total sample was composed of 2085 participants (Table 1). A flowchart showing the causes of participants' exclusion from analyses is presented in Supplemental Figure S1. The most represented participants were women (54.6%), participants aged 50 to 64 years (mean age of 46.3 ± SD 15.8), living in the German-speaking region (65.2%), of Swiss nationality (84.0%), living in households of two people (39.6%), and with primary/secondary education (51.3%). Four questionnaires (0.2%) were not returned.

3.2. Consumption of UPFs according to Characteristics of Participants

Overall, median TEI among participants was 2089 kcal [P25–P75: 1665–2552] (women 1842 vs. men 2417 kcal) and UPFs represented 28.7% of TEI [P25–P75: 19.9–38.9]. Consumption of UPFs was significantly higher among people aged 18 to 29 years (34.8% of TEI) than in older groups (e.g., 26.3% in 65–75-year-olds; $p = 0.001$). Consumption of UPFs was also significantly higher in people living in the German-speaking region (29.6% vs. 28.0% in the Italian-speaking region and 27.2% in the French-speaking region; $p = 0.002$) and among Swiss nationals (29.2% vs. 26.1% for non-Swiss; $p = 0.002$). Associations were also found between UPF consumption (% of TEI) and sex (higher among women, $p = 0.012$), and education (higher among people with lower education, $p = 0.06$). However, no differences in UPF consumption were found according to household size ($p > 0.05$) (Table 1). Seven people did not consume any UPFs during the two recorded days.

3.3. Distribution of Energy Intake (Kcal) from UPFs by Food Group

Table 2 shows the distribution of energy intake from UPFs by food group in the whole sample. In total, the mean ± SD intake of UPFs was 676 ± 440 kcal, representing 31.0% of the mean TEI (2184 kcal) (results slightly different from medians presented in Table 1). Food groups that were the main energy contributors (Columns 1 and 2) were cereal products, legumes & potatoes (564 kcal; 25.6% of TEI); meat, fish & eggs (272 kcal; 12.6% of TEI); and dairy products (269 kcal; 12.4% of TEI).

Salty snacks; confectionary, cakes & biscuits; and other foods, including meat substitutes or added artificial sweeteners were predominantly constituted of UPFs (100.0%, 99.6%, and 94.1%, respectively, Columns 3 and 4). Among UPFs, most calories came from confectionary, cakes & biscuits (204 kcal, 29.5% of total daily intake from UPFs, Column 5); followed by meat, fish & eggs (105 kcal, 14.9%); and cereal products, legumes & potatoes (78 kcal, 12.5%). Together, other foods; ice-creams & milk-based desserts; alcoholic beverages & alcoholic drink substitutes; soups & broth; industrial dishes; and other non-alcoholic beverages accounted for less than 10% of daily UPFs calories. The last two groups (i.e., nuts & seeds; and fruit & vegetables) did not provide ultra-processed energy (Table 2, Column 5).

3.4. Distribution of Weight of Total Diet (Grams) from UPFs by Food Group

On average, participants consumed 3443 g (SD: 981) of foods and beverages per day, 481 g (SD: 463) (14.2%) of which were from UPFs (see Supplemental Table S2). The major contributors to UPF intake were juices & soft drinks (210 g, 26.0%), confectionary, cakes & biscuits (50 g, 15.9%), and dairy products (48 g, 11.1%, Figure 1).

Table 1. Consumption of ultra-processed foods by sociodemographic characteristics. Swiss population aged 18 to 75 years, National Nutrition Survey 2014–2015.

Characteristics	N (%)	TEI (kcal/Day)		UPF Consumption (kcal/Day) [1]		UPF Consumption (%TEI) [2]		p-Value [3]	p-Value [4]
		Medians	P25–P75	Medians	P25–P75	Medians	P25–P75		
All participants	2085 (100.0)	2089	1665–2552	587	364–885	28.7	19.9–38.9		
Sex								0.125	0.012 *
Women	1139 (54.6)	1842	1527–2216	517	325–746	28.4	19.4–38.5		
Men	946 (45.4)	2417	1987–2993	703	445–1056	29.2	20.8–39.9		
Age groups, years								0.001 *	0.001 *
18–29	407 (19.5)	2221	1709–2731	727	478–1060	34.8	24.5–45.0		
30–39	327 (15.7)	2126	1700–2669	646	418–963	31.8	22.3–42.0		
40–49	450 (21.6)	2110	1702–2583	599	380–883	28.2	20.3–37.8		
50–64	562 (27.0)	2021	1640–2507	519	308–811	25.5	16.9–36.6		
65–75	339 (16.3)	1978	1641–2331	495	314–714	26.3	17.1–35.0		
Linguistic region								0.003 *	0.002 *
German	1359 (65.2)	2153	1721–2612	617	399–915	29.6	20.9–39.6		
French	510 (24.5)	1991	1647–2467	526	323–789	27.2	17.7–37.1		
Italian	216 (10.4)	1930	1515–2319	509	298–820	28.0	16.9–39.4		
1st nationality [5]								0.009 *	0.002 *
Swiss	1751 (84.0)	2078	1665–2550	595	379–894	29.2	20.3–39.0		
Non-Swiss	330 (15.8)	2124	1654–2571	557	318–839	26.1	17.5–37.1		
Household size [5]								0.060	0.400
One person	338 (16.2)	1996	1621–2446	573	330–892	29.0	18.5–40.6		
Two people	825 (39.6)	2070	1669–2514	565	353–835	28.1	19.7–37.3		
Three people	336 (16.1)	2103	1728–2522	591	371–901	28.8	19.5–39.7		
Four people and more	582 (27.9)	2132	1688–2678	626	407–945	30.2	21.5–40.1		
Education [5]								0.073	0.060
Primary & secondary	1069 (51.3)	1993	1588–2495	574	355–894	29.1	20.2–39.7		
Tertiary	1012 (48.5)	2160	1762–2617	604	373–870	28.4	19.6–38.4		

[1] Total energy intake from UPFs. [2] Contribution of UPFs from total energy intake. [3] Differences in UPF consumption as the percentage of total energy intake were tested with two-sample Wilcoxon rank-sum (Mann–Whitney) tests for sex and nationality. Kruskal–Wallis equality-of-populations rank tests were used for age, linguistic region, and household size. [4] Differences in UPF consumption as the percentage of total energy intake were tested using multiple quantile regressions. [5] Four questionnaires were not completed (N_{total} = 2081). * $p < 0.05$. TEI: total energy intake. UPFs: ultra-processed food and drink products. P25–P75: 25th and 75th percentiles. CHF: Swiss franc.

Table 2. Distribution of energy intake (kcal) from UPFs by food group, in decreasing order (N = 2085, **bold** = 3 largest numbers, *italic* = 3 smallest numbers, by column).

Food Groups	Total Intake (kcal/Day) Mean (SD)	Contribution to TEI (%TEI) Mean (SD)	UPF Intake (kcal/Day) Mean (SD)	UPF Intake from Total Intake (%) [1] Mean (SD)	UPF Intake from TEI (%TEI) [2] Mean (SD)
Confectionary, cakes & biscuits	204 (216)	9.0 (8.3)	**204 (214)**	**99.6 (4.6)**	**29.5 (23.9)**
Meat, fish & eggs	**272 (218)**	**12.6 (8.8)**	**105 (150)**	35.3 (34.1)	**14.9 (18.6)**
Cereal products, legumes & potatoes	**564 (310)**	**25.6 (10.4)**	**78 (109)**	14.8 (19.3)	**12.5 (16.8)**
Juices & soft drinks	97 (150)	4.1 (5.5)	65 (136)	53.9 (44.4)	8.0 (13.3)
Dairy products	**269 (203)**	**12.4 (8.3)**	50 (86)	16.6 (25.0)	7.9 (13.9)
Seasoning, spices, yeast & herbs	95 (100)	4.4 (4.2)	33 (62)	32.2 (37.5)	5.5 (10.1)
Added fats	182 (152)	8.3 (6.0)	30 (75)	15.8 (25.6)	4.9 (10.0)
Salty snacks	22 (75)	1.0 (2.8)	22 (75)	**100.0 (0.0)**	3.0 (8.5)
Sugar, honey, jam, sweet sauces & syrups	60 (75)	2.7 (3.2)	18 (46)	26.4 (38.0)	2.9 (7.0)
Breakfast cereals	29 (72)	1.2 (3.1)	19 (58)	65.9 (44.9)	2.7 (8.1)
Other foods	*14 (58)*	*0. 7 (2.7)*	13 (58)	**94.1 (23.0)**	2.4 (9.3)
Ice-creams & milk-based desserts	22 (54)	1.0 (2.3)	14 (38)	74.4 (42.0)	2.3 (6.5)
Alcoholic beverages substitutes	107 (159)	4.7 (6.5)	13 (47)	14.4 (29.6)	1.9 (6.4)
Soups & broth	21 (55)	1.0 (2.8)	5 (29)	40.5 (48.3)	0.7 (4.3)
Industrial dishes	*13 (65)*	*0.6 (2.6)*	6 (45)	40.2 (48.3)	0.6 (4.4)
Other non-alcoholic beverages	*15 (33)*	*0.8 (1.5)*	*2 (12)*	*2.26 (13.3)*	*0.2 (2.3)*
Nuts & seeds	39 (84)	1.7 (3.6)	*0 (0)*	*0.0 (0.0)*	*0.0 (0.0)*
Fruit & vegetables	159 (120)	7.8 (6.1)	*0 (0)*	*0.0 (0.0)*	*0.0 (0.0)*
Total	2184 (750)	100.0	676.3 (440.1)	-	100.0

[1] Among consumers only (N varies according to food groups, e.g., N = 2074 for cereal products, legumes & potatoes to N = 155 for industrial dishes). [2] Seven people did not consume any UPFs (N_{total} = 2078).

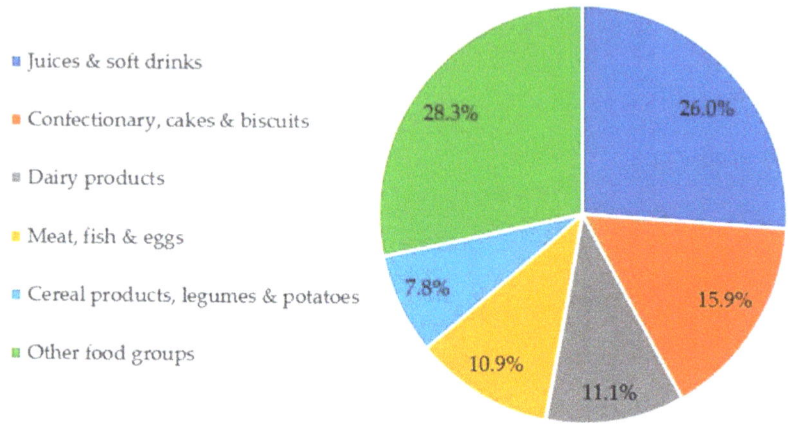

Figure 1. Proportion of UPF intake weight (grams/day) in comparison to the total diet weight, by major food group contributors. Seven people did not consume any UPFs (N_{total} = 2078).

3.5. Contribution of UPFs to Intake of Macro- and Micronutrients

UPFs accounted for 39.3% of the total daily intake of sugars, 32.8% of SFAs, 31.8% of total fats, and 30.7% of total carbohydrates (Figure 2). UPFs accounted for less than 20% of total daily intake for dietary fibre (15.2%). Details on absolute intakes and proportions of missing nutrient values are presented in Supplemental Table S3.

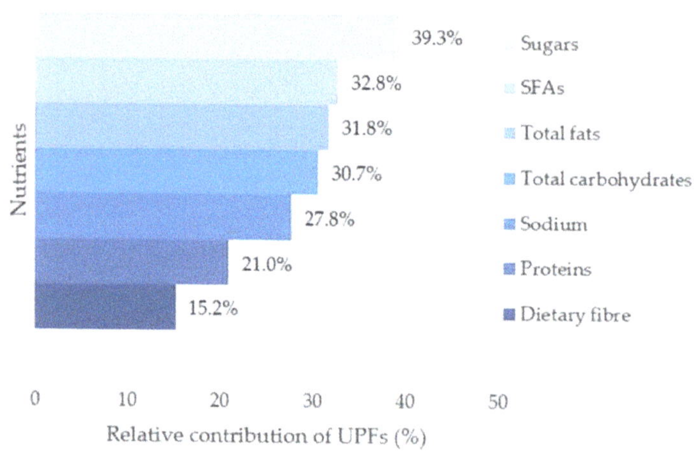

Figure 2. Relative contribution of UPFs to total daily intake (% based on medians) for seven nutrients. Sugars include all mono and disaccharides, e.g., glucose, fructose, lactose, saccharose; SFAs: saturated fatty acids.

4. Discussion

4.1. Principal Findings

UPFs represent a substantial percentage of TEI (29%). We found a higher percentage of energy from UPFs among younger adults, those living in the German-speaking region, and Swiss nationals. Conversely, people aged 50–64 and 65–75 years and non-Swiss nationals were participants who consumed the least UPFs. Major contributors of ultra-processed calories were confectionary, cakes & biscuits; meat, fish & eggs; and cereal products, legumes & potatoes. These three food groups contributed to more than 50% of the energy intake from UPFs. When taking the weight of UPFs in the diet into account, food groups consumed in higher amounts were juices & soft drinks; and confectionary, cakes & biscuits. UPFs provided a large proportion of sugars, SFAs, and total fats. Conversely, the contribution of UPFs was below 20% for dietary fibre.

4.2. Consumption of UPFs according to Countries

A systematic review including several countries showed that the consumption of UPFs greatly varies between Western high-income countries, with the US and UK being the countries with the highest percent of TEI from UPFs (higher than 50%), and Italy being the country with the lowest level (about 10%) [12]. For instance, in Canada, the levels of intake were also elevated (more than 45%). Australia showed levels of UPF consumption ranging from 38.9% to 42.0% of TEI. In Europe, in both Spain and France the consumption varied between 17.0% and more than 30%, depending on the studies. Consumption in Belgium was similar to consumption in Switzerland (means of 30.3% and 31.0%, respectively), while in Portugal the intake was lower (22.2%) but higher than in Italy [12].

4.3. Consumption of UPFs according to Characteristics of Participants

We found that the highest percentage of energy intake from UPFs was in young adults (<30 years) and decreased with age. This trend has already been observed in previous studies [15–17]. Young adults might be attracted by the convenience (limited time spent in the kitchen) of these products [31]. Interestingly, when we related the time required to cook a hot meal at home during a usual week with the consumption of UPFs in menuCH participants, we found that those who spend less than 30 min cooking had a significantly higher percentage of kilocalories from UPFs (Supplementary Table S4). Other authors also showed that time spent on food preparation at home was associated with indicators of diet quality and frequency of fast-food restaurant use [32]. In addition, among adolescents and young adults, the use of social media is high, and greatly promotes the consumption of branded UPFs, such as soft drinks, cakes, crisps, pizzas, and sweets [33].

People from the German-speaking region consumed more UPFs. This is consistent with previous literature showing that people from the German-speaking region less frequently cook hot lunches themselves at home in comparison to people from the French-speaking and Italian-speaking regions [34]. Furthermore, the consumption of UPFs, such as soft drinks (including fruit lemonades and sugar-free soft drinks) or processed meat is higher in the German-speaking part of Switzerland [13].

In the current study, non-Swiss nationals consumed significantly fewer UPFs, even though this group was slightly underrepresented in the sample [13]. The majority of foreigners residing in Switzerland are Italian, German, Portuguese, and French nationals [35]. People from Italy, Portugal, and France may have maintained a diet closer to the Mediterranean diet, which is usually poor in UPFs [36]. Indeed, when the adherence to the Mediterranean diet over 50 years was analysed in 41 countries, Germany ranked 35th and Switzerland 34th, while Portugal, Italy, and France ranked 10th, 14th, and 27th, respectively [37]. Moreover, another study showed that the average household availability of UPFs was lower in Portugal, Italy, and France compared to other European countries such as Germany or Austria (Switzerland not included in this analysis) [38]. Of note, the same phenomenon was found in Australia and Canada, where the intake of UPFs was also significantly lower among immigrants compared to locals [16,17].

Energy intake from UPFs only slightly differed according to education. Other barriers than lower education like taste, daily habits, and lack of time and willpower may play a role in adherence to healthy eating [39]. Furthermore, in this study, the intake from minimally or unprocessed foods was not investigated. It is possible that, even if the consumption of UPFs was similar, foods of NOVA group 1 were more consumed by people with higher education, as demonstrated in Belgium by Vandevijvere et al. [19]. This could be explained by the fact that people with higher education are more health conscious [40–42].

4.4. Distribution of Energy Intake from UPFs by Food Group

Ultra-processed energy came mainly from confectionary, cakes & biscuits; meat, fish & eggs; and cereal products, legumes & potatoes. Comparing our results with other studies is difficult because the way foods are grouped differs from one study to another. However, a study conducted in 22 European countries reported that the two main UPFs consumed among adults were fine bakery wares and sausages [43]. In our study, chocolate, industrial cakes, and cookies are typical UPFs of the group confectionary, cakes & biscuits. Because Swiss people consume the most chocolate per capita worldwide [44], this could explain why confectionary, cakes & biscuits was the food group contributing most to ultra-processed energy.

4.5. Distribution of Intake from UPFs (Grams/Day) by Food Group

The average consumption of UPFs in adults across 22 European countries was estimated at 328 g/day, representing an average share of total weight intake of 12% [43]. In our study, these figures were slightly higher: 481 g/day and 14.2%, respectively. A possible explanation is that alcoholic beverages were not considered in the international study. When the proportion (in weight, g/day) of UPFs in the total diet was analysed, major contributors were juices & soft drinks; confectionary, cakes & biscuits; and dairy products. Across Europe, the most consumed ultra-processed drinks were soft drinks and fruit/vegetable juices [43]. This analysis shows that the UPFs preferred by consumers are similar in Switzerland.

4.6. Nutrition Profile of UPFs

We found that diets rich in UPFs are high in sugars and fats, especially SFAs, and low in fibre, which is in line with other studies [18,45,46]. In this study, UPFs contributed nearly 40% of total sugar and 35% of SFA intake—nutrients that have been associated with a greater risk of chronic diseases [47]. The contribution of sodium was almost 30%, and it is known that a reduction in sodium intake reduces blood pressure [48,49]. In the US diet, the average intake of carbohydrates, added sugars, and SFAs increased significantly with the dietary contribution of UPFs [2]. In the UK, UPFs contributed nearly 65% of all free sugars (different from total sugars) in all age groups [50], and the intake of carbohydrates, free sugars, total fats, SFAs, and sodium increased significantly as UPF consumption increased [51]. In France, UPFs represented most of the total and free sugars and total fats, SFAs, but only a minor part of proteins and fibre [15]. Because of the poor nutritional profile of UPFs, high intake affects people's health, and the risk of several non-communicable diseases is higher [52–54]. Thus, replacing UPFs with less- or un-processed foods could improve the quality of the diet without drastically impacting the intake of proteins [55]. Of note, in our study, values in unsaturated fatty acids and micronutrients were more likely to be missing from the Food Composition Database for UPFs than for non-UPFs, which limited the analysis for these nutrients(Supplementary Table S3).

4.7. Strengths and Limitations

For the assessment of dietary intake we used two 24HDRs, which may have led to misreporting of intake due to social desirability and recall bias [56]. However, 24HDRs are appropriate for estimating average levels of food consumption in nutrition population-based surveys [56] and to describe UPF consumption in a given population [57]. Although we assessed diet in the whole of Switzerland, the number of participants from the Italian-speaking region, a small region in Switzerland, was limited in our sample. The categorization of groups does not always make it possible to distinguish foods within the 18 food groups that are ultra-processed, although Table S1 provides specific examples of ultra-processed products in each group. In addition, food description did not always contain enough information to categorize foods according to the NOVA classifications with certainty; our conservative approach might have underestimated UPF consumption.

Finally, micronutrient content was not available for all foods/beverages, thus limiting the number of nutrients included in our analysis.

Despite these limitations, this is the first study to assess the importance of UPFs in a representative sample of the Swiss population encompassing three linguistic regions. The inclusion of two non-consecutive 24HDR conducted by trained dieticians enabled the estimation of detailed dietary intake (e.g., systematic description of cooking and preservation methods, brand names, etc.), allowing accurate identification of NOVA group 4 foods/beverages. Furthermore, the classification of foods (UPFs vs. non-UPFs) was performed by trained dieticians and discussed in case of discrepancies.

5. Conclusions

Consumption of UPFs accounts for nearly one-third of total calories consumed in Switzerland, and their nutritional profile is unbalanced. Non-communicable disease prevention programs should especially target young adults. Nutritional education messages for reducing UPF consumption should first focus on the highest-contributing food groups, i.e., sugary foods/beverages and processed meat. Additionally, population-based public health measures, such as (i) taxing soft drinks or other UPFs, (ii) front-of-pack warning labels on NOVA 4 products, and (iii) school food policies banning UPFs from school meals, are possible strategies to reduce UPF consumption and prevent non-communicable diseases [58].

Supplementary Materials: The following supporting information can be downloaded at: https://www.mdpi.com/article/10.3390/nu14214486/s1, Table S1: Eighteen foods groups and examples of foods from each food group; Table S2: Distribution of total intake (grams/day) and intake from UPFs (grams/day) by food group in decreasing order; Table S3: Nutrient profile of the overall diet and of ultra-processed products (N = 2085) and missing values from the Food Composition Database, by nutrient; Table S4: Consumption of UPFs according to time to prepare and cook a hot meal at home. Figure S1: Flowchart showing causes of participants' exclusion from analyses.

Author Contributions: Conceptualization: A.C. Methodology: A.C. and V.A.B.M. Project administration: V.A.B.M. and A.C. Data curation: V.A.B.M. and A.C. Formal analysis: V.A.B.M. and A.C. Investigation: A.C. Visualization: V.A.B.M. Supervision: A.C. Writing—original draft: V.A.B.M. and A.C. Writing—Review and Editing: S.B.D.T., C.J.C., F.N.B., S.K.-S., M.K. and P.M.-V. All authors have read and agreed to the published version of the manuscript.

Funding: AC is supported by the Swiss National Science Foundation (Project number 190277). FB is supported by Swiss Cancer Research (KFS-4722-02-2019). SKS is supported by the Australian Research Council Centre of Excellence in Population Ageing Research (Project number CE170100005). The funders had no role in the preparation of the manuscript and decision to publish it.

Institutional Review Board Statement: The study was conducted according to the guidelines laid down in the Declaration of Helsinki. It was approved by all regional ethics committees (lead committee in Lausanne, Protocol 26/13).

Informed Consent Statement: Each participant signed written informed consent.

Data Availability Statement: The whole dataset and relevant documents (e.g., questionnaires, GloboDiet data) are accessible in the repository: https://menuch.iumsp.ch (accessed on 21 October 2022).

Acknowledgments: This article was based on the preliminary results of a Master thesis conducted within the joint Master of Science (MSc) in Health Sciences of HES-SO (University of Applied Sciences and Arts Western Switzerland) and University of Lausanne (UNIL), major in Nutrition and dietetics, at HES-SO Master. We sincerely thank all the researchers who collected and shared the menuCH data and all the volunteers of the study.

Conflicts of Interest: The authors declare no conflict of interest.

References

1. Monteiro, C.A.; Cannon, G.; Levy, R.B.; Moubarac, J.-C.; Louzada, M.L.C.; Rauber, F.; Khandpur, N.; Cediel, G.; Neri, D.; Martinez-Steele, E.; et al. Ultra-Processed Foods: What They Are and How to Identify Them. *Public Health Nutr.* **2019**, *22*, 936–941. [CrossRef]
2. Martinez Steele, E.M.; Popkin, B.M.; Swinburn, B.; Monteiro, C.A. The share of ultra-processed foods and the overall nutritional quality of diets in the US: Evidence from a nationally representative cross-sectional study. *Popul. Health Metr.* **2017**, *15*, 6. [CrossRef]
3. Louzada, M.L.D.C.; Martins, A.P.B.; Canella, D.; Baraldi, L.; Levy, R.B.; Claro, R.; Moubarac, J.-C.; Cannon, G.; Monteiro, C.A. Impact of ultra-processed foods on micronutrient content in the Brazilian diet. *Rev. Saúde Pública* **2015**, *49*, 1–8. [CrossRef]
4. Gupta, S.; Hawk, T.; Aggarwal, A.; Drewnowski, A. Characterizing Ultra-Processed Foods by Energy Density, Nutrient Density, and Cost. *Front. Nutr.* **2019**, *6*, 70. [CrossRef]
5. Fardet, A.; Rock, E. Chronic diseases are first associated with the degradation and artificialization of food matrices rather than with food composition: Calorie quality matters more than calorie quantity. *Eur. J. Nutr.* **2022**, *61*, 2239–2253. [CrossRef]
6. Thorning, T.K.; Bertram, H.C.; Bonjour, J.-P.; De Groot, L.; Dupont, D.; Feeney, E.; Ipsen, R.; Lecerf, J.M.; Mackie, A.; McKinley, M.C.; et al. Whole dairy matrix or single nutrients in assessment of health effects: Current evidence and knowledge gaps. *Am. J. Clin. Nutr.* **2017**, *105*, 1033–1045. [CrossRef]
7. Pagliai, G.; Dinu, M.; Madarena, M.P.; Bonaccio, M.; Iacoviello, L.; Sofi, F. Consumption of ultra-processed foods and health status: A systematic review and meta-analysis. *Br. J. Nutr.* **2020**, *125*, 308–318. [CrossRef]
8. Chen, X.; Zhang, Z.; Yang, H.; Qiu, P.; Wang, H.; Wang, F.; Zhao, Q.; Fang, J.; Nie, J. Consumption of ultra-processed foods and health outcomes: A systematic review of epidemiological studies. *Nutr. J.* **2020**, *19*, 86. [CrossRef]
9. Askari, M.; Heshmati, J.; Shahinfar, H.; Tripathi, N.; Daneshzad, E. Ultra-processed food and the risk of overweight and obesity: A systematic review and meta-analysis of observational studies. *Int. J. Obes.* **2020**, *44*, 2080–2091. [CrossRef]
10. Moradi, S.; Entezari, M.H.; Mohammadi, H.; Jayedi, A.; Lazaridi, A.-V.; Kermani, M.A.H.; Miraghajani, M. Ultra-processed food consumption and adult obesity risk: A systematic review and dose-response meta-analysis. *Crit. Rev. Food Sci. Nutr.* **2021**, 1–12. [CrossRef]
11. Pestoni, G.; Habib, L.; Reber, E.; Rohrmann, S.; Staub, K.; Stanga, Z.; Faeh, D. Ultraprocessed Food Consumption is Strongly and Dose-Dependently Associated with Excess Body Weight in Swiss Women. *Obesity* **2021**, *29*, 601–609. [CrossRef] [PubMed]
12. Marino, M.; Puppo, F.; Del Bo', C.; Vinelli, V.; Riso, P.; Porrini, M.; Martini, D. A Systematic Review of Worldwide Consumption of Ultra-Processed Foods: Findings and Criticisms. *Nutrients* **2021**, *13*, 2778. [CrossRef] [PubMed]
13. Chatelan, A.; Beer-Borst, S.; Randriamiharisoa, A.; Pasquier, J.; Blanco, J.M.; Siegenthaler, S.; Paccaud, F.; Slimani, N.; Nicolas, G.; Camenzind-Frey, E.; et al. Major Differences in Diet across Three Linguistic Regions of Switzerland: Results from the First National Nutrition Survey menuCH. *Nutrients* **2017**, *9*, 1163. [CrossRef] [PubMed]
14. Baraldi, L.G.; Martinez Steele, E.; Canella, D.S.; Monteiro, C.A. Consumption of Ultra-Processed Foods and Associated Sociodemographic Factors in the USA between 2007 and 2012: Evidence from a Nationally Representative Cross-Sectional Study. *BMJ Open* **2018**, *8*, e020574. [CrossRef] [PubMed]
15. Andrade, G.C.; Julia, C.; Deschamps, V.; Srour, B.; Hercberg, S.; Kesse-Guyot, E.; Allès, B.; Chazelas, E.; Deschasaux, M.; Touvier, M.; et al. Consumption of Ultra-Processed Food and Its Association with Sociodemographic Characteristics and Diet Quality in a Representative Sample of French Adults. *Nutrients* **2021**, *13*, 682. [CrossRef] [PubMed]
16. Nardocci, M.; Leclerc, B.-S.; Louzada, M.-L.; Monteiro, C.A.; Batal, M.; Moubarac, J.-C. Consumption of ultra-processed foods and obesity in Canada. *Can. J. Public Health* **2019**, *110*, 4–14. [CrossRef]
17. Marchese, L.; Livingstone, K.M.; Woods, J.L.; Wingrove, K.; Machado, P. Ultra-processed food consumption, socio-demographics and diet quality in Australian adults. *Public Health Nutr.* **2021**, *25*, 94–104. [CrossRef]
18. Moubarac, J.-C.; Batal, M.; Louzada, M.L.; Steele, E.M.; Monteiro, C.A. Consumption of Ultra-Processed Foods Predicts Diet Quality in Canada. *Appetite* **2017**, *108*, 512–520. [CrossRef]
19. Vandevijvere, S.; De Ridder, K.; Fiolet, T.; Bel, S.; Tafforeau, J. Consumption of ultra-processed food products and diet quality among children, adolescents and adults in Belgium. *Eur. J. Nutr.* **2018**, *58*, 3267–3278. [CrossRef]
20. Chatelan, A.; Gaillard, P.; Kruseman, M.; Keller, A. Total, Added, and Free Sugar Consumption and Adherence to Guidelines in Switzerland: Results from the First National Nutrition Survey menuCH. *Nutrients* **2019**, *11*, 1117. [CrossRef]
21. Chatelan, A.; Marques-Vidal, P.; Bucher, S.; Siegenthaler, S.; Metzger, N.; Zuberbühler, C.A.; Camenzind-Frey, E.; Reggli, A.; Bochud, M.; Beer-Borst, S. Lessons Learnt About Conducting a Multilingual Nutrition Survey in Switzerland: Results from menuCH Pilot Survey. *Int. J. Vitam. Nutr. Res.* **2017**, *87*, 25–36. [CrossRef] [PubMed]
22. Slimani, N.; Casagrande, C.; Nicolas, G.; Freisling, H.; Huybrechts, I.; Ocké, M.C.; Niekerk, E.M.; Van Rossum, C.; Bellemans, M.; De Maeyer, M.; et al. The standardized computerized 24-h dietary recall method EPIC-Soft adapted for pan-European dietary monitoring. *Eur. J. Clin. Nutr.* **2011**, *65*, S5–S15. [CrossRef] [PubMed]
23. Crispim, S.P.; De Vries, J.H.M.; Geelen, A.; Souverein, O.W.; Hulshof, P.J.M.; Lafay, L.; Rousseau, A.-S.; Lillegaard, I.T.L.; Andersen, L.F.; Huybrechts, I.; et al. Two non-consecutive 24 h recalls using EPIC-Soft software are sufficiently valid for comparing protein and potassium intake between five European centres—Results from the European Food Consumption Validation (EFCOVAL) study. *Br. J. Nutr.* **2010**, *105*, 447–458. [CrossRef] [PubMed]

24. Camenzind-Frey, E.; Zuberbühler, C. *menuCH—Schweizerisches Fotobuch/Livre Photo Suisse/Manuale Fotografico Svizzero (menuCH Picture Book)*; Federal Office of Public Health & Federal Food Safety and Veterinary Office: Bern, Switzerland, 2014.
25. Federal Food Safety and Veterinary Office FSVO. The Swiss Food Composition Database. Available online: https://www.valeursnutritives.ch/fr/ (accessed on 5 October 2020).
26. Juul, F.; Martinez-Steele, E.; Parekh, N.; Monteiro, C.A.; Chang, V.W. Ultra-processed food consumption and excess weight among US adults. *Br. J. Nutr.* **2018**, *120*, 90–100. [CrossRef]
27. Martínez Steele, E.; Baraldi, L.G.; da Costa Louzada, M.L.; Moubarac, J.-C.; Mozaffarian, D.; Monteiro, C.A. Ultra-Processed Foods and Added Sugars in the US Diet: Evidence from a Nationally Representative Cross-Sectional Study. *BMJ Open* **2016**, *6*, e009892. [CrossRef]
28. Open Food Facts—Suisse. Available online: https://ch-fr.openfoodfacts.org (accessed on 20 April 2022).
29. Swiss Society for Nutrition; Federal Food Safety and Veterinary Office. Pyramide Alimentaire Suisse 2011 (Swiss Food Pyramid). Available online: https://www.sge-ssn.ch/fr/toi-et-moi/boire-et-manger/equilibre-alimentaire/pyramide-alimentaire-suisse/ (accessed on 9 June 2022).
30. Guenther, P.M.; Casavale, K.O.; Reedy, J.; Kirkpatrick, S.I.; Hiza, H.A.B.; Kuczynski, K.J.; Kahle, L.L.; Krebs-Smith, S.M. Update of the healthy eating index: HEI-2010. *J. Acad. Nutr. Diet.* **2013**, *113*, 569–580. [CrossRef] [PubMed]
31. Djupegot, I.L.; Nenseth, C.B.; Bere, E.; Bjørnarå, H.B.T.; Helland, S.H.; Øverby, N.C.; Torstveit, M.K.; Stea, T.H. The association between time scarcity, sociodemographic correlates and consumption of ultra-processed foods among parents in Norway: A cross-sectional study. *BMC Public Health* **2017**, *17*, 447. [CrossRef]
32. Monsivais, P.; Aggarwal, A.; Drewnowski, A. Time Spent on Home Food Preparation and Indicators of Healthy Eating. *Am. J. Prev. Med.* **2014**, *47*, 796–802. [CrossRef]
33. Qutteina, Y.; Hallez, L.; Mennes, N.; De Backer, C.; Smits, T. What Do Adolescents See on Social Media? A Diary Study of Food Marketing Images on Social Media. *Front. Psychol.* **2019**, *10*, 2637. [CrossRef]
34. Bochud, M.; Chatelan, A.; Blanco, J.M.; Beer-Borst, S.M. *Anthropometric Characteristics and Indicators of Eating and Physical Activity Behaviors in the Swiss Adult Population: Results from menuCH 2014–2015*; The Federal Office of Public Health and the Food Safety and Veterinary Office: Bern, Switzerland, 2017. [CrossRef]
35. Federal Statistical Office. Foreign Population. Available online: https://www.bfs.admin.ch/bfs/en/home/statistiken/bevoelkerung/migration-integration/auslaendische-bevoelkerung.html (accessed on 6 May 2022).
36. Dinu, M.; Asensi, M.T.; Pagliai, G.; Lotti, S.; Martini, D.; Colombini, B.; Sofi, F. Consumption of Ultra-Processed Foods Is Inversely Associated with Adherence to the Mediterranean Diet: A Cross-Sectional Study. *Nutrients* **2022**, *14*, 2073. [CrossRef]
37. Vilarnau, C.; Stracker, D.M.; Funtikov, A.; Da Silva, R.; Estruch, R.; Bach-Faig, A. Worldwide adherence to Mediterranean Diet between 1960 and 2011. *Eur. J. Clin. Nutr.* **2018**, *72*, 83–91. [CrossRef] [PubMed]
38. Monteiro, C.A.; Moubarac, J.-C.; Levy, R.B.; Canella, D.S.; da Costa Louzada, M.L.; Cannon, G. Household Availability of Ultra-Processed Foods and Obesity in Nineteen European Countries. *Public Health Nutr.* **2018**, *21*, 18–26. [CrossRef] [PubMed]
39. de Mestral, C.; Khalatbari-Soltani, S.; Stringhini, S.; Marques-Vidal, P. Perceived barriers to healthy eating and adherence to dietary guidelines: Nationwide study. *Clin. Nutr.* **2019**, *39*, 2580–2585. [CrossRef]
40. Lawrence, E.M. Why Do College Graduates Behave More Healthfully Than Those Who Are Less Educated? *J. Health Soc. Behav.* **2017**, *58*, 291–306. [CrossRef] [PubMed]
41. Li, J.; Powdthavee, N. Does more education lead to better health habits? Evidence from the school reforms in Australia. *Soc. Sci. Med.* **2015**, *127*, 83–91. [CrossRef]
42. Raghupathi, V.; Raghupathi, W. The influence of education on health: An empirical assessment of OECD countries for the period 1995–2015. *Arch. Public Health* **2020**, *78*, 1–18. [CrossRef] [PubMed]
43. Mertens, E.; Colizzi, C.; Peñalvo, J.L. Ultra-processed food consumption in adults across Europe. *Eur. J. Nutr.* **2021**, *61*, 1521–1539. [CrossRef]
44. Chocosuisse. Available online: https://www.chocosuisse.ch/en/services/facts-figures (accessed on 7 May 2022).
45. Julia, C.; Martinez, L.; Allès, B.; Touvier, M.; Hercberg, S.; Méjean, C.; Kesse-Guyot, E. Contribution of ultra-processed foods in the diet of adults from the French NutriNet-Santé study. *Public Health Nutr.* **2018**, *21*, 27–37. [CrossRef]
46. de Deus Mendonça, R.; Pimenta, A.M.; Gea, A.; De La Fuente-Arrillaga, C.; Martinez-Gonzalez, M.A.; Lopes, A.C.S.; Bes-Rastrollo, M. Ultraprocessed food consumption and risk of overweight and obesity: The University of Navarra Follow-Up (SUN) cohort study. *Am. J. Clin. Nutr.* **2016**, *104*, 1433–1440. [CrossRef]
47. World Health Organization. *Diet, Nutrition, and the Prevention of Chronic Diseases: Report of a WHO Study Group*; Technical report series; World Health Organization: Geneva, Switzerland, 1995; ISBN 978-92-4-120797-3.
48. Aburto, N.J.; Ziolkovska, A.; Hooper, L.; Elliott, P.; Cappuccio, F.P.; Meerpohl, J.J. Effect of lower sodium intake on health: Systematic review and meta-analyses. *BMJ* **2013**, *346*, f1326. [CrossRef]
49. Leyvraz, M.; Chatelan, A.; Da Costa, B.R.; Taffé, P.; Paradis, G.; Bovet, P.; Bochud, M.; Chiolero, A. Sodium intake and blood pressure in children and adolescents: A systematic review and meta-analysis of experimental and observational studies. *Int. J. Epidemiol.* **2018**, *47*, 1796–1810. [CrossRef] [PubMed]
50. Rauber, F.; Louzada, M.L.D.C.; Steele, E.M.; Rezende, L.; Millett, C.; A Monteiro, C.; Levy, R.B. Ultra-processed foods and excessive free sugar intake in the UK: A nationally representative cross-sectional study. *BMJ Open* **2019**, *9*, e027546. [CrossRef] [PubMed]

51. Rauber, F.; da Costa Louzada, M.L.; Steele, E.M.; Millett, C.; Monteiro, C.A.; Levy, R.B. Ultra-Processed Food Consumption and Chronic Non-Communicable Diseases-Related Dietary Nutrient Profile in the UK (2008–2014). *Nutrients* **2018**, *10*, 587. [CrossRef]
52. Gentry, M. World Cancer Research Fund International (WCRF). *Impact* **2017**, *2017*, 32–33. [CrossRef]
53. Mozaffarian, D. Dietary and Policy Priorities for Cardiovascular Disease, Diabetes, and Obesity: A Comprehensive Review. *Circulation* **2016**, *133*, 187–225. [CrossRef]
54. GBD 2016 Risk Factors Collaborators. Global, regional, and national comparative risk assessment of 84 behavioural, environmental and occupational, and metabolic risks or clusters of risks, 1990–2016: A systematic analysis for the Global Burden of Disease Study 2016. *Lancet* **2017**, *390*, 1345–1422. [CrossRef]
55. Adams, J.; Hofman, K.; Moubarac, J.-C.; Thow, A.M. Public health response to ultra-processed food and drinks. *BMJ* **2020**, *369*, m2391. [CrossRef]
56. DAPA Measurement2 Toolkit. Available online: https://dapa-toolkit.mrc.ac.uk/diet/subjective-methods/24-hour-dietary-recall (accessed on 7 October 2020).
57. FAO. *Guidelines on the Collection of Information on Food Processing through Food Consumption Surveys*; FAO: Rome, Italy, 2015.
58. Popkin, B.M.; Barquera, S.; Corvalan, C.; Hofman, K.J.; Monteiro, C.; Ng, S.W.; Swart, E.C.; Taillie, L.S. Towards unified and impactful policies to reduce ultra-processed food consumption and promote healthier eating. *Lancet Diabetes Endocrinol.* **2021**, *9*, 462–470. [CrossRef]

nutrients

Article

Ultra-Processed Food Consumption Associated with Incident Hypertension among Chinese Adults—Results from China Health and Nutrition Survey 1997–2015

Ming Li [1,*] and Zumin Shi [2]

[1] Centre for Population Health Research, Division of Health Sciences, University of South Australia, Adelaide, SA 5005, Australia
[2] Human Nutrition Department, College of Health Sciences, QU Health, Qatar University, Doha 2713, Qatar
* Correspondence: ming.li@unisa.edu.au

Abstract: Objective: Ultra-processed food (UPF) has been shown to increase the cardiometabolic health risks. We aimed to determine the association between UPF intake based on the NOVA classification and the risk of hypertension incidence during 1997–2015. Methods: Data from 15,054 adults aged $\geq$ 20 years (47.4% males) attending the China Nutrition and Health Survey (CNHS) were used. Food intake at each survey was assessed by a 3-day 24 h dietary recall and weighed food record method between 1997–2011. Cox regression was used to assess the association between UPF intake and incident hypertension. Results: During a mean average of 9.5 years (SD 5.5) of follow up, 4329 hypertension incident cases were identified. The incident rates (per 1000) for non-consumers and 1–49, 50–99, and $\geq$100 g/day of UPF intake were 29.5 and 29.5, 33.4, and 36.3, respectively. Compared with non-consumers, the hazard ratios (95% CI) for UPF intake of 1–49, 50–99, and >100 g/day were 1.00 (0.90–1.12), 1.17 (1.04–1.33), and 1.20 (1.06–1.35), respectively, (p = 0.001) after adjusting for potential confounding factors. There was a significant interaction between UPF intake and age with a higher risk in the younger group (<40 years) than in the older one. Conclusion: UPF consumption was dose-responsively associated with increased risk of hypertension among Chinese adults, especially in younger groups.

Keywords: ultra-processed food; incident hypertension; adults; China

Citation: Li, M.; Shi, Z. Ultra-Processed Food Consumption Associated with Incident Hypertension among Chinese Adults—Results from China Health and Nutrition Survey 1997–2015. *Nutrients* **2022**, *14*, 4783. https://doi.org/10.3390/nu14224783

Academic Editors: Monica Dinu and Daniela Martini

Received: 27 October 2022
Accepted: 9 November 2022
Published: 11 November 2022

1. Introduction

Hypertension is a serious medical condition that significantly increases the risks of heart, brain, and kidney conditions, as well as other diseases. It is the leading preventable risk factor for cardiovascular disease (CVD) and all-cause mortality worldwide [1]. The global prevalence of hypertension in adults aged 30–79 reached 32% in women and 34% in men in 2019 with an increased trend in most low- and middle-income countries [2], while in China, a review of 15 recent epidemiological studies based on national population surveys from 1997–2017 reported that 18–45% of the Chinese adult population ($\geq$18 years of age) had hypertension, and only a limited portion of 4.2–30.1% had it under control [3].

The sharp increasing trend of hypertension, particularly in younger adults, is in line with the dramatic social–economic development observed in China and multidimensional levels of factors associated with hypertension, including environmental, psychosocial, lifestyle, and behavioral factors [4–9].

Among the modifiable dietary factors, certain nutrients, foods, and dietary patterns are associated with high blood pressure/hypertension. For example, high salt consumption has been proven to increase the risk of hypertension substantially in the Chinese population [9]. A meta-analysis of 133 randomized control studies in diverse populations reported that a reduction in sodium decreases systolic blood pressure (SBP) [10]. A recent large 5-year intervention study in Chinese older adults found that using a salt substitute of 70%

sodium chloride and 25% potassium chloride decreases SBP and incidence of stroke, CVD, and death, as compared to the use of regular salt [11]. High sodium intake increases blood pressure by increasing water retention and systemic peripheral resistance, altering the endothelial function and the structure and function of large elastic arteries. High intake can modify sympathetic activity, and autonomic neuronal modulation of the cardiovascular system. In addition, excessive dietary sodium induces alterations in the extracellular matrix of the arterial wall, favoring a process of arterial stiffening [12]. World Health Organization recommends limiting sodium intake to approximately 2.0 g per day (equivalent to approximately 5.0 g salt per day) in the general population [13].

Hypertension is inversely associated with intakes of whole grains, fruits, nuts, and dairy, whereas positively with red meat, processed meat, and sugar-sweetened beverages [14] while in the short term, green tea could lower blood pressure [15]. Overall dietary patterns, such as the Dietary Approaches to Stop Hypertension (DASH) study and both the Nordic diet and Mediterranean diet, are associated with blood pressure [16,17]. Studies in the Chinese population have shown that modern dietary pattern with a high consumption of meat and processed foods is associated with increased cardiometabolic risk [18] and DASH diet can reduce the risk of hypertension induced by air pollution [19].

NOVA classifies foods and drinks based on their processing status into four groups, which allows a novel insight into its health impact [20]. Ultra-processed food (UPF) is the 4th group by this classification that includes products of entirely industrial formulations or made from substances extracted from foods, with minimal whole foods [21]. UPF is commonly high in energy density, sugars, salt, and trans-fats, as well as additives with poor nutrition profiles [20], and it contributes more than half of the total daily energy intake in high-income countries, and its consumption is increasing rapidly in middle-income countries [22,23]. The increased consumption was driven by economic development and urbanization, especially in nutrition transition countries, such as China [24–26]. In addition, food choice at the individual level based not only on nutrients profile but also on taste, convenience, and cost drives the increased trend [27]. Syntheses of observational studies from countries in Europe and the American continents have shown that UPF intake is associated with certain conditions but the association with hypertension is inconsistent [28–30]. For example, the prospective analyses in Mediterranean and Brazilian cohorts demonstrated higher consumption was positively related with the risk of developing hypertension [31–33] while results in Canadian and Lebanon adults showed no evidence of a relationship between UPF consumption and SBP and diastolic blood pressure (DBP) [30].

The mean daily UPF consumption in Chinese adults increased four times between 1997–2011, and higher long-term UPF consumption is associated with increased risk of being overweight/obese and diabetes [34,35]. However, the association between UPF consumption and incident hypertension has not been quantified in China and whether the association interplays with being overweight/obese, having diabetes, dietary patterns, or other behavioral factors remains unknown. This study aimed to fill the knowledge gaps.

2. Research Design and Methods

2.1. Study Design and Sample

This was a prospective follow-up study of UPF intake and incident hypertension between 1997–2015 using data from China Health and Nutrition Survey (CHNS).

The CHNS study is an ongoing household-based cohort study conducted in nine provinces in China [36]. A multistage random-cluster sampling method was applied to select participants in both urban and rural areas. Ten waves of dietary data collection (1989, 1991, 1993, 1997, 2000, 2004, 2006, 2009, 2011, and 2015) have been completed. The overall response rate was >60% based on the first survey in 1989 and >80% based on the previous survey year [36]. A cohort of 15,054 participants meeting the following inclusion criteria were included (Figure 1): aged $\geq$ 20 years; having attended at least two nutrition surveys between 1997–2015; having dietary and blood pressure measures; having plausible energy intake (800–6000 kcal/day for men, and 600–4000 kcal/day for women); being free of hypertension at baseline. The survey was approved by the institutional review committees and informed consent was obtained from all participants [36]. The data used in the current study were de-identified and publicly available.

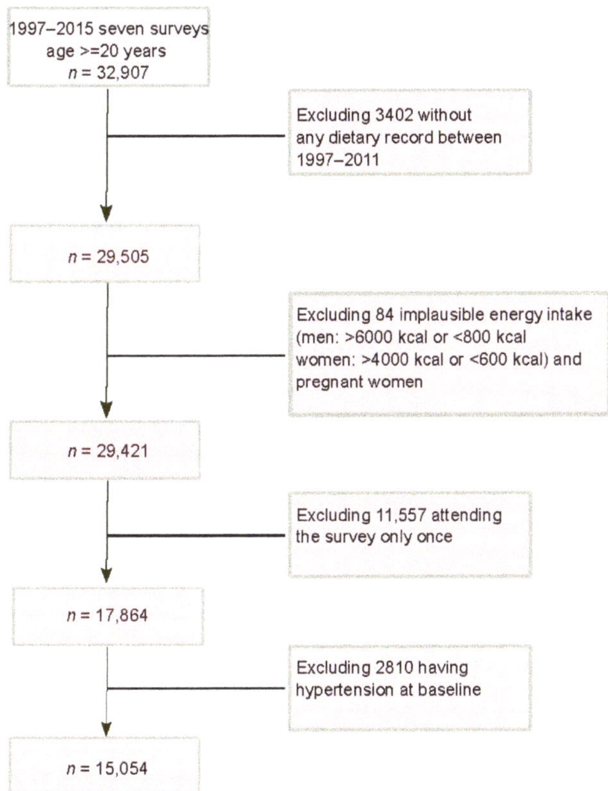

Figure 1. Sample flowchart of participants attending CHNS 1997–2015.

2.2. Outcome Variable: Incident Hypertension

During household visit at each survey, blood pressure was measured by mercury sphygmomanometer based on a standard protocol [36]. Hypertension was defined as having SBP $\geq$ 140 mmHg and/or DBP $\geq$ 90 mmHg or having known hypertension.

2.3. Exposure Variable: UPF Consumption

At each survey, individual food intake data were collected by a trained investigator using a 24 h dietary recall for three consecutive days [36]. Foods and condiments in the home inventory, foods from markets or from gardens, and food waste were weighed and recorded by interviewers at the beginning and end of the three-day survey period. The Chinese food composition tables were used to convert food intake to nutrient intake [37,38]. Around 3000 food items in the food composition tables since 1997 were categorized into four groups based on the NOVA classification [20]. UPF intake for each participant at each survey was categorized into four levels: non-consumers, 1–49 g/day, 50–99 g/day, $\geq$100 g/day. We choose this cut-off based on the fact that the serving size in the context of Chinese food is *Liang* (50 g).

2.4. Covariates

Sociodemographic information was collected at each survey using a structured questionnaire. The following constructed variables were used as indicators of socioeconomic status: education (low: illiterate/primary school; medium: junior middle school; high: high middle school or higher), per capita annual family income (recoded into tertiles as low, medium, and high), urbanization levels (recoded into tertiles as low, medium, and high).

Lifestyle factors from questionnaire included smoking, alcohol drinking, sleep, and physical activity. Smoking status was categorized as non-smokers, ex-smokers, and current smokers. Alcohol consumption was recorded as yes or no. Sleep duration was recorded as ≤6, 7–9, and ≥10 h per day using data collected since 2004. Physical activity level (metabolic equivalent of task MET) was estimated based on self-reported activities (including occupational, domestic, transportation, and leisure time physical activity) and duration using a compendium of physical activities. Tea consumption in each survey wave was categorized into four levels: non-consumers, <2 cups/day, 2–3.9 cups/day, and ≥4 cups/day with one cup being 240 mL.

Height was measured without shoes to the nearest 0.2 cm using a portable stadiometer. Weight was measured without shoes and in light clothing to the nearest 0.1 kg on a calibrated beam scale. Body mass index (BMI) was calculated from weight and height. Overweight/obesity was defined as BMI ≥ 25 kg/m^2.

2.5. Statistical Analysis

Sample characteristics were presented and compared by baseline UPF categories of "None, 1–49, 50–99, ≥100 g/day" using ANOVA for continuous measures or chi-square tests for categorical ones.

The association between UPF intake and incident hypertension was examined using Cox regression with age as the underline time scale [39]. Study entry was the age at baseline. Exit time was the age at incident hypertension or related death or the end of follow-up, whichever occurred first. The proportional hazards assumption was assessed by Schoenfeld residuals. Unadjusted and adjusted hazard ratios (95% CI) were reported from the following models: unadjusted model; adjusted models subsequently adjusted for age, sex, and energy intake; socioeconomic status (income, urbanization, and education), behavioral factors (smoking, alcohol drinking, and physical activity), and BMI; sodium/potassium; intake of fruit and vegetable or green tea; diabetes. All adjusted covariates except sex were treated as time varying measures.

Interaction between UPF intake and other covariates (sociodemographic) on incident hypertension was assessed by introducing a product term in the final regression model (Model 3) and the stratified results were presented. The following sensitivity analysis was conducted: (1) using data from those entering at the first wave (1997) or last wave (2011); (2) data before and after 2004, where UPF increased differently. STATA 17.0 (Stata Corporation, College Station, TX, USA) was used for all the analyses. Statistical significance was considered when $p < 0.05$ (two-sided).

3. Results

3.1. Population Characteristics and UPF Consumption

Among the 15,054 participants included in this study, 6924 entered in 1997, 2160 in 2000, 1406 in 2004, 774 in 2006, 1320 in 2009, and 2470 in 2011. At baseline, the mean age of this sample was 40.2 years (SD 14.4), 47.4% were males, 40.7% resided in highly urbanized area, 29.7% were smokers, and 8.6% were alcohol drinkers. The prevalence of overweight/obesity was 20.1%. The mean daily energy, fat, protein, and carbohydrate intake were 2184 kcal, 67.8 g, 67.9 g, and 321.9 g, respectively.

At baseline, 11,010 (73%) reported no UPF intake, while 1, 276 (8%) reported daily UPF consumption ≥100 g. Compared with non-consumers, those having ≥100 g/day were significantly more likely to be: older aged; males; having higher education and income; living in highly urbanized area; smoking; drinking; having less tea consumption; sleeping <6 h; having less physical activity; having higher intake of energy, fat, protein, and potassium but lower carbohydrates; having higher fruit intake; entering the survey in a more recent survey; and higher prevalence of overweight/obesity. Baseline prevalence of diabetes were no different by levels of UPF intake (Table 1).

Table 1. Baseline sample characteristics by UPF intake (g/day): China Health and Nutrition Survey (n = 15,054).

UPF Intake Level	None	1–49	50–99	≥100	p-Value
n	11,010	1699	1069	1276	
Survey year					<0.001
1997	51.8%	41.1%	25.4%	19.4%	
2000	16.1%	11.1%	10.2%	7.0%	
2004	10.3%	8.1%	6.8%	4.5%	
2006	5.0%	5.5%	5.1%	6.0%	
2009	7.7%	9.1%	12.4%	14.5%	
2011	9.0%	25.3%	40.0%	48.7%	
Age (years), mean (SD)	39.9 (14.4)	40.8 (14.7)	40.7 (14.4)	41.6 (14.3)	<0.001
Sex					<0.001
Men	46.2%	44.2%	49.8%	60.4%	
Women	53.8%	55.8%	50.2%	39.6%	
Income					<0.001
Low	31.4%	21.5%	19.4%	17.9%	
Medium	34.5%	32.3%	33.1%	29.4%	
High	34.1%	46.2%	47.5%	52.7%	
Education					<0.001
Low	41.1%	27.3%	20.8%	19.8%	
Medium	35.5%	34.2%	29.7%	27.9%	
High	23.4%	38.5%	49.5%	52.2%	
Urbanization					<0.001
Low	36.1%	19.3%	15.1%	12.2%	
Medium	30.7%	23.8%	23.2%	21.9%	
High	33.2%	56.9%	61.7%	65.9%	
Energy intake (kcal/d), mean (SD)	2206.6 (653.5)	2037.3 (635.2)	2104.3 (670.7)	2260.7 (722.8)	<0.001
Fat intake (g/d), mean (SD)	65.4 (35.9)	68.7 (34.0)	75.8 (37.1)	81.3 (38.7)	<0.001
Protein intake (g/d), mean (SD)	66.8 (22.7)	67.3 (22.8)	71.2 (23.9)	75.3 (25.6)	<0.001
Carbohydrate intake (g/d), mean (SD)	337.2 (125.0)	285.0 (120.8)	275.5 (117.0)	277.7 (107.4)	<0.001
Sodium intake (mg/d), mean (SD)	5465.8 (6880.9)	5157.6 (4455.0)	4885.6 (4721.0)	5450.1 (5236.1)	0.014
Potassium intake (mg/d), mean (SD)	1611.6 (895.8)	1596.7 (664.2)	1703.7 (757.2)	1872.7 (1174.8)	<0.001
Vegetable intake (g/day), mean (SD)	283.7 (173.4)	262.9 (160.2)	253.9 (156.9)	255.3 (158.2)	<0.001
Fruit intake (g/day), mean (SD)	20.9 (78.0)	49.7 (103.4)	64.9 (112.8)	89.0 (129.2)	<0.001
Tea intake (cup/day)					<0.001
None	64.4%	57.1%	54.4%	51.3%	
<2	12.3%	16.0%	16.0%	17.2%	
2–3.9	12.0%	13.0%	14.6%	12.1%	
4	11.2%	13.8%	15.1%	19.4%	
Smoking					<0.001
Non-smoker	69.8%	70.7%	65.0%	60.1%	
Ex-smokers	1.3%	1.4%	3.2%	3.8%	
Current smokers	29.0%	27.8%	31.8%	36.1%	
Alcohol drinking	31.7%	35.8%	44.1%	51.5%	<0.001
Sleep duration (hours/day)					<0.001
≤6	7.5%	10.3%	9.9%	11.2%	
6–9	80.4%	81.0%	79.6%	80.8%	
>9	12.1%	8.8%	10.5%	8.0%	
Physical activity (MET hours/week), mean (SD)	142.3 (115.2)	127.3 (106.2)	127.8 (104.9)	128.4 (99.7)	<0.001
BMI (kg/m^2), mean (SD)	22.3 (3.1)	22.7 (3.2)	22.9 (3.4)	23.2 (3.4)	<0.001
Diabetes	5.5%	7.6%	4.7%	8.6%	0.43

p from ANOVA for continuous measures or chi-square tests for categorical ones.

The mean daily UPF consumption in this population increased slowly from 10.5 g in 1997 to 14.9 g in 2004, and sharply increased to reach 47.3 g in 2011 (Supplementary Figure S1).

3.2. Incident Hypertension and the Association with UPF Consumption

During a mean average of 9.5 years (median 8.9, SD 5.5) of follow-up (total 142,868 person years), 4329 incident cases were observed. Of them, 689 cases were identified in 2000, 874 in 2004, 575 in 2006, 758 in 2009, 546 in 2011, and 887 in 2015.

The corresponding incident cases for UPF non-consumers, 1–49 g/d, 50–99 g/d, and ≥100 g/d were 3137, 459, 327, and 406, given the unadjusted hazard ratios (HRs) (95% CI) of 1.00, 0.95 (0.86–1.05), 1.08 (0.96–1.21), and 1.12 (1.01–1.25) (p for trend = 0.031). After adjusting for age, sex, total energy intake, education, income, urbanization, smoking, alcohol drinking, physical activity, and BMI, the HRs were not substantially changed, being 1.00, 1.00 (0.90–1.12), 1.17 (1.04–1.33), 1.20 (1.06–1.35) (Model 2, Table 2). Further adjusting for sodium/potassium (Model 3), intake of fruit and vegetable/tea (Model 4), or diabetes (Model 5) did not alter the HRs either.

Table 2. Hazard ratio (95%CI) for hypertension incidence by UPF intake (g/day): China Health and Nutrition Survey (n = 15,054).

UPF Intake Level	None	1–49	50–99	≥100	*p* for Trend
Number of incident cases	3137	459	327	406	
Rate (per 1000 person years)	29.5	29.5	33.4	36.3	
Person years	106,364	15,542	9777	11,186	
Unadjusted model	1.00	0.95 (0.86–1.05)	1.08 (0.96–1.21)	1.12 (1.01–1.25)	0.031
Model 1	1.00	1.03 (0.93–1.13)	1.14 (1.02–1.28)	1.25 (1.13–1.39)	0.000
Model 2	1.00	1.00 (0.90–1.12)	1.17 (1.04–1.33)	1.20 (1.06–1.35)	0.001
Model 3	1.00	1.00 (0.90–1.12)	1.17 (1.04–1.33)	1.20 (1.06–1.35)	0.001
Model 4	1.00	1.00 (0.90–1.12)	1.17 (1.03–1.32)	1.19 (1.06–1.34)	0.001
Model 5	1.00	1.00 (0.90–1.12)	1.17 (1.03–1.33)	1.19 (1.06–1.35)	<0.001

Model 1 adjusted for age, sex, and energy intake. Model 2 further adjusted for income, education, urbanization, smoking, alcohol drinking, physical activity, sleep duration, and BMI. Model 3: model 2 further adjusted for sodium/potassium intake. Model 4: model 2 further adjusted for intake of fruit and vegetables/tea; Model 5: model 4 further adjusted for known diabetes.

Other factors significantly associated with incident hypertension were age, sex, education, income, urbanization, alcohol drinking, and BMI.

The association between UPF and incident hypertension varied by age. Among the younger participants (<40 years), the adjusted HRs (9% CI) were: 1.04 (0.79–1.35) for 1–49 g/d, 1.23 (0.90–1.68) for 50–99 g/d, and 1.54 (1.17–2.04) for ≥100 g/d, compared to non-consumers, significantly higher than in older participants (≥40 years), with corresponding HRs (95% CI) of 0.99 (0.88–1.11), 1.11 (0.97–1.27), 1.15 (1.01–1.32) (p for interaction = 0.017) (Figure 2). There were no significant interactions between UPF and sex, income, education, and urbanization, in relation to the risk of incident hypertension. Sensitivity analysis showed consistent associations (data not shown).

	UPF intake (g/day)					
	None	1-49	50-99	>100	*p*-trend	*p*-interaction
Age (years)						0.017
<40	1.00	1.04 (0.79–1.35)	1.23 (0.90–1.68)	1.54 (1.17–2.04)	0.003	
≥40	1.00	0.99 (0.88–1.11)	1.11 (0.97–1.27)	1.15 (1.01–1.32)	0.023	
Sex						0.626
Men	1.00	0.96 (0.82–1.14)	1.21 (1.02–1.43)	1.17 (1.01–1.36)	0.012	
Women	1.00	1.02 (0.88–1.18)	1.12 (0.93–1.36)	1.27 (1.04–1.54)	0.016	
Income						0.169
Low	1.00	1.01 (0.81–1.27)	0.95 (0.72–1.27)	1.25 (0.96–1.63)	0.249	
Medium	1.00	0.88 (0.72–1.08)	1.22 (0.97–1.52)	1.28 (1.04–1.58)	0.021	
High	1.00	1.09 (0.92–1.28)	1.27 (1.06–1.53)	1.13 (0.95–1.35)	0.026	
Education						0.170
Low	1.00	0.91 (0.77–1.08)	1.04 (0.84–1.30)	1.19 (0.97–1.46)	0.251	
Medium	1.00	1.00 (0.81–1.23)	1.22 (0.98–1.53)	1.08 (0.87–1.34)	0.200	
High	1.00	1.20 (0.97–1.49)	1.38 (1.11–1.73)	1.32 (1.07–1.63)	0.001	
Urbanization						0.143
Low	1.00	0.81 (0.63–1.05)	1.25 (0.93–1.68)	1.07 (0.78–1.45)	0.594	
Medium	1.00	1.12 (0.91–1.38)	0.97 (0.76–1.25)	1.25 (1.00–1.55)	0.094	
High	1.00	1.03 (0.88–1.20)	1.28 (1.08–1.52)	1.23 (1.04–1.45)	0.002	

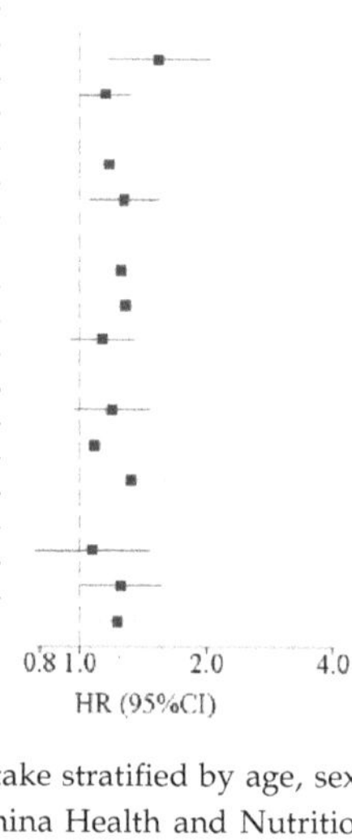

Figure 2. Hazard ratio (95%CI) for hypertension with UPF intake stratified by age, sex, income, education, and urbanization among participants attending China Health and Nutrition Survey (n = 15,054). Model adjusted for age, sex and energy intake, income, education, urbanization, smoking, alcohol drinking, physical activity, and BMI. Stratification variables were not adjusted in the corresponding models.

4. Discussion

In a 10 year follow-up study of 15,054 adults aged ≥ 20 years, UPF consumption was dose-responsively associated with incident hypertension and those having ≥ 100 g/d had an overall increased risk of 15%. There was a significant interaction between UPF and age. In adults aged under 40 years, high UPF intake (≥100 g/d) increased the risk of hypertension by 54% while there was a 15% increased risk in those aged over 40 years.

Our finding of the positive association between UPF intake and hypertension was consistent with three longitudinal studies: the 9-year follow-up Spanish The Seguimiento Universidad de Navarra Project, project, which reported a 21% higher risk among 14,790 university students [31]; the ELSA-Brazil studies among 8754 adults aged 35–74, which reported 23% greater risk of developing hypertension for higher UPF consumption after adjusting for sociodemographic, lifestyle, BMI, and dietary factors [32]; and the 2-year follow up of 1221 graduates in the Cohort of Universities of Minas Gerais, Brazil (CUME Project) Project that reported an increased risk of 35% [33]. Our study confirmed the results of a meta-analysis of prospective association between certain UPF, such as red meat, processed meat, and sugar-sweetened beverages with hypertension [14].

The positive association between UPFs and hypertension can be explained not only by their poor nutrient profile, including high amount of salt, saturated fats, sugar, and energy, but also a lack of whole foods, such as fruits and vegetables [22,24], which were shown in our adjusted model. Plausible biological pathways may include increased energy intake, changes to the gut microbiota, alterations in the gut–brain satiety signalling, and hormonal effects, which may target sodium/potassium balance, endothelial function, oxidation stress, and inflammation [40]. Despite lacking evidence of the long-term effect of non-nutritional bioactive compounds in UPF on human health and food additives, such as artificial sweetener, emulsifiers, thickening and stabilizing agents, and bisphenols, may play roles through the pathways of insulin response or gut microbiota, and/or adipocyte function [41].

In addition to the poor nutrient profile or quality from UPF that poses a risk to health, such as hypertension, growing concerns have emerged with regard to the impact on the food structure characteristics or food matrix during food processing as UPF products are industrial formulations manufactured from substances extracted from foods or synthesized from other organic sources that mostly contain little or no natural complex food [42,43]. Further research is needed to understand the proportional harm associated with the food physical structure, and other attributes of UPF [44].

The impact of UPF intake ≥100 g/day on the risk of developing hypertension among younger adults is of concern. Based on a previous report using CHNS data, the weekly frequency of eating out doubled to 25% between 2004–2011, remarkably higher in younger adults and males [45]. Eating out increases the consumption of UPF, compared with home-prepared meals [46]. Younger adults are heavily exposed to TV advertisements with more than half on food, snacks, and beverages during the times between 20:00 h to 22:00 h [47]. In addition to these environmental changes, it should be noted that younger adults are under pressure from education, jobs, finance, and family. A recent national survey estimated that 16.6% of Chinese adults had experienced mental illness at some point in their lives with the most common being anxiety disorders and the increased prevalence of depression [48]. Further investigation on the UPF consumption and health transition from childhood and adolescence to adulthood is warranted based on our findings that children and adolescents are more likely to have certain UPF that related to being overweight/obese [49] and to the early onset of hypertension in this study population, in addition to the early onset of some cancers, such as colorectal and breast cancers [50]. It is unknown whether early exposure to UPF or its accumulative effect or both can explain the age difference in association with the early onset of hypertension with UPF.

Our result support the Chinese dietary guidelines published in 2022 in which new recommendations have been supplemented. The new guideline emphasizes the needs to avoid UPF, to acquire knowledge and skills to cook, and to select packaged food by reading food labels, in addition to the food-based recommendations [51].

This is the first association study between UPF consumption using NOVA classification and incident hypertension in a large cohort of the Chinese adult population. The study period lasted for ten years and covered the socioeconomic transition, including dietary pattern change. The energy and food intake from the surveys have been proven to be generally valid based on basal metabolic rate [52]. Missing data were low, and no data imputation was needed. In total, 98.6% of the participants were included in the full multivariable model. Hypertension incident cases were ascertained by established international criteria from data using standardized protocols at each survey. Known confounding factors, including sociodemographic, behavioral, health, and dietary factors were adjusted. The consumption of fat, fruit, and vegetables was used as a proxy for diet quality. The statistical analysis was robust, considering the repeated measures of UPF during the follow-up period and using age as a time scale to reduce potential bias [39].

Limitations should be noted. Firstly, misclassification was possible due to incomplete records on food processing methods in the CHNS survey, which was not specifically matched with NOVA classification, and the use of gram for UPF might not be precise for the diverse UPF items (e.g., soft drinks). Secondly, the ascertainment of food items might not be subtle in reflecting the complexity of food processing and variabilities in additive composition between brands for a similar type of product, and, therefore, some food items could only be roughly grouped and the association could be biased. Thirdly, 24 h sodium excretion, which more accurately measures the dietary sodium and metabolism, was not collected in the CHNS but instead we used dietary sodium in the adjusted analysis. Finally, residual confounding was still possible due to the lack of data on ethnicity, which is closely related to culinary culture in China. In addition, stress level as a strong factor of hypertension was not attainable, although daily sleep and alcohol drinking were included as proxies. Further well-designed studies in other populations and settings are warranted to determine causality and identify potential mechanisms.

To conclude, higher UPF consumption was dose-responsively associated with incident hypertension, especially among younger adults aged < 40 years in a 10-year follow-up of Chinese adults between 1997–2011.

Supplementary Materials: The following supporting information can be downloaded at: https://www.mdpi.com/article/10.3390/nu14224783/s1, Figure S1: Age- and sex-adjusted mean intake of UPF in 1997-2011 (n = 15,054).

Author Contributions: M.L. and Z.S. conceived the study, Z.S. analyzed the data, M.L. and Z.S. interpreted the results, M.L. drafted and revised the manuscript. All authors have read and agreed to the published version of the manuscript.

Funding: This research received no external funding.

Institutional Review Board Statement: Not applicable.

Informed Consent Statement: Not applicable.

Data Availability Statement: The current research uses data from the China Health and Nutrition Survey (CHNS). Data described in the manuscript, code book, and analytic code are made publicly and freely available without restriction at https://www.cpc.unc.edu/projects/china accessed on 15 January 2019.

Acknowledgments: This research uses data from China Health and Nutrition Survey. The authors thank the National Institute of Nutrition and Food Safety, China Centre for Disease Control and Prevention, Carolina Population Centre, the University of North Carolina at Chapel Hill, the NIH (R01-HD30880, DK056350, and R01-HD38700) and the Fogarty International Centre for financial support for the CHNS data collection and analysis files from 1989 to 2006, and both parties plus the China-Japan Friendship Hospital, Ministry of Health for support for CHNS 2009 and future surveys.

Conflicts of Interest: The authors declare no conflict of interest.

References

1. Zhou, B.; Perel, P.; Mensah, G.A.; Ezzati, M. Global epidemiology, health burden and effective interventions for elevated blood pressure and hypertension. *Nat. Rev. Cardiol.* **2021**, *18*, 785–802. [CrossRef] [PubMed]
2. NCD Risk Factor Collaboration. Worldwide trends in hypertension prevalence and progress in treatment and control from 1990 to 2019: A pooled analysis of 1201 population-representative studies with 104 million participants. *Lancet* **2021**, *398*, 957–980. [CrossRef]
3. Yin, R.; Yin, L.; Li, L.; Silva-Nash, J.; Tan, J.; Pan, Z.; Zeng, J.; Yan, L.L. Hypertension in China: Burdens, guidelines and policy responses: A state-of-the-art review. *J. Hum. Hypertens* **2022**, *36*, 126–134. [CrossRef] [PubMed]
4. Ma, S.; Yang, L.; Zhao, M.; Magnussen, C.G.; Xi, B. Trends in hypertension prevalence, awareness, treatment and control rates among Chinese adults, 1991–2015. *J. Hypertens* **2021**, *39*, 740–748. [CrossRef]
5. Yang, B.Y.; Qian, Z.; Howard, S.W.; Vaughn, M.G.; Fan, S.J.; Liu, K.K.; Dong, G.H. Global association between ambient air pollution and blood pressure: A systematic review and meta-analysis. *Environ. Pollut.* **2018**, *235*, 576–588. [CrossRef]
6. Hu, B.; Liu, X.; Yin, S.; Fan, H.; Feng, F.; Yuan, J. Effects of psychological stress on hypertension in middle-aged Chinese: A cross-sectional study. *PLoS ONE* **2015**, *10*, e0129163. [CrossRef]
7. Shen, Y.; Chang, C.; Zhang, J.; Jiang, Y.; Ni, B.; Wang, Y. Prevalence and risk factors associated with hypertension and prehypertension in a working population at high altitude in China: A cross-sectional study. *Environ. Health Prev. Med.* **2017**, *22*, 19. [CrossRef]
8. Li, L.; Li, L.; Chai, J.X.; Xiao, L.; Ng, C.H.; Ungvari, G.S.; Xiang, Y.T. Prevalence of Poor Sleep Quality in Patients with Hypertension in China: A Meta-analysis of Comparative Studies and Epidemiological Surveys. *Front. Psychiatry* **2020**, *11*, 591. [CrossRef]
9. Du, S.; Batis, C.; Wang, H.; Zhang, B.; Zhang, J.; Popkin, B.M. Understanding the patterns and trends of sodium intake, potassium intake, and sodium to potassium ratio and their effect on hypertension in China. *Am. J. Clin. Nutr.* **2014**, *99*, 334–343. [CrossRef]
10. Huang, L.; Trieu, K.; Yoshimura, S.; Neal, B.; Woodward, M.; Campbell, N.R.C.; Li, Q.; Lackland, D.T.; Leung, A.A.; Anderson, C.A.M.; et al. Effect of dose and duration of reduction in dietary sodium on blood pressure levels: Systematic review and meta-analysis of randomised trials. *BMJ* **2020**, *368*, m315. [CrossRef]
11. Neal, B.; Wu, Y.; Feng, X.; Zhang, R.; Zhang, Y.; Shi, J.; Zhang, J.; Tian, M.; Huang, L.; Li, Z.; et al. Effect of Salt Substitution on Cardiovascular Events and Death. *N. Engl. J. Med.* **2021**, *385*, 1067–1077. [CrossRef] [PubMed]
12. Grillo, A.; Salvi, L.; Coruzzi, P.; Salvi, P.; Parati, G. Sodium Intake and Hypertension. *Nutrients* **2019**, *11*, 1970. [CrossRef] [PubMed]
13. World Health Organization. *Guideline: Sodium Intake for Adults and Children*; World Health Organization: Geneva, Switzerland, 2012.
14. Schwingshackl, L.; Schwedhelm, C.; Hoffmann, G.; Knuppel, S.; Iqbal, K.; Andriolo, V.; Bechthold, A.; Schlesinger, S.; Boeing, H. Food Groups and Risk of Hypertension: A Systematic Review and Dose-Response Meta-Analysis of Prospective Studies. *Adv. Nutr.* **2017**, *8*, 793–803. [CrossRef] [PubMed]
15. Xu, R.; Yang, K.; Ding, J.; Chen, G. Effect of green tea supplementation on blood pressure: A systematic review and meta-analysis of randomized controlled trials. *Medicine* **2020**, *99*, e19047. [CrossRef] [PubMed]
16. Appel, L.J.; Moore, T.J.; Obarzanek, E.; Vollmer, W.M.; Svetkey, L.P.; Sacks, F.M.; Bray, G.A.; Vogt, T.M.; Cutler, J.A.; Windhauser, M.M.; et al. A clinical trial of the effects of dietary patterns on blood pressure. DASH Collaborative Research Group. *N. Engl. J. Med.* **1997**, *336*, 1117–1124. [CrossRef]
17. Ndanuko, R.N.; Tapsell, L.C.; Charlton, K.E.; Neale, E.P.; Batterham, M.J. Dietary Patterns and Blood Pressure in Adults: A Systematic Review and Meta-Analysis of Randomized Controlled Trials. *Adv. Nutr.* **2016**, *7*, 76–89. [CrossRef]
18. Li, M.; Shi, Z. Dietary Pattern during 1991–2011 and Its Association with Cardio Metabolic Risks in Chinese Adults: The China Health and Nutrition Survey. *Nutrients* **2017**, *9*, 1218. [CrossRef]
19. Xu, H.; Guo, B.; Qian, W.; Ciren, Z.; Guo, W.; Zeng, Q.; Mao, D.; Xiao, X.; Wu, J.; Wang, X.; et al. Dietary Pattern and Long-Term Effects of Particulate Matter on Blood Pressure: A Large Cross-Sectional Study in Chinese Adults. *Hypertension* **2021**, *78*, 184–194. [CrossRef]
20. Monteiro, C.A.; Cannon, G.; Moubarac, J.C.; Levy, R.B.; Louzada, M.L.C.; Jaime, P.C. The UN Decade of Nutrition, the NOVA food classification and the trouble with ultra-processing. *Public Health Nutr.* **2018**, *21*, 5–17. [CrossRef]
21. Monteiro, C.A.; Cannon, G.; Levy, R.B.; Moubarac, J.C.; Louzada, M.L.; Rauber, F.; Khandpur, N.; Cediel, G.; Neri, D.; Martinez-Steele, E.; et al. Ultra-processed foods: What they are and how to identify them. *Public Health Nutr.* **2019**, *22*, 936–941. [CrossRef]
22. Monteiro, C.A.; Lawrence, M.; Millett, C.; Nestle, M.; Popkin, B.M.; Scrinis, G.; Swinburn, B. The need to reshape global food processing: A call to the United Nations Food Systems Summit. *BMJ Glob. Health* **2021**, *6*, e006885. [CrossRef] [PubMed]
23. Monteiro, C.A.; Moubarac, J.C.; Cannon, G.; Ng, S.W.; Popkin, B. Ultra-processed products are becoming dominant in the global food system. *Obes. Rev.* **2013**, *14* (Suppl. S2), 21–28. [CrossRef]
24. Hovhannisyan, V.; Devadoss, S. Effects of urbanization on food demand in China. *Empirl. Econ.* **2020**, *58*, 699–721. [CrossRef]
25. Zhai, F.Y.; Du, S.F.; Wang, Z.H.; Zhang, J.G.; Du, W.W.; Popkin, B.M. Dynamics of the Chinese diet and the role of urbanicity, 1991-2011. *Obes Rev* **2014**, *15* (Suppl. S1), 16–26. [CrossRef]
26. Zhang, J.; Wang, Z.; Du, W.; Huang, F.; Jiang, H.; Bai, J.; Zhang, X.; Zhang, B.; Wang, H. Twenty-Five-Year Trends in Dietary Patterns among Chinese Adults from 1991 to 2015. *Nutrients* **2021**, *13*, 1327. [CrossRef] [PubMed]

27. Drewnowski, A.; Monsivais, P. Chapter 10-Taste, cost, convenience, and food choices. In *Present Knowledge in Nutrition*, 11th ed.; Marriott, B.P., Birt, D.F., Stallings, V.A., Yates, A.A., Eds.; Academic Press: Cambridge, MA, USA, 2020; pp. 185–200.
28. Askari, M.; Heshmati, J.; Shahinfar, H.; Tripathi, N.; Daneshzad, E. Ultra-processed food and the risk of overweight and obesity: A systematic review and meta-analysis of observational studies. *Int. J. Obes.* **2020**, *44*, 2080–2091. [CrossRef] [PubMed]
29. Lane, M.M.; Davis, J.A.; Beattie, S.; Gomez-Donoso, C.; Loughman, A.; O'Neil, A.; Jacka, F.; Berk, M.; Page, R.; Marx, W.; et al. Ultraprocessed food and chronic noncommunicable diseases: A systematic review and meta-analysis of 43 observational studies. *Obes. Rev.* **2021**, *22*, e13146. [CrossRef]
30. Pagliai, G.; Dinu, M.; Madarena, M.P.; Bonaccio, M.; Iacoviello, L.; Sofi, F. Consumption of ultra-processed foods and health status: A systematic review and meta-analysis. *Br. J. Nutr.* **2021**, *125*, 308–318. [CrossRef]
31. Mendonca, R.D.; Lopes, A.C.; Pimenta, A.M.; Gea, A.; Martinez-Gonzalez, M.A.; Bes-Rastrollo, M. Ultra-Processed Food Consumption and the Incidence of Hypertension in a Mediterranean Cohort: The Seguimiento Universidad de Navarra Project. *Am. J. Hypertens* **2017**, *30*, 358–366. [CrossRef]
32. Scaranni, P.; Cardoso, L.O.; Chor, D.; Melo, E.C.P.; Matos, S.M.A.; Giatti, L.; Barreto, S.M.; da Fonseca, M.J.M. Ultra-processed foods, changes in blood pressure and incidence of hypertension: The Brazilian Longitudinal Study of Adult Health (ELSA-Brasil). *Public Health Nutr.* **2021**, *24*, 3352–3360. [CrossRef]
33. Rezende-Alves, K.; Hermsdorff, H.H.M.; Miranda, A.; Lopes, A.C.S.; Bressan, J.; Pimenta, A.M. Food processing and risk of hypertension: Cohort of Universities of Minas Gerais, Brazil (CUME Project). *Public Health Nutr.* **2021**, *24*, 4071–4079. [CrossRef] [PubMed]
34. Li, M.; Shi, Z. Ultra-Processed Food Consumption Associated with Overweight/Obesity among Chinese Adults-Results from China Health and Nutrition Survey 1997–2011. *Nutrients* **2021**, *13*, 2796. [CrossRef] [PubMed]
35. Li, M.; Shi, Z. Association between Ultra-Processed Food Consumption and Diabetes in Chinese Adults-Results from the China Health and Nutrition Survey. *Nutrients* **2022**, *14*, 4241. [CrossRef] [PubMed]
36. Popkin, B.M.; Du, S.; Zhai, F.; Zhang, B. Cohort Profile: The China Health and Nutrition Survey-monitoring and understanding socio-economic and health change in China, 1989–2011. *Int. J. Epidemiol.* **2010**, *39*, 1435–1440. [CrossRef]
37. Institute for Nutrition and Food Safety of the Chinese Center for Disease Control and Prevention. *China Food Composition Table*; Peking University Medical Press: Beijing, China, 2002.
38. Institute for Nutrition and Food Safety of the Chinese Center for Disease Control and Prevention. *China Food Composition Table 2004*; Peking University Medical Press: Beijing, China, 2005.
39. Korn, E.L.; Graubard, B.I.; Midthune, D. Time-to-event analysis of longitudinal follow-up of a survey: Choice of the time-scale. *Am. J. Epidemiol.* **1997**, *145*, 72–80. [CrossRef]
40. Juul, F.; Vaidean, G.; Parekh, N. Ultra-processed Foods and Cardiovascular Diseases: Potential Mechanisms of Action. *Adv. Nutr.* **2021**, *12*, 1673–1680. [CrossRef]
41. Simmons, A.L.; Schlezinger, J.J.; Corkey, B.E. What Are We Putting in Our Food That Is Making Us Fat? Food Additives, Contaminants, and Other Putative Contributors to Obesity. *Curr. Obes. Rep.* **2014**, *3*, 273–285. [CrossRef]
42. Fardet, A.; Rock, E.; Bassama, J.; Bohuon, P.; Prabhasankar, P.; Monteiro, C.; Moubarac, J.C.; Achir, N. Current food classifications in epidemiological studies do not enable solid nutritional recommendations for preventing diet-related chronic diseases: The impact of food processing. *Adv. Nutr.* **2015**, *6*, 629–638. [CrossRef]
43. Fardet, A.; Rock, E. Chronic diseases are first associated with the degradation and artificialization of food matrices rather than with food composition: Calorie quality matters more than calorie quantity. *Eur. J. Nutr.* **2022**, *61*, 2239–2253. [CrossRef]
44. Matos, R.A.; Adams, M.; Sabate, J. Review: The Consumption of Ultra-Processed Foods and Non-communicable Diseases in Latin America. *Front. Nutr.* **2021**, *8*, 622714. [CrossRef]
45. Tian, X.; Zhong, L.; von Cramon-Taubadel, S.; Tu, H.; Wang, H. Restaurants in the Neighborhood, Eating Away from Home and BMI in China. *PLoS ONE* **2016**, *11*, e0167721. [CrossRef] [PubMed]
46. Andrade, G.C.; Gombi-Vaca, M.F.; Louzada, M.; Azeredo, C.M.; Levy, R.B. The consumption of ultra-processed foods according to eating out occasions. *Public Health Nutr.* **2020**, *23*, 1041–1048. [CrossRef] [PubMed]
47. Zhou, Z.; Diao, Q.; Shao, N.; Liang, Y.; Lin, L.; Lei, Y.; Zheng, L. The Frequency of Unhealthy Food Advertising on Mainland Chinese Television (TV) and Children and Adolescents' Risk of Exposure to Them. *PLoS ONE* **2015**, *10*, e0128746. [CrossRef] [PubMed]
48. Huang, Y.; Wang, Y.; Wang, H.; Liu, Z.; Yu, X.; Yan, J.; Yu, Y.; Kou, C.; Xu, X.; Lu, J.; et al. Prevalence of mental disorders in China: A cross-sectional epidemiological study. *Lancet Psychiatry* **2019**, *6*, 211–224. [CrossRef]
49. Li, M.; Dibley, M.J.; Sibbritt, D.W.; Yan, H. Dietary habits and overweight/obesity in adolescents in Xi'an City, China. *Asia Pac. J. Clin. Nutr.* **2010**, *19*, 76–82. [PubMed]
50. Gupta, S.; Harper, A.; Ruan, Y.; Barr, R.; Frazier, A.L.; Ferlay, J.; Steliarova-Foucher, E.; Fidler-Benaoudia, M.M. International Trends in the Incidence of Cancer Among Adolescents and Young Adults. *J. Natl. Cancer Inst.* **2020**, *112*, 1105–1117. [CrossRef] [PubMed]

51. Yang, Y.X.; Wang, X.L.; Leong, P.M.; Zhang, H.M.; Yang, X.G.; Kong, L.Z.; Zhai, F.Y.; Cheng, Y.Y.; Guo, J.S.; Su, Y.X. New Chinese dietary guidelines: Healthy eating patterns and food-based dietary recommendations. *Asia Pac. J. Clin. Nutr.* **2018**, *27*, 908–913. [CrossRef] [PubMed]
52. Wang, Y.; Li, M.; Shi, Z. Higher egg consumption associated with increased risk of diabetes in Chinese adults-China Health and Nutrition Survey. *Br. J. Nutr.* **2021**, *126*, 110–117. [CrossRef]

nutrients

MDPI

Article

Ultra-Processed Food Consumption and Depressive Symptoms in a Mediterranean Cohort

Justyna Godos [1], Marialaura Bonaccio [2], Wahidah H. Al-Qahtani [3], Wolfgang Marx [4], Melissa M. Lane [4], Gian Marco Leggio [1] and Giuseppe Grosso [1,5,*]

1 Department of Biomedical and Biotechnological Sciences, University of Catania, 95123 Catania, Italy
2 Department of Epidemiology and Prevention, IRCCS NEUROMED, 86077 Pozzilli, Italy
3 Department of Food Sciences & Nutrition, College of Food & Agriculture Sciences, King Saud University, Riyadh 11451, Saudi Arabia
4 Food & Mood Centre, The Institute for Mental and Physical Health and Clinical Translation (IMPACT), School of Medicine, Deakin University, Geelong, VIC 3220, Australia
5 Center for Human Nutrition and Mediterranean Foods (NUTREA), University of Catania, 95123 Catania, Italy
* Correspondence: giuseppe.grosso@unict.it

Abstract: Excess consumption of ultra-processed foods (UPFs) is currently under investigation for its potentially detrimental impact on human health. Current evidence demonstrates a substantial association with an increased risk of metabolic disorders, but data on mental health outcomes are just emerging. The aim of this study was to investigate the relationship between the consumption of UPFs and depressive symptoms in a sample of younger Italian adults. A cross-sectional study was conducted on 596 individuals (age 18–35 y) recruited in southern Italy. Food frequency questionnaires and the NOVA classification were used to assess dietary factors; the Center for the Epidemiological Studies of Depression Short Form (CES-D-10) was used to assess presence of depressive symptoms. Individuals in the highest quartile of UPF consumption had higher odds of having depressive symptoms in the energy-adjusted model (odds ratio (OR) = 1.89, 95% confidence interval (CI): 1.06, 3.28); the association remained significant after adjusting for potential confounding factors (OR = 2.04, 95% CI: 1.04, 4.01) and became even stronger after further adjustment for adherence to the Mediterranean diet as a proxy of diet quality (OR = 2.70, 95% CI: 1.32, 5.51). In conclusion, a positive association between UPF consumption and likelihood of having depressive symptoms was found in younger Italian individuals. Given the consistency of the findings after adjustment for diet quality, further studies are needed to understand whether non-nutritional factors may play a role in human neurobiology.

Keywords: ultra-processed foods; NOVA classification; food processing; nutritional psychiatry; depression; depressive symptoms

Citation: Godos, J.; Bonaccio, M.; Al-Qahtani, W.H.; Marx, W.; Lane, M.M.; Leggio, G.M.; Grosso, G. Ultra-Processed Food Consumption and Depressive Symptoms in a Mediterranean Cohort. *Nutrients* **2023**, *15*, 504. https://doi.org/10.3390/nu15030504

Academic Editors: Rosa Casas, Josep A. Tur and Maria Luz Fernandez

Received: 17 November 2022
Revised: 10 December 2022
Accepted: 11 January 2023
Published: 18 January 2023

1. Introduction

Dietary risk factors have been accounted to be responsible for about 10 million deaths due to cardiovascular diseases, metabolic disorders, and certain cancers in 2017 [1]. The issues related to overnutrition, especially in developed countries, depend on several factors that vary from personal choices to exposure to an obesogenic environment, from societal decisions to industry inputs [2]. Younger generations are at the highest risk, being registered as those under the strongest environmental pressure driven by stressful modern lifestyle, lack of time leading to scarce physical activity, poor sleep, and unhealthy behaviors, including low quality dietary habits [3]. In this regard, the globalization of food markets, the growing inputs from the food industry (in terms of quality of available food products), and a general hardship in financial situation are known to promote "Westernized-type diets", as opposed to traditional dietary patterns characterized by minimally-processed, locally produced, plant-based foods [4]. All together, these factors lead to various levels of malnutrition, including low-quality dietary patterns characterized by highly-palatable, convenient, energy-rich, and nutrient-poor foods [5].

Studies investigating the level of processing as a potential variable of interest to predict health outcomes are nowadays using the so-called NOVA classification to identify a category of foods classified as "ultra-processed". Based on the NOVA classification, ultra-processed foods (UPFs) are foods characterized by formulations containing few or no natural ingredients, supplemented with chemical additives and preservatives to prolong shelf life, but also supply intense palatable features and properties (i.e., flavor enhancers, colorants, emulsifiers, artificial sweeteners, thickeners, and foaming/anti-foaming agents) [6]. UPFs are widely consumed in modern societies, although with large differences across countries [7]. Studies show that UPF intake may range from 15–20% of daily energy intake in Mediterranean countries, to up to 80% in US, UK, Canadian, and Australian populations, with a substantially higher rate of consumption among younger individuals [8]. This variation may significantly affect their impact on the general population, as well as the projection of the future burden of disease related to the consumption of UPFs.

There is an ongoing debate about whether the negative health effects of UPFs stem from poor nutritional quality of food processing [9]. Contrary to the idea that UPFs might exert a negative impact on health due to the poor nutritional content, we recently demonstrated that higher risk of mortality associated with high UPF intake was independent from the nutritional quality of the diet [10]. The concerns regarding the consumption of UPFs rely on the observed association with various non-communicable diseases, such as obesity, cardiometabolic diseases, and lately also behavioral disorders [11,12]. Among the most under-studied research topics, diet has been hypothesized to be an independent risk factor for mental disorders [3]. There is growing evidence that various dietary factors may be associated with depressive symptoms, although the nature and direction of this relation are largely unknown [13]. Several hypotheses and mechanisms have been proposed [14,15], which suggest that inflammatory processes related to food intake may explain part of the relationship between dietary factors and brain health [16]. Among the various aspects of the diet that may exert such suggested effects, UPFs are currently under investigation for a potential negative impact toward mental health outcomes [17,18]. The aim of this study was to investigate whether an association between UPF consumption and depressive symptoms could be observed in a cohort of southern Italian younger adults.

2. Materials and Methods

2.1. Study Design and Population

The present study is a cross-sectional analysis of the baseline data from the Mediterranean healthy Eating, Aging and Lifestyle (MEAL) study, an observational study aiming to explore the relation between lifestyle behaviors and non-communicable diseases in a population recruited in the Mediterranean area [19]. Participants were randomly selected between 2014 and 2015 in the main districts of Catania, in southern Italy. The recruitment and data collection were performed through the registered records of local general practitioners stratified by sex and 10-year age groups. Out of 2405 individuals invited, the final sample included 2044 participants with a response rate of 85%. For the purposes of this study, data from individuals under 35 years old were included (n = 735). The goals of the project have been described to the participants prior to acceptance of participation by written informed consent. All the study procedures were conducted in accordance with the Declaration of Helsinki (1989) of the World Medical Association. The study protocol has been reviewed and approved by the concerning ethical committee.

2.2. Background Data

Face-to-face, computer assisted interviews were conducted by trained personnel to collect data on sex, age, educational (the highest educational degree achieved) and occupational (the most important employment during the year before the investigation or before retirement) statuses, smoking status, and physical activity level. Marital status was categorized as (i) unmarried/widowed or (ii) married. Educational status was categorized as (i) low (primary/secondary), (ii) medium (high school), and (iii) high (university). Occupational status was categorized as (i) unemployed, (ii) low (unskilled workers), (iii) medium (partially skilled workers), and (iv) high (skilled workers). The International Physical Activity Questionnaires (IPAQ) [20] were used to evaluate physical activity level and categorized as (i) low, (ii) moderate, and (iii) high. Smoking status was categorized as (i) non-smoker, (ii) ex-smoker, and (iii) current smoker. Eating habits included questions on skipping breakfast, snacking habits, and skipping dinner, with answers categorized as (i) always/often and (ii) seldom/never.

2.3. Dietary Information and UPF Calculation

Validated instruments were used to collect data on dietary consumption over the previous year [21,22]. The food frequency questionnaire (FFQ) included questions on average consumption of 110 foods and beverages with nine response options ranging from "never" to "4–5 times per day". For food items generally consumed over certain periods of the year, the questions referred to seasonal consumption and results were proportionally adjusted. The instrument demonstrated an acceptable relative validity and reliability when validated for the Italian population [21]. Nutrient (macro- and micro-) and non-nutrient (polyphenol) dietary content was estimated by calculating the 24 h intake of foods and beverages (in gr or ml, respectively) and estimating the correspondent daily intake of nutrients from the food composition tables of Council for Research in Agriculture and Analysis of Agricultural Economy (CREA) [23]. Data entries with lacking information or unreliable intakes (<1000 or >6000 kcal/d) were excluded from the analyses (n = 52) leaving a total of 683 individuals included in the analysis.

To provide indication of overall diet quality, adherence to the Mediterranean diet was used as a proxy. The literature-based score [24,25] takes into account the daily consumption of food groups that are considered as key features of the Mediterranean diet, providing positive points for increasing portions (up to 2 points) of fruit, vegetables, legumes, cereals, fish, and olive oil, negative points for increasing portions of meat and dairy foods, and positive points for moderate alcohol intake. The score is ultimately composed of the summary points from 0 to 18 points, with higher scores indicating higher adherence to the Mediterranean diet. For the purposes of this study, the sample was grouped in tertiles and categorized as (i) low, (ii) medium, and (iii) high adherence to the Mediterranean diet.

UPF consumption was calculated by applying the NOVA classification to the major food groups consumed in the study sample [26]. Briefly, 110 food items of the long FFQ were classified as follows: group 1, unprocessed or minimally processed foods (i.e., rice and other cereals, meat, fish, milk, eggs, fruit, vegetables, nuts, etc.), group 2, processed culinary ingredients (i.e., sugar, vegetable oils and butter); group 3, processed foods (i.e., processed breads and cheese); group 4, UPFs (i.e., confectioneries, salty snacks, fast-foods, soft drinks, etc.) [27]. For the purpose of this study, the mean share of the NOVA group 4 (UPFs) to the total daily energy intake was estimated, and participants were categorized into quartiles of energy shares of UPFs as the variable of exposure.

2.4. Depressive Symptoms

Screening for depressive symptoms was performed using the 10-item Center for the Epidemiological Studies of Depression Short Form (CES-D-10) [28]. Briefly, the CES-D-10 is a self-administered tool that includes 10 questions commonly used to test for presence of depressive symptoms in the general population. The frequency of mood/symptoms during the previous week is rated by a 4-point Likert scale ranging from 0 indicating rarely or none of the time (less than 1 day) to 3 indicating most or all of the time (5–7 days). The total score is calculated by summing up the scores of the individual questions and ranges from 0 to 30, with higher scores indicating greater severity of symptoms; conventionally, a score >15 indicates the presence of depressive symptoms. After excluding individuals with incomplete or unreliable questionnaires ($n = 87$), a total sample of 596 was included in the final analysis.

2.5. Statistical Analysis

Categorical variables are presented as frequencies of occurrence and percentages, with a Chi-squared test used to assess differences between quartiles of UPF consumption. Continuous variables are expressed as mean and standard deviations (SDs), with ANOVA test used to test differences between groups. The association between UPF consumption and presence of depressive symptoms was tested by logistic regression analyses through calculation of odds ratios (ORs) and 95% confidence intervals (Cis) for an energy-adjusted model, a multivariate model adjusted for baseline characteristics (age, sex, educational and occupational status, smoking, and physical activity level, marital status, and snacking habits), and a third model with additional adjustment for adherence to the Mediterranean diet (as a proxy for diet quality). All reported p values were based on two-sided tests and compared to a significance level of 5%. SPSS 21 (SPSS Inc., Chicago, IL, USA) software was used for all the statistical calculations.

3. Results

The distribution of baseline characteristics by quartiles of UPF consumption in the study sample is presented in Table 1. There was a significantly different distribution of UPF consumption by marital status ($p = 0.006$) and physical activity level ($p = 0.004$), although with no clear trend across categories, but a tendency of higher consumption in unmarried and medium/highly physically active individuals. In contrast, individuals consuming more UPFs reported significantly lower adherence to the Mediterranean diet, with opposite trends in those reporting lower consumption ($p < 0.001$).

Table 1. Baseline characteristics of the study sample according to quartiles of intake of UPFs ($n = 596$).

	UPF Consumption				*p*-Value
	Q1	Q2	Q3	Q4	
Age, mean (SD)	28.6 (5.7)	29.8 (6.0)	29.6 (5.5)	28.7 (5.9)	0.189
Sex, *n* (%)					0.699
Men	33 (40.2)	46 (35.7)	76 (42.0)	84 (41.2)	
Women	49 (59.8)	83 (64.3)	105 (58.0)	120 (58.8)	
Marital status, *n* (%)					0.006
Unmarried/widowed	51 (62.2)	68 (52.7)	114 (63.0)	146 (71.6)	
Married	31 (37.8)	61 (47.3)	67 (37.0)	58 (28.4)	
Smoking status, *n* (%)					0.587
Never	49 (59.8)	90 (69.8)	120 (66.3)	137 (67.2)	
Current	29 (35.4)	31 (24.0)	53 (29.3)	53 (26.0)	
Former	4 (4.9)	8 (6.2)	8 (4.4)	14 (6.9)	
Educational level, *n* (%)					0.059
Low	17 (20.7)	19 (14.7)	13 (7.2)	32 (15.7)	
Medium	35 (42.7)	49 (38.0)	79 (43.6)	82 (40.2)	
High	30 (36.6)	61 (47.3)	89 (49.2)	90 (44.1)	
Occupational level, *n* (%)					0.153
Unemployed	21 (28.8)	32 (27.8)	37 (25.7)	66 (39.8)	
Low	11 (15.1)	13 (11.3)	15 (10.4)	16 (9.6)	
Medium	16 (21.9)	19 (16.5)	35 (24.3)	25 (15.1)	
High	25 (34.2)	51 (44.3)	57 (39.6)	59 (35.5)	
Physical activity level, *n* (%)					0.004
Low	19 (23.2)	21 (16.3)	19 (10.5)	27 (13.2)	
Medium	28 (34.1)	72 (55.8)	85 (47.0)	110 (53.9)	
High	35 (42.7)	36 (27.9)	77 (42.5)	67 (32.8)	
Eating habits, *n* (%)					
Skipping breakfast	17 (20.7)	31 (24.0)	52 (28.7)	53 (26.0)	0.546
Daily snacking	12 (14.6)	17 (13.2)	23 (12.7)	26 (12.7)	0.975
Skipping dinner	9 (11.0)	12 (9.3)	22 (12.2)	34 (16.7)	0.220
Adherence to Mediterranean diet, *n* (%)					<0.001
Low	30 (36.3)	70 (54.3)	103 (56.9)	147 (72.1)	
Medium	46 (56.1)	46 (35.7)	44 (24.3)	50 (24.5)	
High	6 (7.3)	13 (10.1)	34 (18.8)	7 (3.4)	

When testing for differences in major food groups, micro- and macro-nutrients across quartiles of UPF consumption, most of the macronutrients, sodium, total and processed means were consumed significantly more by those in the highest quartile of UPF intake. In contrast, fiber and certain food groups such as cereals, fruits, vegetables, legumes, dairy products, and olive oil were consumed less among individuals in the highest quartile of UPF intake (Table 2).

Table 2. Mean (and standard deviation) consumption of micro-, macronutrients and major food groups intake according to quartiles of UFP consumption.

	UPF Consumption				*p*-Value
	Q1	Q2	Q3	Q4	
		mean (SD)			
Energy intake (kcal/d)	2045.88 (873.58)	2067.57 (738.74)	2008.67 (821.52)	2283.99 (1105.62) *	0.019
Energy intake (kJ/d)	8230.25 (3651.25)	8378.72 (3084.12)	8186.26 (3446.88)	9275.7 (4593.52) *	0.023
		Macronutrients			
Carbohydrates (g/d)	304.47 (141.20)	315.29 (124.31)	294.34 (128.04)	313.02 (152.21)	0.493
Fiber (g/d)	39.42 (31.57)	33.75 (15.72)	31.34 (15.98)	31.05 (19.5) *	0.008
Protein (g/d)	92.86 (53.88)	89.59 (34.09)	86.08 (40.07)	94.05 (47.70)	0.324
Fat (g/d)	54.51 (24.17)	55.69 (20.12)	59.07 (23.87)	77.84 (41.13) **	<0.001
Cholesterol (mg/d)	173.69 (129.47)	178.7 (80.11)	190.19 (112.56)	244.64 (131.42) **	<0.001
SFA	19.93 (8.74)	21.35 (8.02)	22.49 (8.60)	31.50 (16.31) **	<0.001
MUFA	23.74 (9.41)	24.07 (8.32)	25.23 (10.66)	31.15 (15.85) **	<0.001
PUFA	11.17 (5.89)	10.68 (4.41)	11.36 (7.11)	12.91 (8.23) *	0.019
Total Omega-3	1.73 (0.96)	1.66 (0,78)	1.57 (0.73)	1.61 (0.77)	0.509
		Micronutrients			
Vitamin A (Retinol)	950.29 (571.75)	942.28 (449.58)	868.43 (425.27)	898.99 (491.21)	0.452
Vitamin C (mg/d)	195.19 (185.64)	178.35 (108.56)	164.96 (101.98)	155.92 (106.05)	0.062
Vitamin E (mg/d)	9.30 (4.67)	8.82 (3.53)	8.95 (3.85)	9.87 (5.74)	0.141
Vitamin B12	8.34 (18.02)	6.78 (4.56)	7.01 (6.56)	7.99 (6.79)	0.421
Vitamin D	5.29 (5.52)	5.60 (4.98)	5.94 (7.67)	6.19 (8.41)	0.769
Sodium (mg/d)	2890.45 (1136.79)	3151.52 (1158.34)	2953.90 (985.94)	3302.72 (1599.08) *	0.020
Potassium (mg/d)	4299.95 (3013.97)	3892.82 (1545.45)	3731.17 (1894.24)	3948.06 (2143.45)	0.243
		Foods			
Cereals (g/d)	236.26 (144.97)	256.83 (139.76)	208.58 (112.36)	167.70 (121.70) **	<0.001
Vegetables (g/d)	334.55 (304.85)	283.70 (143.48)	257.64 (143,.60)	256.29 (170.06) *	0.006
Fruit (g/d)	506.21 (458.81)	409.42 (343.75)	406.27 (364.51)	371.04 (320.63) *	0.042
Legumes (g/d)	334.55 (304.85)	283.70 (143.48)	257.64 (143.60)	256.29 (170.06) *	0.009
Nuts (total, g/d)	17.05 (20.58)	16.64 (20.74)	18.35 (22.60)	25.44 (72.74)	0.253
Fish (g/d)	77.90 (120.38)	66.18 (66.88)	78.99 (116.29)	67.80 (88.37)	0.569
Eggs (g/d)	1.89 (3.49)	2.53 (5.12)	2.80 (5.40)	2.13 (4.33)	0.396
Meat (total, g/d)	76.71 (47.87)	74.43 (47.43)	66.97 (37.27)	77.03 (34.53) *	0.077
Red meat (g/d)	34.23 (22.52)	35.28 (25.67)	35.03 (31.88)	35.81 (22.68)	0.973
Processed Meat (g/d)	13.05 (12.45)	17.71 (15.67)	18.03 (14.09)	31.45 (31.46) **	<0.001
Dairy products (g/d)	216.86 (187,86)	238.17 (166,41)	176.12 (155,34)	155,33 (204.18) *	0.019
Alcohol (total, g/d)	8.15 (13.42)	5.06 (7.25)	5.43 (7.54)	6.30 (9.35)	0.078
Olive oil (ml/d)	7.10 (3.10)	7.15 (3.04)	6.44 (2.91)	6.22 (3.08) *	0.018

* indicates $p < 0.05$ for ANOVA analysis, ** indicates $p < 0.001$ for ANOVA analysis.

Table 3 shows the association between UPF consumption and presence of depressive symptoms. Individuals in the highest quartile of UPF consumption had higher odds of having depressive symptoms in the energy-adjusted model (OR = 1.89, 95% CI: 1.06, 3.28); the association remained significant after adjusting for potential confounding factors (including age, sex, energy intake, educational and occupational lever, smoking status, eating habits, and physical activity level) (OR = 2.04, 95% CI: 1.04, 4.01) and became even stronger after further adjustment for adherence to the Mediterranean diet (OR = 2.70, 95% CI: 1.32, 5.51) (Table 3).

Table 3. Association between intake of UPFs and having depressive symptoms in the study sample.

	UPF Consumption			
	Q1	Q2	Q3	Q4
	OR (95% CI)			
Model 1	1	1.05 (0.56, 1.96)	1.17 (0.65, 2.09)	1.87 (1.06, 3.29)
Model 2	1	1.26 (0.62, 2.57)	0.93 (0.46, 1.89)	2.04 (1.04, 4.01)
Model 3	1	1.44 (0.70, 2.97)	1.17 (0.56, 2.43)	2.70 (1.33, 5.51)

Model 1 was adjusted for energy intake. Model 2 was further adjusted for age (mean), sex, marital status, educational level, occupational level, physical activity level, smoking status, eating habits. Model 3 was further adjusted for level of adherence to the Mediterranean diet.

4. Discussion

This study provides cross-sectional evidence of an association between higher UPF consumption and an increase in depressive symptoms. Furthermore, in contrast to the hypothesis that UPF may affect mental health due to the poor nutritional quality of the diet, further adjustment for adherence to the Mediterranean diet (as a proxy of diet quality) increased, rather than reduced, the association between UPF consumption and presence of depressive symptoms, suggesting that components of the diet other than nutritional quality may play a role on the reported association.

Two recent meta-analyses, including mostly cross-sectional investigations, reported that higher consumption of UPFs is associated with increased depressive symptoms [17,18]. With specific reference to the association between UPF consumption and depression, one of the first studies published on this topic has been conducted on about 26,000 French participants within the NutriNet-Santé cohort reporting an average 32% daily energy intake from UPFs; the authors found that a 10% increase in %UPF in the diet was associated with a 21% higher risk of depressive symptoms over a 5-year follow-up period [29]. Another study involving nearly 15,000 Spanish university graduates (mean age 36.7 years) participating in the "Seguimiento Universidad de Navarra" (SUN) Project, reported a 33% higher risk of depression in high UPF consumers (about 400 g/d) after a follow-up of 10 years [30]. In addition, studies with higher mean intake of UPFs reported similar findings. The National Health and Nutrition Examination Survey (NHANES) including nearly 14,000 US adults (with an average 55% of total energy intake from UPFs) showed that individuals in the highest quartile of UPF consumption were more likely (43% higher odds) to have depressive symptoms, compared to the lowest category of consumption [31]. An updated report from the same sample revealed that individuals with the highest level of UPF consumption were significantly more likely to report at least mild depression, more mentally unhealthy and more anxious days per month [32]. Although not specifically quantifying the intake of UPFs, a study conducted on about 3500 participants showed that individuals consuming a "processed food" dietary pattern characterized by high intake of sweetened desserts, fried food, processed meat, refined grains, and high-fat dairy products were more likely to have depressive symptoms compared to those with less consumption [33]. Similar associations have been reported for broader mental health conditions in other cohorts. In a sample of nearly 3000 Brazilian adolescents, higher consumption of UPFs has been associated with higher rates of internalizing symptoms including depression and anxiety [34]. In addition, data from the Adolescent School-Based Health Survey including nearly 100,000 adolescents showed that daily UPF consumption and sedentary behaviors were associated with higher odds for anxiety-induced sleep disturbance [35], which was mediated by loneliness and eating while watching TV or studying [36]. Finally, a cross-sectional study conducted on 1270 Brazilian retail workers showed that UPF consumption was associated with high perceived stress levels [37]. Most studies reported some sort of pattern of background characteristics associated with high UPF consumption: younger age, being unmarried/living alone, frequent out-of-home eating, often high cultural level. In line with our findings, this data suggest that younger individuals might be a more susceptible group of the population at higher risk of mood disorders due to a number of potential factors (work-related stress, lack of time, financial instability, etc.). This suggests that the rise in UPF consumption may be driven not only due to their highly palatable nature, but also due to economic and practical convenience.

From a mechanistic point of view, several hypotheses have been suggested and supported by scientific literature to explain the detrimental effects of UPF consumption on mental health outcomes. UPF consumption, as well as various dietary factors, may affect systemic inflammation with a consequently higher risk of non-communicable diseases, including mental disorders [38,39]. High UPF consumption has been demonstrated to be characterized by a rise in intake of refined sugars (such as high-fructose corn syrup) and saturated/trans fatty acids, accompanied with lower intake of fiber [7]. The high energy density of UPFs may lead to an imbalance of regulation and homeostatic maintenance of cells, causing an impairment of their microenvironment and finally compromising their functionality and integrity [40]. High intracellular glucose derived from high-free sugar food products increases intermediate metabolites of oxidative metabolism, mitochondrial dysfunction, and subsequently increases reactive oxygen species (ROS) production [41]. Similarly (albeit with totally different mechanisms), a high consumption of saturated and trans fatty acids induces a suffering of the endoplasmic reticulum at an intracellular level, modification of cellular membranes, and activation of transcription factors related to oxidative stress and proinflammatory pathways, including nuclear factor kB, related to the production of proinflammatory cytokines and the mTOR, JNK, and AKT pathways [42]. Finally, high consumption of UPFs has been reported to be often associated with lower intake of fiber, which may represent an additional mechanism related to disruption of homeostasis, immune regulation, and establishment of mental health issues [43]. Specifically, high UPF consumption as well as a lack of dietary fiber may induce an imbalance of the gut microbiota and lead to dysbiosis [44]; this condition is characterized by changes in their functional composition and metabolic activities, including a reduction in short chain fatty acids (SCFAs) and a rise in lipopolysaccharides producing bacteria, intestinal barrier dysfunction, and bacterial translocation into the bloodstream, tissues, and organs causing systemic immune system activation and inflammation [45]. Moreover, gut microbiota may communicate with the central nervous system through interaction with enteroendocrine and enterochromaffins cells, which are able to transmit signals via vagal or afferent nerve fibers and induce responses into the brain (i.e., serotonin release) [46]. Besides this indirect mechanism of central nervous system involvement, gut microbiota modifications also impact gut peptides and hormones (i.e., neuropeptide Y, glucagone-like peptide-1, cholecystokinin, ghrelin, corticotropin-releasing factor) which are all involved, to a various extent and through different mechanisms, in the complex gut–brain axis communication [47]. Long-term exposure to highly-palatable UPFs, and production of pro-inflammatory cytokines and secondary products of oxidative stress at brain level may also play a role in the alteration of the physiological feeding patterns, leading to food-anticipatory and binge-type behaviors, potential failure in self-control [48,49], which in turn are associated with anxiety/depressive symptoms [50] and alteration of sleep quality [51].

Concerns also arise regarding food additives that have been shown to exert neurotoxicity and clinical manifestation of depression, cognitive decline, and eating disorders [52]. Common food additives are generally used in UPFs for a variety of purposes, including in the alteration of organoleptic properties such as non-caloric sweeteners, flavor and color enhancers, emulsifiers, foaming/anti-foaming and anti-caking agents [53]; these compounds have been shown to affect human physiology in various ways, including oral processing, alteration of the gut microbiota homeostasis, uncoupling between predicted calories and consequent response from the digestive system, and further development of oxidative stress and the pro-inflammatory actions as the main mechanisms of toxicity [54]. The promotion of inflammatory processes associated with the consumption of UPFs may potentially affect the functioning of common neuronal signaling systems (i.e., serotonergic and dopaminergic systems) and of certain brain regions (i.e., amygdala) implicated in mental health disorders [55]. Some additives may contain nanoparticles that exert higher toxicity when compared to the bulk material because they are absorbed, cross various biological barriers, and may accumulate in tissues and organs [56]. These compounds may have direct interactions at the cellular level, exerting local toxicity by increasing the reactive oxygen and nitrogen species production, inducing mitochondrial and DNA oxidation, and activating pro-inflammatory cellular pathways [57]. Finally, the transformation processes of UPFs may lead to the production of substances, such as acrylamide, acrolein, polycyclic aromatic hydrocarbons, and furan that are known to be toxic to the human organism [58]. These compounds have been shown to potentially exert neurotoxicity via microbiota–gut–brain axis signaling and inflammasome-related neuroinflammation [59–61].

The findings of this study should be considered in light of some methodological limitations. First, due to the observational nature of the study design, the results may be affected by reverse causation, and the study design does not allow us to define cause–effect relationships, only association. However, the findings from this study do not necessarily imply that UPFs must necessarily cause depression, but that a mutual relation may exist, that UPFs might be consumed as comfort foods by an at-risk population (i.e., younger individuals with emerging mood disorders), and that it can establish a vicious cycle by further enhancing detrimental effects on brain health related to depression. Second, although the most common potential confounding factors have been taken into account when adjusting for multivariate analyses, the existence of residual unmeasured confounders cannot be ruled out. Third, the assessment of dietary intakes (FFQ) is limited in its nature by recall bias, portion size uncertainty, and social desirability. Finally, the general low consumption of UPF, especially among older individuals, did not allow us to provide data for all age groups, and results for older participants were null possible due to lack of statistical power (data not shown); although it is important to distinguish the findings by age groups due to generational differences in exposure and risk factors, we missed the opportunity to generalize the results also among older groups of individuals.

5. Conclusions

In conclusion, a positive association between UPF consumption and likelihood of having depressive symptoms was found in younger southern Italian adults. Further studies are needed to corroborate this association, also among other populations. It is crucial to understand whether non-nutritional factors may also play a role in human neurobiology. The specific involvement of brain regions involved in behavioral disorders needs to be further investigated to better understand the impact of food additives on human mental health.

Author Contributions: Conceptualization, J.G. and G.G.; methodology, G.G.; formal analysis, G.G.; writing—original draft preparation, J.G.; writing—review and editing, M.B., W.H.A.-Q., W.M., M.M.L., G.M.L. and G.G.; supervision, G.G.; funding acquisition, W.H.A.-Q. All authors have read and agreed to the published version of the manuscript.

Funding: This work was supported by the Distinguished Scientist Fellowship Program (DSFP) at King Saud University, Riyadh, Saudi Arabia. J.G. was supported by the co-financing of the European Union—FSE-REACT-EU, PON Research and Innovation 2014–2020 DM1062/2021; CUP: E65F21002560001. Wolfgang Marx is currently funded by an NHMRC Investigator Grant (#2008971) and a Multiple Sclerosis Research Australia early-career fellowship. Wolfgang has previously received funding from the Cancer Council Queensland and university grants/fellowships from La Trobe University, Deakin University, University of Queensland, and Bond University. Wolfgang has received industry funding and/or has attended events funded by Cobram Estate Pty. Ltd. and Bega Dairy and Drinks Pty Ltd. Wolfgang has received travel funding from the Nutrition Society of Australia. Wolfgang has received consultancy funding from Nutrition Research Australia and ParachuteBH. Wolfgang has received speaker honoraria from The Cancer Council Queensland and the Princess Alexandra Research Foundation.

Institutional Review Board Statement: The study was conducted in accordance with the Declaration of Helsinki, and approved by the Institutional Review Board (or Ethics Committee) of CE Catania 2 (protocol code 802/23 December 2014).

Informed Consent Statement: Informed consent was obtained from all subjects involved in the study.

Data Availability Statement: The data that support the findings of this study are available upon reasonable request.

Conflicts of Interest: The authors declare no conflict of interest.

References

1. GBD 2019 Risk Factors Collaborators. Global burden of 87 risk factors in 204 countries and territories, 1990–2019: A systematic analysis for the Global Burden of Disease Study 2019. *Lancet* **2020**, *396*, 1223–1249. [CrossRef]
2. Mozaffarian, D. Dietary and policy priorities for cardiovascular disease, diabetes, and obesity: A comprehensive review. *Circulation* **2016**, *133*, 187–225. [CrossRef]
3. Godos, J.; Currenti, W.; Angelino, D.; Mena, P.; Castellano, S.; Caraci, F.; Galvano, F.; Del Rio, D.; Ferri, R.; Grosso, G. Diet and mental health: Review of the recent updates on molecular mechanisms. *Antioxidants* **2020**, *9*, 346. [CrossRef]
4. GBD 2017 Diet Collaborators. Health effects of dietary risks in 195 countries, 1990–2017: A systematic analysis for the Global Burden of Disease Study 2017. *Lancet* **2019**, *393*, 1958–1972. [CrossRef]
5. Imamura, F.; Micha, R.; Khatibzadeh, S.; Fahimi, S.; Shi, P.; Powles, J.; Mozaffarian, D. Global Burden of Diseases Nutrition and Chronic Diseases Expert Group (NutriCoDE) Dietary quality among men and women in 187 countries in 1990 and 2010: A systematic assessment. *Lancet Glob. Health* **2015**, *3*, e132–e142. [CrossRef] [PubMed]
6. Monteiro, C.A.; Cannon, G.; Moubarac, J.-C.; Levy, R.B.; Louzada, M.L.C.; Jaime, P.C. The UN Decade of Nutrition, the NOVA food classification and the trouble with ultra-processing. *Public Health Nutr.* **2018**, *21*, 5–17. [CrossRef] [PubMed]
7. Martini, D.; Godos, J.; Bonaccio, M.; Vitaglione, P.; Grosso, G. Ultra-Processed Foods and Nutritional Dietary Profile: A Meta-Analysis of Nationally Representative Samples. *Nutrients* **2021**, *13*, 3390. [CrossRef]
8. Marino, M.; Puppo, F.; Del Bo', C.; Vinelli, V.; Riso, P.; Porrini, M.; Martini, D. A Systematic Review of Worldwide Consumption of Ultra-Processed Foods: Findings and Criticisms. *Nutrients* **2021**, *13*, 2778. [CrossRef]
9. Poti, J.M.; Braga, B.; Qin, B. Ultra-processed Food Intake and Obesity: What Really Matters for Health-Processing or Nutrient Content? *Curr. Obes. Rep.* **2017**, *6*, 420–431. [CrossRef] [PubMed]
10. Bonaccio, M.; Di Castelnuovo, A.; Ruggiero, E.; Costanzo, S.; Grosso, G.; De Curtis, A.; Cerletti, C.; Donati, M.B.; de Gaetano, G.; Iacoviello, L. Joint association of food nutritional profile by Nutri-Score front-of-pack label and ultra-processed food intake with mortality: Moli-sani prospective cohort study. *BMJ* **2022**, e070688. [CrossRef]
11. Elizabeth, L.; Machado, P.; Zinöcker, M.; Baker, P.; Lawrence, M. Ultra-Processed Foods and Health Outcomes: A Narrative Review. *Nutrients* **2020**, *12*, 1955. [CrossRef]
12. Lane, M.M.; Davis, J.A.; Beattie, S.; Gómez-Donoso, C.; Loughman, A.; O'Neil, A.; Jacka, F.; Berk, M.; Page, R.; Marx, W.; et al. Ultraprocessed food and chronic noncommunicable diseases: A systematic review and meta-analysis of 43 observational studies. *Obes. Rev.* **2021**, *22*, e13146. [CrossRef] [PubMed]
13. Marx, W.; Moseley, G.; Berk, M.; Jacka, F. Nutritional psychiatry: The present state of the evidence. *Proc. Nutr. Soc.* **2017**, *76*, 427–436. [CrossRef] [PubMed]
14. Adan, R.A.H.; van der Beek, E.M.; Buitelaar, J.K.; Cryan, J.F.; Hebebrand, J.; Higgs, S.; Schellekens, H.; Dickson, S.L. Nutritional psychiatry: Towards improving mental health by what you eat. *Eur. Neuropsychopharmacol.* **2019**, *29*, 1321–1332. [CrossRef] [PubMed]
15. Marx, W.; Lane, M.; Hockey, M.; Aslam, H.; Berk, M.; Walder, K.; Borsini, A.; Firth, J.; Pariante, C.M.; Berding, K.; et al. Diet and depression: Exploring the biological mechanisms of action. *Mol. Psychiatry* **2021**, *26*, 134–150. [CrossRef]
16. González Olmo, B.M.; Butler, M.J.; Barrientos, R.M. Evolution of the human diet and its impact on gut microbiota, immune responses, and brain health. *Nutrients* **2021**, *13*, 196. [CrossRef] [PubMed]
17. Lane, M.M.; Gamage, E.; Travica, N.; Dissanayaka, T.; Ashtree, D.N.; Gauci, S.; Lotfaliany, M.; O'Neil, A.; Jacka, F.N.; Marx, W. Ultra-Processed Food Consumption and Mental Health: A Systematic Review and Meta-Analysis of Observational Studies. *Nutrients* **2022**, *14*, 2568. [CrossRef]
18. Mazloomi, S.N.; Talebi, S.; Mehrabani, S.; Bagheri, R.; Ghavami, A.; Zarpoosh, M.; Mohammadi, H.; Wong, A.; Nordvall, M.; Kermani, M.A.H.; et al. The association of ultra-processed food consumption with adult mental health disorders: A systematic review and dose-response meta-analysis of 260,385 participants. *Nutr. Neurosci.* **2022**, 1–19. [CrossRef]
19. Grosso, G.; Marventano, S.; D'Urso, M.; Mistretta, A.; Galvano, F. The Mediterranean healthy eating, ageing, and lifestyle (MEAL) study: Rationale and study design. *Int. J. Food Sci. Nutr.* **2017**, *68*, 577–586. [CrossRef]
20. Craig, C.L.; Marshall, A.L.; Sjöström, M.; Bauman, A.E.; Booth, M.L.; Ainsworth, B.E.; Pratt, M.; Ekelund, U.; Yngve, A.; Sallis, J.F.; et al. International physical activity questionnaire: 12-country reliability and validity. *Med. Sci. Sports Exerc.* **2003**, *35*, 1381–1395. [CrossRef]
21. Marventano, S.; Mistretta, A.; Platania, A.; Galvano, F.; Grosso, G. Reliability and relative validity of a food frequency questionnaire for Italian adults living in Sicily, Southern Italy. *Int. J. Food Sci. Nutr.* **2016**, *67*, 857–864. [CrossRef]
22. Buscemi, S.; Rosafio, G.; Vasto, S.; Massenti, F.M.; Grosso, G.; Galvano, F.; Rini, N.; Barile, A.M.; Maniaci, V.; Cosentino, L.; et al. Validation of a food frequency questionnaire for use in Italian adults living in Sicily. *Int. J. Food Sci. Nutr.* **2015**, *66*, 426–438. [CrossRef] [PubMed]

23. Tabelle Composizione Degli Alimenti CREA. Available online: https://www.crea.gov.it/-/tabella-di-composizione-degli-alimenti (accessed on 12 May 2021).
24. Marventano, S.; Godos, J.; Platania, A.; Galvano, F.; Mistretta, A.; Grosso, G. Mediterranean diet adherence in the Mediterranean healthy eating, aging and lifestyle (MEAL) study cohort. *Int. J. Food Sci. Nutr.* **2018**, *69*, 100–107. [CrossRef] [PubMed]
25. Sofi, F.; Macchi, C.; Abbate, R.; Gensini, G.F.; Casini, A. Mediterranean diet and health status: An updated meta-analysis and a proposal for a literature-based adherence score. *Public Health Nutr.* **2014**, *17*, 2769–2782. [CrossRef] [PubMed]
26. Godos, J.; Giampieri, F.; Al-Qahtani, W.H.; Scazzina, F.; Bonaccio, M.; Grosso, G. Ultra-Processed Food Consumption and Relation with Diet Quality and Mediterranean Diet in Southern Italy. *Int. J. Environ. Res. Public Health* **2022**, *19*, 11360. [CrossRef] [PubMed]
27. Moubarac, J.-C.; Parra, D.C.; Cannon, G.; Monteiro, C.A. Food classification systems based on food processing: Significance and implications for policies and actions: A systematic literature review and assessment. *Curr. Obes. Rep.* **2014**, *3*, 256–272. [CrossRef]
28. Radloff, L.S. The use of the Center for Epidemiologic Studies Depression Scale in adolescents and young adults. *J. Youth Adolesc.* **1991**, *20*, 149–166. [CrossRef]
29. Adjibade, M.; Julia, C.; Allès, B.; Touvier, M.; Lemogne, C.; Srour, B.; Hercberg, S.; Galan, P.; Assmann, K.E.; Kesse-Guyot, E. Prospective association between ultra-processed food consumption and incident depressive symptoms in the French NutriNet-Santé cohort. *BMC Med.* **2019**, *17*, 78. [CrossRef]
30. Gómez-Donoso, C.; Sánchez-Villegas, A.; Martínez-González, M.A.; Gea, A.; Mendonça, R.d.D.; Lahortiga-Ramos, F.; Bes-Rastrollo, M. Ultra-processed food consumption and the incidence of depression in a Mediterranean cohort: The SUN Project. *Eur. J. Nutr.* **2020**, *59*, 1093–1103. [CrossRef]
31. Zheng, L.; Sun, J.; Yu, X.; Zhang, D. Ultra-Processed Food Is Positively Associated with Depressive Symptoms Among United States Adults. *Front. Nutr.* **2020**, *7*, 600449. [CrossRef]
32. Hecht, E.M.; Rabil, A.; Martinez Steele, E.; Abrams, G.A.; Ware, D.; Landy, D.C.; Hennekens, C.H. Cross-sectional examination of ultra-processed food consumption and adverse mental health symptoms. *Public Health Nutr.* **2022**, *25*, 3225–3234. [CrossRef] [PubMed]
33. Akbaraly, T.N.; Brunner, E.J.; Ferrie, J.E.; Marmot, M.G.; Kivimaki, M.; Singh-Manoux, A. Dietary pattern and depressive symptoms in middle age. *Br. J. Psychiatry* **2009**, *195*, 408–413. [CrossRef] [PubMed]
34. Faisal-Cury, A.; Leite, M.A.; Escuder, M.M.L.; Levy, R.B.; Peres, M.F.T. The relationship between ultra-processed food consumption and internalising symptoms among adolescents from São Paulo city, Southeast Brazil. *Public Health Nutr.* **2022**, *25*, 2498–2506. [CrossRef]
35. Werneck, A.O.; Vancampfort, D.; Oyeyemi, A.L.; Stubbs, B.; Silva, D.R. Joint association of ultra-processed food and sedentary behavior with anxiety-induced sleep disturbance among Brazilian adolescents. *J. Affect. Disord.* **2020**, *266*, 135–142. [CrossRef] [PubMed]
36. Werneck, A.O.; Hoare, E.; Silva, D.R. Do TV viewing and frequency of ultra-processed food consumption share mediators in relation to adolescent anxiety-induced sleep disturbance? *Public Health Nutr.* **2021**, *24*, 5491–5497. [CrossRef] [PubMed]
37. Lopes Cortes, M.; Andrade Louzado, J.; Galvão Oliveira, M.; Moraes Bezerra, V.; Mistro, S.; Souto Medeiros, D.; Arruda Soares, D.; Oliveira Silva, K.; Nicolaevna Kochergin, C.; Honorato Dos Santos de Carvalho, V.C.; et al. Unhealthy Food and Psychological Stress: The Association between Ultra-Processed Food Consumption and Perceived Stress in Working-Class Young Adults. *Int. J. Environ. Res. Public Health* **2021**, *18*, 3863. [CrossRef]
38. Grosso, G.; Laudisio, D.; Frias-Toral, E.; Barrea, L.; Muscogiuri, G.; Savastano, S.; Colao, A. Anti-Inflammatory Nutrients and Obesity-Associated Metabolic-Inflammation: State of the Art and Future Direction. *Nutrients* **2022**, *14*, 1137. [CrossRef]
39. Lane, M.M.; Lotfaliany, M.; Forbes, M.; Loughman, A.; Rocks, T.; O'Neil, A.; Machado, P.; Jacka, F.N.; Hodge, A.; Marx, W. Higher Ultra-Processed Food Consumption Is Associated with Greater High-Sensitivity C-Reactive Protein Concentration in Adults: Cross-Sectional Results from the Melbourne Collaborative Cohort Study. *Nutrients* **2022**, *14*, 3309. [CrossRef]
40. Gatineau, E.; Polakof, S.; Dardevet, D.; Mosoni, L. Similarities and interactions between the ageing process and high chronic intake of added sugars. *Nutr. Res. Rev.* **2017**, *30*, 191–207. [CrossRef]
41. DiNicolantonio, J.J.; Mehta, V.; Onkaramurthy, N.; O'Keefe, J.H. Fructose-induced inflammation and increased cortisol: A new mechanism for how sugar induces visceral adiposity. *Prog. Cardiovasc. Dis.* **2018**, *61*, 3–9. [CrossRef]
42. Duan, Y.; Zeng, L.; Zheng, C.; Song, B.; Li, F.; Kong, X.; Xu, K. Inflammatory links between high fat diets and diseases. *Front. Immunol.* **2018**, *9*, 2649. [CrossRef]
43. Kumar, J.; Rani, K.; Datt, C. Molecular link between dietary fibre, gut microbiota and health. *Mol. Biol. Rep.* **2020**, *47*, 6229–6237. [CrossRef] [PubMed]
44. Lane, M.; Howland, G.; West, M.; Hockey, M.; Marx, W.; Loughman, A.; O'Hely, M.; Jacka, F.; Rocks, T. The effect of ultra-processed very low-energy diets on gut microbiota and metabolic outcomes in individuals with obesity: A systematic literature review. *Obes. Res. Clin. Pract.* **2020**, *14*, 197–204. [CrossRef] [PubMed]
45. Martin, C.R.; Osadchiy, V.; Kalani, A.; Mayer, E.A. The Brain-Gut-Microbiome Axis. *Cell. Mol. Gastroenterol. Hepatol.* **2018**, *6*, 133–148. [CrossRef] [PubMed]
46. Gershon, M.D. 5-Hydroxytryptamine (serotonin) in the gastrointestinal tract. *Curr. Opin. Endocrinol. Diabetes Obes.* **2013**, *20*, 14–21. [CrossRef] [PubMed]
47. Lach, G.; Schellekens, H.; Dinan, T.G.; Cryan, J.F. Anxiety, depression, and the microbiome: A role for gut peptides. *Neurotherapeutics* **2018**, *15*, 36–59. [CrossRef] [PubMed]

48. Bake, T.; Morgan, D.G.A.; Mercer, J.G. Feeding and metabolic consequences of scheduled consumption of large, binge-type meals of high fat diet in the Sprague-Dawley rat. *Physiol. Behav.* **2014**, *128*, 70–79. [CrossRef]
49. Bake, T.; Murphy, M.; Morgan, D.G.A.; Mercer, J.G. Large, binge-type meals of high fat diet change feeding behaviour and entrain food anticipatory activity in mice. *Appetite* **2014**, *77*, 60–71. [CrossRef]
50. Nakajima, S.; Fukasawa, K.; Gotoh, M.; Murakami-Murofushi, K.; Kunugi, H. Saturated fatty acid is a principal cause of anxiety-like behavior in diet-induced obese rats in relation to serum lysophosphatidyl choline level. *Int. J. Obes.* **2020**, *44*, 727–738. [CrossRef]
51. Andrich, D.E.; Melbouci, L.; Ou, Y.; Leduc-Gaudet, J.-P.; Chabot, F.; Lalonde, F.; Lira, F.S.; Gaylinn, B.D.; Gouspillou, G.; Danialou, G.; et al. Altered Feeding Behaviors and Adiposity Precede Observable Weight Gain in Young Rats Submitted to a Short-Term High-Fat Diet. *J. Nutr. Metab.* **2018**, *2018*, 1498150. [CrossRef]
52. Medina-Reyes, E.I.; Rodríguez-Ibarra, C.; Déciga-Alcaraz, A.; Díaz-Urbina, D.; Chirino, Y.I.; Pedraza-Chaverri, J. Food additives containing nanoparticles induce gastrotoxicity, hepatotoxicity and alterations in animal behavior: The unknown role of oxidative stress. *Food Chem. Toxicol.* **2020**, *146*, 111814. [CrossRef] [PubMed]
53. Silva, M.M.; Reboredo, F.H.; Lidon, F.C. Food colour additives: A synoptical overview on their chemical properties, applications in food products, and health side effects. *Foods* **2022**, *11*, 379. [CrossRef] [PubMed]
54. Rinninella, E.; Cintoni, M.; Raoul, P.; Gasbarrini, A.; Mele, M.C. Food additives, gut microbiota, and irritable bowel syndrome: A hidden track. *Int. J. Environ. Res. Public Health* **2020**, *17*, 8816. [CrossRef]
55. Contreras-Rodriguez, O.; Solanas, M.; Escorihuela, R.M. Dissecting ultra-processed foods and drinks: Do they have a potential to impact the brain? *Rev. Endocr. Metab. Disord.* **2022**, *23*, 697–717. [CrossRef] [PubMed]
56. Mantovani, A.; Aquilina, G.; Cubadda, F.; Marcon, F. Risk-Benefit Assessment of Feed Additives in the One Health Perspective. *Front. Nutr.* **2022**, *9*, 843124. [CrossRef]
57. Lau, K.; McLean, W.G.; Williams, D.P.; Howard, C.V. Synergistic interactions between commonly used food additives in a developmental neurotoxicity test. *Toxicol. Sci.* **2006**, *90*, 178–187. [CrossRef]
58. Konduracka, E. A link between environmental pollution and civilization disorders: A mini review. *Rev. Environ. Health* **2019**, *34*, 227–233. [CrossRef]
59. Zhao, M.; Zhang, B.; Deng, L. The Mechanism of Acrylamide-Induced Neurotoxicity: Current Status and Future Perspectives. *Front. Nutr.* **2022**, *9*, 859189. [CrossRef]
60. Igarashi, K.; Uemura, T.; Kashiwagi, K. Acrolein toxicity at advanced age: Present and future. *Amino Acids* **2018**, *50*, 217–228. [CrossRef]
61. Muscogiuri, G.; Barrea, L.; Laudisio, D.; Savastano, S.; Colao, A. Obesogenic endocrine disruptors and obesity: Myths and truths. *Arch. Toxicol.* **2017**, *91*, 3469–3475. [CrossRef]

nutrients

MDPI

Article

Adherence to the Mediterranean Diet and Health-Related Quality of Life during the COVID-19 Lockdown: A Cross-Sectional Study including Preschoolers, Children, and Adolescents from Brazil and Spain

Desirée Victoria-Montesinos [1], Almudena Tárraga-Marcos [2], Javier Brazo-Sayavera [3,4], Estela Jiménez-López [5], Héctor Gutiérrez-Espinoza [6,*], Josefa María Panisello Royo [7], Pedro J. Tárraga-López [2,*] and José Francisco López-Gil [5,8]

1 Faculty of Pharmacy and Nutrition, UCAM Universidad Católica San Antonio de Murcia, 30107 Murcia, Spain
2 Departamento de Ciencias Médicas, Facultad de Medicina, Universidad Castilla-La Mancha (UCLM), 02006 Albacete, Spain
3 PDU EFISAL, Centro Universitario Regional Noreste, Universidad de la República (UdelaR), Rivera 40000, Uruguay
4 Department of Sports and Computer Science, Universidad Pablo de Olavide (UPO), 41013 Seville, Spain
5 Health and Social Research Center, Universidad de Castilla-La Mancha, 16071 Cuenca, Spain
6 Escuela de Fisioterapia, Universidad de las Américas, Quito 170504, Ecuador
7 Fundación para el Fomento de la Salud, 28006 Madrid, Spain
8 Department of Environmental Health, Harvard University T.H. Chan School of Public Health, Boston, MA 02138, USA
* Correspondence: pjtarraga@sescam.jccm.es (P.J.T.-L.); hector.gutierrez@udla.edu.ec (H.G.-E.)

Citation: Victoria-Montesinos, D.; Tárraga-Marcos, A.; Brazo-Sayavera, J.; Jiménez-López, E.; Gutiérrez-Espinoza, H.; Panisello Royo, J.M.; Tárraga-López, P.J.; López-Gil, J.F. Adherence to the Mediterranean Diet and Health-Related Quality of Life during the COVID-19 Lockdown: A Cross-Sectional Study including Preschoolers, Children, and Adolescents from Brazil and Spain. *Nutrients* **2023**, *15*, 677. https://doi.org/10.3390/nu15030677

Academic Editors: Monica Dinu and Daniela Martini

Received: 6 January 2023
Revised: 23 January 2023
Accepted: 24 January 2023
Published: 29 January 2023

Abstract: Scientific literature has suggested positive associations between the Mediterranean diet (MD) and the health-related quality of life (HRQoL) in young populations. However, to our knowledge, this relationship is unexplored during a situation of social isolation (i.e., lockdown). The objective of the current study is to examine the relationship between the MD and HRQoL during the COVID-19 lockdown among preschoolers, children, and adolescents from Brazil and Spain. This cross-sectional study includes a sample of 1099 three- to seventeen-year-old participants (47.6% girls) who were recruited via social networks. The HRQoL was assessed with the EQ-5D-Y. The Quality Index for Children and Teenagers (KIDMED) questionnaire was applied to evaluate the relationship between the MD and HRQoL. The highest prevalence of reported problems was found for worried, sad, or unhappy participants (39.8%). Furthermore, the lowest proportion of HRQoL problems was observed for "mobility" (2.5%). The proportion of high adherence to the MD was 44.3%. Participants with greater MD adherence reported higher HRQoL mean scores when compared with those who did not adhere to the MD (83.7 ± 0.6 vs. 85.6 ± 0.7, respectively; $p < 0.05$). Adherence to the MD and especially daily fruit intake were related to higher HRQoL during the COVID-19 lockdown among Brazilian and Spanish young people aged three to seventeen years.

Keywords: Mediterranean dietary patterns; eating healthy; lifestyle; pandemic; youths; children; adolescents; preschoolers

1. Introduction

In March 2020, the World Health Organization (WHO) declared the global COVID-19 outbreak a pandemic [1]. An extreme increase in the number of infections led to the declaration of several restraint measures in most countries, such as lockdowns [2]. During these lockdowns, most countries (e.g., Brazil and Spain) established compulsory limitations, such as use of face masks, restricted international travel, and completely closed schools [1]. This social isolation, in combination with physical distancing, produced changes in the daily routines of citizens (e.g., physical activity levels, eating habits) [3,4]. For instance, a greater consumption of unhealthy foods was identified, especially among vulnerable populations such as children and adolescents [5]. Furthermore, a recent systematic review conducted by Pourghazi et al. [6] has pointed out uneven changes before and during the COVID-19 pandemic in young people (i.e., an increase in sweet and snack intake and a decrease in fruit, vegetables, and fast-food intake consumption). Supporting this notion, an additional aspect of concern from a public health perspective is the increase in the prevalence of malnutrition (e.g., excess weight) among children and adolescents during the COVID-19 pandemic [7]. Conversely, the COVID-19 lockdown has also lead to more family meals and a higher degree of resilience among family members [8], which has been related to greater psychological and physical health among this population [9].

Childhood and adolescence are periods characterized by an increase in food requirements due to higher nutritional demands for optimal physiological and psychological maturation [10]. Nutritional deficiencies among children and adolescents can lead to poorer development and future health problems [10]. In this line, the WHO states that healthy eating is a relevant factor in protecting against malnutrition and non-communicable diseases [11]. Among the healthiest dietary patterns, the Mediterranean diet (MD) is characterized by a dietary pattern rich in plant-based foods (fruits, vegetables, legumes, cereals, seeds, nuts, and olives), a moderate-to-high consumption of fish/seafood, a moderate intake of eggs, dairy products (preferably yogurt and cheese) and poultry, and a low intake of red meat, with olive oil as a main source of added fat [12]. Thus, this dietary pattern is one of the most well-known for its health benefits worldwide [13]. Although there is a belief that the MD is a characteristic eating habit of people living in countries around the Mediterranean Sea, other Mediterranean-type ecosystems can be placed from the 30° parallel to the 45° parallel of the southern or northern latitudes, with their coasts oriented towards the west [14]. Similarly, researchers from several non-Mediterranean countries (i.e., Colombia [15] and Brazil [16]) have adapted and validated the Mediterranean Diet Quality Index for Children and Teenagers (KIDMED) to assess adherence to the MD. The main reason for this is that this eating pattern is linked with a decrease in inflammatory status, as well as a reduced likelihood of being overweight and obese among young people [17,18], whereas a low adherence to the MD has been related to overweight and obesity in these age phases [19]. Despite its potential benefits, a low adherence to the MD seems to be prevalent in this population [20–22]. For instance, prior to the COVID-19 pandemic, the prevalence of children and adolescents having a low adherence to the Mediterranean diet was 4.2% in Spain [14], 14.9% and 27% for children and adolescents in Greece [15,16], and between 23.0% and 33.0% in Italy [17,18]. Based on these results, geographical region seems to play an important role in higher adherence, whereas Mediterranean countries do not always show the highest adherence to the Mediterranean diet.

Another relevant factor from early childhood to adolescence is the health-related quality of life (HRQoL), which plays a key role in physical and psychological well-being [23]. HRQoL is defined as the individual's perception of his or her position in life and the evaluation of different dimensions of this perception, along with their influence on health status [24]. Furthermore, HRQoL is a concept that has been widely documented in recent years to assess children and their health status in a comprehensive manner, including in its assessment physical, cognitive, social, and psychological functioning. It is crucial to prevent unhealthy behaviors [23]. Among young people, physical and psychological wellbeing declined significantly during the COVID-19 lockdown [25]. In fact, a recent systematic review by Nobari et al. [26] that included 3177 children and adolescents indicated a decrease in HRQoL among young people during the COVID-19 pandemic. Three articles showed that the COVID-19 pandemic significantly impacted the HRQoL of children and adolescents. Although another did not report a comparison between the pre-pandemic period and during the COVID-19 pandemic, a reduction in the HRQoL can be observed. Relevant HRQoL factors include dietary habits (i.e., frequency, composition, and amount of beverages/food consumed), which are associated with higher HRQoL scores in mental and physical dimensions [27]. In addition, these dietary habits have been associated with overweight and obesity in this population [28], which are related to lower HRQoL [29].

According to the results of some studies, the lockdown and physical distancing that occurred as a consequence of the COVID-19 pandemic may have contributed to a modification in some dietary habits and the HRQoL among young populations [30,31]. In this line, a systematic review by Della Valle et al. [4] indicated that adherence to the MD might have increased during COVID-19 lockdown. Furthermore, some systematic reviews have concluded that the COVID-19 pandemic has led to a decline in the HRQoL of children and adolescents [26,32]. A recent systematic review analyzed the association of adherence to the MD with the HRQoL in young people [27]. However, all studies included in that review were conducted before [33–36] and other was performed after [37] the COVID-19 lockdown, indicating an association between a greater adherence to the MD and a higher HRQoL among young populations [27,37]. To our knowledge, this is the first study that has analyzed this relationship in young people during the situation of social isolation (i.e., lockdown) due to the COVID-19 pandemic. Thus, the objective of the current study is to examine the association between the MD (i.e., adherence to this eating pattern and its specific components) and HRQoL during the COVID-19 lockdown among preschoolers, children, and adolescents aged 3−17 years from Brazil and Spain.

2. Materials and Methods

2.1. Participants and Study Design

Out of the 1263 participants from Brazil and Spain, 143 were eliminated since they were too young (aged < 3 years) or too old (aged > 17 years) for the study. Additionally, 21 participants were removed due to missing data. This left 1099 participants whose data were included in the final analysis. As in-person contact was not allowed, these participants were recruited through social networks. Through a snowball sampling technique, an online survey was designed and delivered in both Brazil and Spain. The survey took approximately 15 min to complete and was filled out by the participants' parents or guardians. Before carrying out the survey, the objective of this study was explained and informed consent was obtained. The data was collected over a period of 15 days in both countries (from 29 March to 13 April 2020). During this period, the entire Brazilian and Spanish populations were expected to remain at home (excepting essential workers) and were only allowed to go out for healthcare, basic food shopping, and some justified exceptions. As criteria, only parents/legal guardians of the young Brazilian and Spanish populations aged 3–17 years who signed the informed consent were involved. Regarding the exclusion criteria, participants were not included when their parents/guardians did not totally complete the online survey. This study had the approval of the ethical committees of the *Universidade Tecnologica do Paraná* (CAAE: 32023220.8.0000.5547; approval number: 4.275.232 and the *Universidad Católica de Murcia* (code: CE112001).

2.2. Procedures

2.2.1. Health-Related Quality of Life (Dependent Variable)

The HRQoL (health-related quality of life) was evaluated by the EQ-5D-Y (proxy version 1), a tool specifically designed for evaluating HRQoL in young people [38]. The EQ-5D-Y assesses five different aspects of HRQoL (mobility, self-care, ability to carry out usual activities, presence of pain or discomfort, and emotional well-being) on a three-point scale (no problems, some problems, or a lot of problems). The results of the EQ-5D-Y are represented by a three-digit code, with each digit representing the level of severity in each of the five dimensions. The dimensions of HRQoL were also grouped into two categories: "no problems" or "any problems" (combining "some problems" and "a lot of problems"). In addition, the EQ-5D-Y includes a visual analogue scale (VAS) with scores ranging from 0 (the lowest HRQoL) to 100 (the highest HRQoL). The reliability and validity of the EQ-5D-Y have been established in the previous research [39].

2.2.2. Adherence to the Mediterranean Diet (Independent Variable)

Adherence to the Mediterranean diet (MD) was evaluated using the KIDMED [40]. The KIDMED is a sixteen-question test that ranges from −4 to 12 points. All the items included in the KIDMED and its scoring system can be found in Table S1. Questions about unhealthy aspects of the MD were scored as −1 (e.g., "Takes sweets and candy several times every day"), and questions about healthy aspects were scored as +1 (e.g., "Takes a fruit or fruit juice every day"). The total scores from the KIDMED were then divided into three categories: high MD ($\geq$ 8 points), moderate MD (4–7 points), and low MD ($\leq$ 3 points).

2.2.3. Covariates

The following information was collected from the parents or guardians of the participants: age, sex (male or female), nationality, educational level of the primary breadwinner (non-university or university), socioeconomic status, excess weight (overweight or obese), and daily movement behaviors. These variables were all considered to be covariates. The family affluence scale-III (FAS-III) score ranges from 0 to 13 points, with higher scores indicating a higher socioeconomic status based on the responses to six questions about the family's resources (e.g., number of vehicles, bedrooms, computers, etc.) [41]. Anthropometric data (e.g., height, weight) were also collected for the children, and the body mass index (BMI) z-score and the proportion of excess weight (i.e., overweight and obesity) were computed following the World Health Organization (WHO) criteria [42,43]. Daily movement behaviors, including physical activity, recreational screen time, and sleep duration, were also measured. Physical activity was evaluated by asking parents how many days their child was physically active for at least 60 min in the previous week. Recreational screen time was measured by asking parents about the time their child spent on various sedentary activities (e.g., watching TV, playing video games, or using electronic devices) as follows: (a) "How many hours a day, during the COVID-19 lockdown, does your child usually spend using electronic devices such as computers, tablets or smartphones for other purposes (e.g., homework, emailing, tweeting, Facebook, chatting, surfing the internet)?"; (b) (b) "How many hours a day, during the COVID-19 lockdown, does your child spend playing games on a computer, games console, tablet, smartphone or other electronic device (not including moving or fitness games)?"; (c) "How many hours a day, during the COVID-19 lockdown, does your child spend watching TV, videos (including YouTube or similar services), DVDs, and other entertainment on a screen?". The three answers were summed considering a week distribution of five weekdays and two weekend days. Sleep duration was determined by asking parents about the bedtimes and wake times (for weekdays and weekend days separately) of their child as follows: "What time does your child usually go to bed?" and "What time does your usually get up?". The average daily sleep duration was computed for each participant as follows: [(average nocturnal sleep duration on weekdays $\times$ 5) + (average nocturnal sleep duration on weekends $\times$ 2)]/7.

2.3. *Statistical Analysis*

The descriptive data for the study were presented in two ways: as the mean and standard deviation in the case of continuous data, and as the number and percentage in case of categorical data. Differences between adherence to the MD and problems in the HRQoL domains were tested using the Pearson's chi-square test (χ^2) or Fisher's exact test (when expected values in any of the cells of a contingency table were below five participants). Preliminary analyses showed no interaction between adherence to the MD and sex (p = 0.351) or age group (p = 0.174) in relation to HRQoL. Thus, all the samples were analyzed together to increase the statistical power. Analyses of covariance (ANCOVAs) were performed to assess the differences in HRQoL in relation to adherence to the MD, as well as to determine the specific MD components associated with HRQoL. Furthermore, linear regression analyses were performed to test the individual MD components associated with a higher HRQoL. Sex, age, nationality, socioeconomic status, breadwinner's educational level, excess weight, and daily movement behaviors were included as covariates. Statistical analyses were performed by the software Statistical Package for Social Sciences (SPSS, IBM Corp., Armonk, NY, USA) in its version 28 for Windows. A p-value < 0.05 denoted statistical significance.

3. Results

Table 1 shows the descriptive data for study participants. The proportion of high adherence to the MD was 44.3%. The VAS mean score was 84.5 ± 15.9. The lowest proportion of HRQoL problems was observed for "mobility" (2.5%). Conversely, the highest proportion was found for "feeling worried, sad, or unhappy" (39.8%).

Table 1. Characteristics of the study participants (N = 1099).

Variables	M/n	SD/%
Age (years)	11.5	4.5
Preschoolers (3–5 years)	146	13.3
Children (6–12 years)	444	40.4
Adolescents (13–17 years)	509	46.3
Sex		
Boys	576	52.4
Girls	523	47.6
Nationality		
Brazilian	495	45.0
Spanish	604	55.0
Breadwinner's educational level		
University studies	538	49.0
Non-university studies	561	51.0
Socioeconomic status		
FAS-III (score)	7.5	2.3
Anthropometric data		
Weight (kg)	43.0	19.5
Height (cm)	145.5	24.8
BMI (z-score)	0.9	2.0
Overweight/Obesity [a]	467	42.5
Daily movement behaviors		
Physical activity (days/week)	4.0	2.3
Recreational screen time (h/day)	7.6	5.9
Sleep duration (h/day)	9.9	1.5
Adherence to the MD		
KIDMED (score)	7.0	2.4
Low/Moderate MD	612	55.7
High MD	487	44.3
HRQoL		
VAS (score)	84.5	15.9
Mobility (%, any problem)	28	2.5
Looking after myself (%, any problem)	112	10.2
Doing usual activities (%, any problem)	113	10.3
Having pain or discomfort (%, any problem)	181	16.5
Feeling worried, sad, or unhappy (%, any problem)	437	39.8

Data expressed as a mean and standard deviation for continuous variables or numbers and percentage for categorical variables. BMI—body mass index; FAS—family affluence scale; HRQoL—health-related quality of life; KIDMED—Mediterranean Diet Quality Index for Children and Teenagers; MD—Mediterranean diet; and VAS—visual analogue scale. [a] According to the World Health Organization criteria [43].

In Table 2, we assessed the association between adherence to the MD and the different domains of HRQoL. Despite a higher proportion of problems across all HRQoL domains in participants with low/moderate MD, only significant differences were obtained for the "having pain or discomfort" domain ($p = 0.020$). Sub-group analyses by sex and age group can be found in Table S2.

Table 2. Association between Mediterranean diet adherence and problems in different dimensions of health-related quality of life.

HRQoL Domain [a]	Low/Moderate MD	High MD	*p*
Mobility	18 (2.9)	10 (2.1)	0.442
Looking after myself	62 (10.1)	50 (10.3)	0.941
Doing usual activities	71 (11.6)	42 (8.6)	0.106
Having pain or discomfort	115 (18.8)	66 (13.6)	0.020
Feeling worried, sad, or unhappy	246 (40.2)	191 (39.2)	0.743

[a] Data expressed as proportion of participants reporting any problem in the different health-related quality of life domains, according to adherence to the Mediterranean diet. HRQoL—health-related quality of life; MD—Mediterranean diet. *p* values obtained by Pearson's chi-square test or Fisher's exact test.

Figure 1 depicts the VAS mean score in relation to adherence to the MD. After adjusting by several covariates, participants with high MD adherence reported greater HRQoL mean scores (M = 83.7; SE = 0.6) than those participants with low/moderate MD adherence (M = 85.6; SE = 0.7). In both the unadjusted and adjusted models, the differences between groups were statistically significant ($p < 0.05$ for both). Figures S1 and S2 show sub-group analyses by sex and age group, respectively.

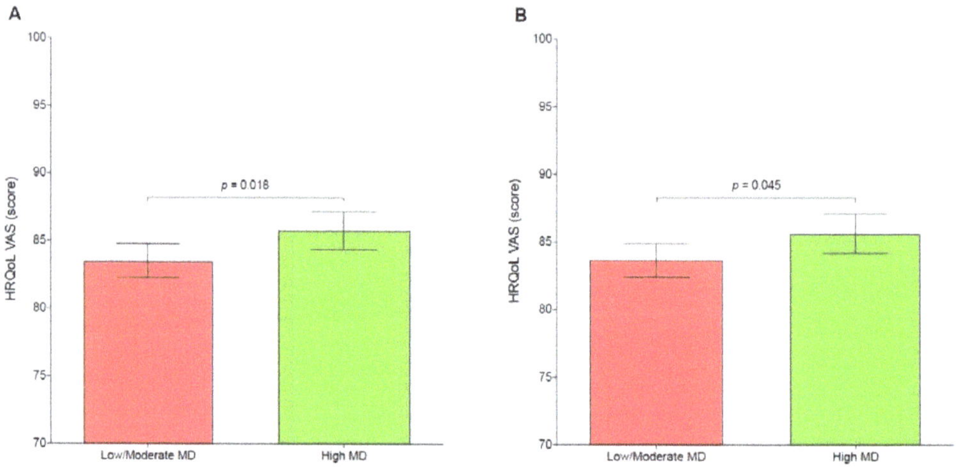

Figure 1. Mean differences of health-related quality of life according to the degree of adherence to the Mediterranean diet during the COVID-19 lockdown. HRQoL—health-related quality of life; MD—Mediterranean diet; VAS—visual analogue scale. (**A**): unadjusted; (**B**): adjusted by sex, age, nationality, socioeconomic status, breadwinner's educational level, excess weight, and daily movement behaviors. *p* values obtained by analysis of covariance.

Table 3 indicates the association between different MD components of the KIDMED and the HRQoL scores. After adjusting by several covariates, eating fruit or consuming fruit juice every day was significantly associated with a greater HRQoL score ($p < 0.005$).

Table 3. Association between different Mediterranean diet components and health-related quality of life score.

Items	*B*	SE	LLCI	ULCI	*p*
Takes a fruit or fruit juice every day	3.59	1.28	1.09	6.09	0.005
Has a second fruit every day	0.29	1.13	−1.94	2.51	0.800
Has fresh or cooked vegetables regularly once a day	−0.05	1.20	−2.41	2.31	0.968
Has fresh or cooked vegetables more than once a day	−0.04	1.15	−2.29	2.21	0.973
Consumes fish regularly (at least 2–3 times per week)	−1.84	1.14	−4.07	0.39	0.106
Goes more than once a week to a fast-food (hamburger) restaurant	0.99	1.55	−2.05	4.02	0.522
Likes pulses and eats them more than once a week	0.00	1.41	−2.78	2.77	0.999
Consumes pasta or rice almost every day (5 or more times per week)	−0.35	1.19	−2.68	1.98	0.765
Has cereals or grains (bread, etc.) for breakfast	−0.06	1.20	−2.41	2.28	0.958
Consumes nuts regularly (at least 2–3 times per week)	−0.13	1.06	−2.22	1.96	0.905
Uses olive oil at home	−0.76	1.75	−4.20	2.68	0.665
Skips breakfast	−2.28	2.15	1.94	−6.49	0.289
Has a dairy product for breakfast (yogurt, milk, etc.)	−0.36	1.54	−3.37	2.66	0.816
Has commercially baked goods or pastries for breakfast	−0.59	1.45	−3.43	2.26	0.686
Takes two yogurts and/or some cheese (40 g) daily	0.10	1.00	−1.86	2.06	0.921
Takes sweets and candy several times every day	0.48	1.23	−1.94	2.90	0.698

Data expressed as non-standardized beta coefficient, standard error, and 95% confidence interval. Adjusted by sex, age, nationality, socioeconomic status, breadwinner's educational level, excess weight, and daily movement behaviors. *p* values obtained by multiple linear regression.

4. Discussion

To our knowledge, this is the first study to analyze associations between the MD (overall adherence and its specific components) and the HRQoL in preschoolers, children, and adolescents from Brazil and Spain during the COVID-19 lockdown. Overall, our results showed that a higher adherence to the MD was related to higher levels of the HRQoL and lower instances of HRQoL problems in the "having pain or discomfort" domain. When the different MD components were considered, daily fruit intake or fruit use was associated with higher scores of HRQoL.

Our findings depict that a high adherence to the MD was linked with higher HRQoL scores. This result is in accordance with most of the studies conducted on this matter, both before [27] and after the COVID-19 lockdown [44]. Nevertheless, divergent results have also been found [27]. These divergent results could be explained by the age of the included samples, with different levels of psychological maturity [45], or by the consumption of different proportions of food groups [46]. Although the mechanism by which the MD may improve the HRQoL is not fully explained, it could be related to the high amount of antioxidants, polyphenols, vitamins, and dietary fiber present in some foods (i.e., fruits). The daily intake of fruit (an eating habit characteristic of the MD) could lead to benefits in mental well-being and normal brain function [47,48]. An additional possible reason for the association between the MD and HRQoL could be (at least partially) explained due to the possible beneficial influence of a healthy diet on certain psychological factors (i.e., positive affect, life satisfaction, moods, and emotions) [49,50]. Another possible explanation could be related to the association between eating habits, sleep-related problems, and the HRQoL. In this sense, López-Gil et al. [51] observed a relationship among healthy eating habits and sleep-related problems, which has been related to a lower HRQoL [52]. Their results showed that eating habits characteristic of the MD (i.e., a higher fruit, vegetable, and bean intake, a lower sweet consumption, having breakfast more often, and having family meals) were related to lower sleep-related problems. It is possible that the consumption of certain MD components facilitates an adequate sleep duration (diminishing sleep-related problems) and, therefore, a higher HRQoL [53,54]. Lastly, breakfast (an important component of the MD) has been associated with greater psychosocial health in Spanish preschoolers, children, and adolescents [55,56]. In this sense, social context appears to be crucial in this relationship [57] as family meals at home may help to increase the quality of the meals, which could improve well-being and decrease psychosocial behavioral problems, increasing the HRQoL in this population [58].

In addition, another finding of the current study was that a higher adherence to the MD was related to lower pain or discomfort symptoms during the COVID-19 lockdown (especially in children and adolescents). Previous studies have depicted that a higher adherence to the MD is related to lesser pain in children and adolescents [59,60]. Although the mechanisms by which a healthy diet may increase the HRQoL are unknown, the intake of some type of foods (e.g., fruits, vegetables, and olive oil) typical of the MD could offer benefits against oxidative stress and inflammation [61,62], which have been hypothesized to be pain contributors [63,64]. Therefore, it is possible that participants with a higher adherence to the MD have lower levels of oxidative stress and inflammation that could contribute to perceiving less pain.

On the other hand, by specifically analyzing the components of the MD, our results showed that a daily consumption of fruit/fruit juice was related to a higher HRQoL. Despite being a period characterized by a decrease in healthy behaviors [3,65–67], young people who had this healthy eating habit reported a higher HRQoL in comparison with those who did not. These results are in line with those obtained in different cross-sectional pre-pandemic studies [68–70], in which a higher fruit consumption was related to a higher HRQoL. This finding could be explained due to the antioxidant and anti-inflammatory effects of the bioactive compounds contained in fruits such as folate, vitamin C, polyphenols, fiber, etc. [71]. In addition to physical health, the adequate consumption of fruits and their biocomponents also seems to improve mental health [48]. This may be because of the role of essential nutrients provided by healthy food (i.e., fruits) in neurotransmitter formation related to mental health, such as serotonin, dopamine, and oxytocin [72], which could lead to a higher HRQoL.

This study has some strengths that should be acknowledged, including the use of large samples of preschoolers, children, and adolescents from Brazil and Spain, and being the first to examine the specific association of the MD and its specific components with HRQoL during the COVID-19 lockdown in this age group. Conversely, this study has several limitations that should be taken into account. First, the cross-sectional design of the study means that it is not possible to determine cause-and-effect associations. Thus, we do not have information from prior to the COVID-19 lockdown on some variables that could influence the results obtained (e.g., previous adherence to the MD or previous diets). For a better understanding of the link between adherence to the MD and the HRQoL, it would be necessary to conduct longitudinal studies. Second, the information obtained was based on self-reported questionnaires that may introduce social desirability and recall biases. Third, the KIDMED does not provide exact information on types of food and the frequency of consumption. This means that, for example, in evaluating the consumption of dairy products, we do not know whether skimmed or whole milk was consumed, which may influence neurotransmitter formation such as serotonin and dopamine, since whole milk has a higher quantity of these precursors (tryptophan and tyrosine) [73]. Fourth, the compulsory limitations during the COVID-19 pandemic could differently influence the adherence to the MD and the HRQoL according to the type of lockdown imposed by each country's government (e.g., a total or partial lockdown). Fifth, although several covariates were adjusted, residual confounding is still possible (e.g., type of family and frequency of family meals).

5. Conclusions

Adherence to the MD, especially daily fruit intake, were related to higher HRQoL scores during the COVID-19 lockdown among Brazilian and Spanish young people aged 3 to 17 years. This finding is clinically significant, as the pandemic appears to have deteriorated the HRQoL in the young population and a better understanding of the associated factors (i.e., healthy diet) could help to establish concrete measures in case of situations of social isolation. Further studies with different designs (i.e., interventions) are required to establish cause-and-effect relationships and to verify whether greater adherence to the MD produces improvements in the HRQoL in the young population.

Supplementary Materials: The following are available online at https://www.mdpi.com/article/10.3390/nu15030677/s1: Figure S1. Mean differences of health-related quality of life according to the degree of adherence to the Mediterranean diet during the COVID-19 lockdown by sex; HRQoL—health-related quality of life; MD—Mediterranean diet; VAS—visual analogue scale; A—boys, B—girls; Adjusted by age, nationality, socioeconomic status, breadwinner's educational level, excess weight, and daily movement behaviors. *p* values obtained by analysis of covariance. Figure S2. Mean differences of health-related quality of life according to the degree of adherence to the Mediterranean diet during the COVID-19 lockdown by age group. HRQoL—health-related quality of life; MD—Mediterranean diet; VAS—visual analogue scale; A—preschoolers; B—children; C—adolescents. Adjusted by age, nationality, socioeconomic status, breadwinner's educational level, excess weight, and daily movement behaviors. *P* values obtained by analysis of covariance. Table S1. Mediterranean Diet Quality Index for Children and Teenagers. Table S2. Association between Mediterranean Diet adherence and problems in different dimensions of health-related quality of life by sex or age group.

Author Contributions: Conceptualization, J.F.L.-G. and J.B.-S.; methodology, J.F.L.-G. and J.B.-S.; validation, J.F.L.-G.; formal analysis, J.F.L.-G.; investigation, J.F.L.-G. and J.B.-S.; resources, J.F.L.-G. and J.B.-S.; data curation, J.F.L.-G.; writing—original draft preparation, D.V.-M. and J.F.L.-G.; writing—review and editing, D.V.-M., J.F.L.-G., A.T.-M., E.J.-L., H.G.-E., J.M.P.R., P.J.T.-L. and J.B.-S.; visualization, J.F.L.-G.; supervision, J.F.L.-G.; project administration, J.F.L.-G. All authors have read and agreed to the published version of the manuscript.

Funding: This research received no external funding.

Institutional Review Board Statement: The study was conducted in accordance with the Declaration of Helsinki and approved by the Institutional Ethics Committee of the *Universidade Tecnológica do Paraná* (UTFPR; protocol code 4.275.232) and the *Universidad Católica de Murcia* (UCAM; protocol code CE112001).

Informed Consent Statement: Informed consent was obtained from all subjects involved in the study.

Data Availability Statement: Not applicable.

Acknowledgments: The authors would like to acknowledge the help from Hallyne Bergamini Silva Caetano, Cristiano Israel Caetano, Rafael Gomes Sentone, Anderson Caetano Paulo, Antonio García-Hermoso, Edina Maria de Camargo, Cezane P. Reuter, Anelise R. Gaya, and Noelia González Gálvez, and the contribution of the Ayuntamiento de Archena (Spain). J.F.L.-G. is a Margarita Salas Fellow (*Universidad de Castilla-La Mancha*—2021-MS-20563).

Conflicts of Interest: The authors declare no conflict of interest.

References

1. World Health Organization. *Coronavirus Disease 2019 (COVID-19): Situation Report—51*; WHO: Geneva, Switzerland, 2020.
2. Adil, M.T.; Rahman, R.; Whitelaw, D.; Jain, V.; Al-Taan, O.; Rashid, F.; Munasinghe, A.; Jambulingam, P. SARS-CoV-2 and the Pandemic of COVID-19. *Postgrad. Med. J.* **2021**, *97*, 110–116. [CrossRef]
3. Neville, R.D.; Lakes, K.D.; Hopkins, W.G.; Tarantino, G.; Draper, C.E.; Beck, R.; Madigan, S. Global Changes in Child and Adolescent Physical Activity During the COVID-19 Pandemic: A Systematic Review and Meta-Analysis. *JAMA Pediatr.* **2022**, *176*, 886–894. [CrossRef] [PubMed]
4. Della Valle, P.G.; Mosconi, G.; Nucci, D.; Vigezzi, G.P.; Gentile, L.; Gianfredi, V.; Bonaccio, M.; Gianfagna, F.; Signorelli, C.; Iacoviello, L.; et al. Adherence to the Mediterranean Diet during the COVID-19 National Lockdowns: A Systematic Review of Observational Studies. *Acta Biomed.* **2021**, *92*, e2021440. [CrossRef] [PubMed]
5. Marino, M.; Puppo, F.; Del Bo', C.; Vinelli, V.; Riso, P.; Porrini, M.; Martini, D. A Systematic Review of Worldwide Consumption of Ultra-Processed Foods: Findings and Criticisms. *Nutrients* **2021**, *13*, 2778. [CrossRef] [PubMed]
6. Pourghazi, F.; Eslami, M.; Ehsani, A.; Ejtahed, H.-S.; Qorbani, M. Eating Habits of Children and Adolescents during the COVID-19 Era: A Systematic Review. *Front. Nutr.* **2022**, *9*, 1004953. [CrossRef] [PubMed]
7. Fore, H.H.; Dongyu, Q.; Beasley, D.M.; Ghebreyesus, T.A. Child Malnutrition and COVID-19: The Time to Act Is Now. *Lancet* **2020**, *396*, 517–518. [CrossRef]

8. Scapaticci, S.; Neri, C.R.; Marseglia, G.L.; Staiano, A.; Chiarelli, F.; Verduci, E. The Impact of the COVID-19 Pandemic on Lifestyle Behaviors in Children and Adolescents: An International Overview. *Ital. J. Pediatr.* **2022**, *48*, 22. [CrossRef]
9. Fulkerson, J.A.; Strauss, J.; Neumark-Sztainer, D.; Story, M.; Boutelle, K. Correlates of Psychosocial Well-Being among Overweight Adolescents: The Role of the Family. *J. Consult. Clin. Psychol.* **2007**, *75*, 181–186. [CrossRef]
10. Magriplis, E.; Zampelas, A. Chapter 3—Mediterranean Diet in Children and Adolescents. In *The Mediterranean Diet*, 2nd ed.; Preedy, V.R., Watson, R.R., Eds.; Academic Press: Cambridge, MA, USA, 2020; pp. 25–39. ISBN 978-0-12-818649-7.
11. World Health Organization. *Healthy Diet*; WHO: Geneva, Switzerland, 2020.
12. Bach-Faig, A.; Berry, E.M.; Lairon, D.; Reguant, J.; Trichopoulou, A.; Dernini, S.; Medina, F.X.; Battino, M.; Belahsen, R.; Miranda, G.; et al. Mediterranean Diet Pyramid Today. Science and Cultural Updates. *Public Health Nutr.* **2011**, *14*, 2274–2284. [CrossRef]
13. Dinu, M.; Pagliai, G.; Casini, A.; Sofi, F. Mediterranean Diet and Multiple Health Outcomes: An Umbrella Review of Meta-Analyses of Observational Studies and Randomised Trials. *Eur. J. Clin. Nutr.* **2018**, *72*, 30–43. [CrossRef]
14. Köppen, V. *Das Geographische System Der Klimate*; Gebrüder Borntraeger: Berlin, Germany, 1936.
15. Carrillo, H.A.; Ramírez-Vélez, R. Adherence to the Mediterranean Diet in a Sample of Colombian Schoolchildren: An Evaluation of the Psychometric Properties of the Kidmed Questionnaire [Adherencia a La Dieta Mediterránea En Una Población Escolar Colombiana: Evaluación de Las Propiedades Psicométricas Del Cuestionario Kidmed]. *Nutr. Hosp.* **2020**, *37*, 73–79. [CrossRef] [PubMed]
16. Simon, M.I.S.D.S.; Forte, G.C.; Marostica, P.J.C. Translation and Cultural Adaptation of the Mediterranean Diet Quality Index in Children and Adolescents [Tradução E Adaptação Cultural Do Mediterranean Diet Quality Index In Children And Adolescents]. *Rev. Paul. Pediatr.* **2020**, *38*, e2018242. [CrossRef] [PubMed]
17. Bawaked, R.A.; Schröder, H.; Ribas-Barba, L.; Izquierdo-Pulido, M.; Pérez-Rodrigo, C.; Fíto, M.; Serra-Majem, L. Association of Diet Quality with Dietary Inflammatory Potential in Youth. *Food Nutr. Res.* **2017**, *61*, 1328961. [CrossRef]
18. Mistretta, A.; Marventano, S.; Antoci, M.; Cagnetti, A.; Giogianni, G.; Nolfo, F.; Rametta, S.; Pecora, G.; Marranzano, M. Mediterranean Diet Adherence and Body Composition among Southern Italian Adolescents. *Obes. Res. Clin. Pract.* **2017**, *11*, 215–226. [CrossRef] [PubMed]
19. Pereira-da-Silva, L.; Rêgo, C.; Pietrobelli, A. The Diet of Preschool Children in the Mediterranean Countries of the European Union: A Systematic Review. *Int. J. Environ. Res. Public Health* **2016**, *13*, 572. [CrossRef]
20. Cabrera, S.G.; Fernández, N.H.; Hernández, C.R.; Nissensohn, M.; Román-Viña, B.; Serra-Majem, L. KIDMED Test; Prevalence of Low Adherence to the Mediterranean Diet in Children and Young: A Systematic Review. *Nutr. Hosp.* **2015**, *32*, 2390–2399. [CrossRef]
21. Gil, J.F.L.; Camargo, E.; Lucas, J.L.Y. Adherencia a la dieta mediterránea en escolares de Educación Primaria partícipes en actividad física: Una revisión sistemática. *Cult. Cienc. Y Deporte* **2020**, *15*, 267–275.
22. Grosso, G.; Galvano, F. Mediterranean Diet Adherence in Children and Adolescents in Southern European Countries. *NFS J.* **2016**, *3*, 13–19. [CrossRef]
23. Haverman, L.; Limperg, P.F.; Young, N.L.; Grootenhuis, M.A.; Klaassen, R.J. Paediatric Health-Related Quality of Life: What Is It and Why Should We Measure It? *Arch. Dis. Child.* **2017**, *102*, 393–400. [CrossRef]
24. World Health Organisation. WHOQOL—Measuring Quality of Life | The World Health Organization. Available online: https://www.who.int/tools/whoqol (accessed on 9 January 2023).
25. Lehmann, S.; Haug, E.; Bjørknes, R.; Sandal, G.M.; Fadnes, L.T.; Skogen, J.C. Quality of Life among Young People in Norway during the COVID-19 Pandemic. A Longitudinal Study. *Eur. Child Adolesc. Psychiatry* **2022**, 1–11. [CrossRef]
26. Nobari, H.; Fashi, M.; Eskandari, A.; Villafaina, S.; Murillo-Garcia, Á.; Pérez-Gómez, J. Effect of COVID-19 on Health-Related Quality of Life in Adolescents and Children: A Systematic Review. *Int. J. Environ. Res. Public Health* **2021**, *18*, 4563. [CrossRef] [PubMed]
27. Romero-Robles, M.A.; Ccami-Bernal, F.; Ortiz-Benique, Z.N.; Pinto-Ruiz, D.F.; Benites-Zapata, V.A.; Patiño, D.C. Adherence to Mediterranean Diet Associated with Health-Related Quality of Life in Children and Adolescents: A Systematic Review. *BMC Nutr.* **2022**, *8*, 57. [CrossRef]
28. Leech, R.M.; McNaughton, S.A.; Timperio, A. The Clustering of Diet, Physical Activity and Sedentary Behavior in Children and Adolescents: A Review. *Int. J. Behav. Nutr. Phys. Act.* **2014**, *11*, 4. [CrossRef]
29. Ul-Haq, Z.; Mackay, D.F.; Fenwick, E.; Pell, J.P. Meta-Analysis of the Association between Body Mass Index and Health-Related Quality of Life among Children and Adolescents, Assessed Using the Pediatric Quality of Life Inventory Index. *J. Pediatr.* **2013**, *162*, 280–286.e1. [CrossRef] [PubMed]
30. Ravens-Sieberer, U.; Kaman, A.; Erhart, M.; Devine, J.; Schlack, R.; Otto, C. Impact of the COVID-19 Pandemic on Quality of Life and Mental Health in Children and Adolescents in Germany. *Eur. Child Adolesc. Psychiatry* **2022**, *31*, 879–889. [CrossRef]
31. Titis, E. Parental Perspectives of the Impact of COVID-19 Lockdown on Food-Related Behaviors: Systematic Review. *Foods* **2022**, *11*, 2851. [CrossRef]
32. Ahn, S.N. The Potential Impact of COVID-19 on Health-Related Quality of Life in Children and Adolescents: A Systematic Review. *Int. J. Environ. Res. Public Health* **2022**, *19*, 14740. [CrossRef]
33. Evaristo, O.S.; Moreira, C.; Lopes, L.; Abreu, S.; Agostinis-Sobrinho, C.; Oliveira-Santos, J.; Póvoas, S.; Oliveira, A.; Santos, R.; Mota, J. Associations between Physical Fitness and Adherence to the Mediterranean Diet with Health-Related Quality of Life in Adolescents: Results from the LabMed Physical Activity Study. *Eur. J. Public Health* **2018**, *28*, 631–635. [CrossRef]

34. Muros, J.J.; Pérez, F.S.; Ortega, F.Z.; Sánchez, V.M.G.; Knox, E. The Association between Healthy Lifestyle Behaviors and Health-Related Quality of Life among Adolescents. *J. Pediatr. (Rio. J.)* **2017**, *93*, 406–412. [CrossRef]
35. Costarelli, V.; Koretsi, E.; Georgitsogianni, E. Health-Related Quality of Life of Greek Adolescents: The Role of the Mediterranean Diet. *Qual. Life Res.* **2013**, *22*, 951–956. [CrossRef]
36. Ferrer-Cascales, R.; Albaladejo-Blázquez, N.; Ruiz-Robledillo, N.; Clement-Carbonell, V.; Sánchez-SanSegundo, M.; Zaragoza-Martí, A. Higher Adherence to the Mediterranean Diet Is Related to More Subjective Happiness in Adolescents: The Role of Health-Related Quality of Life. *Nutrients* **2019**, *11*, 698. [CrossRef]
37. Papadaki, S.; Carayanni, V.; Notara, V.; Chaniotis, D. Anthropometric, Lifestyle Characteristics, Adherence to the Mediterranean Diet, and COVID-19 Have a High Impact on the Greek Adolescents' Health-Related Quality of Life. *Foods* **2022**, *11*, 2726. [CrossRef] [PubMed]
38. Wille, N.; Badia, X.; Bonsel, G.; Burström, K.; Cavrini, G.; Devlin, N.; Egmar, A.-C.; Greiner, W.; Gusi, N.; Herdman, M.; et al. Development of the EQ-5D-Y: A Child-Friendly Version of the EQ-5D. *Qual. Life Res.* **2010**, *19*, 875–886. [CrossRef]
39. Gusi, N.; Perez-Sousa, M.A.; Gozalo-Delgado, M.; Olivares, P.R. Validez y fiabilidad de la versión proxy del EQ-5D-Y en español. *An. Pediatría* **2014**, *81*, 212–219. [CrossRef]
40. Serra-Majem, L.; Ribas, L.; Ngo, J.; Ortega, R.M.; García, A.; Pérez-Rodrigo, C.; Aranceta, J. Food, Youth and the Mediterranean Diet in Spain. Development of KIDMED, Mediterranean Diet Quality Index in Children and Adolescents. *Public Health Nutr.* **2004**, *7*, 931–935. [CrossRef]
41. Currie, C.; Molcho, M.; Boyce, W.; Holstein, B.; Torsheim, T.; Richter, M. Researching Health Inequalities in Adolescents: The Development of the Health Behaviour in School-Aged Children (HBSC) Family Affluence Scale. *Soc. Sci. Med.* **2008**, *66*, 1429–1436. [CrossRef]
42. World Health Organization. WHO Child Growth Standards: Length/Height-for-Age, Weight-for-Age, Weight-for-Length, Weight-for-Height and Body Mass Index-for-Age: Methods and Development. Available online: https://www.who.int/publications-detail-redirect/924154693X (accessed on 2 January 2023).
43. de Onis, M.; Onyango, A.W.; Borghi, E.; Siyam, A.; Nishida, C.; Siekmann, J. Development of a WHO Growth Reference for School-Aged Children and Adolescents. *Bull. World Health Organ.* **2007**, *85*, 660–667. [CrossRef] [PubMed]
44. Pastor, R.; Bouzas, C.; Albertos, I.; García, C.; García, Á.; Prieto, T.; Velázquez, J.; Sánchez-Jiménez, E.; Rodríguez, R.; Martín, F.J.; et al. Health-Related Quality of Life in Spanish Schoolchildren and Its Association with the Fitness Status and Adherence to the Mediterranean Diet. *Nutrients* **2022**, *14*, 2322. [CrossRef]
45. Arain, M.; Haque, M.; Johal, L.; Mathur, P.; Nel, W.; Rais, A.; Sandhu, R.; Sharma, S. Maturation of the Adolescent Brain. *Neuropsychiatr. Dis. Treat.* **2013**, *9*, 449–461. [CrossRef]
46. Nabbosa, S.; Lee, S. Associations between Food Groups and Health-Related Quality of Life in Korean Adults. *Nutrients* **2022**, *14*, 3643. [CrossRef] [PubMed]
47. Azupogo, F.; Seidu, J.A.; Issaka, Y.B. Higher Vegetable Intake and Vegetable Variety Is Associated with a Better Self-Reported Health-Related Quality of Life (HR-QoL) in a Cross-Sectional Survey of Rural Northern Ghanaian Women in Fertile Age. *BMC Public Health* **2018**, *18*, 920. [CrossRef] [PubMed]
48. Muscaritoli, M. The Impact of Nutrients on Mental Health and Well-Being: Insights From the Literature. *Front. Nutr.* **2021**, *8*, 656290. [CrossRef] [PubMed]
49. Gaspar, T.; Ribeiro, J.P.; de Matos, M.G.; Leal, I.; Ferreira, A. Health-Related Quality of Life in Children and Adolescents: Subjective Well Being. *Span. J. Psychol.* **2012**, *15*, 177–186. [CrossRef] [PubMed]
50. López-Gil, J.F.; García-Hermoso, A. Adherence to the Mediterranean Diet and Subjective Well-Being among Chilean Children. *Appetite* **2022**, *172*, 105974. [CrossRef] [PubMed]
51. López-Gil, J.F.; Gaya, A.R.; Reuter, C.P.; Caetano, C.I.; Sentone, R.G.; Caetano, H.B.S.; Brazo-Sayavera, J. Sleep-Related Problems and Eating Habits during COVID-19 Lockdown in a Southern Brazilian Youth Sample. *Sleep Med.* **2021**, *85*, 150–156. [CrossRef] [PubMed]
52. Combs, D.; Goodwin, J.L.; Quan, S.F.; Morgan, W.J.; Shetty, S.; Parthasarathy, S. Insomnia, Health-Related Quality of Life and Health Outcomes in Children: A Seven Year Longitudinal Cohort. *Sci. Rep.* **2016**, *6*, 27921. [CrossRef]
53. López-Gil, J.F.; Brazo-Sayavera, J.; de Campos, W.; Yuste Lucas, J.L. Meeting the Physical Activity Recommendations and Its Relationship with Obesity-Related Parameters, Physical Fitness, Screen Time, and Mediterranean Diet in Schoolchildren. *Children* **2020**, *7*, 263. [CrossRef]
54. García-de-Miguel, M.; Huertas-Hoyas, E.; Pérez-Corrales, J.; Rodríguez-Rivas, C.; García-Bravo, C.; García-Bravo, S.; Camacho-Montaño, L.R. Food Intake Changes and Their Impact on Quality of Life in Spanish Citizens with and without COVID-19 during Lockdown. *Healthcare* **2022**, *10*, 1414. [CrossRef]
55. Zahedi, H.; Djalalinia, S.; Sadeghi, O.; Garizi, F.Z.; Asayesh, H.; Payab, M.; Zarei, M.; Qorbani, M. Breakfast Consumption and Mental Health: A Systematic Review and Meta-Analysis of Observational Studies. *Nutr. Neurosci.* **2022**, *25*, 1250–1264. [CrossRef]
56. Ahadi, Z.; Kelishadi, R.; Qorbani, M.; Zahedi, H.; Aram, M.; Motlagh, M.E.; Ardalan, G.; Shafiee, G.; Arzaghi, S.M.; Asayesh, H.; et al. Association of Breakfast Intake with Psychiatric Distress and Violent Behaviors in Iranian Children and Adolescents: The CASPIAN- IV Study. *Indian J. Pediatr.* **2016**, *83*, 922–929. [CrossRef] [PubMed]
57. Mullan, B.A.; Singh, M. A Systematic Review of the Quality, Content, and Context of Breakfast Consumption. *Nutr. Food Sci.* **2010**, *40*, 81–114. [CrossRef]

58. Wu, X.Y.; Zhuang, L.H.; Li, W.; Guo, H.W.; Zhang, J.H.; Zhao, Y.K.; Hu, J.W.; Gao, Q.Q.; Luo, S.; Ohinmaa, A.; et al. The influence of diet quality and dietary behavior on health-related quality of life in the general population of children and adolescents: A systematic review and meta-analysis. *Qual. Life Res. Int. J. Qual. Life Asp. Treat. Care Rehabil.* **2019**, *28*, 1989–2015. [CrossRef] [PubMed]
59. Cenni, S.; Sesenna, V.; Boiardi, G.; Casertano, M.; Di Nardo, G.; Esposito, S.; Strisciuglio, C. The Mediterranean Diet in Paediatric Gastrointestinal Disorders. *Nutrients* **2023**, *15*, 79. [CrossRef]
60. El Amrousy, D.; Elashry, H.; Salamah, A.; Maher, S.; Abd-Elsalam, S.M.; Hasan, S. Adherence to the Mediterranean Diet Improved Clinical Scores and Inflammatory Markers in Children with Active Inflammatory Bowel Disease: A Randomized Trial. *J. Inflamm. Res.* **2022**, *15*, 2075–2086. [CrossRef]
61. Hosseini, B.; Berthon, B.S.; Saedisomeolia, A.; Starkey, M.R.; Collison, A.; Wark, P.A.B.; Wood, L.G. Effects of Fruit and Vegetable Consumption on Inflammatory Biomarkers and Immune Cell Populations: A Systematic Literature Review and Meta-Analysis. *Am. J. Clin. Nutr.* **2018**, *108*, 136–155. [CrossRef] [PubMed]
62. Sanchez-Rodriguez, E.; Biel-Glesson, S.; Fernandez-Navarro, J.R.; Calleja, M.A.; Espejo-Calvo, J.A.; Gil-Extremera, B.; de la Torre, R.; Fito, M.; Covas, M.-I.; Vilchez, P.; et al. Effects of Virgin Olive Oils Differing in Their Bioactive Compound Contents on Biomarkers of Oxidative Stress and Inflammation in Healthy Adults: A Randomized Double-Blind Controlled Trial. *Nutrients* **2019**, *11*, 561. [CrossRef]
63. Dai, J.; Jones, D.P.; Goldberg, J.; Ziegler, T.R.; Bostick, R.M.; Wilson, P.W.; Manatunga, A.K.; Shallenberger, L.; Jones, L.; Vaccarino, V. Association between Adherence to the Mediterranean Diet and Oxidative Stress. *Am. J. Clin. Nutr.* **2008**, *88*, 1364–1370. [CrossRef] [PubMed]
64. Kaushik, A.S.; Strath, L.J.; Sorge, R.E. Dietary Interventions for Treatment of Chronic Pain: Oxidative Stress and Inflammation. *Pain* **2020**, *9*, 487–498. [CrossRef]
65. Ruiz-Roso, M.B.; de Carvalho Padilha, P.; Mantilla-Escalante, D.C.; Ulloa, N.; Brun, P.; Acevedo-Correa, D.; Ferreira Peres, W.A.; Martorell, M.; Aires, M.T.; de Oliveira Cardoso, L.; et al. COVID-19 Confinement and Changes of Adolescent's Dietary Trends in Italy, Spain, Chile, Colombia and Brazil. *Nutrients* **2020**, *12*, 1807. [CrossRef]
66. Madigan, S.; Eirich, R.; Pador, P.; McArthur, B.A.; Neville, R.D. Assessment of Changes in Child and Adolescent Screen Time During the COVID-19 Pandemic: A Systematic Review and Meta-Analysis. *JAMA Pediatr.* **2022**, *176*, 1188–1198. [CrossRef]
67. Hartman-Munick, S.M.; Lin, J.A.; Milliren, C.E.; Braverman, P.K.; Brigham, K.S.; Fisher, M.M.; Golden, N.H.; Jary, J.M.; Lemly, D.C.; Matthews, A.; et al. Association of the COVID-19 Pandemic With Adolescent and Young Adult Eating Disorder Care Volume. *JAMA Pediatr.* **2022**, *176*, 1225–1232. [CrossRef] [PubMed]
68. Regan, C.; Walltott, H.; Kjellenberg, K.; Nyberg, G.; Helgadóttir, B. Investigation of the Associations between Diet Quality and Health-Related Quality of Life in a Sample of Swedish Adolescents. *Nutrients* **2022**, *14*, 2489. [CrossRef]
69. Marques, A.; Peralta, M.; Santos, T.; Martins, J.; de Matos, M.G. Self-Rated Health and Health-Related Quality of Life Are Related with Adolescents' Healthy Lifestyle. *Public Health* **2019**, *170*, 89–94. [CrossRef] [PubMed]
70. Shariati-Bafghi, S.-E.; Rashidkhani, B.; Fadardi, J.S.; Safarian, M.; Edalatian, J.; Ranjbar, G.; Nematy, M. Dietary Patterns and Health-Related Quality of Life among Iranian Adolescents. *Qual. Life Res.* **2022**, *31*, 789–802. [CrossRef] [PubMed]
71. Pano, O.; Martínez-Lapiscina, E.H.; Sayón-Orea, C.; Martinez-Gonzalez, M.A.; Martinez, J.A.; Sanchez-Villegas, A. Healthy Diet, Depression and Quality of Life: A Narrative Review of Biological Mechanisms and Primary Prevention Opportunities. *World J. Psychiatry* **2021**, *11*, 997–1016. [CrossRef]
72. Holder, M.D. The Contribution of Food Consumption to Well-Being. *ANM* **2019**, *74*, 44–52. [CrossRef]
73. Górska-Warsewicz, H.; Rejman, K.; Laskowski, W.; Czeczotko, M. Milk and Dairy Products and Their Nutritional Contribution to the Average Polish Diet. *Nutrients* **2019**, *11*, 1771. [CrossRef]

nutrients

MDPI

Article

Adherence to the Mediterranean Diet and Ultra-Processed Foods Consumption in a Group of Italian Patients with Celiac Disease

Marta Tristan Asensi [1,*], Giuditta Pagliai [1], Sofia Lotti [1], Abigail Corrao [2], Barbara Colombini [1], Ilaria Giangrandi [1,3], Francesco Sofi [1,3] and Monica Dinu [1]

1 Department of Experimental and Clinical Medicine, University of Florence, 50134 Florence, Italy
2 Department of Human Science, Georgetown University, Washington, DC 20057, USA
3 Unit of Clinical Nutrition, Careggi University Hospital, 50134 Florence, Italy
* Correspondence: marta.tristanasensi@unifi.it; Tel.: +39-3791439612

Abstract: Evidence on the consumption of ultra-processed foods (UPF) in adults with celiac disease (CD) and its impact on Mediterranean Diet (MD) adherence is still limited. Our aim was to determine UPF consumption and its relationship with MD adherence in a group of adults, according to the presence of CD. This case-control study included 103 adults with CD and 312 without CD. UPF intake was assessed using the NOVA Food Frequency Questionnaire (NFFQ), while MD adherence was assessed using the Medi-Lite score. UPF represented 14.5% of the diet of participants with CD (246 g/day) and came mainly from cereals-based products (29%) and sweets (24.2%). UPF consumption did not differ with the presence of CD, but participants with CD had significantly ($p < 0.05$) higher consumption of precooked pasta and pre-packaged breads. Participants with CD also reported a significantly lower MD adherence than participants without CD (9.4 vs. 10.4), with higher intake of meat and dairy products, and lower consumption of vegetables and fish. An inverse trend was found between UPF consumption and MD adherence in adults with CD, although not statistically significant. These findings highlight the importance of improving nutrition education for subjects with CD, which should not only focus on gluten exclusion.

Keywords: celiac disease; ultra-processed foods; NOVA classification; NFFQ; Mediterranean diet; Medi-Lite; gluten free diet

Citation: Tristan Asensi, M.; Pagliai, G.; Lotti, S.; Corrao, A.; Colombini, B.; Giangrandi, I.; Sofi, F.; Dinu, M. Adherence to the Mediterranean Diet and Ultra-Processed Foods Consumption in a Group of Italian Patients with Celiac Disease. *Nutrients* **2023**, *15*, 938. https://doi.org/10.3390/nu15040938

Academic Editor: David S. Sanders

Received: 10 January 2023
Revised: 8 February 2023
Accepted: 10 February 2023
Published: 13 February 2023

1. Introduction

Celiac disease (CD) is a chronic autoimmune disease that affects the small intestine in genetically susceptible individuals. This condition is triggered by exposure to gluten, a protein complex found in wheat, rye, and barley [1]. Although the prevalence of CD is highest in childhood, being one of the most common chronic diseases in children, it is well known that it affects all age groups [1,2]. Moreover, CD is more prevalent in females than in males [2]. The most common presenting symptoms are malabsorption, diarrhea, steatorrhea, and weight loss resulting from damage to the intestinal mucosa. Other less common symptoms include abdominal pain, reflux esophagitis, or depression [1].

The mainstay of treatment for CD is strict adherence to a gluten-free diet [3]. Although the only restriction of the diet is the exclusion of gluten, the nutritional adequacy of this diet remains controversial, as some evidence suggests that the gluten-free diet is unbalanced due to lower fiber intake and greater intakes of sugar, total fat, and saturated fat [4]. A potential cause has been attributed to the quality of gluten-free products, which are generally associated with processed foods with low nutritional density [5–7]. Indeed, the so-called ultra-processed foods (UPF), known as "industrial formulations typically with five or more and usually many ingredients" [8], are associated with unbalanced diets due to their formulation rich in free sugars, fats, and salt [9–11].

In addition, some studies also suggest that people with CD do not adequately consume the different food groups that are part of a healthy diet [12]. One of the dietary patterns with robust evidence of its beneficial effect on health status is the Mediterranean diet (MD) [13]. To date, several studies in Mediterranean countries have attempted to investigate whether people with CD follow the MD, particularly in the pediatric population [12,14–16]. However, scientific evidence of UPF consumption in adults with CD and the possible impact it may have on MD adherence is still limited. For this reason, the aim of the present study was to determine the consumption of UPF and its possible relationship with MD adherence in a group of adults with CD compared to a group of adults without this condition.

2. Materials and Methods

2.1. Study Design and Data Collection

All study participants were recruited sequentially from patients referred to the Clinical Nutrition Unit of Azienda Ospedaliera Universitaria Careggi, Florence, Italy, during the period from January to December 2021. The cases consisted of all patients with CD who made a first nutritional visit to our unit during the study period. The controls were also patients who had a first nutritional visit at our unit, but without a diagnosis of CD or other disease. For optimal comparison, each case was matched with 3 controls, with age and gender matching. All subjects who were asked to participate in the study agreed. Data were collected through a self-administered questionnaire created with the online survey platform SurveyMonkey (www.surveymonkey.com, accessed 31 September 2022) [17]. Prior to starting the questionnaire, participants were asked to read the study project information sheet and to sign the informed consent form. Afterwards, participants were asked to fill a brief questionnaire on sociodemographic characteristics (age, gender, weight, height, civil status, and educational level) and two validated questionnaires to collect data on adherence to MD and UPF intake. Body mass index (BMI) was calculated as weight (kg)/height (m^2). Civil status was classified as single, married/partnered, and divorced/widowed, while education level was classified as secondary school, high school, and university. This observational study was conducted in conformity with the guidelines set out in the Declaration of Helsinki and was approved by the Ethics Committee of the Tuscany Region, Azienda Ospedaliera Universitaria Careggi, Florence, Italy [CEAVC 18353/OSS].

2.2. NOVA Food Frequency Questionnaire (NFFQ)

Food consumption was assessed using the NOVA Food Frequency Questionnaire (NFFQ), a validated questionnaire aimed to evaluate the food consumption of Italian adults according to NOVA group classification [18]. Participants were asked to fill the 94 items of the NFFQ by indicating their typical frequency of consumption and portion size in their diet for a typical month within the last 12 months. Participants could choose 1 of 10 consumption frequency options, ranging from never or less than once a month to daily consumption. Portion sizes included 6 options, ranging from 0.5 to 3 portions, identified according to Italian reference portions and the portions indicated on food labels. The 94 items are subdivided into 9 main food groups: (1) fruits and nuts; (2) vegetables and legumes; (3) cereals and tubers; (4) meat and fish; (5) milk, dairy products, and eggs; (6) oils, fats, and seasonings; (7) sweets and sweeteners; (8) beverages; and (9) other. The amount consumed for each item, and the larger food group, was calculated in grams per week and grams per day.

All food and beverages included in the NFFQ were categorized according to the level of food processing, as unprocessed and minimally processed (MPF); processed culinary ingredients (PCI); processed (PF); and UPF. For the analysis of NFFQ data, PCI products were grouped with PF, as they are to be consumed with other foods, not alone as an independent food group. The weight ratio of foods within each level of processing was calculated, rather than the energy ratio, to more effectively represent processed foods that do not provide calories and non-nutritional components of processed foods.

2.3. The Medi-Lite Adherence Score

Participants also completed the Medi-Lite score [19], an evidence-based tool developed in 2014 and validated in 2017 [20], to determine individual adherence to MD. This questionnaire assigns points from 0 to 2 to daily and/or weekly consumption of 9 food groups, in accordance with the MD, to generate a final score ranging from 0 (lowest adherence to MD) to 18 (highest adherence to MD). Point values are assigned according to the frequency of consumption, using reference portions; the quantities chosen as cut-offs for each item were calculated on the basis of available literature linking the consumption of typical and non-typical Mediterranean foods to health indicators [19]. For foods typical of the MD, including fruits, vegetables, cereals, legumes, and fish, 2 points are assigned to the highest level of consumption, 1 point to the intermediate level, and 0 points to the lowest level of intake. Similarly for olive oil, 2 points are awarded for regular use, 1 for frequent use, and 0 for occasional use. Further, 2 points are assigned to the lowest intake level, 1 to the intermediate level, and 0 to the highest level of consumption for meat and meat products and dairy products, which are foods non-typical of the MD. For alcohol, 2 points are given to the intermediate level of consumption, 1 for the lowest, and 0 for the highest intake level.

In this study, the optimal intake for individual food groups was defined as the choice that produced > 2 points and corresponded to the following consumption levels: fruit > 2 portions per day, vegetables > 2.5 portions per day, cereals > 1.5 portions per day, dairy products < 1 portion per day, meat and meat products < 1 portion per day, fish > 2.5 portions per week, legumes > 2 portions per week, alcohol 1–2 alcohol units per day, and regular use of olive oil.

2.4. Statistical Analysis

Statistical analysis was performed using the statistical package IBM SPSS Statistics for Macintosh, version 28.0 (IBM Corp., Armonk, N.Y., USA). Continuous variables are expressed as mean ± standard deviation (SD) or median and range (min–max), as appropriate. Categorical variables are reported as frequencies and percentages. The Mann–Whitney test was used to compare participants with CD and participants without CD, while the Chi-square test was used to test proportions.

A general linear model adjusted for age, gender, BMI, education level, civil status, and daily food consumed was conducted to assess daily UPF consumption, according to the presence of CD. Since this test assumes a normal distribution of data, non-normally distributed data were transformed into logs, and further analyses were performed with the processed data. However, to facilitate interpretation, the log data were again converted to the original scale (antilog) and presented as geometric means with 95% confidence intervals (CIs). To assess the possible relationship between MD adherence and UPF consumption in participants with CD and participants without CD, they were divided into tertiles (1st tertile ≤ 10%; 2nd tertile = 10–17%; 3rd tertile ≥ 17%) according to their level of UPF intake (proportion by weight). A p-value < 0.05 was considered statistically significant.

3. Results

From January to December 2021, a total of 112 adults with CD were enrolled in the study as cases and matched with 336 adults without CD as controls. Subjects who did not answer all test questions were excluded, leaving a total of 103 cases with CD (92%) and 312 controls without CD (93%), for a total of 415 participants. The mean age of study participants was 39.6 ± 13.1 years, 84.3% of whom were women. About 25% of the subjects was overweight or obese (mean BMI 23.1 ± 4.2 kg/m^2). Participants with CD had an average of 11.4 ± 8.6 years of CD progression since diagnosis. The sociodemographic characteristics of the subjects according to the presence of CD are shown in Table 1. More than half of the study population reported being married or with a partner. Overall, the study population was highly educated; a significant difference was found amongst participants with CD reporting secondary school as the highest educational level achieved as compared to participants without CD. No other significant differences between the two groups were observed.

Table 1. Sociodemographic characteristics of the study population.

	Adults with Celiac Disease (n = 103)	Adults without Celiac Disease (n = 312)	p-Value
Age (years)	40.4 ± 12.6	39.3 ± 13.3	0.471
Gender, n (% women)	89 (86.4)	261 (83.7)	0.505
Body weight (kg)	63.4 ± 12.1	64.8 ± 14	0.823
BMI (kg/m^2)	23.2 ± 4.2	23.1 ± 4.1	0.740
BMI ≥ 25, n (%)	25 (24)	78 (25.0)	0.882
Education level			
Secondary school, n (%)	10 (9.7)	14 (4.5)	0.049
High school, n (%)	36 (35)	133 (42.6)	0.169
University degree, n (%)	57 (55.3)	165 (52.9)	0.206
Civil status			
Single, n (%)	37 (35.9)	131 (42)	0.276
Married/partner, n (%)	61 (59.2)	150 (48.1)	0.050
Divorced/Widowed, n (%)	5 (4.9)	31(10)	0.112
Years of disease	11.4 ± 8.6	-	-

Legend: BMI = Body Mass Index. Data are reported as mean ± standard deviation (SD) or number and percentage (%), as appropriate.

3.1. Ultra-Processed Foods' Consumption

UPF represented 14.5 ± 8% of the diet of participants with CD, which was equivalent to 246.3 ± 139.2 g/day. These values did not differ significantly from those reported in participants without CD, which showed a dietary percentage of 15.7 ± 9.5% of UPF equivalent to 264.4 ± 163.5 g/day. Table 2 reports the consumption of single UPF categories in the two groups. After adjustment for possible confounding factors, such as gender, age, and total food consumed (g/day), significant differences between participants with CD and without CD were found in the consumption of vegetables and legumes (p = 0.021) and meat and fish (p = 0.016), with lower consumption in participants without CD compared to participants with CD (−47.2% and −28.7%, respectively). Although no significant differences in overall consumption of ultra-processed cereals and tubers were found between the two groups, participants with CD were significantly more likely to consume ready-to-heat pasta (p = 0.033) and pre-packaged breads and bread alternatives (p = 0.012) than participants without CD (+32% and+25.5%, respectively). No significant differences were found in the consumption of the other categories of UPF between the two groups.

Table 2. UPF intake (g/day) in adults according to the presence of celiac disease.

	Adults with Celiac Disease (n = 103)	Adults without Celiac Disease (n = 312)	p-Value
Vegetables and legumes UPF	15.9 (5.4–26.4)	30.1 (24.1–36.2)	0.021
Ready-to-heat vegetables and legumes (with added ingredients)	15.9 (5.4–26.4)	30.1 (24.1–36.2)	0.021
Cereals and tubers UPF	66.5 (57.0–75.9)	56.2 (50.8–61.6)	0.066
Ready-to-heat pasta/gnocchi dishes	9.7 (7.2–12.1)	6.6 (5.1–8.0)	0.033
Pre-packaged breads and bread alternatives	26.3 (21.7–30.9)	19.6 (17.0–22.2)	0.012
Pre-packaged pizza, focaccia, sandwich, and savory pies	15.9 (11.6–20.2)	13.0 (10.5–15.5)	0.254

Table 2. *Cont.*

	Adults with Celiac Disease (*n* = 103)	Adults without Celiac Disease (*n* = 312)	*p*-Value
Breakfast cereals and energy bars (with added sugar)	5.1 (3.0–7.1)	5.6 (4.4–6.8)	0.667
Pre-packaged potatoes, croquets, and instant soups	9.2 (6.0–12.3)	11.1 (9.3–13.0)	0.292
Meat and fish UPF	8.7 (6.1–11.2)	12.2 (10.8–13.8)	0.016
Nuggets, sticks, sausages, burgers, and other reconstituted meat products	7.5 (5.3–9.8)	10.2 (8.9–11.5)	0.043
Fish nuggets, fish sticks, and other reconstituted fish products	1.1 (0.3–1.9)	2.1 (1.6–2.5)	0.043
Milk and dairy products UPF	28.6 (19.9–37.3)	33.4 (28.4–38.4)	0.346
Milk beverages (e.g., probiotic milk with added sugar)	6.4 (1.7–11.1)	7.9 (5.2–10.6)	0.579
Fruit or flavored yogurts (e.g., vanilla flavored)	20.6 (13.4–27.8)	23.9 (19.8–28.1)	0.431
Melted cheese (also used as a sandwich filling)	1.6 (1.0–2.2)	1.6 (1.2–1.9)	0.949
Fats and seasoning UPF	4.9 (3.5–6.2)	5.2 (4.5–6.0)	0.654
Margarines and other spreads, or instant sauces (e.g., mayonnaise, ketchup)	4.9 (3.5–6.2)	5.2 (4.5–6.0)	0.654
Sweets and Sweeteners UPF	51.4 (44.4–58.4)	45.8 (41.7–49.8)	0.173
Pre-packaged biscuits, cakes, snacks, and ice-cream	38.4 (32.1–44.7)	33.3 (29.7–36.9)	0.169
Chocolate, spreads (e.g., nut spread), and candies	13.0 (10.2–15.8)	12.4 (10.8–14.0)	0.740
Beverages UPF	36.9 (24.9–49.0)	46.3 (39.4–53.2)	0.186
Alcoholic beverages, soft and energy drinks (e.g., iced tea, coke)	36.9 (24.9–49.0)	46.3 (39.4–53.2)	0.186
Other UPF	26.6 (17.3–35.8)	19.9 (14.6–25.1)	0.216
Plant-based dairy substitutes (e.g., soy drinks)	25.0 (16.3–33.7)	16.9 (11.9–21.9)	0.112
Plant-based meat substitutes (e.g., veggie burger)	1.6 (0.2–3.0)	3.0 (2.2–3.8)	0.078

Legend: UPF = ultra-processed foods. Data are reported as geometric mean and 95% confidence interval (CI). General linear model adjusted for age, gender, BMI, education level, civil status, and daily food consumption (g/day).

The food categories that contribute the most to UPF consumption in participants with CD were cereals and tubers (29%), sweets and sweeteners (24.2%), and beverages (12.7%).

3.2. Adherence to the Mediterranean Diet

Participants with CD reported a significantly ($p < 0.001$) lower adherence score to the MD (9.4 ± 2.2) than participants without CD (10.4 ± 2.5). Figure 1 shows the results obtained for the different food groups considered in the Medi-Lite questionnaire, according to whether the optimal recommendations of the MD were met. A significantly larger number of participants with CD did not meet the recommended portions of meat and meat products (61.2% vs. 40.1%), vegetables (77.7% vs. 58.3%), fish (93.2% vs. 79.8%), and dairy products (50.5% vs. 37.8%) than participants without CD, which indicates a higher consumption of non-traditional Mediterranean foods, such as meat and dairy products, and lower consumption of traditional Mediterranean, such as vegetables and fish, in participants with CD compared to participants without CD.

3.3. Ultra-Processed Foods' Consumption and Adherence to the Mediterranean Diet

Figure 2 shows the Medi-Lite Score according to the level of UPF consumption in the diet (1st tertile $\leq$ 10%; 2nd tertile 10–17%; 3rd tertile $\geq$ 17%) in adults with and without CD, adjusted for age, gender, BMI, education level, civil status, and daily food consumption (g/day). Both participants with and without CD who had a higher intake of UPF reported a lower adherence to MD [9.0 (8.3–9.7) and 9.6 (9.2–10.1), respectively] than participants with medium [9.3 (8.6–10.1) and 10.7 (10.3–11.2), respectively] or low [9.9 (9.3–10.6) and 11.0 (10.6–11.5), respectively] consumption of UPF. This trend was statistically significant only in participants without CD ($p<0.001$).

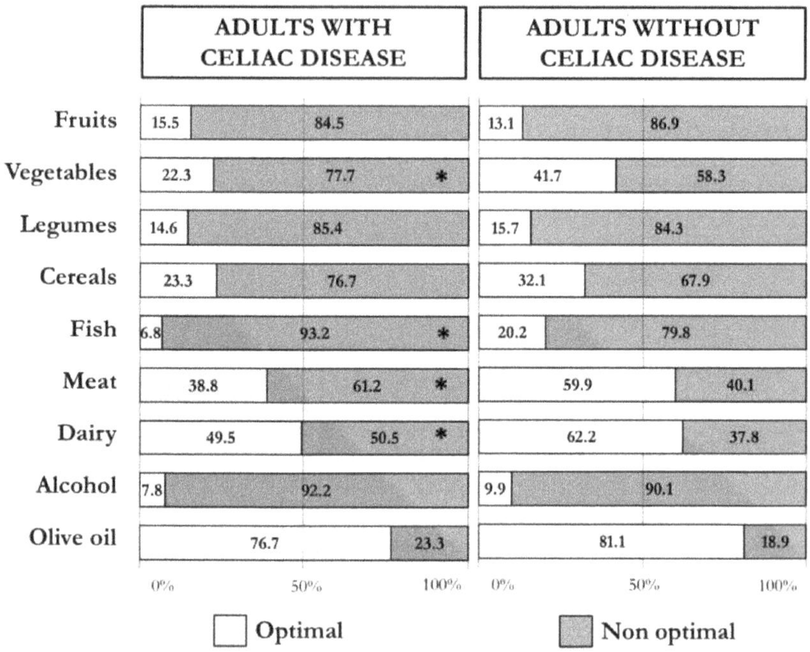

Figure 1. Percentages of adults with and without celiac disease who reported an optimal (2 points) and non-optimal (≤1 points) intake of the single food groups of the Medi-Lite score (* p-value < 0.05).

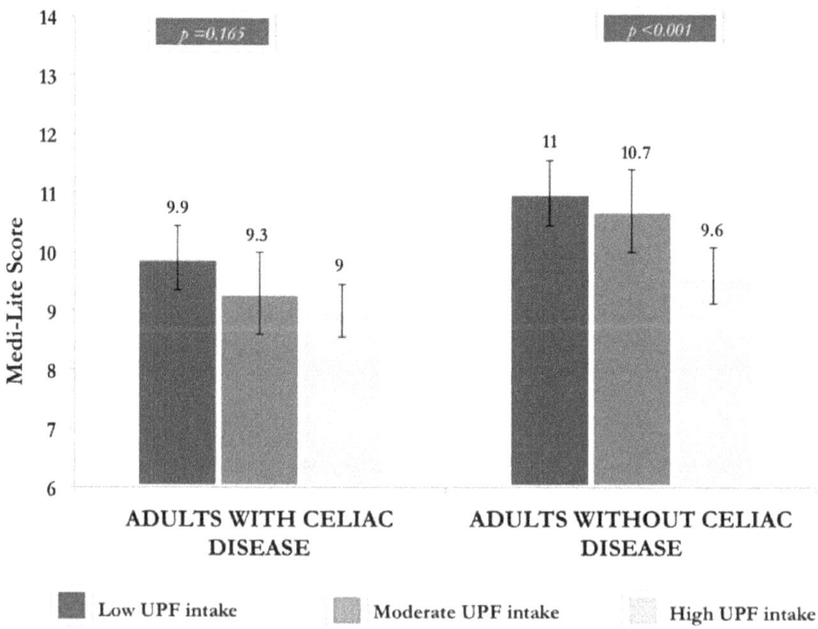

Figure 2. Medi-Lite Score according to the level of UPF consumption in adults with and without celiac disease. Legend: UPF: ultra-processed foods. Vertical bars represent the 95% confidence interval (CI).

4. Discussion

The present is the first study to assessed UPF consumption in relation to MD adherence in a group of adults with CD, in comparison to adults without CD, in Italy. In our study, UPF consumption did not differ according to the presence of CD. However, subjects with CD reported higher consumption of precooked pasta and pre-packaged breads than subjects without CD, with cereal-based products and sweets being the most common UPF in the diets of adults with CD. On the other hand, participants with CD had significantly lower MD adherence compared to participants without CD, with higher consumption of non-traditional Mediterranean foods, such as meat and dairy products, and lower consumption of traditional foods, such as vegetables and fish. Furthermore, an inverse trend was found between UPF consumption and MD adherence in adults with CD, although not statistically significant.

CD is a chronic autoimmune disease of the small intestine, for which the only current treatment is to follow a gluten-free diet [21]. Recent studies have reported a poorer nutritional quality associated with the gluten-free diet due to higher total fat and saturated fat contents and lower fiber content [16,22]. One of the possible causes has been attributed to the quality of gluten-free products, which are generally associated with processed products with low nutritional density [5–7]. Using the validated NFFQ questionnaire specifically designed to assess UPF consumption based on the NOVA classification, we analyzed UPF consumption in a group of adults with CD compared to a group of adults without CD. The higher consumption of ultra-processed pasta and bread we found in participants with CD is not surprising, as these are cereal-based products that naturally contain gluten. In fact, the UPF most consumed by the participants with CD were found to be agglutinated versions of cereal-based products and sweets, including snacks and biscuits, commonly known as gluten-free products. Various studies have shown the importance of these products, not only for their contribution to daily energy intake, but also as a source of dietary carbohydrates for patients with CD. In children and adolescents with CD, gluten-free products contributed between 24 and 36% of the total daily caloric intake and provided 49.5–73% of the total carbohydrate intake [6,16,23]. Although gluten-free products are clearly an important part of UPF consumption among people with CD, the total UPF consumption in our study did not differ from the intake of adults without CD. One possible hypothesis could be related to the use of gluten as an additive ingredient in UPF to improve their organoleptic properties [8], which limits the choice of UPF products among people with CD. For example, the use of different additives in the production of processed meat products [24] can explain the fact that the consumption of ultra-processed meat and fish-based foods in our study was lower in participants with CD than in participants without CD. Nevertheless, in contrast to our findings, a previous study conducted in Spain reported a higher UPF consumption in subjects with CD compared to subjects without CD. These differences can be determinate by the different methods used to evaluate UPF consumption, as well as the different age of participants. In fact, the research conducted in Spain included in the study children and adolescents, who tend to have a higher UPF consumption than adults [25], even in people with CD [26].

The gluten-free diet, despite being a restrictive dietary profile due to the total elimination of gluten-containing foods, should be a varied and balanced diet based on a healthy eating pattern [1]. The countries bordering the Mediterranean Sea are characterized by following the lifestyle promoted by the MD, a dietary pattern associated with robust scientific evidence on its health benefits [13]. MD is characterized by plant-based foods, with high consumption of vegetables, fruits, whole grains, legumes, nuts, and seeds; olive oil as the main source of fat; moderate consumption of fish, dairy products, and eggs; and low consumption of meat. All the food groups mentioned above could be included in a gluten-free diet, with only the exception of whole grains, which should be substituted by naturally gluten-free cereals, such as rice, corn, millet, or teff. For this reason, MD could also be a suitable dietary pattern for people with CD. However, in line with the study published by Morreale et al., [12] our study shows significantly lower MD adherence in adults with CD compared to adults without CD. Specifically, adults with CD reported a higher consumption of non-traditional Mediterranean foods, such as meat, milk, and dairy products, and a lower consumption of traditional foods, such as vegetables and fish. In both studies, healthy adults also reported low values for the consumption of typical Mediterranean foods, highlighting the decrease in MD adherence already observed in several studies [27–29].

As the current literature suggests, the change in eating habits due to the introduction of UPF may replace the consumption of fresh foods that form the basis of traditional diets [30]. In a previous study conducted on Italian adults, UPF consumption was indeed associated with lower adherence to MD [31]. Data on subjects with CD, however, are more limited. To date, only a Spanish study conducted in children and adolescents has analyzed this, finding that patients with CD who consumed higher energy intake from UPF had lower adherence to MD and vice versa [14]. In our study, we also observed a relationship between higher UPF consumption and lower adherence to MD, although the result reached statistical significance only in controls. This could be due to the fact that the case population was limited and underlines the need for further studies on this topic. In fact, due to the increased risk of non-communicable diseases associated with high UPF consumption, such as cardiovascular diseases, cerebrovascular disease, and depression [32], it is of great importance to promote a healthy dietary pattern, such as MD and a limitation of UPF. Furthermore, considering the presence of gluten-free products as a source of UPF in the diets of people with CD, it should also be relevant to promote the consumption of naturally gluten-free cereals to improve the diet quality of people with CD. In fact, some studies have already demonstrated the efficacy of nutrition education in people with CD, reporting an improvement in UPF consumption and better adherence to MD in children, adolescents [15,33], and adults [26].

The present study has some limitations that need to be underlined. Firstly, due to its design, we cannot establish any cause–effect relationship. Secondly, we cannot exclude a sample bias where the participants with and without CD selected for our study may not accurately represent the members of the population. The use of a self-administered online survey is also a limitation of this study, because although it is a very useful and low-cost tool, at the same time, it may lead to recall bias and misclassification errors. However, we used a validated questionnaire specifically designed to estimate UPF intake according to NOVA classification, thus avoiding the misclassification of foods according to the extent and purpose of food processing. Moreover, the evaluation of food intake was made with the proportions, by weight, of UPF in the diet. This method is more suitable than using the energy proportion, to avoid underestimating non-caloric beverages, such as artificially sweetened beverages, as well as various of the non-nutritional components related to food processing, such as additives, neoformed contaminants, food contact materials, or alterations of the food matrix.

5. Conclusions

In conclusion, our study suggests higher consumption of ultra-processed pasta and bread-based foods and lower MD adherence in adults with CD compared to adults without CD. In addition, higher UPF consumption was observed in adults with CD with lower adherence to MD. These findings highlight the importance of improving the nutrition education of people with CD, which should not only focus on gluten exclusion, but also on promoting a balanced and healthy dietary pattern. Special attention should be paid to the intake of gluten-free products, while encouraging the consumption of naturally gluten-free cereal products in order to limit UPF consumption by people with CD.

Author Contributions: Conceptualization, M.T.A., M.D. and F.S.; methodology, M.T.A. and M.D.; software, M.T.A.; formal analysis, M.T.A. and G.P.; investigation, M.T.A., G.P., S.L., A.C. and I.G., data curation, M.T.A. and G.P.; writing—original draft preparation, M.T.A., G.P. and A.C.; writing—review and editing, M.D., B.C. and F.S.; visualization, M.T.A. and G.P.; supervision, B.C., F.S. and M.D. All authors have read and agreed to the published version of the manuscript.

Funding: This research did not receive any specific grant from any funding agency, commercial, or non-profit sector.

Institutional Review Board Statement: All procedures involving human subjects were approved by the Ethics Committee of the Tuscany Region, Careggi University Hospital, Florence, Italy [CEAVC 18353/OSS].

Informed Consent Statement: Informed consent was obtained from all subjects participating in the study.

Data Availability Statement: Additional information is available on request from the corresponding author.

Conflicts of Interest: The authors declare no conflict of interest.

References

1. Al-Toma, A.; Volta, U.; Auricchio, R.; Castillejo, G.; Sanders, D.S.; Cellier, C.; Mulder, C.J.; Lundin, K.E.A. European Society for the Study of Coeliac Disease (ESsCD) guideline for coeliac disease and other gluten-related disorders. *United Eur. Gastroenterol. J.* **2019**, *7*, 583–613. [CrossRef] [PubMed]
2. Singh, P.; Arora, A.; Strand, T.A.; Leffler, D.A.; Catassi, C.; Green, P.H.; Kelly, C.P.; Ahuja, V.; Makharia, G.K. Global Prevalence of Celiac Disease: Systematic Review and Meta-analysis. *Clin. Gastroenterol. Hepatol.* **2018**, *16*, 823–836. [CrossRef] [PubMed]
3. Ludvigsson, J.F.; Bai, J.C.; Biagi, F.; Card, T.R.; Ciacci, C.; Ciclitira, P.J.; Green, P.H.; Hadjivassiliou, M.; Holdoway, A.; van Heel, D.A.; et al. Diagnosis and management of adult coeliac disease: Guidelines from the British Society of Gastroenterology. *Gut* **2014**, *63*, 1210–1228. [CrossRef] [PubMed]
4. Cardo, A.; Churruca, I.; Lasa, A.; Navarro, V.; Vázquez-Polo, M.; Perez-Junkera, G.; Larretxi, I. Nutritional Imbalances in Adult Celiac Patients Following a Gluten-Free Diet. *Nutrients* **2021**, *13*, 2877. [CrossRef]
5. Fry, L.; Madden, A.M.; Fallaize, R. An investigation into the nutritional composition and cost of gluten-free versus regular food products in the UK. *J. Hum. Nutr. Diet.* **2018**, *31*, 108–120. [CrossRef]
6. Larretxi, I.; Simon, E.; Benjumea, L.; Miranda, J.; Bustamante, M.A.; Lasa, A.; Eizaguirre, F.J.; Churruca, I. Gluten-free-rendered products contribute to imbalanced diets in children and adolescents with celiac disease. *Eur. J. Nutr.* **2019**, *58*, 775–783. [CrossRef]
7. Babio, N.; Lladó Bellette, N.; Besora-Moreno, M.; Castillejo, G.; Guillén, N.; Martínez-Cerezo, F.; Vilchez, E.; Roger, E.; Hernández-Alonso, P.; Salas Salvadó, J. A comparison of the nutritional profile and price of gluten-free products and their gluten-containing counterparts available in the Spanish market. *Nutr. Hosp.* **2020**, *37*, 814–822. [CrossRef]
8. Monteiro, C.A.; Cannon, G.; Levy, R.B.; Moubarac, J.C.; Louzada, M.L.; Rauber, F.; Khandpur, N.; Cediel, G.; Neri, D.; Martinez-Steele, E.; et al. Ultra-processed foods: What they are and how to identify them. *Public Health Nutr.* **2019**, *22*, 936–941. [CrossRef]
9. Martínez Steele, E.; Baraldi, L.G.; da Costa Louzada, M.; Moubarac, J.C.; Mozaffarian, D.; Monteiro, C.A. Ultra-processed foods and added sugars in the US diet: Evidence from a nationally representative cross-sectional study. *BMJ Open* **2016**, *6*, e009892. [CrossRef]
10. Steele, E.M.; Marrón Ponce, J.A.; Cediel, G.; Louzada, M.L.C.; Khandpur, N.; Machado, P.; Moubarac, J.C.; Rauber, F.; Corvalán, C.; Levy, R.B.; et al. Potential reductions in ultra-processed food consumption substantially improve population cardiometabolic-related dietary nutrient profiles in eight countries. *Nutr. Metab. Cardiovasc. Dis.* **2022**, *32*, 2739–2750. [CrossRef]
11. Batal, M.; Johnson-Down, L.; Moubarac, J.C.; Ing, A.; Fediuk, K.; Sadik, T.; Tikhonov, C.; Chan, L.; Willows, N. Quantifying associations of the dietary share of ultra-processed foods with overall diet quality in First Nations peoples in the Canadian provinces of British Columbia, Alberta, Manitoba and Ontario. *Public Health Nutr.* **2018**, *21*, 103–113. [CrossRef] [PubMed]
12. Morreale, F.; Agnoli, C.; Roncoroni, L.; Sieri, S.; Lombardo, V.; Mazzeo, T.; Elli, L.; Bardella, M.T.; Agostoni, C.; Doneda, L.; et al. Are the dietary habits of treated individuals with celiac disease adherent to a Mediterranean diet? *Nutr. Metab. Cardiovasc. Dis.* **2018**, *28*, 1148–1154. [CrossRef] [PubMed]
13. Dinu, M.; Pagliai, G.; Casini, A.; Sofi, F. Mediterranean diet and multiple health outcomes: An umbrella review of meta-analyses of observational studies and randomised trials. *Eur. J. Clin. Nutr.* **2018**, *72*, 30–43. [CrossRef]
14. Nestares, T.; Martín-Masot, R.; Flor-Alemany, M.; Bonavita, A.; Maldonado, J.; Aparicio, V.A. Influence of Ultra-Processed Foods Consumption on Redox Status and Inflammatory Signaling in Young Celiac Patients. *Nutrients* **2021**, *13*, 156. [CrossRef]
15. Suárez-González, M.; Bousoño García, C.; Jiménez Treviño, S.; Iglesias Cabo, T.; Díaz Martín, J.J. Influence of nutrition education in paediatric coeliac disease: Impact of the role of the registered dietitian: A prospective, single-arm intervention study. *J. Hum. Nutr. Diet.* **2020**, *33*, 775–785. [CrossRef]
16. Lionetti, E.; Antonucci, N.; Marinelli, M.; Bartolomei, B.; Franceschini, E.; Gatti, S.; Catassi, G.N.; Verma, A.K.; Monachesi, C.; Catassi, C. Nutritional Status, Dietary Intake, and Adherence to the Mediterranean Diet of Children with Celiac Disease on a Gluten-Free Diet: A Case-Control Prospective Study. *Nutrients* **2020**, *12*, 143. [CrossRef]
17. SurveyMonkey. Available online: https://www.surveymonkey.com/ (accessed on 31 September 2022).
18. Dinu, M.; Bonaccio, M.; Martini, D.; Madarena, M.P.; Vitale, M.; Pagliai, G.; Esposito, S.; Ferraris, C.; Guglielmetti, M.; Rosi, A.; et al. Reproducibility and validity of a food-frequency questionnaire (NFFQ) to assess food consumption based on the NOVA classification in adults. *Int. J. Food. Sci. Nutr.* **2021**, *72*, 861–869. [CrossRef]

19. Sofi, F.; Macchi, C.; Abbate, R.; Gensini, G.F.; Casini, A. Mediterranean diet and health status: An updated meta-analysis and a proposal for a literature-based adherence score. *Public Health Nutr.* **2014**, *17*, 2769–2782. [CrossRef]
20. Sofi, F.; Dinu, M.; Pagliai, G.; Marcucci, R.; Casini, A. Validation of a literature-based adherence score to Mediterranean diet: The MEDI-LITE score. *Int. J. Food. Sci. Nutr.* **2017**, *68*, 757–762. [CrossRef]
21. Caio, G.; Volta, U.; Sapone, A.; Leffler, D.A.; De Giorgio, R.; Catassi, C.; Fasano, A. Celiac disease: A comprehensive current review. *BMC Med.* **2019**, *17*, 142. [CrossRef]
22. Ferrara, P.; Cicala, M.; Tiberi, E.; Spadaccio, C.; Marcella, L.; Gatto, A.; Calzolari, P.; Castellucci, G. High fat consumption in children with celiac disease. *Acta Gastro-Enterol. Belg.* **2009**, *72*, 296–300.
23. Zuccotti, G.; Fabiano, V.; Dilillo, D.; Picca, M.; Cravidi, C.; Brambilla, P. Intakes of nutrients in Italian children with celiac disease and the role of commercially available gluten-free products. *J. Hum. Nutr. Diet.* **2013**, *26*, 436–444. [CrossRef]
24. Spychaj, A.; Pospiech, E.; Iwanska, E.; Montowska, M. Detection of allergenic additives in processed meat products. *J. Sci. Food. Agric.* **2018**, *98*, 4807–4815. [CrossRef]
25. Ruggiero, E.; Esposito, S.; Costanzo, S.; Di Castelnuovo, A.; Cerletti, C.; Donati, M.B.; de Gaetano, G.; Iacoviello, L.; Bonaccio, M.; INHES Study Investigators. Ultra-processed food consumption and its correlates among Italian children, adolescents and adults from the Italian Nutrition & Health Survey (INHES) cohort study. *Public Health Nutr.* **2021**, *24*, 6258–6271. [CrossRef]
26. Perez-Junkera, G.; Vázquez-Polo, M.; Eizagirre, F.J.; Benjumea, L.; Tutau, C.; Esteban, B.; Miranda, J.; Larretxi, I.; Navarro, V.; Churruca, I.; et al. Application of a Platform for Gluten-Free Diet Evaluation and Dietary Advice: From Theory to Practice. *Sensors* **2022**, *22*, 732. [CrossRef] [PubMed]
27. Veronese, N.; Notarnicola, M.; Cisternino, A.M.; Inguaggiato, R.; Guerra, V.; Reddavide, R.; Donghia, R.; Rotolo, O.; Zinzi, I.; Leandro, G.; et al. Trends in adherence to the Mediterranean diet in South Italy: A cross sectional study. *Nutr. Metab. Cardiovasc. Dis.* **2020**, *30*, 410–417. [CrossRef]
28. Da Silva, R.; Bach-Faig, A.; Quintana, B.R.; Buckland, G.; de Almeida, M.D.V.; Serra-Majem, L. Worldwide variation of adherence to the Mediterranean diet, in 1961-1965 and 2000-2003. *Public Health Nutr.* **2009**, *12*, 1676–1684. [CrossRef]
29. Vilarnau, C.; Stracker, D.M.; Funtikov, A.; da Silva, R.; Estruch, R.; Bach-Faig, A. Worldwide adherence to Mediterranean Diet between 1960 and 2011. *Eur. J. Clin. Nutr.* **2019**, *72*, 83–91. [CrossRef]
30. Monteiro, C.A.; Cannon, G.; Moubarac, J.C.; Levy, R.B.; Louzada, M.L.C.; Jaime, P.C. The UN Decade of Nutrition, the NOVA food classification and the trouble with ultra-processing. *Public Health Nutr.* **2018**, *21*, 5–17. [CrossRef]
31. Dinu, M.; Asensi, M.T.; Pagliai, G.; Lotti, S.; Martini, D.; Colombini, B.; Sofi, F. Consumption of Ultra-Processed Foods Is Inversely Associated with Adherence to the Mediterranean Diet: A Cross-Sectional Study. *Nutrients* **2022**, *14*, 2073. [CrossRef]
32. Pagliai, G.; Dinu, M.; Madarena, M.P.; Bonaccio, M.; Iacoviello, L.; Sofi, F. Consumption of ultra-processed foods and health status: A systematic review and meta-analysis. *Br. J. Nutr.* **2021**, *125*, 308–318. [CrossRef] [PubMed]
33. Suárez-González, M.; Bousoño-García, C.; Jiménez-Treviño, S.; Díaz-Martín, J.J. Gluten-Free Diet: Nutritional Strategies to Improve Eating Habits in Children with Celiac Disease: A Prospective, Single-arm Intervention Study. *Nutrients* **2021**, *13*, 1108. [CrossRef] [PubMed]

nutrients

MDPI

Article

Association between Late-Eating Pattern and Higher Consumption of Ultra-Processed Food among Italian Adults: Findings from the INHES Study

Marialaura Bonaccio [1,*], Emilia Ruggiero [1], Augusto Di Castelnuovo [2], Claudia Francisca Martínez [1], Simona Esposito [1], Simona Costanzo [1], Chiara Cerletti [1], Maria Benedetta Donati [1], Giovanni de Gaetano [1] and Licia Iacoviello [1,3,†] on behalf of the INHES Study Investigators

1 Department of Epidemiology and Prevention, IRCCS NEUROMED, 86077 Pozzilli, Italy
2 Mediterranea Cardiocentro, 80122 Napoli, Italy
3 Department of Medicine and Surgery, Research Center in Epidemiology and Preventive Medicine (EPIMED), University of Insubria, 21100 Varese, Italy
* Correspondence: marialaura.bonaccio@moli-sani.org
† Listed in the Appendix A.

Abstract: Late eating is reportedly associated with adverse metabolic health, possibly through poor diet quality. We tested the hypothesis that meal timing could also be linked to food processing, an independent predictor of health outcomes. We analysed data on 8688 Italians (aged > 19years) from the Italian Nutrition & HEalth Survey (INHES) established in 2010–2013 throughout Italy. Dietary data were collected through a single 24 h dietary recall, and the NOVA classification was used to categorize foods according to increasing levels of processing: (1) minimally processed foods (e.g., fruits); (2) culinary ingredients (e.g., butter); (3) processed foods (e.g., canned fish); (4) ultra-processed foods (UPFs; e.g., carbonated drinks, processed meat). We then calculated the proportion (%) of each NOVA group on the total weight of food eaten (g/d) by creating a weight ratio. Subjects were classified as early or late eaters based on the population's median timing for breakfast, lunch and dinner. In multivariable-adjusted regression models, late eaters reported a lower intake of minimally processed food (β = −1.23; 95% CI −1.75 to −0.71), a higher intake of UPF (β = 0.93; 0.60 to 1.25) and reduced adherence to a Mediterranean Diet (β = −0.07; −0.12 to −0.03) as compared to early eaters. Future studies are warranted to examine whether increased UPF consumption may underpin the associations of late eating with adverse metabolic health reported in prior cohorts.

Keywords: meal timing; late eating; food processing; ultra-processed food; NOVA classification

Citation: Bonaccio, M.; Ruggiero, E.; Di Castelnuovo, A.; Martínez, C.F.; Esposito, S.; Costanzo, S.; Cerletti, C.; Donati, M.B.; de Gaetano, G.; Iacoviello, L. Association between Late-Eating Pattern and Higher Consumption of Ultra-Processed Food among Italian Adults: Findings from the INHES Study. *Nutrients* **2023**, *15*, 1497. https://doi.org/10.3390/nu15061497

Academic Editor: Megan A. McCrory

Received: 2 February 2023
Revised: 1 March 2023
Accepted: 16 March 2023
Published: 20 March 2023

1. Introduction

Obesity and associated cardiometabolic diseases continue to rise worldwide despite extensive public health efforts to reverse this trend [1]. Unhealthy diets, i.e., diets not meeting nutritional requirements, are major risk factors for obesity and associated diseases [2,3], and therefore, common strategies to tackle obesity and diet-related diseases have been almost exclusively focused on food composition, leading to recommendations to reduce sugar, salt and fat while emphasizing high intakes of foods that are natural sources of fibre, vitamins and minerals [4].

Among the factors that possibly contribute to the rise in obesity and cardiometabolic diseases, growing attention has been paid to the timing of food intake (i.e., the time when meals are usually consumed), which has been associated with various indicators of adiposity, possibly, but not entirely, through higher energy intake [5–12].

Population studies suggest that late eating, which refers to a delay in the timing of meals (usually the main meal of the day or the last meal, i.e., dinner) [12] may be a factor implicated in obesity and other non-communicable diseases related to nutrition [13–15].

Potential mechanistic links through which meal timing may promote obesity and associated diseases include, among others, the lower diet quality and higher calorie intake observed in late eaters [16–18]. However, no prior studies to date have evaluated the possible association of meal timing with the intake of foods with different degrees of processing. Actually, it has been suggested that obesity prevalence continues to increase concomitantly with the increased consumption of ultra-processed foods (UPFs) [19]. According to the NOVA classification, UPFs are industrial formulations of ingredients, containing little or no whole food and typically including flavouring and colouring agents, emulsifiers and other cosmetic additives [20]. Consistently, population-based cohorts support a direct association of a large dietary share of UPFs with obesity [21,22] and cardiometabolic diseases [23], as well as with the incidence of major chronic diseases, regardless of the overall diet quality [24].

To fill this knowledge gap, we tested the hypothesis that the meal timing pattern is differentially associated with the intake of foods that have different food processing levels according to the NOVA classification. This study was conducted using a large dataset of adults recruited throughout Italy in 2010–2013.

2. Materials and Methods

2.1. Study Population

The data are from the Italian Nutrition & HEalth Survey (INHES), which was a 3-year telephone-based survey on nutrition and health designed to collect information on dietary habits (i.e., quality, quantity, food and meal patterns), food choice determinants, and food health awareness of the Italian population according to geographical distribution, age, gender and socioeconomic status. A total of 9422 men and women aged ≥4 years throughout Italy were enrolled between November 2010 and November 2013. Details about this cohort have been previously described [25].

To capture an adequate proportion of weekdays and weekends, a survey calendar was organized at a group level accordingly in order to distribute the sample subjects across four seasons (excluding Christmas, Easter and mid-August periods).

During the recruitment phase, the computer-assisted telephone interview method was used to collect dietary data (dietary habits and behaviour), the health status of the subjects, risk factors, anthropometric measurements (for example, height and weight) and health perception. Given the study objective, participants were excluded for the following reasons: subjects below 20 years of age (n = 571), missing data on diet (n = 2), extreme energy intakes reported (<800 kcal/d in men and <500 kcal/d in women or >4000 kcal/d in men and >3500 kcal/d in women; n = 159) and missing data on meal timing (n = 2). Therefore, a total of 8688 subjects were analysed.

2.2. Assessment of Dietary Data

A self-recorded diary, using computer-based 1-day 24-h dietary recall interview (24-HDR) software, and an Italian version of the European Food Propensity Questionnaire were used to record dietary data [26,27].

Subjects were instructed to recall and record the following data for each meal consumed: (a) time and place of food intake; (b) detailed description of foods (or beverages) and (c) the quantity of intake and the food brand chosen (for manufactured foods). Further, a picture booklet was used as a reference by the subjects to report portion sizes. Lastly, participants answered whether they were currently on any diet and whether their consumption differed from their habitual diet.

Individual food items and recipes reported by the participants were later matched with those available in the food list of the data management system INRAN-DIARIO 3.1 [26,28] by a nutritionist during the interviews.

Finally, a total of 2000 single food items extracted from the final output database were included in the software food list.

The NOVA classification [29] was used to categorize each food item into one of the following categories according to the extent and purpose of food processing: (1) fresh or minimally processed foods (e.g., fruit, meat, milk); (2) processed culinary ingredients (e.g., oils, butter, sugar); (3) processed food items (e.g., canned fish, unpackaged freshly made breads); or (4) UPFs containing predominantly industrial substances and little or no whole foods (e.g., carbonated drinks, processed meat, sweet or savoury packaged snacks). Consumption (in g/d) in each of the four NOVA groups and the percentage they represented with respect to the total amount of food eaten were determined in order to obtain a weight ratio. We used this approach instead of the energy ratio because total food amounts better account for non-nutritional factors related to food processing (e.g., neo-formed contaminants, additives and alterations to the structure of raw foods) [30]. The full list of individual foods and food groups categorized according to the NOVA classification is available in Table 1. For analyses on individual meal types, we calculated the consumption in each NOVA group separately for breakfast, lunch and dinner. Adherence to the Mediterranean Diet was evaluated by the Mediterranean Diet Score (MDS) as proposed by Trichopoulou et al. [31]. Briefly, we assigned 1 point to healthy foods (i.e., fruits and nuts, vegetables, legumes, fish, cereals, monounsaturated-to-saturated fat ratio) whose consumption was above the sex-specific medians of intake in the adult population of the whole INHES cohort; foods presumed to be detrimental (i.e., meat and dairy products) were given a positive score if their consumption was below the median. All other intakes received 0 points. For alcohol intake (ethanol), participants who consumed alcohol (men: 10–50 g/d; women: 5–25 g/d) scored 1 point; otherwise, the score was 0. The Mediterranean Diet Score potentially ranges from 0 to 9 (the latter reflecting maximum adherence).

Table 1. Classification of individual food items and food groups by degree of food processing according to NOVA in the INHES study, Italy, 2010–2013.

NOVA Food Category	Food Items
Group 1: Unprocessed or minimally processed foods	Water; fresh, squeezed or dried fruits and leafy and root vegetables; nuts; fresh legumes; wheat; rice; pasta; flour; potatoes; meat; poultry; fish and seafood; milk; plain yogurts without added sugar; eggs; spices; tea and coffee.
Group 2: Processed culinary ingredients	Vinegars; creams; vegetable oils; butter; lard; sugar and honey.
Group 3: Processed foods	Jam; cured traditional ham; olives; canned fruits; salted or sugared nuts; canned or bottled vegetables and legumes; breads; artisanal pizza; smoked and canned fish; cheese; wine and beer.
Group 4: Ultra-processed food	Processed meat (e.g., salami, mortadella, sausages, hamburger, chicken nuggets); fish products (e.g., fish sticks); packaged breads and buns; bread substitutes (e.g., crackers, rusks, breadstick); breakfast cereals and bars; fruit yogurt; fruit drinks; carbonated soft drinks; cocoa drinks; alcoholic drinks (e.g., rum, gin, whisky); energy drinks and bars; milk substitutes (e.g., soy drinks); margarine; mayonnaise and similar; sliced cheese; sweet packaged snacks; plant-based meat alternatives (e.g., veggie burgers); non-sugar sweeteners; sweet biscuits; cakes, croissant and other non-handmade pastries; ice-cream; chocolate; candies and gums; non-sugar sweeteners; baby food.

To evaluate overall diet quality, we also calculated the Food Standards Agency Nutrient Profiling System (FSAm-NPS) dietary index, which is used to compute the Nutri-Score front-of-pack labelling system that ranks food items according to their nutritional value [32].

The FSAm-NPS score was calculated as previously implemented in other population cohorts [24,33] as follows: for all foods and beverages consumed, based on composition for each 100 g of content, 0 to 40 points were allocated for nutrients that should be consumed in limited amounts (A points), i.e., total sugars (g), saturated fats (g), sodium (mg) and energy (kJ), and 0 to 15 points were given for nutrients or components that should be promoted, i.e., dietary fibre (g) and protein (g), and for fruit, vegetables, legumes and nuts (%) (C points). The total score of the product was calculated by subtracting the sum of C points from the sum of A points. Thus, the final FSAm-NPS score for each food/beverage was based on a scale that could theoretically range from −15 (healthiest food) to +40 (least healthy food). Based on this overall FSAm-NPS score, the Nutri-Score labelling system categorizes food products into five colours, associated with letters A (dark green) to E (dark orange), reflecting their nutritional quality [32]. The FSAm-NPS dietary index (DI) was computed at the individual level as an energy-weighted mean of the FSAm-NPS scores of all foods and beverages consumed by each participant using the following equation:

$$\text{FSA} - \text{NPS DI} = \frac{\sum_{i=1}^{n} \text{FS}_i \text{E}_i}{\sum_{i=1}^{n} \text{E}_i}$$

FS_i represents the score of food/beverage 'i', E_i is the energy intake from food/beverage 'i' specific to each participant, and 'n' is the total number of foods/beverages consumed. An increase in the FSAm-NPS dietary index values therefore reflects a decrease in the overall diet quality value.

2.3. Assessment of Meal Timing

The timing of main meals (i.e., breakfast, lunch and dinner) was obtained by using information provided by participants during the 24 h dietary recall, where they were asked to indicate the time of each eating occasion. For each main meal, we calculated the study population sample's median time and assigned 1 point to those participants reporting having (a) breakfast after 7 am (study sample median time); (b) lunch after 1 p.m. (study sample median time); and (c) dinner after 8 p.m. (study sample median time). Individuals consuming meals before the median time were given 0 points. Participants scoring ≥2 points were considered to have a late meal timing pattern; otherwise, people were classified as having an early meal timing pattern. For simplification, we called them late eaters and early eaters, respectively.

2.4. Ascertainment of Covariates

Education was based on the highest qualification attained and was categorized as up to elementary school (corresponding to ≤5 years of study), lower secondary (>5–≤8 years), upper secondary (>8–≤13 years) and postsecondary (>13 years). Present occupations were categorized into six groups: manual, non-manual, housewife, retired, student and unemployed. Marital status was defined as married/living in a couple, single, separated/divorced and widowed. The definition of urban or rural environments was based on the urbanization level described by the European Institute of Statistics (EUROSTAT definition)—obtained by the tool 'Atlante Statistico dei Comuni' provided by the Italian National Institute of Statistics [34]. Subjects were classified as never (one who has never smoked, or who has smoked less than 100 cigarettes in the lifetime), current (smoking one or more cigarettes per day at the time of the interview), former (one who had quit smoking at the time of interview) or occasional smokers (smoking less than 1 cigarette per day at the time of interview). History of cardiovascular disease and cancer and a previous diagnosis of diabetes, hyperlipidaemia or hypertension were self-reported and categorized as yes/no. Body mass index (BMI) was calculated by using self-reported measurements of height and weight, calculated as kg/m^2 and grouped into three categories: normal (≤25 kg/m^2), overweight (>25–<30 kg/m^2) or obese (≥30 kg/m^2). Self-reported sport activity was used as a categorical variable (yes/no).

2.5. Statistical Analysis

The general characteristics of the analytic sample according to early and late-eating patterns are presented as numbers and percentages for categorical variables and means with standard deviations (SDs) for continuous traits. Differences in the distribution of baseline covariates were calculated using generalized linear models adjusted for age, sex and energy intake (GENMOD procedure for categorical variables and GLM procedure for continuous variables in SAS software).

Beta coefficients with 95% confidence intervals (95% CI) from multivariable-adjusted linear regression analyses were used to evaluate the association between the meal timing pattern (independent variable) and each category of NOVA (continuous dependent variable) or dietary index (i.e., the Mediterranean Diet Score and the FSAm-NPS dietary index; continuous dependent variables). Each dietary variable was standardized to one standard deviation to allow comparison. An *a priori* approach was used to select potential covariates instead of statistical criteria [35]. Two models were ultimately fitted: model 1 was adjusted for age, sex and energy intake, and multivariable model 2 was model 1 but further adjusted for education, geographical area, place of residence, sport activity, occupation, marital status, smoking, BMI, cardiovascular disease, cancer, hypertension, diabetes and hyperlipidaemia. To maximize data availability, missing data on covariates were handled using multiple imputation (SAS PROC MI, followed by PROC MIANALYZE; n = 10 imputed datasets).

We conducted subgroup analyses to test the robustness of the findings by analysing the potential effect modification of the association of the meal timing pattern with each dietary score by various risk factors, such as age (19–50 years; 51–65 years and 66–97 years) and sex. We used SAS/STAT software, version 9.4 (SAS Institute Inc., Cary, NC, USA), for the analysis.

3. Results

The analytic sample consists of 4053 men (46.7%) and 4635 women (53.3%) with a mean age of 56.9 years (±14.6). The average (SD) weight contributions of unprocessed/minimally processed foods, culinary ingredients, processed foods and UPFs to the diet were 73.7% (±12.0), 2.6% (±1.2), 15.9% (±10.7) and 7.8% (±7.0), respectively. More than half (58.1%) of the total calories came from unprocessed/minimally processed foods and culinary ingredients, while 24.6% came from processed food, and 17.3% were from UPFs.

The characteristics of the study participants according to the meal timing pattern are presented in Table 2. As compared to early eaters, late eaters were younger, were more likely to live in Southern Italy and urban environments, had a higher educational level and were prevalently non-manual workers. Additionally, late eaters were less likely to report chronic diseases (e.g., CVD) or other health conditions (e.g., hypertension and hyperlipidaemia). No relevant differences in BMI, diabetes or history of cancer were found. Differences in dietary factors were also observed between meal timing patterns. Specifically, late eaters tended to consume less energy from carbohydrates while reporting higher energy from fats (Table 3).

In multivariable-adjusted regression analyses, we found that late eaters were less likely to consume unprocessed/minimally processed foods as compared to early eaters (β = −0.10; 95% CI −0.14 to −0.06) while reporting the increased consumption of UPFs (β = 0.13; 95% CI 0.09 to 0.18) and processed culinary ingredients (β = 0.05; 95% CI 0.01 to 0.10); eating late was also found to be inversely associated with adherence to the Mediterranean Diet (β = −0.07; 95% CI −0.12 to −0.03) and directly associated with the FSAm-NPS dietary index (β = 0.10; 95% CI 0.05 to 0.14) (Table 4; Model 2). The direction and strengths of these associations were substantially confirmed in all age groups and in men and women, especially for UPF consumption and diet quality indices

(Supplementary Tables S1 and S2); however, the relationships of late eating with unprocessed/minimally processed food or processed food intake were stronger in the young group than in the elderly (Supplementary Table S1). Additionally, an effect modification by sex was observed in relation to the consumption of unprocessed/minimally processed foods and culinary ingredients (Supplementary Table S2).

Table 2. Characteristics of 8688 participants (20–97 years) in the INHES study, Italy, 2010–2013.

	Meal Timing Pattern			
	All	**Early Eaters**	**Late Eaters**	***p*-Value**
N of subjects, %	8688 (100.0)	5781 (66.5)	2907 (33.5)	-
Sex				0.44
Men	4053 (46.7)	2680 (46.4)	1373 (47.3)	
Women	4635 (53.3)	3101 (53.6)	1534 (52.8)	
Age (years; mean ± SD)	56.9 ± 14.6	58.9 ± 14.5	52.9 ± 13.9	<0.0001
Age groups, years				<0.0001
19–50	2967 (34.2)	1718 (29.7)	1249 (43.0)	
51–65	2863 (32.9)	1799 (31.1)	1064 (36.6)	
66–97	2858 (32.9)	2264 (39.2)	594 (20.4)	
Geographical area				<0.0001
Northern	3556 (40.9)	2932 (50.7)	624 (21.5)	
Centre	1407 (16.2)	886 (15.3)	521 (17.9)	
Southern	3725 (42.9)	1963 (34.0)	1762 (60.6)	
Place of residence				<0.0001
Rural	1178 (13.6)	861 (14.9)	317 (10.9)	
Urban	7510 (86.4)	4920 (85.1)	2590 (89.1)	
Educational level				<0.0001
Up to elementary	1540 (17.7)	1252 (21.7)	288 (9.9)	
Lower secondary	2268 (26.1)	1589 (27.5)	679 (23.4)	
Upper secondary	3430 (39.5)	2142 (37.0)	1288 (44.3)	
Postsecondary	1385 (15.9)	747 (12.9)	638 (21.9)	
Missing data	65 (0.8)	51 (0.9)	14 (0.5)	
Occupation				0.0001
Non-manual workers	2658 (30.6)	1525 (26.4)	1133 (39.0)	
Manual workers	1537 (17.7)	1006 (17.4)	531 (18.3)	
Housewife	958 (11.0)	623 (10.9)	325 (11.2)	
Retired	3129 (36.0)	2406 (41.6)	723 (24.9)	
Student	142 (1.6)	61 (1.1)	81 (2.7)	
Unemployed	251 (2.9)	145 (2.5)	106 (3.6)	
Missing data	13 (0.2)	5 (0.1)	8 (0.3)	
Marital status				0.19
Married/in couple	6533 (75.2)	4382 (75.8)	2151 (74.0)	
Single	1244 (14.3)	707 (12.2)	537 (18.5)	
Separated/divorced	270 (3.1)	185 (3.2)	85 (2.9)	
Widowed	616 (7.1)	492 (8.5)	124 (4.3)	
Missing data	25 (0.3)	15 (0.3)	10 (0.3)	
Smoking habit				<0.0001
No	5180 (59.6)	3533 (61.1)	1647 (56.7)	
Current	1390 (16.0)	888 (15.4)	502 (17.3)	
Ex	1925 (22.2)	1244 (21.5)	681 (23.4)	
Occasional	163 (1.9)	96 (1.7)	67 (2.3)	
Missing data	30 (0.3)	20 (0.3)	10 (0.3)	
Sport activity				0.067
No	7096 (81.7)	4835 (83.6)	2261 (77.8)	
Yes	1585 (18.2)	943 (16.3)	642 (22.1)	
Missing data	7 (0.1)	3 (0.1)	4 (0.1)	
Cardiovascular disease				0.095
No	8397 (96.7)	5576 (96.6)	2821 (97.0)	
Yes	291 (3.3)	205 (3.4)	86 (3.0)	

Table 2. *Cont.*

	Meal Timing Pattern			
	All	**Early Eaters**	**Late Eaters**	***p*-Value**
Cancer				0.73
No	8397 (96.6)	5570 (96.3)	2827 (97.2)	
Yes	291 (3.4)	211 (3.7)	80 (2.8)	
Hypertension				0.026
No	5859 (67.4)	3762 (65.1)	2097 (72.1)	
Yes	2809 (32.4)	2008 (34.7)	801 (27.6)	
Missing data	20 (0.2)	11 (0.2)	9 (0.3)	
Hyperlipidaemia				0.010
No	6756 (77.8)	4456 (77.2)	2300 (79.0)	
Yes	1902 (21.9)	1307 (22.5)	595 (20.6)	
Missing data	30 (0.3)	18 (0.3)	12 (0.4)	
Diabetes				0.33
No	7997 (92.1)	5281 (91.4)	2716 (93.4)	
Yes	661 (7.6)	482 (8.3)	179 (6.2)	
Missing data	30 (0.3)	18 (0.3)	12 (0.4)	
Body mass index				0.13
Normal weight	4168 (48.0)	2727 (47.2)	1441 (49.6)	
Overweight	3333 (38.3)	2250 (38.9)	1083 (37.2)	
Obese	1172 (13.5)	795 (13.8)	377 (13.0)	
Missing data	15 (0.2)	9 (0.1)	6 (0.2)	

Values are reported as numbers and percentages unless otherwise stated. Means were adjusted for age, sex and energy intake. *p*-values were obtained using generalized linear models for both continuous and categorical dependent variables adjusted for age, sex and energy intake.

Table 3. Dietary factors associated with meal timing pattern in 8688 participants (20–97 years) from the INHES study, Italy, 2010–2013.

	Meal Timing Pattern		
	Early Eaters	**Late Eaters**	***p*-Value**
Energy intake (kcal/d)	1889 ± 578	1913 ± 592	0.093
Alcohol intake (g/d)	9.0 ± 14.5	8.9 ± 13.9	0.87
Carbohydrate (% TEI)	49.1 ± 9.9	48.5 ± 9.7	0.018
Sugar (g/d)	70.1 ± 29.9	69.4 ± 30.2	0.30
Fibre intake (g/d)	18.0 ± 7.8	18.1 ± 8.1	0.48
Protein (% TEI)	16.0 ± 3.8	16.1 ± 3.8	0.27
Fat (% TEI)	34.6 ± 7.9	35.1 ± 7.8	0.0079
Saturated fat (% TEI)	10.1 ± 3.8	10.3 ± 3.7	0.085
Saturated fat (g/d)	21.6 ± 11.4	21.9 ± 11.5	0.11
MUFA (% TEI)	10.1 ± 3.8	10.3 ± 3.7	0.085
PUFA (% TEI)	4.2 ± 1.6	4.2 ± 1.6	0.26
Dietary cholesterol (mg/d)	235.7 ± 168.4	232.1 ± 168.9	0.34
Sodium (mg/d)	1620 ± 1095	1600 ± 1063	0.36
Minimally processed food (Group 1)	74.0 ± 11.8	72.7 ± 12.2	<0.0001
Culinary ingredients (Group 2)	2.6 ±1.2	2.6 ± 1.2	0.12
Processed food (Group 3)	15.9 ± 10.6	16.4 ± 10.7	0.033
Ultra-processed food (Group 4)	7.5 ± 6.7	8.3 ± 7.3	<0.0001

TEI = total energy intake. MUFA = monounsaturated fats. PUFA = polyunsaturated fats. Means and *p*-values obtained from general linear regression models adjusted for sex, age and energy intake.

Analyses separated by meal type showed that late breakfast eating was associated with the reduced consumption of unprocessed/minimally processed foods and processed foods and a higher intake of UPFs at breakfast, as well as with lower adherence to the Mediterranean Diet and a higher FSAm-NPS dietary index. Similarly, participants who had delayed dinners were more likely to eat processed foods or UPFs and tended to reduce the intake of unprocessed/minimally processed foods, and also reported less adherence to a Mediterranean Diet and a larger dietary share of foods with poor nutritional quality. Finally, late lunch eaters reported a higher intake of processed culinary ingredients (Figure 1).

Table 4. Association of food processing according to NOVA classification with meal timing pattern in 8688 participants (20–97 years) from the INHES study, Italy 2010–2013.

	Meal Timing Pattern
	Late vs. Early Eaters
NOVA Groups	β (95% CI)
Minimally processed food (Group 1)	
Model 1	−0.11 (−0.15 to −0.07)
Model 2	−0.10 (−0.14 to −0.06)
Culinary ingredients (Group 2)	
Model 1	0.03 (−0.01 to 0.08)
Model 2	0.05 (0.01 to 0.10)
Processed food (Group 3)	
Model 1	0.04 (0.003 to 0.08)
Model 2	0.02 (−0.02 to 0.06)
Ultra-processed food (Group 4)	
Model 1	0.11 (0.07 to 0.15)
Model 2	0.13 (0.09 to 0.18)
Mediterranean Diet Score	
Model 1	−0.03 (−0.07 to 0.01)
Model 2	−0.07 (−0.12 to −0.03)
FSAm-NPS dietary index	
Model 1	0.05 (0.01 to 0.10)
Model 2	0.10 (0.05 to 0.14)

Model 1: Multivariable-adjusted linear regression including age, sex and energy intake. Model 2: Multivariable-adjusted linear regression including age, sex, energy intake, place of residence, educational level, occupation, marital status, smoking status, sport activity, body mass index, history of cardiovascular disease, history of cancer, diabetes, hyperlipidaemia and hypertension. FSAm-NPS = Food Standards Agency Nutrient Profiling System. Each dietary variable was standardized to allow comparison.

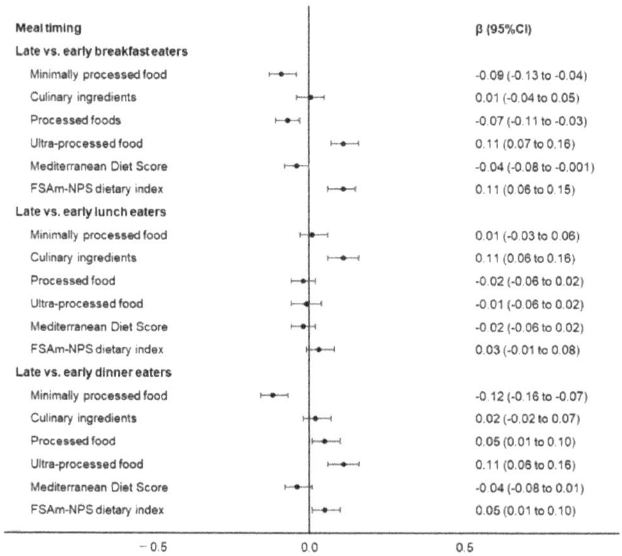

Figure 1. Timing of food intake for individual meals (late vs. early eaters) associated with food processing according to NOVA classification, adherence to the Mediterranean Diet and the Food Standards Agency Nutrient Profiling System (FSAm-NPS) dietary index in 8688 participants (20–97 years) from the INHES study, Italy, 2010–2013. Regression coefficients β with 95% CIs from a multivariable-adjusted linear regression including age, sex, energy intake, place of residence, educational level, occupation, marital status, smoking status, sport activity, body mass index, history of cardiovascular disease, history of cancer, diabetes, hyperlipidaemia and hypertension. Each dietary variable was standardized to allow comparison.

4. Discussion

In this large cohort of 8688 adults from the general Italian population, a late-eating pattern was associated with both a higher consumption of UPFs and a lower intake of unprocessed/minimally processed foods, as well as with poorer diet quality. Evidence from population studies has consistently suggested that the timing of meal intake is a reliable predictor of cardiometabolic health outcomes, with late eating being reportedly associated with obesity and glucose intolerance in observational studies [10,36]. The key role of timed meals has been also supported by animal [37] and intervention studies in humans showing that late eating may adversely impact the success of weight-loss therapy [38].

Mechanistic hypotheses to support the association of late eating with adverse cardiometabolic health are likely multifactorial and include the fact that late eating may contribute to circadian misalignment, i.e., a lack of synchrony of light/dark cycles and behavioural rhythms with the endogenous circadian system [38–40], which was found to adversely impact both energy balance and glycaemic control [41] and changes in the diversity of the microbiota [42].

A number of studies indicate that late eaters tend to have a lower overall diet quality and higher energy intake [16,17,43,44], which may in part explain the adverse cardiometabolic health associated with delaying meals to later in the day; this was also confirmed by our analyses showing that late eating was associated with reduced adherence to a traditional Mediterranean Diet and higher values of the FSAm-NPS dietary index, which is used to compute the Nutri-Score front-of-pack labels and reflects the consumption of less-nutrient-dense foods. However, others reported that energy intake and overall diet quality were not found to vary significantly across eating times [39].

As all prior studies were focused on the nutritional composition of diets, regardless of food processing levels, we used a complementary approach by examining whether meal timing is differentially associated with the food intakes with different levels of processing according to the NOVA classification.

UPF intake is on arise worldwide and constitutes more than half of the total calories eaten in the US, UK and Canada [45–47] while being less consumed in Mediterranean countries, such as Italy [48] and Spain [49]. An increasing number of large-scale population studies indicate that elevated intakes of UPFs can be a major threat to human health, being directly associated with an increased risk of cardiovascular disease, cancer and diabetes, as well as reduced survival [23,24]. A systematic review summarizing the evidence for the association between food processing and cardiometabolic factors in adults found that a large dietary share of UPFs is positively associated with worse cardiometabolic health, as reflected by increased levels of overweight and obesity, metabolic syndrome and high blood pressure [50]. Additionally, a high proportion of UPFs in the diet was linked to altered levels of inflammation [51], which was found to be increased in association with mistimed meals in both animals [52] and humans [53].

Both the direct association of the meal timing pattern with UPFs and its inverse relationship with unprocessed/minimally processed foods observed in our study suggest that the degree of food processing could be among the potential mechanisms/factors that link mistimed meals to impaired cardiometabolic outcomes. Besides being nutrient-poor (e.g., rich in fat, sodium and salt, and low in fibre and nutrients), UPFs are a major dietary source of chemicals (e.g., endocrine-disrupting chemicals such as bisphenol and phthalates commonly used in food packaging) and neo-formed compounds (e.g., acrylamide), which may have severe implications for health, as suggested by robust research, ranging from laboratory-based to prospective epidemiological studies [54].

Most importantly, food processing impacts both the nutritional composition (e.g., decreased antioxidant potential of some foods resulting from removing germ and bran) and food matrix (i.e., the 'architecture' of the food, which derives from nutrient interactions), which is crucial to the food's overall health potential, specifically in satiety and glycaemic responses, as well as in determining nutrient bioavailability [55].

While complex, natural, minimally or unprocessed foods have more or less intact structures, and their nutritional properties are substantially unaltered [55], highly processed foods are typically unstructured, fractionated and usually heavily supplemented with free glucose and sucrose, which renders glucose more available for absorption, thereby increasing blood glycaemic response [56]. Diets with a large share of foods with a high glycaemic index are well-established risk factors for cardiometabolic diseases and mortality [57].

Interestingly, in our study late eating was associated with an approximately absolute 1% higher proportion of UPF intake relative to the total food eaten; prior cohort studies showed that even such a small increment possibly leads to a higher risk of mortality both in general populations [24] and among people with pre-existing cardiovascular disease [58]. Despite consuming more UPFs, late eaters also tended to report lower diet quality overall, and in this regard, it is worth noting that most highly processed foods are typically less nutrient-dense [59]. In addition, diets high in UPFs were found to have a higher impact on mortality than the overall diet quality [24].

Lastly, a late meal pattern in our study was associated with younger age, a higher educational level and being single; all these characteristics were reportedly associated with a higher consumption of UPFs in previous cohort studies [48,60], while unmarried individuals were also found to have lower diet quality overall [61,62]. However, our estimates were from multivariable-adjusted models that also account for these socioeconomic and demographic factors, and other drivers for UPF consumption need consideration (e.g., heavy marketing, availability, low cost, attractiveness, high palatability and domination of food supply chains) [20].

Strengths and Limitations

To the best of our knowledge, this is the first study that analysed meal timing in association with food processing and also with the dietary index underpinning the Nutri-Score front-of-pack label. The major strengths of this study include a large sample size representative of the Italian population, with a complete assessment of diet, lifestyle and other covariates used to minimize, at least in part, confounding. Moreover, the use of 24 h recall is more advantageous than other tools (e.g., food frequency questionnaires) to assess participants' diets and to classify foods based on the extent of processing according to NOVA [63]. Despite its strengths, among its limitations, we acknowledge the observational nature of our study and the cross-sectional design of the analyses, which limits causal inference. Further, errors in the visual display of foods and potential bias could have been introduced by the interviewer in the telephone-based survey. Additionally, the decline in the use of landline phones may have resulted in an under-representation of respondents. Another weakness is that the study relied on self-reported dietary data, which are susceptible to bias and error, including social desirability and recall bias, imprecision in assessing portion sizes and inadequacies in food composition tables; however, data were collected by trained interviewers, and each participant received by mail, beforehand, a short photograph atlas and guidance notes to estimate food portion sizes. It was not possible to include some unmeasured factors as confounders due to their unavailability; however, it is a weakness in any observational study. Limitations also include that we dichotomized our population into early and late eaters using the population median timing, as a consensus on the most suitable approach to quantifying food timing is still lacking [39]. We also acknowledge that the NOVA classification remains controversial, mainly due to its equivocal definition of ultra-processed food and multiple revisions and refinements over time [64]; however, its utility value in nutrition epidemiology research has been widely acknowledged allowing comparison with previous studies. Finally, the generalizability of our findings might be limited to the Italian population.

5. Conclusions

As well as reporting poor diet quality overall, late eaters are prone to consume more UPFs and fewer minimally processed food than early eaters. These findings contribute to increased knowledge on the mechanisms underpinning the association between late eating and adverse cardiometabolic health previously reported in several experimental and observational studies [12,13,39]. Anticipating the timing of meals may provide a complementary strategy for reducing UPF consumption and increasing unprocessed or minimally processed food intakes, which typically require more time and effort than ready-to-eat/heat meals. Undeniably, mistimed meals are strongly influenced by several factors, especially socioeconomic conditions that are difficult to tackle. Further research is warranted to test whether the consumption of UPFs could be a mediator of the association between mistimed meals and adverse cardiometabolic health.

Supplementary Materials: The following supporting information can be downloaded at https://www.mdpi.com/article/10.3390/nu15061497/s1. Table S1: Association of food processing according to NOVA classification with meal timing pattern across age groups from the INHES Study, Italy 2010–2013; Table S2: Association of food processing according to NOVA classification with meal timing pattern in men and women from the INHES Study, Italy 2010–2013.

Author Contributions: M.B., E.R. and A.D.C. conceived the present study, contributed to its design and to the interpretation of data; C.F.M. contributed to data interpretation and critically revised the manuscript; S.C. and A.D.C. managed data collection; E.R. and S.E. analysed the data; M.B. and C.F.M. wrote the manuscript; M.B.D., C.C., G.d.G. and L.I. originally inspired the INHES study and critically reviewed the manuscript. All authors have read and agreed to the published version of the manuscript.

Funding: The enrolment phase of the INHES study was supported by a research grant from Barilla SpA. The present analyses were partially supported by the Italian Ministry of Health—Ricerca Corrente 2022–2024. The present analyses were funded under the National Recovery and Resilience Plan (NRRP), Mission 4 Component 2 Investment 1.3—Call for tender No. 341 of 15/03/2022 of the Italian Ministry of University and Research funded by the European Union—NextGenerationEU. Project code PE0000003, Concession Decree No. 1550 of 11/10/2022 adopted by the Italian Ministry of University and Research, CUP D93C22000890001, Project title "Research and innovation network on food and nutrition Sustainability, Safety and Security–Working ON Foods" (ONFoods). The funders had no role in the study design or the collection, analysis or interpretation of data; in the writing of the report; or in the decision to submit this article for publication.

Institutional Review Board Statement: This study was conducted according to the guidelines laid down in the Declaration of Helsinki, and all procedures involving human subjects were approved by the Ethical Committee of the Catholic University of Rome.

Informed Consent Statement: Verbal informed consent was obtained from all subjects. Verbal consent was witnessed and formally recorded.

Data Availability Statement: The data underlying this article will be shared on reasonable request to the corresponding author. The data are stored in an institutional repository (https://repository.neuromed.it) and access is restricted by ethical approval and the legislation of the European Union.

Acknowledgments: We thank the Research Group of the Cardiovascular Epidemiology Observatory/Health Examination Survey 2008 for making available the list of subjects recruited in their survey. ER was supported by Fondazione Umberto Veronesi, who is gratefully acknowledged. CFM was financially supported by the Joint Platform Laboratory of Umberto Veronesi Foundation—Department of Epidemiology and Prevention at IRCCS Neuromed in Pozzilli, Italy.

Conflicts of Interest: All authors of the present manuscript declare that they have no conflicts of interest to disclose.

Appendix A

INHES Study Investigators
Principal Investigator: Licia Iacoviello
Study coordinator: Americo Bonanni

Scientific Committee: Marialaura Bonaccio, Francesca Bracone, Chiara Cerletti, Simona Costanzo, Augusto Di Castelnuovo, Mariarosaria Persichillo, Maria Benedetta Donati, Giovanni de Gaetano.

Dietary questionnaire validation: Mariarosaria Persichillo and Francesco Zito.

Questionnaire administration: Lucia Aurisano, Paola Barisciano, Valentina Bonaccio, Francesca De Lucia, Giovanna Galuppo, Filippo Petrucci, Anna Sciarretta, Angelita Verna.

Data management: Simona Costanzo, Augusto Di Castelnuovo, Marco Olivieri.

References

1. Roth, G.A.; Mensah, G.A.; Johnson, C.O.; Addolorato, G.; Ammirati, E.; Baddour, L.M.; Barengo, N.C.; Beaton, A.Z.; Benjamin, E.J.; Benziger, C.P. GBD-NHLBI-JACC Global Burden of Cardiovascular Diseases Writing Group. Global Burden of Cardiovascular Diseases and Risk Factors, 1990–2019: Update From the GBD 2019 Study. *J. Am. Coll. Cardiol.* **2020**, *76*, 2982–3021, Erratum in: *J. Am. Coll. Cardiol.* **2021**, *77*, 1958–1959. [CrossRef]
2. Swinburn, B.A.; Kraak, V.I.; Allender, S.; Atkins, V.J.; Baker, P.I.; Bogard, J.R.; Brinsden, H.; Calvillo, A.; De Schutter, O.; Devarajan, R.; et al. The Global Syndemic of Obesity, Undernutrition, and Climate Change: The Lancet Commission report. *Lancet* **2019**, *393*, 791–846, Erratum in: *Lancet* **2019**, *393*, 746. [CrossRef] [PubMed]
3. GBD 2017 Diet Collaborators. Health effects of dietary risks in 195 countries, 1990–2017: A systematic analysis for the Global Burden of Disease Study 2017. *Lancet* **2019**, *393*, 1958–1972, Erratum in: *Lancet* **2021**, *397*, 2466. [CrossRef]
4. Waijers, P.M.; Feskens, E.J.; Ocké, M.C. A critical review of predefined diet quality scores. *Br. J. Nutr.* **2007**, *97*, 219–231. [CrossRef]
5. Garaulet, M.; Gómez-Abellán, P. Timing of food intake and obesity: A novel association. *Physiol. Behav.* **2014**, *134*, 44–50. [CrossRef] [PubMed]
6. Davis, R.; Rogers, M.; Coates, A.M.; Leung, G.K.W.; Bonham, M.P. The Impact of Meal Timing on Risk of Weight Gain and Development of Obesity: A Review of the Current Evidence and Opportunities for Dietary Intervention. *Curr. Diab. Rep.* **2022**, *22*, 147–155. [CrossRef]
7. Xiao, Q.; Garaulet, M.; Scheer, F.A.J.L. Meal timing and obesity: Interactions with macronutrient intake and chronotype. *Int. J. Obes* **2019**, *43*, 1701–1711. [CrossRef]
8. Baron, K.G.; Reid, K.J.; Horn, L.V.; Zee, P.C. Contribution of evening macronutrient intake to total caloric intake and body mass index. *Appetite* **2013**, *60*, 246–251. [CrossRef]
9. Baron, K.G.; Reid, K.J.; Kern, A.S.; Zee, P.C. Role of sleep timing in caloric intake and BMI. *Obesity* **2011**, *19*, 1374–1381. [CrossRef]
10. Wang, J.B.; Patterson, R.E.; Ang, A.; Emond, J.A.; Shetty, N.; Arab, L. Timing of energy intake during the day is associated with the risk of obesity in adults. *J. Hum. Nutr. Diet.* **2014**, *27* (Suppl. 2), 255–262. [CrossRef]
11. Kahleova, H.; Lloren, J.I.; Mashchak, A.; Hill, M.; Fraser, G.E. Meal Frequency and Timing Are Associated with Changes in Body Mass Index in Adventist Health Study 2. *J. Nutr.* **2017**, *147*, 1722–1728. [CrossRef] [PubMed]
12. Martínez-Lozano, N.; Tvarijonaviciute, A.; Ríos, R.; Barón, I.; Scheer, F.A.J.L.; Garaulet, M. Late Eating Is Associated with Obesity, Inflammatory Markers and Circadian-Related Disturbances in School-Aged Children. *Nutrients* **2020**, *12*, 2881. [CrossRef] [PubMed]
13. St-Onge, M.P.; Ard, J.; Baskin, M.L.; Chiuve, S.E.; Johnson, H.M.; Kris-Etherton, P.; Varady, K.; American Heart Association Obesity Committee of the Council on Lifestyle and Cardiometabolic Health; Council on Cardiovascular Disease in the Young; Council on Clinical Cardiology; et al. Meal Timing and Frequency: Implications for Cardiovascular Disease Prevention: A Scientific Statement from the American Heart Association. *Circulation* **2017**, *135*, e96–e121. [CrossRef]
14. Leung, G.K.W.; Huggins, C.E.; Bonham, M.P. Effect of meal timing on postprandial glucose responses to a low glycemic index meal: A crossover trial in healthy volunteers. *Clin. Nutr.* **2019**, *38*, 465–471. [CrossRef]
15. Wehrens, S.M.T.; Christou, S.; Isherwood, C.; Middleton, B.; Gibbs, M.A.; Archer, S.N.; Skene, D.J.; Johnston, J.D. Meal Timing Regulates the Human Circadian System. *Curr. Biol.* **2017**, *27*, 1768–1775. [CrossRef]
16. Gallant, A.; Lundgren, J.; Drapeau, V. Nutritional Aspects of Late Eating and Night Eating. *Curr. Obes. Rep.* **2014**, *3*, 101–107. [CrossRef]
17. Gontijo, C.A.; Balieiro, L.C.T.; Teixeira, G.P.; Fahmy, W.M.; Crispim, C.A.; Maia, Y.C.P. Effects of timing of food intake on eating patterns, diet quality and weight gain during pregnancy. *Br. J. Nutr.* **2020**, *123*, 922–933. [CrossRef]
18. Silva, C.M.; Teixeira, B.S.; Wright, K.P., Jr.; Maia, Y.C.P.; Crispim, C.A. Time-Related Eating Patterns Are Associated with the Total Daily Intake of Calories and Macronutrients in Day and Night Shift Workers. *Nutrients* **2022**, *14*, 2202. [CrossRef]
19. Monteiro, C.A.; Moubarac, J.C.; Levy, R.B.; Canella, D.S.; Louzada, M.L.D.C.; Cannon, G. Household availability of ultra-processed foods and obesity in nineteen European countries. *Public Health Nutr.* **2018**, *21*, 18–26. [CrossRef]
20. Monteiro, C.A.; Cannon, G.; Levy, R.B.; Moubarac, J.C.; Louzada, M.L.; Rauber, F.; Khandpur, N.; Cediel, G.; Neri, D.; Martinez-Steele, E.; et al. Ultra-processed foods: What they are and how to identify them. *Public Health Nutr.* **2019**, *22*, 936–941. [CrossRef] [PubMed]
21. Rauber, F.; Chang, K.; Vamos, E.P.; da Costa Louzada, M.L.; Monteiro, C.A.; Millett, C.; Levy, R.B. Ultra-processed food consumption and risk of obesity: A prospective cohort study of UK Biobank. *Eur. J. Nutr.* **2021**, *60*, 2169–2180. [CrossRef] [PubMed]

22. Sandoval-Insausti, H.; Jiménez-Onsurbe, M.; Donat-Vargas, C.; Rey-García, J.; Banegas, J.R.; Rodríguez-Artalejo, F.; Guallar-Castillón, P. Ultra-Processed Food Consumption Is Associated with Abdominal Obesity: A Prospective Cohort Study in Older Adults. *Nutrients* **2020**, *12*, 2368. [CrossRef] [PubMed]
23. Pagliai, G.; Dinu, M.; Madarena, M.P.; Bonaccio, M.; Iacoviello, L.; Sofi, F. Consumption of ultra-processed foods and health status: A systematic review and meta-analysis. *Br. J. Nutr.* **2021**, *125*, 308–318. [CrossRef]
24. Bonaccio, M.; Di Castelnuovo, A.; Ruggiero, E.; Costanzo, S.; Grosso, G.; De Curtis, A.; Cerletti, C.; Donati, M.B.; de Gaetano, G.; Iacoviello, L.; et al. Joint association of food nutritional profile by Nutri-Score front-of-pack label and ultra-processed food intake with mortality: Moli-sani prospective cohort study. *BMJ.* **2022**, *378*, e070688. [CrossRef] [PubMed]
25. Pounis, G.; Bonanni, A.; Ruggiero, E.; Di Castelnuovo, A.; Costanzo, S.; Persichillo, M.; Bonaccio, M.; Cerletti, C.; Riccardi, G.; Donati, M.B.; et al. INHES Investigators. Food group consumption in an Italian population using the updated food classification system FoodEx2: Results from the Italian Nutrition & HEalth Survey (INHES) study. *Nutr. Metab. Cardiovasc. Dis.* **2017**, *27*, 307–328. [CrossRef] [PubMed]
26. Leclercq, C.; Arcella, D.; Piccinelli, R.; Sette, S.; Le Donne, C.; Turrini, A.; INRAN-SCAI 2005–06 Study Group. The Italian National Food Consumption Survey INRAN-SCAI 2005–06: Main results in terms of food consumption. *Public Health Nutr.* **2009**, *12*, 2504–2532. [CrossRef] [PubMed]
27. Illner, A.K.; Harttig, U.; Tognon, G.; Palli, D.; Salvini, S.; Bower, E.; Amiano, P.; Kassik, T.; Metspalu, A.; Engeset, D.; et al. Feasibility of innovative dietary assessment in epidemiological studies using the approach of combining different assessment instruments. *Public Health Nutr.* **2011**, *14*, 1055–1063. [CrossRef]
28. Sette, S.; Le Donne, C.; Piccinelli, R.; Arcella, D.; Turrini, A.; Leclercq, C. INRAN-SCAI 2005–6 Study Group. The third Italian National Food Consumption Survey, INRAN-SCAI 2005–06–part 1: Nutrient intakes in Italy. *Nutr. Metab. Cardiovasc. Dis.* **2011**, *21*, 922–932. [CrossRef]
29. Monteiro, C.A.; Cannon, G.; Levy, R.; Moubarac, J.C.; Jaime, P.; Martins, A.P.; Canella, D.; Louzada, M.; Parra, D. NOVA. The star shines bright. Food classification. *Public Health World Nutr.* **2016**, *7*, 28–38.
30. Srour, B.; Fezeu, L.K.; Kesse-Guyot, E.; Allès, B.; Méjean, C.; Andrianasolo, R.M.; Chazelas, E.; Deschasaux, M.; Hercberg, S.; Galan, P.; et al. Ultra-processed food intake and risk of cardiovascular disease: Prospective cohort study (NutriNet-Santé). *BMJ* **2019**, *365*, l1451. [CrossRef]
31. Trichopoulou, A.; Costacou, T.; Bamia, C.; Trichopoulos, D. Adherence to a Mediterranean diet and survival in a Greek population. *N. Engl. J. Med.* **2003**, *348*, 2599–2608. [CrossRef] [PubMed]
32. Julia, C.; Hercberg, S. Development of a new front-of-pack nutrition label in France: The five-Colour Nutri-Score. *Public Health Panorama.* **2017**, *3*, 712–725.
33. Deschasaux, M.; Huybrechts, I.; Julia, C.; Hercberg, S.; Egnell, M.; Srour, B.; Kesse-Guyot, E.; Latino-Martel, P.; Biessy, C.; Casagrande, C.; et al. Association between nutritional profiles of foods underlying Nutri-Score front-of-pack labels and mortality: EPIC cohort study in 10 European countries. *BMJ* **2020**, *370*, m3173. [CrossRef] [PubMed]
34. ISTAT: Istituto Nazionale di Statistica; Atlante Statistico dei Comuni. Edizione 2014. Available online: https://www.istat.it/it/archivio/113712 (accessed on 20 January 2023).
35. Hernán, M.A.; Hernández-Díaz, S.; Werler, M.M.; Mitchell, A.A. Causal knowledge as a prerequisite for confounding evaluation: An application to birth defects epidemiology. *Am. J. Epidemiol.* **2002**, *155*, 176–184. [CrossRef]
36. Jakubowicz, D.; Barnea, M.; Wainstein, J.; Froy, O. High caloric intake at breakfast vs. dinner differentially influences weight loss of overweight and obese women. *Obesity* **2013**, *21*, 2504–2512. [CrossRef]
37. Arble, D.M.; Bass, J.; Laposky, A.D.; Vitaterna, M.H.; Turek, F.W. Circadian timing of food intake contributes to weight gain. *Obesity* **2009**, *17*, 2100–2102. [CrossRef]
38. Garaulet, M.; Gómez-Abellán, P.; Alburquerque-Béjar, J.J.; Lee, Y.-C.; Ordovás, J.M.; Scheer, F.A.J.L. Timing of food intake predicts weight loss effectiveness. *Int. J. Obes.* **2013**, *37*, 604–611. [CrossRef]
39. Dashti, H.S.; Gómez-Abellán, P.; Qian, J.; Esteban, A.; Morales, E.; Scheer, F.A.J.L.; Garaulet, M. Late eating is associated with cardiometabolic risk traits, obesogenic behaviors, and impaired weight loss. *Am. J. Clin. Nutr.* **2020**, *113*, 154–161. [CrossRef]
40. Dashti, H.S.; Scheer, F.A.J.L.; Saxena, R.; Garaulet, M. Timing of Food Intake: Identifying Contributing Factors to Design Effective Interventions. *Adv. Nutr.* **2019**, *10*, 606–620. [CrossRef]
41. Morris, C.J.; Yang, J.N.; Garcia, J.I.; Myers, S.; Bozzi, I.; Wang, W.; Buxton, O.M.; Shea, S.A.; Scheer, F.A. Endogenous circadian system and circadian misalignment impact glucose tolerance via separate mechanisms in humans. *Proc. Natl. Acad. Sci. USA* **2015**, *112*, E2225–E2234. [CrossRef]
42. Lopez-Minguez, J.; Gómez-Abellán, P.; Garaulet, M. Timing of Breakfast, Lunch, and Dinner. Effects on Obesity and Metabolic Risk. *Nutrients* **2019**, *11*, 2624. [CrossRef] [PubMed]
43. Lima, M.T.M.; Nunes, F.S.M.; Custódio, I.D.D.; Carvalho, K.P.; Canto, P.P.L.; Paiva, C.E.; Crispim, C.A.; Paiva Maia, Y.C. Eating Earlier and More Frequently Is Associated With Better Diet Quality in Female Brazilian Breast Cancer Survivors Using Tamoxifen. *J. Acad. Nutr. Diet.* **2022**, *122*, 1688–1702. [CrossRef] [PubMed]
44. de Castro, J.M. The time of day of food intake influences overall intake in humans. *J. Nutr.* **2004**, *134*, 104–111. [CrossRef] [PubMed]

45. Baraldi, L.G.; Martinez Steele, E.; Canella, D.S.; Monteiro, C.A. Consumption of ultra-processed foods and associated sociodemographic factors in the USA between 2007 and 2012: Evidence from a nationally representative cross-sectional study. *BMJ Open* **2018**, *8*, e020574. [CrossRef]
46. Rauber, F.; da Costa Louzada, M.L.; Steele, E.M.; Millett, C.; Monteiro, C.A.; Levy, R.B. Ultra-Processed Food Consumption and Chronic Non-Communicable Diseases-Related Dietary Nutrient Profile in the UK (2008–2014). *Nutrients* **2018**, *10*, 587. [CrossRef]
47. Polsky, J.Y.; Moubarac, J.C.; Garriguet, D. Consumption of ultra-processed foods in Canada. *Health Rep.* **2020**, *31*, 3–15. [CrossRef]
48. Ruggiero, E.; Esposito, S.; Costanzo, S.; Di Castelnuovo, A.; Cerletti, C.; Donati, M.B.; de Gaetano, G.; Iacoviello, L.; Bonaccio, M. INHES Study Investigators. Ultra-processed food consumption and its correlates among Italian children, adolescents and adults from the Italian Nutrition & Health Survey (INHES) cohort study. *Public Health Nutr.* **2021**, *24*, 6258–6271. [CrossRef]
49. Blanco-Rojo, R.; Sandoval-Insausti, H.; López-Garcia, E.; Graciani, A.; Ordovás, J.M.; Banegas, J.R.; Rodríguez-Artalejo, F.; Guallar-Castillón, P. Consumption of Ultra-Processed Foods and Mortality: A National Prospective Cohort in Spain. *Mayo. Clin. Proc.* **2019**, *94*, 2178–2188. [CrossRef]
50. Santos, F.S.D.; Dias, M.D.S.; Mintem, G.C.; Oliveira, I.O.; Gigante, D.P. Food processing and cardiometabolic risk factors: A systematic review. *Rev. Saude. Publica.* **2020**, *54*, 70. [CrossRef]
51. Mignogna, C.; Costanzo, S.; Di Castelnuovo, A.; Ruggiero, E.; Shivappa, N.; Hebert, J.R.; Esposito, S.; De Curtis, A.; Persichillo, M.; Cerletti, C.; et al. Moli-sani Study Investigators. The inflammatory potential of the diet as a link between food processing and low-grade inflammation: An analysis on 21,315 participants to the Moli-sani study. *Clin. Nutr.* **2022**, *41*, 2226–2234. [CrossRef]
52. Fonken, L.K.; Lieberman, R.A.; Weil, Z.M.; Nelson, R.J. Dim light at night exaggerates weight gain and inflammation associated with a high-fat diet in male mice. *Endocrinology* **2013**, *154*, 3817–3825. [CrossRef] [PubMed]
53. Morris, C.J.; Purvis, T.E.; Hu, K.; Scheer, F.A. Circadian misalignment increases cardiovascular disease risk factors in humans. *Proc. Natl. Acad. Sci. USA* **2016**, *113*, E1402–E1411. [CrossRef] [PubMed]
54. Juul, F.; Vaidean, G.; Parekh, N. Ultra-processed Foods and Cardiovascular Diseases: Potential Mechanisms of Action. *Adv. Nutr.* **2021**, *12*, 1673–1680. [CrossRef] [PubMed]
55. Fardet, A. Minimally processed foods are more satiating and less hyperglycemic than ultra-processed foods: A preliminary study with 98 ready-to-eat foods. *Food Funct.* **2016**, *7*, 2338–2346. [CrossRef]
56. Fardet, A. Characterization of the Degree of Food Processing in Relation With Its Health Potential and Effects. *Adv. Food Nutr. Res.* **2018**, *85*, 79–129. [CrossRef]
57. Jenkins, D.J.A.; Dehghan, M.; Mente, A.; Bangdiwala, S.I.; Rangarajan, S.; Srichaikul, K.; Mohan, V.; Avezum, A.; Díaz, R.; Rosengren, A. PURE Study Investigators. Glycemic Index, Glycemic Load, and Cardiovascular Disease and Mortality. *N. Engl. J. Med.* **2021**, *384*, 1312–1322. [CrossRef]
58. Bonaccio, M.; Costanzo, S.; Di Castelnuovo, A.; Persichillo, M.; Magnacca, S.; De Curtis, A.; Cerletti, C.; Donati, M.B.; de Gaetano, G.; Iacoviello, L. Ultra-processed food intake and all-cause and cause-specific mortality in individuals with cardiovascular disease: The Moli-sani Study. *Eur. Heart J.* **2022**, *43*, 213–224. [CrossRef] [PubMed]
59. Romero Ferreiro, C.; Lora Pablos, D.; Gómez de la Cámara, A. Two Dimensions of Nutritional Value: Nutri-Score and NOVA. *Nutrients* **2021**, *13*, 2783. [CrossRef] [PubMed]
60. Rico-Campà, A.; Martínez-González, M.A.; Alvarez-Alvarez, I.; Mendonça, R.D.; de la Fuente-Arrillaga, C.; Gómez-Donoso, C.; Bes-Rastrollo, M. Association between consumption of ultra-processed foods and all cause mortality: SUN prospective cohort study. *BMJ* **2019**, *365*, l1949. [CrossRef]
61. Alkerwi, A.; Vernier, C.; Sauvageot, N.; Crichton, G.E.; Elias, M.F. Demographic and socioeconomic disparity in nutrition: Application of a novel Correlated Component Regression approach. *BMJ Open* **2015**, *5*, e006814. [CrossRef]
62. Papadaki, A.; Wood, L.; Sebire, S.J.; Jago, R. Adherence to the Mediterranean diet among employees in South West England: Formative research to inform a web-based, work-place nutrition intervention. *Prev. Med. Rep.* **2015**, *2*, 223–228. [CrossRef] [PubMed]
63. Jung, S.; Park, S.; Kim, J.Y. Comparison of dietary share of ultra-processed foods assessed with a FFQ against a 24-h dietary recall in adults: Results from KNHANES 2016. *Public Health Nutr.* **2022**, *25*, 1166–1175. [CrossRef] [PubMed]
64. Gibney, M.J. Ultra-processed foods: Definitions and policy issues. *Curr. Dev. Nutr.* **2018**, *3*, nzy077. [CrossRef] [PubMed]

nutrients

MDPI

Review

Low-Grade Inflammation and Ultra-Processed Foods Consumption: A Review

Marta Tristan Asensi [1], Antonia Napoletano [1], Francesco Sofi [1,2] and Monica Dinu [1,*]

[1] Department of Experimental and Clinical Medicine, University of Florence, 50134 Florence, Italy
[2] Unit of Clinical Nutrition, Careggi University Hospital, 50134 Florence, Italy
* Correspondence: monica.dinu@unifi.it; Tel.: +39-3791439612

Abstract: Low-grade inflammation alters the homeostasis of the organism and favors the onset of many chronic diseases. The global growth in the prevalence of noncommunicable diseases in recent years has been accompanied by an increase in the consumption of ultra-processed foods (UPF). Known to be hyperpalatable, economic and ready-to-eat, increased consumption of UPF has already been recognized as a risk factor for several chronic diseases. Different research groups have tried to investigate whether UPF consumption could promote low-grade inflammation and thus favor the development of noncommunicable diseases. Current evidence highlights the adverse health effects of UPF characteristics, not only due to the nutrients provided by a diet rich in UPF, but also due to the non-nutritive components present in UPF and the effect they may have on gut health. This review aims to summarize the available evidence on the possible relationship between excessive UPF consumption and modulation of low-grade inflammation, as potential promoters of chronic disease.

Keywords: ultra-processed foods; NOVA classification; low-grade inflammation; chronic diseases

Citation: Tristan Asensi, M.; Napoletano, A.; Sofi, F.; Dinu, M. Low-Grade Inflammation and Ultra-Processed Foods Consumption: A Review. *Nutrients* **2023**, *15*, 1546. https://doi.org/10.3390/nu15061546

Academic Editor: Samuel Fernández-Tomé

Received: 28 February 2023
Revised: 14 March 2023
Accepted: 21 March 2023
Published: 22 March 2023

1. Introduction

Inflammation is an immunosurveillance response essential for host defense, which serves to repair damaged tissues and eliminate toxic agents [1]. However, when this response becomes chronic, it results in the presence of immune system cells for an increasing period of time. This state of low-grade inflammation can lead to dysmetabolic conditions that disrupt homeostasis, favoring the development of a wide range of noncommunicable diseases such as cancer, diabetes and cardiovascular diseases [2].

Current evidence highlights diet among the modifiable behavioral risk factors for the development of noncommunicable diseases [3]. In recent years, particular attention has been paid to the increased consumption of ultra-processed foods (UPF) worldwide [4]. Characterized by being hyperpalatable, affordable and ready-to-eat, UPF have led to a worsening of the diet quality due to their nutritional composition [5] and have already been recognized as a risk factor for diet-related diseases [6].

Recent scientific research has sought to investigate whether UPF consumption could promote low-grade inflammation and thus favor the development of noncommunicable diseases. Emerging evidence attributes the negative effects of UPF consumption not only to the nutrients provided by a diet rich in UPF, but also to the non-nutritive components and the effect they may have on the gut microbiota. This review aims to summarize the available evidence on the possible relationship between excessive UPF consumption and modulation of low-grade inflammation as potential promoters of chronic diseases.

2. Low-Grade Inflammation

The inflammatory response is a defense mechanism of the innate immune system [7] that protects the host from harmful stimuli such as viruses, bacteria, toxins and infections by eliminating pathogens and promoting the repair of damaged tissues [1]. At the onset of inflammation, the innate immune cells perceive pathogen invasion or cell damage and initiate the inflammatory cascade by actively releasing soluble proinflammatory mediators. These signals also activate leukocytes and microvascular changes, such as increased vasodilation and vascular permeability, allowing leukocytes to reach the affected tissues from the blood [8]. Such inflammatory activity should resolve once the threat is overcome, becoming temporarily restricted and self-limiting to maintain homeostasis [9,10]. However, failure of immune resolution or continued exposure to environmental and biological factors that promote the activation of the inflammatory response can lead to a chronic inflammatory process. This results in the presence of immune cells such as lymphocytes, macrophages and plasma cells in the tissue for long periods of time, as well as of proinflammatory cytokines, chemokines and other proinflammatory molecules [11,12]. Although this condition recognized as low-grade inflammation has minimal or no clinical manifestations, the prolonged inflammatory response can cause consequences for tissue health, which can develop into tissue fibrosis and possible loss of function [13].

The presence of low-grade inflammation disrupts the homeostatic balance, altering the crosstalk between immune and metabolic responses and promoting chronic metabolic inflammation. This so-called "metainflammation" is primarily caused by metabolic and nutrient excess and triggers immune cell infiltration and the secretion of inflammatory cytokines into the tissue environment, which may inhibit glucose uptake or alter lipid metabolism [2,14]. As a result, chronic metabolic inflammation is particularly associated with an increased risk of noncommunicable diseases, such as cancer, diabetes and cardiovascular disease. An example is insulin resistance caused by chronic exposure to inflammatory biomarkers, which often lead to diabetes [15]. Low-grade inflammation plays an important role also in the development of cardiovascular diseases, due to its involvement in atheroprogression [16], and may favor the progression of different types of cancer by promoting cell proliferation, decreasing apoptosis and increasing angiogenesis and metastasis [17]. At present, it is not well-established which biomarkers can best represent low-grade inflammation, although among the most widely used in scientific studies are soluble mediators (chemokines and cytokines), acute-phase proteins (fibrinogen and C-Reactive Protein (CRP)) or blood cellular markers (granulocytes and total white blood cells) [18].

Diet as a Risk Factor for Low-Grade Inflammation

Among the environmental and lifestyle factors that can promote or intensify inflammation, increasing scientific evidence supports the role of diet. Potential nutritional compounds influencing inflammation processes include macro- and micronutrients, bioactive molecules such as polyphenols and specific food components [19]. Overall, plant-based dietary patterns with a high consumption of vegetables, fruits and whole grains, a moderate consumption of legumes and fish and a low consumption of red meat have been associated with a greater anti-inflammatory potential (Figure 1). These include several traditional healthy diets, such as the Mediterranean or the Nordic diet, which are usually based on minimally processed or unprocessed foods [20,21]. A meta-analysis that evaluated a total of 2300 subjects from 17 clinical trials showed that greater adherence to the Mediterranean diet was associated with lower levels of inflammatory biomarkers, particularly CRP and interleukin-6 (IL-6) [22]. These findings were confirmed in a recent meta-analysis assessing the effect of multiple dietary patterns on inflammatory biomarkers [23]. The authors concluded that the Mediterranean diet appeared as the dietary pattern with the most significant reductions in inflammatory biomarkers, including IL-6 and CRP [23]. Similar results were observed for the Nordic diet, with a review of intervention and observational studies revealing its beneficial influence on low-grade inflammation amelioration [24].

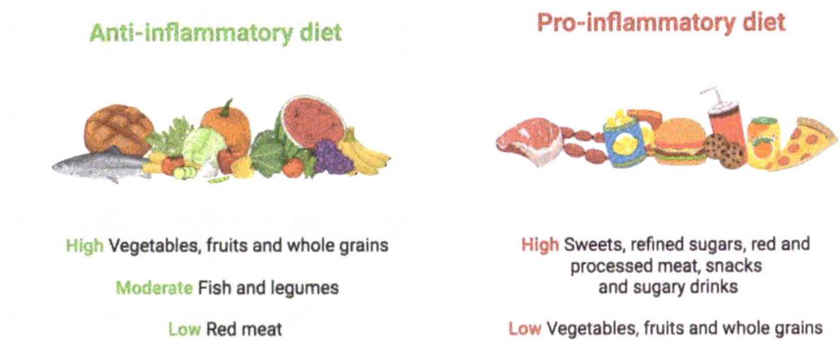

Figure 1. Dietary patterns and inflammation.

A growing number of studies show that the protective effects of these dietary patterns against inflammation are related to the dietary pattern as a whole, not just to its individual components [19]. All these dietary models share the presence of whole grains, fiber, vegetables, fruits, fish, polyunsaturated fatty acids (PUFAs), particularly marine n-3 PUFAs, vitamin C, vitamin E and carotenoids. In contrast, dietary factors that promote inflammation are oxidized lipids, saturated fatty acids (SFAs) and trans fatty acids, which are present at high levels in Western dietary patterns. Unfortunately, in recent years, the increased availability and variety of foods has led to a change in traditional dietary patterns, favoring a nutritional transition and a globalization of the diet towards a Western dietary pattern [25]. This dietary pattern, characterized by a high caloric intake and a high consumption of sweets, refined cereals, red and processed meats, snacks and sugary drinks, has been associated with an increased pro-inflammatory potential and higher levels of CRP and IL-6 [26].

To further investigate the role of diet in modulating inflammation, several literature-based indices have been developed. The energy-adjusted dietary inflammatory index (E-DII) analyzes the potential effect of 45 dietary elements on 6 inflammatory markers, both pro-inflammatory (IL-1b, IL-6, tumor necrosis factor (TNF)-α and CRP) and anti-inflammatory (IL-4, IL-10). The Empirical Diet Inflammatory Pattern (EDIP) is based on food group consumption and divides the dietary intake into nine inflammatory and nine anti-inflammatory food groups according to their impact on the CRP, IL-6 and TNF-αR2 biomarkers of inflammation [27]. Using these indices, many studies have assessed the potential inflammatory effect of diet on the health status. Recently, an umbrella review was conducted on DII and human health [28]. Umbrella reviews are overviews of systematic reviews and meta-analyses that provide a comprehensive and systematic evaluation of the scientific literature available for a specific research topic and offer the possibility to understand the strength of the evidence and the extent of potential biases [29]. In their umbrella review [28], authors found strong evidence supporting the relationship between a high dietary inflammatory index and an increased risk of myocardial infarction. They also found highly suggestive evidence for increased risk of cancer, in particular oral, respiratory, pancreatic and colorectal cancer, and all-cause mortality [28]. As for EDIP, several observational studies have associated a higher score with increased fasting blood sugar and decreased high-density lipoprotein (HDL) cholesterol levels, as well as with an increased risk of weight gain, metabolic syndrome, nonalcoholic fatty liver disease, heart failure and depression [30–36].

3. Ultra-Processed Foods (UPF)

One of the cornerstones of the Western diet are UPF, widely available and increasingly consumed in the contemporary society [4,37]. The possible role of UPF in the nutrition–

health relationship was first highlighted by Monteiro et al. in 2009, with the introduction of the NOVA classification [38]. NOVA is a system that groups foods according to the nature, extent and purpose of the industrial processes they undergo, rather than in terms of the nutrients they contain [38]. In this classification, foods are assigned to one of four groups: Group 1 contains unprocessed or minimally processed foods, i.e., the edible parts of plants or animals taken directly from nature or minimally modified/preserved; Group 2 contains processed culinary ingredients, such as salt, sugar, oil or starch, produced from Group 1 foods; Group 3 contains processed foods such as canned vegetables or freshly baked bread, produced by combining Group 1 and Group 2 foods; Group 4 contains UPFs, defined as "formulations of ingredients, mostly of exclusive industrial use, that have little or none of the food intact and are typically created by a range of industrial techniques and processes" [38]. UPFs are identified by a long list of ingredients, are ready-to-eat, highly palatable, and usually inexpensive. The most commonly consumed UPFs include soft and sweetened beverages, processed bread, refined breakfast cereals, confectionery products, pre-packaged sauces, ready-to-heat meals and processed meats products [39]. Possible mechanisms behind their link with the health status may involve both their nutritional composition and "processing". Indeed, in terms of nutritional composition, UPF are typically nutritionally unbalanced due to their ingredients [40]. Most UPF are energy-dense products high in added sugars, saturated and trans fatty acids and sodium and low in protein, fiber and certain micronutrients including potassium, magnesium, vitamin C, vitamin D, zinc, phosphorus, vitamin B12 and niacin [40].

UPF are also characterized by the presence of non-nutritive components, such as additives and chemicals. Additives are frequently added to make the final product more palatable, with better sensory qualities and longer shelf life. Commonly used additives in the manufacture of UPF include flavorings, emulsifiers and sweeteners such as aspartame, cyclamate or stevia-derived compounds [41]. As to the supposed presence of harmful chemicals in UPF, it has been suggested that they may derive from the processing or packaging of these products [42]. Processing could also alter the physical properties of food products, leading to a higher glycemic load and a reduced gut–brain satiety signaling, both responsible for overconsumption [43].

According to previous studies, all these aspects could explain the reason why the incidence of several chronic noncommunicable diseases is increasing along with UPF consumption [41]. Among adults, multiple meta-analyses found that a higher UPF consumption is significantly associated with an increased risk of overweight and obesity, metabolic syndrome, hypertension, diabetes and cardiovascular disease [6,44–47]. A higher UPF consumption has also been associated with a higher risk of cancer, particularly breast cancer [6,48], anxiety and depression [49] and all-cause mortality [50,51]. In children and adolescents, significant relationships were found with overweight and obesity [25,52].

4. UPF and Low-Grade Inflammation

The number of human studies investigating whether the consumption of UPF could promote low-grade inflammation, so favoring the development of noncommunicable diseases, is still limited. The available studies have focused mainly on two aspects: how excessive UPF consumption may affect the presence of biomarkers of inflammation, and how the nutritional composition or non-nutritional components of UPF may influence the development of chronic inflammation and gut dysbiosis, previously correlated with a pro-inflammatory state (Figure 2).

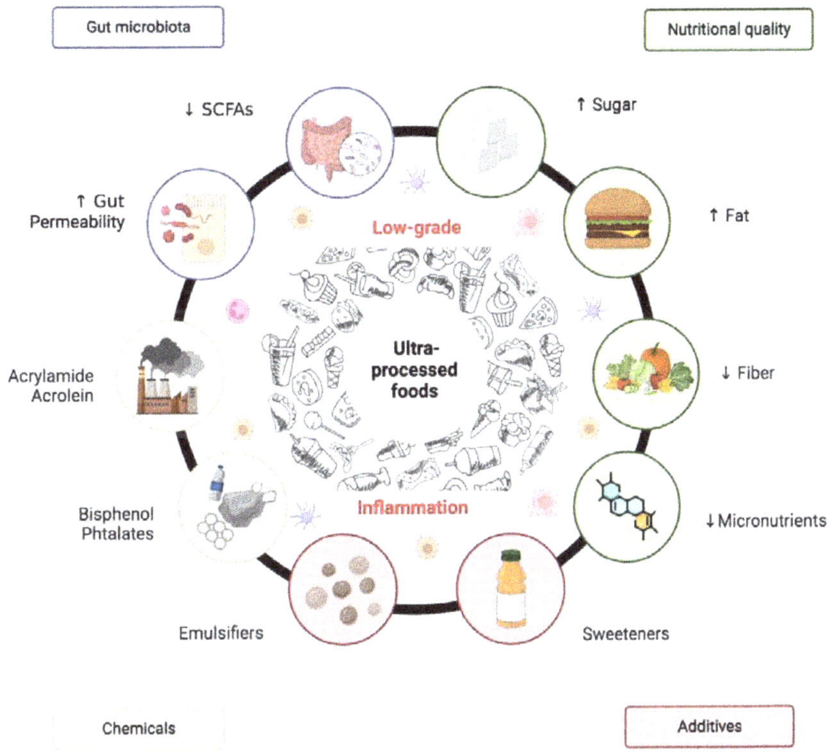

Figure 2. Possible mechanisms explaining the relationship between UPF and low-grade inflammation. ↑ increased; ↓ reduced.

The vast majority of studies that have examined the relationship between UPF consumption and inflammation are observational, either cross-sectional or cohort studies (Table 1), with only one clinical trial currently available [53]

Table 1. Observational studies assessing the relationship between UPF consumption and inflammatory biomarkers.

Author, Year	Study Design	Country	Participants, n	Gender	Age	Study Population	Outcome	Main Results
Lopes et al., 2019 [54]	Cross-sectional analysis of Longitudinal Study of Adult Health (ELSA-Brasil) baseline cohort	Brazil	8468	M/F	35–74	General population	CRP	A higher tertile of UPF intake was associated with a 14% increase in CRP levels only among women. Significance was lost when adjusting for BMI.
Lane et al., 2022 [55]	Cross-sectional analysis of Melbourne Collaborative Cohort	Australia	2018	M/F	57 ± 9	General population	hs-CRP	A 100 g increase in UPF consumption was associated with a 4% increase in hs-CRP concentration, independently of BMI.
Martins et al., 2022 [56]	Cross-sectional	Brazil	391	M/F	17–18	General population	Leptin, IL-6, IL-8, CRP TNF- α	The highest tertiles of UPF intake showed higher levels of CRP and serum leptin and a 79% increase in IL-8 levels. No association was found for IL-6 and TNF- α
Silva Dos Santos et al., 2022 [57]	Cross-sectional analysis of EPITeen Cohort and Pelotas Birth Cohort	Brazil, Portugal	3412	M/F	27–30	General population	IL-6	A positive association between levels of IL-6 and UPF intake was found among females from the Portugal cohort and males from the Brazil cohort.
Kesley et al., 2022 [58]	Cross-sectional analysis of Norwegian Mother, Father and Child Cohort	Norway	2984	F	30 ± 4	Pregnant women	CRP	An increase UPF intake was associated with a 5.4% increase in CRP levels, even after adjustment for pre-pregnancy BMI
Mignogna et al., 2022 [59]	Cross-sectional analysis of Moli-sani cohort	Italy	21,315	M/F	55 ± 3	General population	INFLA-score E-DII score	INFLA-score was associated with higher E-DII score and UPF intake. When adjusting for E-DII, the association of UPF with the INFLA-score was mitigated by 32.6%
Silva et al., 2019 [60]	Cross-sectional	Brazil	784	F	28 ± 5	Pregnant women	E-DII score	E-DII score was positively associated with consumption of UPF when adjusting for covariates including pre-pregnancy BMI

UPF: ultra-processed foods; CRP: C-reactive protein; BMI: body mass index; hs-CRP: high-sensitivity C-reactive protein; IL: interleukin; TNF: tumor necrosis factor; INFLA: low-grade inflammation; E-DII: energy-adjusted dietary inflammatory index.

CRP is the most investigated inflammatory biomarker to date in relation to UPF consumption. In the only available clinical trial, subjects assigned to a diet based on unprocessed foods showed a significant reduction in hs-CRP levels, while subjects on a diet rich in UPF did not report significant changes [53]. The authors suggested that these results might indicate that the subjects were already regularly consuming a large amount of UPF, as already observed in the US population [53]. As for data from observational studies, they are not consistent and suggest that the relationship may depend on gender and body mass index (BMI). For example, in the ELSA-Brasil study, a significant association between high UPF consumption and higher CRP levels was found in women, but the association lost its significance when adjusting for BMI [54]. Similarly, in the Melbourne Collaborative Cohort Study, the association between high UPF consumption and CRP levels remained significant only in men, after adjustment for BMI [55]. In adolescents, Martins et al. found that subjects consuming more UPF in their diet had higher CRP and IL-8 values, but the association was significant only for IL-8 [56]. Other biomarkers studied to a lesser extent are some proinflammatory cytokines such as IL-6. Dos Santos et al. investigated the possible relationship between UPF consumption and IL-6 concentrations in two cohorts, showing an association only in women in the Portuguese cohort and only in men in the Brazilian cohort [57]. The conclusion was that the UPF intake could be associated with higher IL-6 levels, although the relation was not explained by adiposity [57].

As to the E-DII score, a cross-sectional study in Brazil found a direct relationship between a higher dietary energy intake from UPF and a higher rate of dietary inflammation in pregnant women [58]. Similar findings were obtained in the Italian cohort Moli-Sani, where a higher consumption of UPF was related to a higher pro-inflammatory potential of the adults' diet [59]. In this cohort, further analyses were performed using the low-grade inflammation (INFLA)-Score, which allows the assessment of the possible intensity of low-grade inflammation through the effects of biomarkers of inflammation (platelets, white blood cell (WBC), CRP and granulocyte-to-lymphocyte ratio), obtaining the same association [59].

5. Possible Mechanisms Explaining the Relationship between UPF and Low-Grade Inflammation

5.1. Nutritional Aspects

UPF consumption could contribute to an inflammatory state through several mechanisms. First, it could be the high intake of sugars, salt, saturated fats and trans fatty acids typical of a UPF-rich diet that directly promotes the development of chronic inflammation [61]. When high intakes of these nutrients and their possible relationship to the modulation of inflammation are considered individually, the results to date are mixed. UPF are usually high in simple sugars, in the form of either sucrose or a high-fructose syrup, so they tend to be foods that raise the blood glucose markedly and rapidly, i.e., with a high glycemic index/glycemic load [62]. This postprandial increase in the glucose levels in turn causes an increase in insulin levels, which promotes a proinflammatory state [63]. Although these mechanisms appear to play an important role in diet and the promotion of low-grade inflammation, intervention studies are not very clear in this regard. In the TOSCA.IT study, an association was found between the intake of added sugars $\geq$10% of the daily energy intake and increased CRP levels in adults with diabetes [64]. Other observational studies associated a higher consumption of sugar-sweetened beverages with increased levels of CRP and IL-6 in adults and children [65–67]. Regarding the glycemic response, although an intervention study found a positive association between glycemic load and plasma hs-CRP in healthy middle-aged women [68], a recent meta-analysis including 28 randomized controlled trials found no association between the glycemic index and different markers of inflammation in adults [69].

UPF also have a high salt content, contributing to a high sodium intake. Several cross-sectional studies associated a higher salt intake with higher CRP levels in adults and elderly people [70,71], although this association was not found in adolescents [72]. A recent meta-analysis also found no associations between dietary sodium level and markers of inflammation, although it should be noted that the researchers pointed out that their findings were likely due to methodological errors [73].

As for the fat content of UPF, their inflammatory potential derives not only from a higher consumed quantity with respect to other foods, but also from a poorer quality. In fact, trans fatty acids resulting from the industrial process are associated with a higher presence of low-grade inflammation. Specifically, they have been related to higher levels of hs-CRP, IL-6 and TNF-α [74–76]. Diets with a high processed-food content have also been associated with a higher intake of omega-6 fatty acids, resulting in a higher omega-6/omega-3 ratio and the potential promotion of low-grade inflammation [77].

Finally, consuming large amounts of UPF sometimes results in the replacement of foods that are the basis of a healthy and balanced diet. Examples are fruits and vegetables, which are correlated with an anti-inflammatory effect thanks to the presence of numerous phytocompounds [78,79]. Recent studies clearly show how people consuming more UPF have a lower intake of fruit and vegetables [80] and consequently ingest less substances with an anti-inflammatory effect. A low fruit and vegetable consumption also results in a low dietary fiber intake. In the E-DIITM, fiber is considered one of the factors that reduce diet-related inflammation. In previous studies, an adequate fiber intake was shown to be important in maintaining low CRP levels and in maintaining homeostasis of the gut microbiota [81]. A high UPF consumption can also lead to deficiencies of micronutrients considered to be anti-inflammatory factors in the diet, such as magnesium, vitamin C, vitamin D, zinc and niacin [82].

5.2. Non-Nutritional Aspects

Results from an Italian cohort study suggested that only part of the proinflammatory effect of a high UPF consumption can be directly attributed to the nutritional components of the diet, while the rest could be attributed to non-nutritional factors that may promote low-grade inflammation [59]. One of the non-nutritional factors present in UPF are additives, which are added to mimic or intensify the sensory qualities of foods [83]. Among the most studied are sweeteners, especially non-caloric ones such as acesulfame potassium, sucralose or aspartame, due to their widespread use in soft drinks to provide a sweet taste without the energy value of sugars [84]. Recently, there has also been growing interest in the harmful effect of emulsifiers used to improve the shelf life and texture of food products. Although scientific evidence to date is limited, animal and in vitro studies suggest that sweeteners and emulsifiers may contribute to the inflammatory cascade [85–87]. One of the hypothesized mechanisms is the modulation of the microbiota, but data are inconsistent, and further studies are needed to investigate these mechanisms [88,89]. It has also been hypothesized that the non-caloric sweeteners' harmful effect might be due to an acute metabolic response [90]. However, data from two recent meta-analyses do not support this hypothesis, as they found no association between the consumption of non-caloric sweetened beverages and an increased insulinemic effect or acute glycemic response [91,92].

Non-nutrient components such as bisphenol or phthalates may also be present in UPF due to the migration of chemical substances that are part of food packaging. In fact, several cross-sectional studies reported higher levels of both substances in the urine of people with a high UPF consumption [42,93–96]. Because of their structure, bisphenol and phthalates can disrupt various aspects of the hormonal action and are therefore called endocrine disruptors. They can interfere with the synthesis, secretion, transport, signaling and metabolism of hormones; therefore, they have been associated with adverse health consequences, including the development of diseases such as obesity, diabetes and cardiovascular disease [97,98].

A recent meta-analysis investigating the role of different endocrine disruptors on the inflammatory response showed that increased exposure to Bisphenol A (BPA) is significantly associated with higher levels of IL-6 and CRP, while increased exposure to phthalates is associated with higher levels of CRP, IL-6 and IL-10 [99]. Although the adverse effects of BPA have led to various restrictions on its use, the analogs that replaced it appear to have similar effects [100]. On the other hand, UPF may contain chemicals derived from food processing, especially due to the heat treatment to which food is subjected. One example is acrylamide as a result of the Maillard reaction between amino acids and sugars, exposure to which in adults has been associated with an increased presence of biomarkers of inflammation such as CRP or Mean Platelet Volume (MPV) [101]. Another chemical instead derived from lipid oxidation is acrolein, high exposure to which has been associated with a higher concentration of Hs-CRP in adults in the United States [102] and of CRP in adults in China [103].

5.3. Gut Microbiota Modulation

The human gut microbiota is a dynamic and complex network composed of hundreds of thousands of microorganisms, including bacteria, fungi, archaea, viruses and protozoa [104]. When in its normal state of homeostasis, the gut microbiota plays a key role in host health through the immune system function and protection against pathogens. However, when the gut microbiota is altered compared to the community found in healthy individuals, gut dysbiosis occurs [84]. This dysbiosis is associated with a high degree of inflammation, caused by a lower presence of short- chain-fatty-acids-(SCFAs)-producing bacteria, and increased permeability of the gut [105]. Both diet quality and the presence of the additives previously described may influence intestinal dysbiosis, offering a possible explanation for the mechanism linking an increased consumption of UPF with the presence of low-grade inflammation.

In fact, it has been suggested that a diet rich in fiber can decrease the systemic inflammatory response by improving the intestinal barrier function and modulating the intestinal microbiota [81]. This is because dietary fiber is essential for the formation of SCFAs, which are thought to play a key role in neuroimmunoendocrine regulation [106]. In fact, SCFAs are associated with a lower concentration of CRP and plasma lipopolysaccharide, an endotoxin used as a marker to assess intestinal permeability linked to increased low-grade inflammation [107–110]. In contrast, Western diets with a high fat content have been associated with increased intestinal permeability due to a greater presence of lipopolysaccharides in humans and mice [111,112]. Similar results were observed in mice fed a diet rich in refined sugar, also associated with an atypical composition of the intestinal microbiota [113]. In a cross-sectional study conducted in the U.S.A., the increased consumption of highly processed food was associated with intestinal permeability biomarkers [114]. Also in a study conducted in Italy, intestinal permeability tended to increase in subjects with low adherence to the Mediterranean diet, who also reported a high intake of food high in fat and sugar, referred to as junk food [115]. Finally, a French study involving 862 healthy adults found that the regular consumption of foods such as soft drinks, fatty sweet products, fried foods, processed meats, ready-to-eat meals, cheese and desserts, most of them recognized as UPF, was associated with reduced bacterial diversity, indicating an altered microbiota composition [116]. In contrast, the PREDIMED-PLUS study in older adults found no such association and suggested that perhaps the contradictory results with the previous study were due to the lower UPF consumption of the studied population [117].

Several studies have also highlighted additives as possible factors affecting the microbiota. Studies in murine models suggested different mechanisms through which emulsifying additives could contribute to intestinal dysbiosis, increasing intestinal permeability and promoting a proinflammatory state [89,118]. However, these studies remain limited, and the results in humans are contrasting. For example, a double-blind controlled study comparing seven adults on an emulsifier-rich diet to nine adults on an emulsifier-free diet observed changes in the gut microbiome and metabolome that may be related to chronic inflammatory diseases [119]. In contrast, a cross-sectional study involving 588 adults found no association with biomarkers related to increased intestinal permeability, although it found an association with increased levels of systemic inflammation [114]. Similarly, studies in murine models suggested that artificial sweeteners can alter the intestinal microbiota, favoring the enrichment of proinflammatory bacteria that promote the formation of endotoxins such as lipopolysaccharides [85,86,120]. However, the results to date are inconsistent, and further research will be needed to investigate these mechanisms.

6. Conclusions and Future Perspectives

Low-grade inflammation plays a pivotal role in the pathogenesis of noncommunicable diseases, which are becoming increasingly prevalent worldwide. In recent years, diet has been highlighted as one of the main risk factors for these diseases, together with the increased consumption of UPF, which through different mechanisms, may contribute to promote a proinflammatory state. Although the evidence on the association between UPF consumption and inflammation is still limited and, in some cases, the results are discordant, considering the potential impact of their excessive consumption on the health status, as well as their potential role in favoring the presence of chronic inflammation, public policies that limit their consumption are required. These public policies should also include the promotion of traditional diets based on unprocessed or minimally processed foods, in order to modulate low-grade inflammation and improve people's health status. Future human research evaluating clusters of inflammation markers instead of individual biomarkers may help to better understand the mechanism involved in the modulation of low-grade inflammation by a high consumption of UPF. This information could also be useful in establishing policies that promote the reformulation of UPF to minimize their adverse health effects.

Author Contributions: F.S., M.D. and M.T.A., conceptualization; M.T.A. and A.N., revision of the literature; M.T.A. and A.N., writing—original draft preparation; F.S. and M.D., critical revision; F.S. and M.D., writing—review and editing; F.S. and M.D., supervised the manuscript. All authors have read and agreed to the published version of the manuscript.

Funding: This research received no specific grant from any funding agency, commercial or not-for-profit sectors.

Institutional Review Board Statement: Not applicable.

Informed Consent Statement: Not applicable.

Data Availability Statement: Additional data are available from the corresponding author on reasonable request.

Conflicts of Interest: The authors declare no conflict of interest.

References

1. Furman, D.; Campisi, J.; Verdin, E.; Carrera-Bastos, P.; Targ, S.; Franceschi, C.; Ferrucci, L.; Gilroy, D.W.; Fasano, A.; Miller, G.W.; et al. Chronic inflammation in the etiology of disease across the life span. *Nat. Med.* **2019**, *25*, 1822–1832. [CrossRef] [PubMed]
2. Hotamisligil, G.S. Foundations of Immunometabolism and Implications for Metabolic Health and Disease. *Immunity* **2017**, *47*, 406–420. [CrossRef]
3. GBD 2017 Diet Collaborators. Health effects of dietary risks in 195 countries, 1990–2017: A systematic analysis for the Global Burden of Disease Study 2017. *Lancet* **2019**, *393*, 1958–1972. [CrossRef]
4. Marino, M.; Puppo, F.; Del Bo', C.; Vinelli, V.; Riso, P.; Porrini, M.; Martini, D. A Systematic Review of Worldwide Consumption of Ultra-Processed Foods: Findings and Criticisms. *Nutrients* **2021**, *13*, 2778. [CrossRef]
5. Monteiro, C.A.; Levy, R.B.; Claro, R.M.; Castro, I.R.; Cannon, G. A new classification of foods based on the extent and purpose of their processing. *Cad. Saude. Publica* **2010**, *26*, 2039–2049. [CrossRef]
6. Pagliai, G.; Dinu, M.; Madarena, M.P.; Bonaccio, M.; Iacoviello, L.; Sofi, F. Consumption of ultra-processed foods and health status: A systematic review and meta-analysis. *Br. J. Nutr.* **2021**, *125*, 308–318. [CrossRef] [PubMed]
7. Marshall, J.S.; Warrington, R.; Watson, W.; Kim, H.L. An introduction to immunology and immunopathology. *Allergy Asthma Clin. Immunol.* **2018**, *14*, 49. [CrossRef] [PubMed]
8. Medzhitov, R. Inflammation 2010: New adventures of an old flame. *Cell* **2010**, *140*, 771–776. [CrossRef] [PubMed]
9. Kotas, M.E.; Medzhitov, R. Homeostasis, inflammation, and disease susceptibility. *Cell* **2015**, *160*, 816–827. [CrossRef] [PubMed]
10. Fullerton, J.N.; Gilroy, D.W. Resolution of inflammation: A new therapeutic frontier. *Nat. Rev. Drug Discov.* **2016**, *15*, 551–567. [CrossRef]

11. Calder, P.C.; Ahluwalia, N.; Albers, R.; Bosco, N.; Bourdet-Sicard, R.; Haller, D.; Holgate, S.T.; Jönsson, L.S.; Latulippe, M.E.; Marcos, A.; et al. A consideration of biomarkers to be used for evaluation of inflammation in human nutritional studies. *Br. J. Nutr.* **2013**, *109*, S1–S34. [CrossRef] [PubMed]
12. Serhan, C.N.; Brain, S.D.; Buckley, C.D.; Gilroy, D.W.; Haslett, C.; O'Neill, L.A.; Perretti, M.; Rossi, A.G.; Wallace, J.L. Resolution of inflammation: State of the art, definitions and terms. *FASEB J.* **2007**, *21*, 325–332. [CrossRef] [PubMed]
13. Hotamisligil, G.S. Inflammation, metaflammation and immunometabolic disorders. *Nature* **2017**, *542*, 177–185. [CrossRef]
14. Hotamisligil, G.S. Inflammation and metabolic disorders. *Nature* **2006**, *444*, 860–867. [CrossRef] [PubMed]
15. Liu, C.; Feng, X.; Li, Q.; Wang, Y.; Li, Q.; Hua, M. Adiponectin, TNF-α and inflammatory cytokines and risk of type 2 diabetes: A systematic review and meta-analysis. *Cytokine* **2016**, *86*, 100–109. [CrossRef]
16. Lawler, P.R.; Bhatt, D.L.; Godoy, L.C.; Lüscher, T.F.; Bonow, R.O.; Verma, S.; Ridker, P.M. Targeting cardiovascular inflammation: Next steps in clinical translation. *Eur. Heart J.* **2021**, *42*, 113–131. [CrossRef]
17. Fouad, Y.A.; Aanei, C. Revisiting the hallmarks of cancer. *Am. J. Cancer Res.* **2017**, *7*, 1016–1036.
18. Calder, P.C.; Ahluwalia, N.; Brouns, F.; Buetler, T.; Clement, K.; Cunningham, K.; Esposito, K.; Jönsson, L.S.; Kolb, H.; Lansink, M.; et al. Dietary factors and low-grade inflammation in relation to overweight and obesity. *Br. J. Nutr.* **2011**, *106*, S5–S78. [CrossRef]
19. Ramos-Lopez, O.; Martinez-Urbistondo, D.; Vargas-Nuñez, J.A.; Martinez, J.A. The Role of Nutrition on Meta-inflammation: Insights and Potential Targets in Communicable and Chronic Disease Management. *Curr. Obes. Rep.* **2022**, *11*, 305–335. [CrossRef]
20. Shin, P.K.; Park, S.J.; Kim, M.S.; Kwon, D.Y.; Kim, M.J.; Kim, K.; Chun, S.; Lee, H.J.; Choi, S.W. A Traditional Korean Diet with a Low Dietary Inflammatory Index Increases Anti-Inflammatory IL-10 and Decreases Pro-Inflammatory NF-κB in a Small Dietary Intervention Study. *Nutrients* **2020**, *12*, 2468. [CrossRef]
21. Bonaccio, M.; Costanzo, S.; Di Castelnuovo, A.; Gialluisi, A.; Ruggiero, E.; De Curtis, A.; Persichillo, M.; Cerletti, C.; Donati, M.B.; de Gaetano, G.; et al. Increased Adherence to a Mediterranean Diet Is Associated with Reduced Low-Grade Inflammation after a 12.7-Year Period: Results from the Moli-sani Study. *J. Acad. Nutr. Diet.* **2022**. [CrossRef]
22. Schwingshackl, L.; Hoffmann, G. Mediterranean dietary pattern, inflammation and endothelial function: A systematic review and meta-analysis of intervention trials. *Nutr. Metab. Cardiovasc. Dis.* **2014**, *24*, 929–939. [CrossRef]
23. Koelman, L.; Egea Rodrigues, C.; Aleksandrova, K. Effects of Dietary Patterns on Biomarkers of Inflammation and Immune Responses: A Systematic Review and Meta-Analysis of Randomized Controlled Trials. *Adv. Nutr.* **2022**, *13*, 101–115. [CrossRef] [PubMed]
24. Lankinen, M.; Uusitupa, M.; Schwab, U. Nordic Diet and Inflammation—A Review of Observational and Intervention Studies. *Nutrients* **2019**, *11*, 1369. [CrossRef] [PubMed]
25. Costa, C.S.; Del-Ponte, B.; Assunção, M.C.; Santos, I.S. Consumption of ultra-processed foods and body fat during childhood and adolescence: A systematic review. *Public Health Nutr.* **2018**, *21*, 148–159. [CrossRef]
26. Lopez-Garcia, E.; Schulze, M.B.; Fung, T.T.; Meigs, J.B.; Rifai, N.; Manson, J.E.; Hu, F.B. Major dietary patterns are related to plasma concentrations of markers of inflammation and endothelial dysfunction. *Am. J. Clin. Nutr.* **2004**, *80*, 1029–1035. [CrossRef] [PubMed]
27. Tabung, F.K.; Smith-Warner, S.A.; Chavarro, J.E.; Wu, K.; Fuchs, C.S.; Hu, F.B.; Chan, A.T.; Willett, W.C.; Giovannucci, E.L. Development and Validation of an Empirical Dietary Inflammatory Index. *J. Nutr.* **2016**, *146*, 1560–1570. [CrossRef] [PubMed]
28. Marx, W.; Veronese, N.; Kelly, J.T.; Smith, L.; Hockey, M.; Collins, S.; Trakman, G.L.; Hoare, E.; Teasdale, S.B.; Wade, A.; et al. The Dietary Inflammatory Index and Human Health: An Umbrella Review of Meta-Analyses of Observational Studies. *Adv. Nutr.* **2021**, *12*, 1681–1690. [CrossRef]
29. Aromataris, E.; Fernandez, R.; Godfrey, C.M.; Holly, C.; Khalil, H.; Tungpunkom, P. Summarizing systematic reviews: Methodological development, conduct and reporting of an umbrella review approach. *Int. J. Evid. Based Healthc.* **2015**, *13*, 132–140. [CrossRef]
30. Parker, W.R.; Annabathula, R.V.; Skipina, T.M.; Soliman, E.Z. Associations of empirical dietary inflammatory index with heart failure in adults from the United States. *Eur. J. Clin. Nutr.* **2023**, *77*, 112–115. [CrossRef]
31. Farhadnejad, H.; Tehrani, A.N.; Jahromi, M.K.; Teymoori, F.; Mokhtari, E.; Salehi-Sahlabadi, A.; Mirmiran, P. The association between dietary inflammation scores and non-alcoholic fatty liver diseases in Iranian adults. *BMC Gastroenterol.* **2022**, *22*, 267. [CrossRef]
32. Pang, T.; Alman, A.C.; Gray, H.L.; Basu, A.; Shi, L.; Snell-Bergeon, J.K. Empirical dietary inflammatory pattern and metabolic syndrome: Prospective association in participants with and without type 1 diabetes mellitus in the coronary artery calcification in type 1 diabetes (CACTI) study. *Nutr. Res.* **2021**, *94*, 1–9. [CrossRef]
33. Cong, X.; Tracy, M.; Edmunds, L.S.; Hosler, A.S.; Appleton, A.A. The relationship between inflammatory dietary pattern in childhood and depression in early adulthood. *Brain Behav. Immun. Health* **2020**, *2*, 100017. [CrossRef]
34. Tabung, F.K.; Satija, A.; Fung, T.T.; Clinton, S.K.; Giovannucci, E.L. Long-Term Change in both Dietary Insulinemic and Inflammatory Potential Is Associated with Weight Gain in Adult Women and Men. *J. Nutr.* **2019**, *149*, 804–815. [CrossRef] [PubMed]
35. Soltani, S.; Moslehi, N.; Hosseini-Esfahani, F.; Vafa, M. The Association between Empirical Dietary Inflammatory Pattern and Metabolic Phenotypes in Overweight/Obese Adults. *Int. J. Endocrinol. Metab.* **2018**, *16*, e60048. [CrossRef] [PubMed]

36. Shakeri, Z.; Mirmiran, P.; Khalili-Moghadam, S.; Hosseini-Esfahani, F.; Ataie-Jafari, A.; Azizi, F. Empirical dietary inflammatory pattern and risk of metabolic syndrome and its components: Tehran Lipid and Glucose Study. *Diabetol. Metab. Syndr.* **2019**, *11*, 16. [CrossRef] [PubMed]
37. Monteiro, C.A.; Moubarac, J.C.; Cannon, G.; Ng, S.W.; Popkin, B. Ultra-processed products are becoming dominant in the global food system. *Obes. Rev.* **2013**, *14*, 21–28. [CrossRef]
38. Monteiro, C.A. Nutrition and health. The issue is not food, nor nutrients, so much as processing. *Public Health Nutr.* **2009**, *12*, 729–731. [CrossRef]
39. Wang, L.; Du, M.; Wang, K.; Khandpur, N.; Rossato, S.L.; Drouin-Chartier, J.; Steele, E.M.; Giovannucci, E.; Song, M.; Zhang, F.F. Association of ultra-processed food consumption with colorectal cancer risk among men and women: Results from three prospective US cohort studies. *BMJ* **2022**, *378*, e068921. [CrossRef]
40. Martini, D.; Godos, J.; Bonaccio, M.; Vitaglione, P.; Grosso, G. Ultra-Processed Foods and Nutritional Dietary Profile: A Meta-Analysis of Nationally Representative Samples. *Nutrients* **2021**, *13*, 3390. [CrossRef]
41. Monteiro, C.A.; Cannon, G.; Lawrence, M.; Costa Louzada, M.L.; Pereira Machado, P. *Ultra-Processed Foods, Diet Quality, and Health Using the NOVA Classification System*; FAO: Rome, Italy, 2019.
42. Buckley, J.P.; Kim, H.; Wong, E.; Rebholz, C.M. Ultra-processed food consumption and exposure to phthalates and bisphenols in the US National Health and Nutrition Examination Survey, 2013–2014. *Environ. Int.* **2019**, *131*, 105057. [CrossRef] [PubMed]
43. Small, D.M.; DiFeliceantonio, A.G. Processed foods and food reward. *Science* **2019**, *363*, 346–347. [CrossRef]
44. Lane, M.M.; Davis, J.A.; Beattie, S.; Gómez-Donoso, C.; Loughman, A.; O'Neil, A.; Jacka, F.; Berk, M.; Page, R.; Marx, W.; et al. Ultraprocessed food and chronic noncommunicable diseases: A systematic review and meta-analysis of 43 observational studies. *Obes. Rev.* **2021**, *22*, e13146. [CrossRef]
45. Wang, M.; Du, X.; Huang, W.; Xu, Y. Ultra-processed Foods Consumption Increases the Risk of Hypertension in Adults: A Systematic Review and Meta-analysis. *Am. J. Hypertens.* **2022**, *35*, 892–901. [CrossRef]
46. Moradi, S.; Entezari, M.H.; Mohammadi, H.; Jayedi, A.; Lazaridi, A.V.; Kermani, M.A.; Miraghajani, M. Ultra-processed food consumption and adult obesity risk: A systematic review and dose-response meta-analysis. *Crit. Rev. Food Sci. Nutr.* **2023**, *63*, 249–260. [CrossRef] [PubMed]
47. Moradi, S.; Hojjati Kermani, M.A.; Bagheri, R.; Mohammadi, H.; Jayedi, A.; Lane, M.M.; Asbaghi, O.; Mehrabani, S.; Suzuki, K. Ultra-Processed Food Consumption and Adult Diabetes Risk: A Systematic Review and Dose-Response Meta-Analysis. *Nutrients* **2021**, *13*, 4410. [CrossRef] [PubMed]
48. Fiolet, T.; Srour, B.; Sellem, L.; Kesse-Guyot, E.; Allès, B.; Méjean, C.; Deschasaux, M.; Fassier, P.; Latino-Martel, P.; Beslay, M.; et al. Consumption of ultra-processed foods and cancer risk: Results from NutriNet-Santé prospective cohort. *BMJ* **2018**, *360*, k322. [CrossRef] [PubMed]
49. Lane, M.M.; Gamage, E.; Travica, N.; Dissanayaka, T.; Ashtree, D.N.; Gauci, S.; Lotfaliany, M.; O'Neil, A.; Jacka, F.N.; Marx, W. Ultra-Processed Food Consumption and Mental Health: A Systematic Review and Meta-Analysis of Observational Studies. *Nutrients* **2022**, *14*, 2568. [CrossRef]
50. Taneri, P.E.; Wehrli, F.; Roa-Díaz, Z.M.; Itodo, O.A.; Salvador, D.; Raeisi-Dehkordi, H.; Bally, L.; Minder, B.; Kiefte-de Jong, J.C.; Laine, J.E.; et al. Association Between Ultra-Processed Food Intake and All-Cause Mortality: A Systematic Review and Meta-Analysis. *Am. J. Epidemiol.* **2022**, *191*, 1323–1335. [CrossRef]
51. Suksatan, W.; Moradi, S.; Naeini, F.; Bagheri, R.; Mohammadi, H.; Talebi, S.; Mehrabani, S.; Hojjati Kermani, M.A.; Suzuki, K. Ultra-Processed Food Consumption and Adult Mortality Risk: A Systematic Review and Dose-Response Meta-Analysis of 207,291 Participants. *Nutrients* **2021**, *14*, 174. [CrossRef]
52. Louzada, M.L.; Costa, C.D.; Souza, T.N.; Cruz, G.L.; Levy, R.B.; Monteiro, C.A. Impact of the consumption of ultra-processed foods on children, adolescents and adults' health: Scope review. *Cad. Saude. Publica* **2022**, *37*, e00323020. [CrossRef]
53. Hall, K.D.; Ayuketah, A.; Brychta, R.; Cai, H.; Cassimatis, T.; Chen, K.Y.; Chung, S.T.; Costa, E.; Courville, A.; Darcey, V.; et al. Ultra-Processed Diets Cause Excess Calorie Intake and Weight Gain: An Inpatient Randomized Controlled Trial of Ad Libitum Food Intake. *Cell Metab.* **2019**, *30*, 67–77. [CrossRef] [PubMed]
54. Lopes, A.E.; Araújo, L.F.; Levy, R.B.; Barreto, S.M.; Giatti, L. Association between consumption of ultra-processed foods and serum C-reactive protein levels: Cross-sectional results from the ELSA-Brasil study. *Sao Paulo Med. J.* **2019**, *137*, 169–176. [CrossRef]
55. Lane, M.M.; Lotfaliany, M.; Forbes, M.; Loughman, A.; Rocks, T.; O'Neil, A.; Machado, P.; Jacka, F.N.; Hodge, A.; Marx, W. Higher Ultra-Processed Food Consumption Is Associated with Greater High-Sensitivity C-Reactive Protein Concentration in Adults: Cross-Sectional Results from the Melbourne Collaborative Cohort Study. *Nutrients* **2022**, *14*, 3309. [CrossRef]
56. Martins, G.M.; França, A.K.; Viola, P.C.; Carvalho, C.A.; Marques, K.D.; Santos, A.M.; Batalha, M.A.; Alves, J.D.; Ribeiro, C.C. Intake of ultra-processed foods is associated with inflammatory markers in Brazilian adolescents. *Public Health Nutr.* **2022**, *25*, 591–599. [CrossRef] [PubMed]
57. Silva Dos Santos, F.; Costa Mintem, G.; de Oliveira, I.O.; Horta, B.L.; Ramos, E.; Lopes, C.; Gigante, D.P. Consumption of ultra-processed foods and interleukin-6 in two cohorts from high- and middle-income countries. *Br. J. Nutr.* **2022**, 1–11. [CrossRef]
58. Kelsey, P.T.; Papadopoulou, E.; Borge, T.C.; Dahl, C.; Brantsæter, A.L.; Erlund, I.; Meltzer, H.M.; Haug, L.S.; Caspersen, I.H. Ultra-processed food consumption and associations with biomarkers of nutrition and inflammation in pregnancy: The Norwegian Environmental Biobank. *Front. Nutr.* **2022**, *9*, 1052001. [CrossRef]

59. Mignogna, C.; Costanzo, S.; Di Castelnuovo, A.; Ruggiero, E.; Shivappa, N.; Hebert, J.R.; Esposito, S.; De Curtis, A.; Persichillo, M.; Cerletti, C.; et al. The inflammatory potential of the diet as a link between food processing and low-grade inflammation: An analysis on 21,315 participants to the Moli-sani study. *Clin. Nutr.* **2022**, *41*, 2226–2234. [CrossRef]
60. Silva, C.A.; Santos, I.D.S.; Shivappa, N.; Hebert, J.R.; Crivellenti, L.C.; Sartorelli, D.S. The role of food processing in the inflammatory potential of diet during pregnancy. *Rev. Saude. Publica* **2019**, *53*, 113. [CrossRef] [PubMed]
61. Christ, A.; Lauterbach, M.; Latz, E. Western Diet and the Immune System: An Inflammatory Connection. *Immunity* **2019**, *51*, 794–811. [CrossRef]
62. Atkinson, F.S.; Brand-Miller, J.C.; Foster-Powell, K.; Buyken, A.E.; Goletzke, J. International tables of glycemic index and glycemic load values 2021: A systematic review. *Am. J. Clin. Nutr.* **2021**, *114*, 1625–1632. [CrossRef] [PubMed]
63. Dror, E.; Dalmas, E.; Meier, D.T.; Wueest, S.; Thévenet, J.; Thienel, C.; Timper, K.; Nordmann, T.M.; Traub, S.; Schulze, F.; et al. Postprandial macrophage-derived IL-1β stimulates insulin, and both synergistically promote glucose disposal and inflammation. *Nat. Immunol.* **2017**, *18*, 283–292. [CrossRef]
64. Vitale, M.; Masulli, M.; Rivellese, A.A.; Babini, A.C.; Boemi, M.; Bonora, E.; Buzzetti, R.; Ciano, O.; Cignarelli, M.; Cigolini, M.; et al. Influence of dietary fat and carbohydrates proportions on plasma lipids, glucose control and low-grade inflammation in patients with type 2 diabetes-The TOSCA.IT Study. *Eur. J. Nutr.* **2016**, *55*, 1645–1651. [CrossRef]
65. de Koning, L.; Malik, V.S.; Kellogg, M.D.; Rimm, E.B.; Willett, W.C.; Hu, F.B. Sweetened beverage consumption, incident coronary heart disease, and biomarkers of risk in men. *Circulation* **2012**, *125*, 1735–1741. [CrossRef] [PubMed]
66. Kosova, E.C.; Auinger, P.; Bremer, A.A. The Relationships between Sugar-Sweetened Beverage Intake and Cardiometabolic Markers in Young Children. *J. Acad. Nutr. Diet.* **2013**, *113*, 219–227. [CrossRef]
67. Hert, K.A.; Fisk, P.S.; Rhee, Y.S.; Brunt, A.R. Decreased consumption of sugar-sweetened beverages improved selected biomarkers of chronic disease risk among US adults: 1999 to 2010. *Nutr. Res.* **2014**, *34*, 58–65. [CrossRef] [PubMed]
68. Liu, S.; Manson, J.E.; Buring, J.E.; Stampfer, M.J.; Willett, W.C.; Ridker, P.M. Relation between a diet with a high glycemic load and plasma concentrations of high-sensitivity C-reactive protein in middle-aged women. *Am. J. Clin. Nutr.* **2002**, *75*, 492–498. [CrossRef] [PubMed]
69. Milajerdi, A.; Saneei, P.; Larijani, B.; Esmaillzadeh, A. The effect of dietary glycemic index and glycemic load on inflammatory biomarkers: A systematic review and meta-analysis of randomized clinical trials. *Am. J. Clin. Nutr.* **2018**, *107*, 593–606. [CrossRef]
70. Li, K.; Song, H.; Wei, F.; Liu, D.; Zhao, Y.; Yin, H.; Cui, Y.; Zhang, H.; Liu, Z. High salt intake damages myocardial viability and induces cardiac remodeling via chronic inflammation in the elderly. *Front. Cardiovasc. Med.* **2022**, *9*, 952691. [CrossRef]
71. Yilmaz, R.; Akoglu, H.; Altun, B.; Yildirim, T.; Arici, M.; Erdem, Y. Dietary salt intake is related to inflammation and albuminuria in primary hypertensive patients. *Eur. J. Clin. Nutr.* **2012**, *66*, 1214–1218. [CrossRef]
72. Zhu, H.; Pollock, N.K.; Kotak, I.; Gutin, B.; Wang, X.; Bhagatwala, J.; Parikh, S.; Harshfield, G.A.; Dong, Y. Dietary sodium, adiposity, and inflammation in healthy adolescents. *Pediatrics* **2014**, *133*, e635–e642. [CrossRef]
73. Basdeki, E.D.; Kollias, A.; Mitrou, P.; Tsirimiagkou, C.; Georgakis, M.K.; Chatzigeorgiou, A.; Argyris, A.; Karatzi, K.; Manios, Y.; Sfikakis, P.P.; et al. Does Sodium Intake Induce Systemic Inflammatory Response? A Systematic Review and Meta-Analysis of Randomized Studies in Humans. *Nutrients* **2021**, *13*, 2632. [CrossRef] [PubMed]
74. Pranger, I.G.; Muskiet, F.A.; Kema, I.P.; Singh-Povel, C.; Bakker, S.J. Potential Biomarkers for Fat from Dairy and Fish and Their Association with Cardiovascular Risk Factors: Cross-sectional Data from the LifeLines Biobank and Cohort Study. *Nutrients* **2019**, *11*, 1099. [CrossRef] [PubMed]
75. Hadj Ahmed, S.; Kharroubi, W.; Kaoubaa, N.; Zarrouk, A.; Batbout, F.; Gamra, H.; Najjar, M.F.; Lizard, G.; Hininger-Favier, I.; Hammami, M. Correlation of trans fatty acids with the severity of coronary artery disease lesions. *Lipids Health Dis.* **2018**, *17*, 52. [CrossRef]
76. Antoniazzi, L.; Arroyo-Olivares, R.; Bittencourt, M.S.; Tada, M.T.; Lima, I.; Jannes, C.E.; Krieger, J.E.; Pereira, A.C.; Quintana-Navarro, G.; Muñiz-Grijalvo, O.; et al. Association of dietary components with dyslipidemia and low-grade inflammation biomarkers in adults with heterozygous familial hypercholesterolemia from different countries. *Eur. J. Clin. Nutr.* **2019**, *73*, 1622–1625. [CrossRef]
77. DiNicolantonio, J.J.; O'Keefe, J.H. Importance of maintaining a low omega-6/omega-3 ratio for reducing inflammation. *Open Heart* **2018**, *5*, e000946. [CrossRef] [PubMed]
78. Neale, E.P.; Batterham, M.J.; Tapsell, L.C. Consumption of a healthy dietary pattern results in significant reductions in C-reactive protein levels in adults: A meta-analysis. *Nutr. Res.* **2016**, *36*, 391–401. [CrossRef]
79. Bahrampour, N.; Shiraseb, F.; Noori, S.; Clark, C.C.; Mirzaei, K. Is there any putative mediatory role of inflammatory markers on the association between ultra-processed foods and resting metabolic rate? *Front. Nutr.* **2022**, *9*, 932225. [CrossRef]
80. Dinu, M.; Asensi, M.T.; Pagliai, G.; Lotti, S.; Martini, D.; Colombini, B.; Sofi, F. Consumption of Ultra-Processed Foods Is Inversely Associated with Adherence to the Mediterranean Diet: A Cross-Sectional Study. *Nutrients* **2022**, *14*, 2073. [CrossRef]
81. Liu, T.; Wang, C.; Wang, Y.Y.; Wang, L.L.; Ojo, O.; Feng, Q.Q.; Jiang, X.S.; Wang, X.H. Effect of dietary fiber on gut barrier function, gut microbiota, short-chain fatty acids, inflammation, and clinical outcomes in critically ill patients: A systematic review and meta-analysis. *JPEN J. Parenter. Enteral Nutr.* **2022**, *46*, 997–1010. [CrossRef]
82. Hébert, J.R.; Shivappa, N.; Wirth, M.D.; Hussey, J.R.; Hurley, T.G. Perspective: The Dietary Inflammatory Index (DII)-Lessons Learned, Improvements Made, and Future Directions. *Adv. Nutr.* **2019**, *10*, 185–195. [CrossRef]

83. Monteiro, C.A.; Cannon, G.; Moubarac, J.C. The UN Decade of Nutrition, the NOVA food classification and the trouble with ultra-processing. *Public Health Nutr.* **2018**, *21*, 5–17. [CrossRef] [PubMed]
84. Debras, C.; Chazelas, E.; Sellem, L.; Porcher, R.; Druesne-Pecollo, N.; Esseddik, Y.; de Edelenyi, F.S.; Agaësse, C.; De Sa, A.; Lutchia, R.; et al. Artificial sweeteners and risk of cardiovascular diseases: Results from the prospective NutriNet-Santé cohort. *BMJ* **2022**, *378*, e071204. [CrossRef]
85. Bian, X.; Chi, L.; Gao, B.; Tu, P.; Ru, H.; Lu, K. Gut Microbiome Response to Sucralose and Its Potential Role in Inducing Liver Inflammation in Mice. *Front. Physiol.* **2017**, *8*, 487. [CrossRef]
86. Bian, X.; Tu, P.; Chi, L.; Gao, B.; Ru, H.; Lu, K. Saccharin induced liver inflammation in mice by altering the gut microbiota and its metabolic functions. *Food Chem. Toxicol.* **2017**, *107*, 530–539. [CrossRef] [PubMed]
87. Viennois, E.; Merlin, D.; Gewirtz, A.T.; Chassaing, B. Dietary Emulsifier-Induced Low-Grade Inflammation Promotes Colon Carcinogenesis. *Cancer Res.* **2017**, *77*, 27–40. [CrossRef]
88. Murali, A.; Giri, V.; Cameron, H.J.; Sperber, S.; Zickgraf, F.M.; Haake, V.; Driemert, P.; Walk, T.; Kamp, H.; Rietjens, I.M.; et al. Investigating the gut microbiome and metabolome following treatment with artificial sweeteners acesulfame potassium and saccharin in young adult Wistar rats. *Food Chem. Toxicol.* **2022**, *165*, 113123. [CrossRef] [PubMed]
89. Chassaing, B.; Koren, O.; Goodrich, J.K.; Poole, A.C.; Srinivasan, S.; Ley, R.E.; Gewirtz, A.T. Dietary emulsifiers impact the mouse gut microbiota promoting colitis and metabolic syndrome. *Nature* **2015**, *519*, 92–96. [CrossRef]
90. Khan, T.A.; Sievenpiper, J.L. Low-energy sweeteners and cardiometabolic health: Is there method in the madness? *Am. J. Clin. Nutr.* **2020**, *112*, 917–919. [CrossRef]
91. Greyling, A.; Appleton, K.M.; Raben, A.; Mela, D.J. Acute glycemic and insulinemic effects of low-energy sweeteners: A systematic review and meta-analysis of randomized controlled trials. *Am. J. Clin. Nutr.* **2020**, *112*, 1002–1014. [CrossRef]
92. Zhang, R.; Noronha, J.C.; Khan, T.A.; McGlynn, N.; Back, S.; Grant, S.M.; Kendall, C.W.; Sievenpiper, J.L. The Effect of Non-Nutritive Sweetened Beverages on Postprandial Glycemic and Endocrine Responses: A Systematic Review and Network Meta-Analysis. *Nutrients* **2023**, *15*, 1050. [CrossRef] [PubMed]
93. Huang, Y.C.; Huang, P.R.; Lo, Y.C.; Sun, C.W.; Pan, W.H.; Wang, S.L.; Huang, H.B. Food Processing and Phthalate Exposure: The Nutrition and Health Survey in Taiwan (1993–1996 and 2005–2008). *Front. Nutr.* **2021**, *8*, 766992. [CrossRef]
94. Dubeau, C.; Aker, A.; Caron-Beaudoin, É.; Ayotte, P.; Blanchette, C.; McHugh, N.G.; Lemire, M. Perfluoroalkyl acid and bisphenol-A exposure via food sources in four First Nation communities in Quebec, Canada. *Public Health Nutr.* **2023**, *26*, 106–121. [CrossRef] [PubMed]
95. Naspolini, N.F.; Machado, P.P.; Moreira, J.C.; Asmus, C.I.; Meyer, A. Maternal consumption of ultra-processed foods and newborn exposure to perfluoroalkyl substances (PFAS). *Cad. Saude. Publica* **2021**, *37*, e00152021. [CrossRef]
96. Martínez Steele, E.; Khandpur, N.; da Costa Louzada, M.L.; Monteiro, C.A. Association between dietary contribution of ultra-processed foods and urinary concentrations of phthalates and bisphenol in a nationally representative sample of the US population aged 6 years and older. *PLoS ONE* **2020**, *15*, e0236738. [CrossRef]
97. Moon, S.; Yu, S.H.; Lee, C.B.; Park, Y.J.; Yoo, H.J.; Kim, D.S. Effects of bisphenol A on cardiovascular disease: An epidemiological study using National Health and Nutrition Examination Survey 2003–2016 and meta-analysis. *Sci. Total. Environ.* **2021**, *763*, 142941. [CrossRef]
98. Gore, A.C.; Chappell, V.A.; Fenton, S.E.; Flaws, J.A.; Nadal, A.; Prins, G.S.; Toppari, J.; Zoeller, R.T. EDC 2: The Endocrine Society's Second Scientific Statement on Endocrine-Disrupting Chemicals. *Endocr. Rev.* **2015**, *36*, E1–E150. [CrossRef]
99. Liu, Z.; Lu, Y.; Zhong, K.; Wang, C.; Xu, X. The associations between endocrine disrupting chemicals and markers of inflammation and immune responses: A systematic review and meta-analysis. *Ecotoxicol. Environ. Saf.* **2022**, *234*, 113382. [CrossRef]
100. Ye, X.; Wong, L.Y.; Kramer, J.; Zhou, X.; Jia, T.; Calafat, A.M. Urinary Concentrations of Bisphenol A and Three Other Bisphenols in Convenience Samples of U.S. Adults during 2000–2014. *Environ. Sci. Technol.* **2015**, *49*, 11834–11839. [CrossRef] [PubMed]
101. Wang, B.; Wang, X.; Yu, L.; Liu, W.; Song, J.; Fan, L.; Zhou, M.; Yang, M.; Ma, J.; Cheng, M.; et al. Acrylamide exposure increases cardiovascular risk of general adult population probably by inducing oxidative stress, inflammation, and TGF-β1: A prospective cohort study. *Environ. Int.* **2022**, *164*, 107261. [CrossRef]
102. Feng, X.; Liang, R.; Shi, D.; Wang, D.; Xu, T.; Chen, W. Urinary acrolein metabolites, systemic inflammation, and blood lipids: Results from the National Health and Nutrition Examination Survey. *Chemosphere* **2022**, *286*, 131791. [CrossRef]
103. Wang, B.; Yu, L.; Liu, W.; Yang, M.; Fan, L.; Zhou, M.; Ma, J.; Wang, X.; Nie, X.; Cheng, M. Cross-sectional and longitudinal associations of acrolein exposure with pulmonary function alteration: Assessing the potential roles of oxidative DNA damage, inflammation, and pulmonary epithelium injury in a general adult population. *Environ. Int.* **2022**, *167*, 107401. [CrossRef]
104. Sommer, F.; Bäckhed, F. The gut microbiota—Masters of host development and physiology. *Nat. Rev. Microbiol.* **2013**, *11*, 227–238. [CrossRef]
105. Petersen, C.; Round, J.L. Defining dysbiosis and its influence on host immunity and disease. *Cell. Microbiol.* **2014**, *16*, 1024–1033. [CrossRef] [PubMed]
106. Silva, Y.P.; Bernardi, A.; Frozza, R.L. The Role of Short-Chain Fatty Acids from Gut Microbiota in Gut-Brain Communication. *Front. Endocrinol.* **2020**, *11*, 25. [CrossRef]
107. Delzenne, N.M.; Olivares, M.; Neyrinck, A.M.; Beaumont, M.; Kjølbæk, L.; Larsen, T.M.; Benítez-Páez, A.; Romaní-Pérez, M.; Garcia-Campayo, V.; Bosscher, D.; et al. Nutritional interest of dietary fiber and prebiotics in obesity: Lessons from the MyNewGut consortium. *Clin. Nutr.* **2020**, *39*, 414–424. [CrossRef]

108. Makki, K.; Deehan, E.C.; Walter, J.; Bäckhed, F. The Impact of Dietary Fiber on Gut Microbiota in Host Health and Disease. *Cell Host Microbe* **2018**, *23*, 705–715. [CrossRef] [PubMed]
109. Sonnenburg, E.D.; Sonnenburg, J.L. Starving our microbial self: The deleterious consequences of a diet deficient in microbiota-accessible carbohydrates. *Cell Metab.* **2014**, *20*, 79–786. [CrossRef] [PubMed]
110. Kaczmarczyk, O.; Dąbek-Drobny, A.; Woźniakiewicz, M.; Paśko, P.; Dobrowolska-Iwanek, J.; Woźniakiewicz, A.; Targosz, A.; Ptak-Belowska, A.; Piątek-Guziewicz, A.; Zagrodzki, P.; et al. Association between fecal levels of Short-Chain Fatty Acids and serum Pro and Anti-Inflammatory Cytokines in patients with Inflammatory Bowel Disease. *Folia Med. Cracov.* **2022**, *62*, 43–55. [CrossRef]
111. Guerville, M.; Leroy, A.; Sinquin, A.; Laugerette, F.; Michalski, M.C.; Boudry, G. Western-diet consumption induces alteration of barrier function mechanisms in the ileum that correlates with metabolic endotoxemia in rats. *Am. J. Physiol. Endocrinol. Metab.* **2017**, *313*, E107–E120. [CrossRef]
112. López-Moreno, J.; García-Carpintero, S.; Jimenez-Lucena, R.; Haro, C.; Rangel-Zúñiga, O.A.; Blanco-Rojo, R.; Yubero-Serrano, E.M.; Tinahones, F.J.; Delgado-Lista, J.; Pérez-Martínez, P.; et al. Effect of Dietary Lipids on Endotoxemia Influences Postprandial Inflammatory Response. *J. Agric. Food Chem.* **2017**, *65*, 7756–7763. [CrossRef]
113. Montrose, D.C.; Nishiguchi, R.; Basu, S.; Staab, H.A.; Zhou, X.K.; Wang, H.; Meng, L.; Johncilla, M.; Cubillos-Ruiz, J.R.; Morales, D.K.; et al. Dietary Fructose Alters the Composition, Localization, and Metabolism of Gut Microbiota in Association with Worsening Colitis. *Cell. Mol. Gastroenterol. Hepatol.* **2021**, *11*, 525–550. [CrossRef] [PubMed]
114. Um, C.Y.; Hodge, R.A.; Tran, H.Q.; Campbell, P.T.; Gewirtz, A.T.; McCullough, M.L. Association of Emulsifier and Highly Processed Food Intake with Circulating Markers of Intestinal Permeability and Inflammation in the Cancer Prevention Study-3 Diet Assessment Sub-Study. *Nutr. Cancer* **2022**, *74*, 1701–1711. [CrossRef] [PubMed]
115. Di Palo, D.M.; Garruti, G.; Di Ciaula, A.; Molina-Molina, E.; Shanmugam, H.; De Angelis, M.; Portincasa, P. Increased Colonic Permeability and Lifestyles as Contributing Factors to Obesity and Liver Steatosis. *Nutrients* **2020**, *12*, 564. [CrossRef] [PubMed]
116. Partula, V.; Mondot, S.; Torres, M.J.; Kesse-Guyot, E.; Deschasaux, M.; Assmann, K.; Latino-Martel, P.; Buscail, C.; Julia, C.; Galan, P.; et al. Associations between usual diet and gut microbiota composition: Results from the Milieu Intérieur cross-sectional study. *Am. J. Clin. Nutr.* **2019**, *109*, 1472–1483. [CrossRef] [PubMed]
117. Atzeni, A.; Martínez, M.Á.; Babio, N.; Konstanti, P.; Tinahones, F.J.; Vioque, J.; Corella, D.; Fitó, M.; Vidal, J.; Moreno-Indias, I.; et al. Association between ultra-processed food consumption and gut microbiota in senior subjects with overweight/obesity and metabolic syndrome. *Front. Nutr.* **2022**, *9*, 976547. [CrossRef]
118. Swidsinski, A.; Ung, V.; Sydora, B.C.; Loening-Baucke, V.; Doerffel, Y.; Verstraelen, H.; Fedorak, R.N. Bacterial overgrowth and inflammation of small intestine after carboxymethylcellulose ingestion in genetically susceptible mice. *Inflamm. Bowel Dis.* **2009**, *15*, 359–364. [CrossRef]
119. Chassaing, B.; Compher, C.; Bonhomme, B.; Liu, Q.; Tian, Y.; Walters, W.; Nessel, L.; Delaroque, C.; Hao, F.; Gershuni, V.; et al. Randomized Controlled-Feeding Study of Dietary Emulsifier Carboxymethylcellulose Reveals Detrimental Impacts on the Gut Microbiota and Metabolome. *Gastroenterology* **2022**, *162*, 743–756. [CrossRef] [PubMed]
120. Bian, X.; Chi, L.; Gao, B.; Tu, P.; Ru, H.; Lu, K. The artificial sweetener acesulfame potassium affects the gut microbiome and body weight gain in CD-1 mice. *PLoS ONE* **2017**, *12*, e0178426. [CrossRef]

nutrients MDPI

Article

Processing and Nutritional Quality of Breakfast Cereals Sold in Italy: Results from the Food Labelling of Italian Products (FLIP) Study

Donato Angelino [1], Monica Dinu [2], Barbara Gandossi [3], Nicoletta Pellegrini [4,*] and Daniela Martini [3,*,†] on behalf of the SINU Young Working Group

1 Department of Bioscience and Technology for Food, Agriculture and Environment, University of Teramo, 64100 Teramo, Italy
2 Department of Experimental and Clinical Medicine, University of Florence, 50134 Florence, Italy
3 Department of Food, Environmental and Nutritional Sciences (DeFENS), University of Milan, 20133 Milan, Italy
4 Department of Agricultural, Food, Environmental and Animal Sciences, University of Udine, 33100 Udine, Italy
* Correspondence: nicoletta.pellegrini@uniud.it (N.P.); daniela.martini@unimi.it (D.M.); Tel.: +39-0432-558183 (N.P.); +39-02-50316727 (D.M.)
† Collaborators of the SINU Young Working Group are indicated in Appendix A (Table A1).

Abstract: This study aimed to compare the level of processing (as assessed by the NOVA classification) and the nutritional quality (as assessed by nutrition values, Nutri-Score and NutrInform battery) of breakfast cereals currently on the Italian market. A total of 349 items were found, mostly belonging to the NOVA 4 group (66.5%) and to Nutri-Score C and A (40% and 30%, respectively). The NOVA 4 products showed the highest energy, total fat, saturates, and sugar content per 100 g and had the highest number of items with Nutri-Score C (49%) and D (22%). Conversely, NOVA 1 products had the highest content of fibre and protein, the lowest amounts of sugars and salt, and 82% of them were Nutri-Score A, while few Nutri-Score B and C were found. Differences were attenuated when products were compared for their NutrInform battery, with NOVA 4 items showing only slightly fuller batteries for saturated fats, sugar, and salt than NOVA 1 and NOVA 3 products. Overall, these results suggest that the NOVA classification partially overlaps with systems based on the nutritional quality of foods. The lower nutritional quality of NOVA 4 foods may at least partially explain the association found between the consumption of ultra-processed foods and the risk of chronic diseases.

Keywords: food label; NOVA system; ultra-processed foods; front-of-pack labelling

Citation: Angelino, D.; Dinu, M.; Gandossi, B.; Pellegrini, N.; Martini, D., on behalf of the SINU Young Working Group. Processing and Nutritional Quality of Breakfast Cereals Sold in Italy: Results from the Food Labelling of Italian Products (FLIP) Study. *Nutrients* **2023**, *15*, 2013. https://doi.org/10.3390/nu15082013

Academic Editors: Robert Hamlin and Manohar Garg

Received: 6 March 2023
Revised: 17 April 2023
Accepted: 19 April 2023
Published: 21 April 2023

1. Introduction

In recent years, increasing attention has been paid to the level of food processing [1]. NOVA is a classification system that groups foods according to the nature, extent, and purpose of the industrial processes they undergo, rather than in terms of nutrients [2]. In the NOVA classification, foods are assigned to one of the following four groups: NOVA 1 contains unprocessed or minimally processed foods, i.e., the edible parts of plants or animals taken directly from nature or minimally modified/preserved; NOVA 2 contains processed culinary ingredients, such as salt, sugar, oil, or starch, produced from NOVA 1 foods; NOVA 3 contains processed foods such as canned vegetables or freshly baked bread, produced by combining NOVA 1 and NOVA 2 foods; NOVA 4 contains ultra-processed foods, i.e., industrially formulated ready-to-eat products that are predominantly or entirely composed of food-derived substances and additives, with few or no intact foods from NOVA 1.

Up to date, many studies have shown an association between the consumption of NOVA 4 foods and health status, particularly regarding body weight [3], mortality [4], and chronic non-communicable diseases such as cardiovascular disease and depression [5,6]. Among the mechanisms hypothesized to explain these associations is that higher consumption of NOVA 4 foods leads to diets high in calories, free sugars, fat, and salt and low in dietary fibre [7]. Unlike the NOVA system, front-of-pack nutrition labels (FOPNLs) distinguish foods and beverages according to their energy and nutrient contribution to the overall diet. There are currently more than thirty different FOPNLs (proposed or implemented) in the world, many of them in use in multiple countries [8]. Some FOPNLs express the overall nutritional value of a food by using some or all the information from the nutrition declaration and/or other nutritional elements (e.g., the Nutri-Score, a graphic scale that divides the nutritional score into five classes expressed by a color and a letter). Other FOPNLs repeat specific numerical information from the mandatory nutrition declaration in a neutral manner (e.g., the NutrInform battery proposed by Italy).

Given that NOVA classification is not based on nutrient content, it is interesting to understand whether the classification of a food in NOVA 4 coincides with a worse classification by FOPNLs. In theory, NOVA and FOPNLs are complementary systems that focus on different aspects, so their application to a specific food may not necessarily lead to the same conclusions. An example is plant substitutes for animal foods that have a Nutri-Score A (indicating high nutritional quality) while being NOVA 4 [9]. This kind of discrepancy also emerged for other food groups, especially in a survey of foods on the Spanish market where only 75.5% of NOVA 4 foods were classified as having medium-low nutritional quality (C, D, and E) in the Nutri-Score [10]. Since there are no such comparisons made on products found in the Italian market, our aim was to compare the processing (as assessed by the NOVA classification) and the nutritional quality (as assessed by nutrition values retrieved in the nutritional declaration, Nutri-Score and NutrInform battery) of breakfast cereals currently on the Italian market, using data from the Food Labelling of Italian Products (FLIP) database. The choice of breakfast cereals comes from the fact that they can belong to different NOVA groups based on whether they are cereal-only, have added sugar or salt, or have many other ingredients not typically used domestically.

2. Materials and Methods

2.1. Data Collection and Extraction

Breakfast cereals included in the present work were selected as described in a previous study performed within the FLIP project [11]. The online search for the information was performed in July 2022.

The information retrieved for each product was the same as previously collected in a previous study [11], while the NOVA group was assigned to each item considering the processing classification system based on the NOVA classification [2,12].

The Nutri-Score was calculated for each item in accordance with the rules reported in the specific document (https://www.santepubliquefrance.fr/media/files/02-determinants-de-sante/nutrition-et-activite-physique/nutri-score/qr-scientifique-technique-en, Last access 1 February 2023).

Data for NutrInform battery (i.e., energy, fat, saturates, sugar, and salt content) were calculated by considering the standard serving size of 30 g [13] as defined in the manual of use (https://www.nutrinformbattery.it/Manuale_uso_NutrInform_Battery.pdf, Last access 1 February 2023).

Two researchers (DM and MD) extracted the data and double-checked the accuracy of data extraction, while inaccuracies were solved by a third researcher (DA).

All the retrieved data were collected in a Microsoft Excel database and sub-grouped for specific comparisons, i.e., tertiles of sugar, fibre, and salt, according to the NOVA group. Items were classified on the basis of the descriptive name as follows: (i) muesli, (ii) flakes, (iii) bran cereals, (iv) puffed cereals, and (v) others (e.g., cereals with honey or cream-filled cereals). Conversely, based on the presence of wholegrain ingredients, items were classified into refined, partially wholegrain (i.e., at least one ingredient), and wholegrain.

2.2. Data Analysis

Data were organized and statistically analyzed by using the Statistical Package for Social Sciences software (IBM SPSS Statistics, Version 28.0, IBM Corp., Chicago, IL, USA). The significance level was set at $p < 0.05$. Data of energy, nutrient, and fibre contents were expressed as median (interquartile range). In the descriptive analysis of the number of items, variables were expressed as absolute values. The Kolmogorov–Smirnov test was used to assess the normality of data distribution that was rejected.

Data were analyzed by means of the Mann–Whitney non-parametric test to allow comparisons of two independent groups or the Kruskal–Wallis test for independent samples with multiple pairwise comparisons. In addition, the variability of the nutritional values—as energy, nutrient, and fibre contents per 100 g of products—across the different items was assessed by means of a Principal Component (PC) analysis with varimax rotation, also considering the Nutri-Score and the NOVA group categorizations.

3. Results

3.1. Number and Characteristics of Food Items

Table 1 reports the number and the main characteristics of the retrieved items classified on the basis of the three NOVA groups (i.e., NOVA 1, 3, and 4). No items were classified as NOVA 2 since this group includes culinary ingredients. A total of 349 single items of breakfast cereals were included in the final evaluation, mostly belonging to the NOVA 4 group (66.5% out of the total). NOVA 4 prevailed in all the types and mostly in muesli (82%) and other cereals (95%), with the exception of puffed cereals in which 21 items (49%) were classified as NOVA 1 and 19 (44%) as NOVA 4.

Table 1. Number and characteristics of retrieved breakfast cereals stratified based on the NOVA group.

		NOVA Groups		
		NOVA 1	NOVA 3	NOVA 4
Total		60	57	232
Type	Muesli	2	17	85
	Flakes	29	33	63
	Bran cereals	8	1	9
	Puffed cereals	21	3	19
	Other cereals	0	3	56
Organic	No	13	28	176
	Yes	47	29	56
Branded	No	17	30	118
	Yes	43	27	114
Nutrition claim	No	24	14	52
	Yes	36	43	180
Fibre claim	No	28	33	114
	Yes	32	24	118
Fat claim	No	54	42	207
	Yes	6	15	25
Salt claim	No	54	55	229
	Yes	6	2	3
Vitamin and mineral claim	No	56	42	126
	Yes	4	15	106
Sugar claim	No	50	51	218
	Yes	10	6	14
Protein claim	No	47	54	221
	Yes	13	3	11

Table 1. *Cont.*

		NOVA Groups		
		NOVA 1	**NOVA 3**	**NOVA 4**
Health claim	No	54	51	172
	Yes	6	6	60
	No	37	30	95
Wholegrain	Partially #	4	13	121
	Yes	19	14	16

Legend: NOVA 1: minimally processed foods; NOVA 3, processed foods; NOVA 4, ultra-processed foods. # Items partially produced with wholegrain ingredients (i.e., at least one).

Regarding nutrition claims (NCs), among the 232 products classified as NOVA 4, 78% of the items showed at least 1 NC, while within the 60 items in the NOVA 1 group, 60% showed an NC. Health claims were instead reported by 26% of items in the NOVA 4 group and in 10% of those in the NOVA 1 group.

Fibre-related NCs were similarly distributed in NOVA 1 and NOVA 4 groups, with 47% and 49% of products, respectively, carrying this type of claim. Conversely, NCs related to minerals and/or vitamins were mainly present in the NOVA 4 group, with 93% of the items showing this type, while only 7% of items within the NOVA 1 group had this NC. Other NCs were displayed in a few items across the NOVA groups.

Regarding the presence of wholegrain ingredients, 58% and 87% of items made with refined ingredients or only partially with wholegrain ingredients (i.e., at least one) were in the NOVA 4 group, respectively. Conversely, items made with wholegrains were almost equally present in the NOVA 1 (39%), NOVA 3 (28%), and NOVA 4 (33) groups.

The nutritional quality of breakfast cereals belonging to the three NOVA groups stratified by cereal types is reported in Table 2. By considering all the retrieved breakfast cereals, NOVA 4 products showed the highest energy and sugar content per 100 g, while NOVA 1 products were characterized by the highest content of fibre and protein and the lowest amount of sugars and salt. Intriguingly, NOVA 1 and NOVA 4 cereals showed higher total and saturated fats and lower carbohydrate amounts than NOVA 3 products. Concerning the different types of cereals, a high variability in results of the nutritional characteristics of products grouped for NOVA classification was found. In fact, bran cereals and other cereals do not show any difference for energy and saturate contents within the different NOVA groups, while for all the other products NOVA 4 is almost always higher in energy, total carbohydrates, sugar, and salt than NOVA 1 and, in some cases, also NOVA 3 products. When salt is taken into consideration, almost all the NOVA 3 and NOVA 4 groups contain a consistent higher amount of salt—up to ten to twenty times more—than NOVA 1 products. Except for muesli cereals, all the other products labelled as NOVA 1 resulted significantly higher in protein content than NOVA 4 and in most cases also compared to NOVA 3.

Table 2. Energy, nutrients, and salt content of retrieved breakfast cereals stratified based on the NOVA group.

	NOVA	Energy (kJ/100 g)	Energy (kcal/100 g)	Total Fat (g/100 g)	Sfa (g/100 g)	Carbohydrates (g/100 g)	Sugar (g/100 g)	Fibre (g/100 g)	Protein (g/100 g)	SALT (g/100 g)
All	NOVA 1	1549 (1514–1596) c	366 (358–378) c	5.0 (2.6–7.0) a	0.9 (0.5–1.3) a	61.0 (57.7–69.0) b	1.1 (0.7–2.0) c	9.4 (6.5–12.1) a	12.1 (11.0–13.8) a	0.0 (0.0–0.0) B
	NOVA 3	1584 (1564–1634) b	375 (370–385) b	2.1 (1.0–7.3) b	0.5 (0.3–1.1) b	73.0 (60.7–82.0) a	15.0 (6.6–20.4) b	6.0 (3.0–8.8) b	8.5 (7.3–11.0) b	0.4 (0.2–1.0) A
	NOVA 4	1668 (1601–1837) a	396 (378–438) a	5.7 (2.6–15.0) a	1.6 (0.6–4.0) a	69.0 (61.0–78.8) b	20.0 (14.0 -25.0) a	6.0 (4.3–8.0) b	8.5 (7.4–10.0) b	0.5 (0.2–0.9) A
Muesli	NOVA 1	1499 (1467–1531) b	356 (348–364) a,b	6.6 (5.9–7.3) b	1.6 (1.0–2.1) a	58.5 (58.0–59.0) a	18.5 (11.0–26.0) a	11.0 (11.0–11.0) a	10.2 (9.3–11.0) a	0.03 (0.02–0.03) A,B
	NOVA 3	1756 (1565–1820) a,b	418 (372–434) b	12.7 (7.3–16.3) a	4.3 (1.1–5.3) a	59.0 (58.0–63.2) a	18.0 (16.0–22.0) a	8.0 (7.1–9.0) a,b	10.0 (9.3–11.0) a	0.2 (0.02–0.2) B
	NOVA 4	1834 (1746–1917) a	437 (416–458) a	16.0 (13.0–19.0) a	4.1 (1.9–5.1) a	61.1 (58.0–63.2) a	20.0 (16.0–24.0) a	7.2 (6.2–8.4) b	9.0 (8.5–11.0) a	0.2 (0.1–0.4) A
Flakes	NOVA 1	1549 (1517–1566) b	366 (360–372) b	7.0 (5.4–7.0) a	1.2 (1.0–1.3) a	59.1 (58.6–63.0) b	1.1 (0.7–1.4) b	9.9 (8.0–10.0) a	12.0 (11.6–13.0) a	0.01 (0.01–0.03) B
	NOVA 3	1578 (1564–1604) a	372 (370–378) a	1.0 (1.0–1.5) c	0.3 (0.2–0.5) b	81.0 (81.0–83.0) a	8.0 (6.2–16.0) a	3.3 (3.0–6.6) b	8.0 (7.3–8.5) b	0.9 (0.5–1.9) A
	NOVA 4	1605 (1574–1662) a	379 (371–390) a	2.0 (1.5–5.0) b	0.5 (0.3–1.5) b	78.0 (73.0–81.0) a	14.0 (7.8–17.8) a	4.7 (63.5–6.3) b	8.1 (7.4–11.6) b	0.8 (0.5–1.4) A
Bran cereals	NOVA 1	1412 (1291–1526) a	336 (309–363) a	6.1 (4.5–7.5) a	1.1 (0.8–1.3) a	47.5 (37.0–52.1) a	2.0 (1.2–2.7) b	22.1 (16.0–30.0) c	14.6 (13.5–15.4) a	0.01 (0.002–0.01) B
	NOVA 3	1436 (1436–1436) a	343 (343–343) a	4.5 (4.5–4.5) a,b	0.9 (0.9–0.9) a	46.0 (46.0–46.0) a	18.0 (18.0–18.0) a	29.0 (29.0–29.0) b	15.0 (15.0–15.0) a,b	1.2 (1.2–1.2) A
	NOVA 4	1342 (1318–1342) a	321 (316–321) a	3.9 (3.5–3.9) b	0.7 (0.7–1.0) a	41.0 (38.0–41.0) a	17.0 (13.0–18.0) a	35.0 (33.0–35.0) a	14.0 (13.0–14.6) b	1.3 (1.0–1.3) A
Puffed cereals	NOVA 1	1600 (1541–1611) b	378 (365–381) b	2.5 (1.1–3.1) a	0.5 (0.4–0.6) b	71.0 (68.0–85.0) b	0.6 (0.5–1.7) b	6.8 (0.8–8.5) a	11.5 (7.1–14.0) a	0.01 (0.003–0.01) B
	NOVA 3	1620 (1570–1673) a,b	382 (375–396) a,b	4.2 (1.5–5.7) a	0.7 (0.5–0.8) a	75.8 (73.0–85.0) a,b	25.0 (22.0–45.4) a	2.8 (2.2–7.5) a	9.6 (6.0–12.6) b	0.02 (0.000–0.3) A
	NOVA 4	1651 (1609–1690) a	390 (385–400) b	3.2 (1.9–5.1) a	0.6 (0.5–1.0) a	79.0 (78.0–85.0) a	33.0 (23.0–41.0) a	4.6 (2.5–5.5) a	7.0 (5.5–8.5) b	0.1 (0.01–0.7) A
Other cereals	NOVA 1	-	-	-	-	-	-	-	-	-
	NOVA 3	1992 (1531–2016)	476 (362–481)	22.0 (2.0–22.0)	8.9 (0.6–11.0)	60.0 (60.0–69.0)	19.0 (4.2–27.0)	5.7 (4.5–10.0)	8.0 (7.2–12.0)	0.6 (0.3–0.9)
	NOVA 4	1652 (1617–1864)	390 (382–443)	4.6 (2.3–14.1)	1.8 (0.9–4.0)	73.9 (69.4–79.0) *	25.0 (22.7–29.0)	5.0 (3.9–6.8)	7.7 (6.7–8.7)	0.7 (0.5–0.9)

Legend: For each group, different letters or asterisks in the same column after parenthesis indicate significant differences among types (Kruskal–Wallis test for independent samples with multiple pairwise comparisons or Mann–Whitney non-parametric test for two independent samples, $p < 0.05$). SFA, saturates.

Nutritional information of products belonging to the different NOVA groups and stratified on the basis of whole grain ingredients is shown in Supplementary Table S1. Data evidenced a few differences, only in terms of sugar and salt contents, among whole grain products belonging to different NOVA groups. Concerning products made with refined grains and partially produced with wholegrain, data evidenced significantly lower contents of energy, carbohydrate, sugar, fibre, and protein in NOVA 1 products compared to the NOVA 3 and 4 ones. It is worth underlining that the three product categories (refined grain, partially produced with wholegrain, and wholegrain products) have a different number of items, which may impact the intra-product variability of the data (Table 1). Data were also grouped on the basis of the tertiles of sugar, fibre, and salt amounts and then compared on the basis of the NOVA group (Supplementary Table S2). Concerning tertilization based on the sugar amount, products did not reveal particular differences between the NOVA groups, except for NOVA 1 products in the first tertile, which were lower in energy, total and saturated fats, carbohydrates, sugars, and salt and higher in fibre and protein compared to NOVA 2 and 3 products. When products were stratified for tertiles of fibre content, a generally worse nutritional profile in NOVA 4 compared to NOVA 1 items was observed, but here, again, the different number of items should be carefully considered. Finally, the salt tertilization did not show any result worth being highlighted, except for sugar amounts, which were much higher in NOVA 3 and NOVA 4 compared to NOVA 1 group.

3.2. *Nutri-Score and NutrInform Battery of Breakfast Cereals*

When only the Nutri-Score of the products was taken into consideration, regardless of the technological process, data showed that 141 (40%) items were labelled as C, 105 (30%) as A, 58 (17%) as D, and the remaining 45 (13%) items had a Nutri-Score B.

Figure 1 reports the distribution of the retrieved breakfast cereals according to NOVA categorization and Nutri-Score. As shown, NOVA 1 breakfast cereals displayed the highest proportion of products with Nutri-Score A (*n* = 49; 82%), followed by products belonging to NOVA 3 (*n* = 18; 32%) and NOVA 4 (*n* = 38; 16%) groups. The NOVA 1 group also showed the lowest proportion of products with Nutri-Score C (*n* = 3; 5%). Conversely, the Nutri-Score C prevailed in both NOVA 3 (*n* = 24; 42%) and NOVA 4 (*n* = 114; 49%) groups.

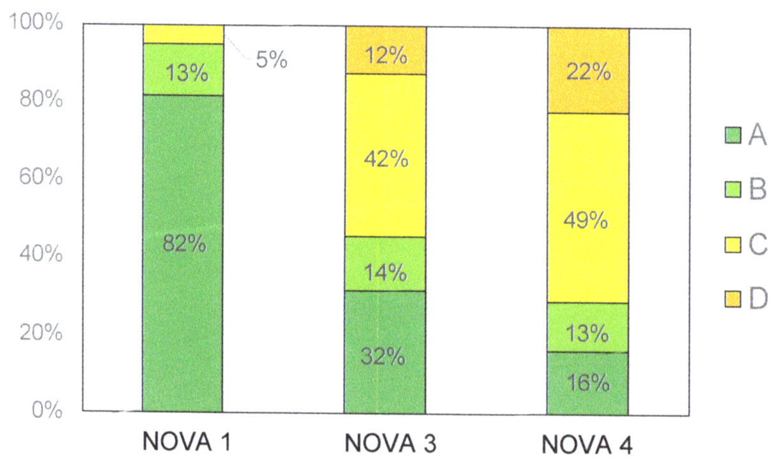

Figure 1. Percentage distribution of breakfast cereals based on NOVA group and Nutri-Score.

Finally, no products with Nutri-Score D were found in the NOVA 1 group, while 7 items (12%) and 51 items (22%) with Nutri-Score D were found in the NOVA 3 and NOVA 4 groups, respectively. No products displayed a Nutri-Score E.

Figure 2 reports the distribution of energy and some nutrient contents in breakfast cereals, classified according to the NOVA group and Nutri-Score. On the whole, a wide variability in terms of energy, nutrients, and fibre has been found regardless of the NOVA group and Nutri-Score values. Almost all the products, irrespective of the Nutri-Score value and the NOVA group, fell within the range of energy 300–500 kcal/100 g; data showed that NOVA 3 and NOVA 4 products had the largest variability in terms of total and saturated fats, sugar, and salt, with no exceptions among the Nutri-Score values. NOVA 1 products showed lower values of total and saturated fats than NOVA 3 and 4 ones, with some exceptions. Concerning fibre values, only Nutri-Score D products had lower than 10 g/100 g fibre content; all the other products showed a wide range of values, thus not allowing a clear grouping of products on the basis of their NOVA group or Nutri-Score.

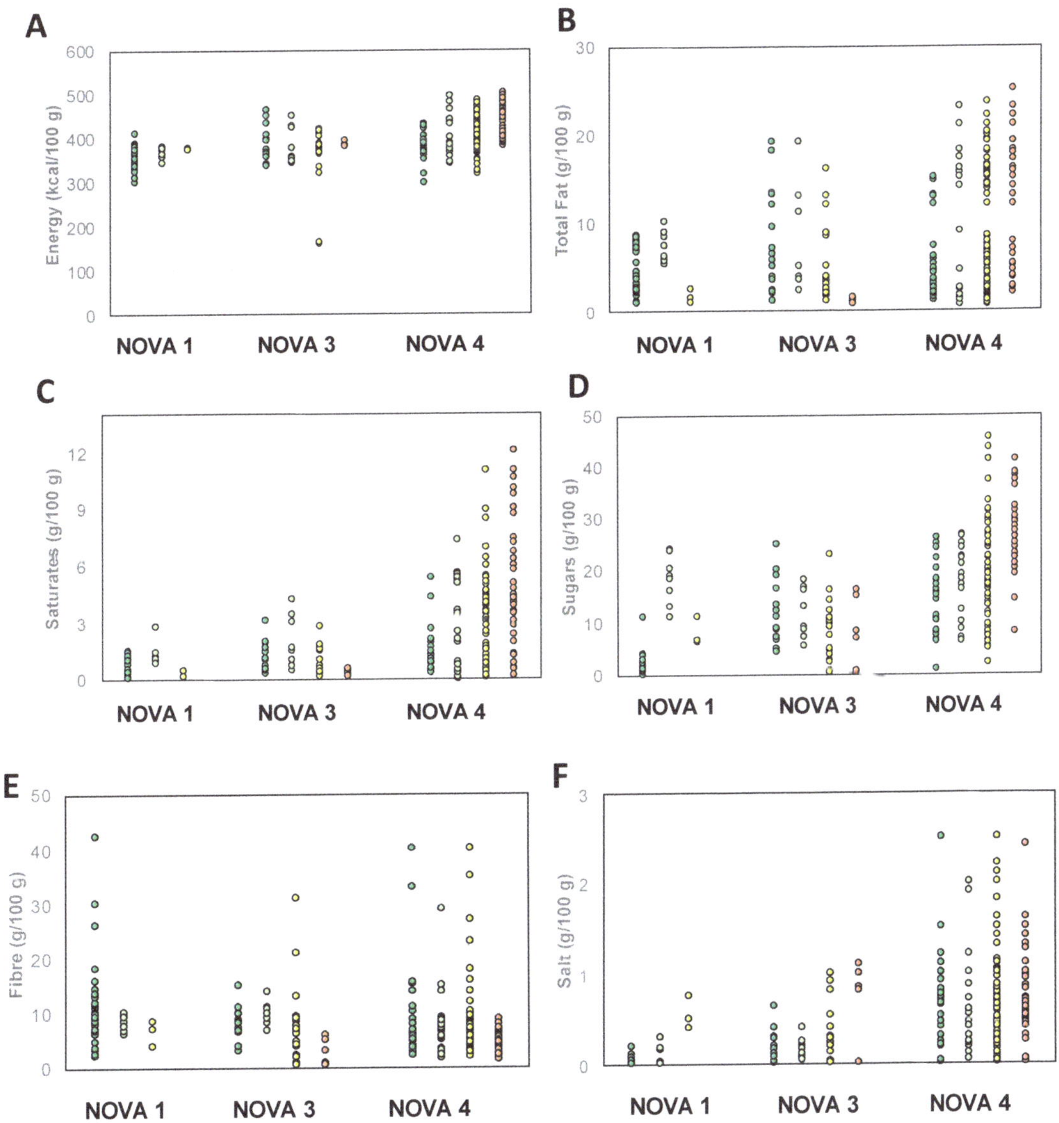

Figure 2. Energy (**A**) and nutrient (**B**–**F**) content of breakfast cereals grouped on the basis of NOVA group and Nutri-Score. Legend of Nutri-Scores: ● = A; ○ = B; ● = C; ● = D.

The variability in the nutritional composition of the breakfast cereals was explored by PC analysis (Figure 3). Two PCs explained 68.7% of the total variability: (i) PC1—which accounted for 36.6% of the variability—was mainly positively loaded by energy, total and saturated fats, and sugars, while being negatively loaded by total carbohydrates, salt, fibre, and protein; (ii) PC2—explaining the 32.1% of the total variability—was mainly positively loaded by total carbohydrates, sugars, and salt and negatively loaded by fibre and protein (Figure 3A). The score plot in Figure 3B confirms a high variability that did not allow products to be grouped on the basis of the NOVA and Nutri-Score values. On the whole, most of the products with B, C, and D Nutri-Scores were described as having high energy, total carbohydrates, salt, total and saturated fats, and sugars, with no distinctions for NOVA groups. Then, a main characterization of the A products by fibre and protein was slightly evidenced, but no distinction among NOVA groups could be pointed out.

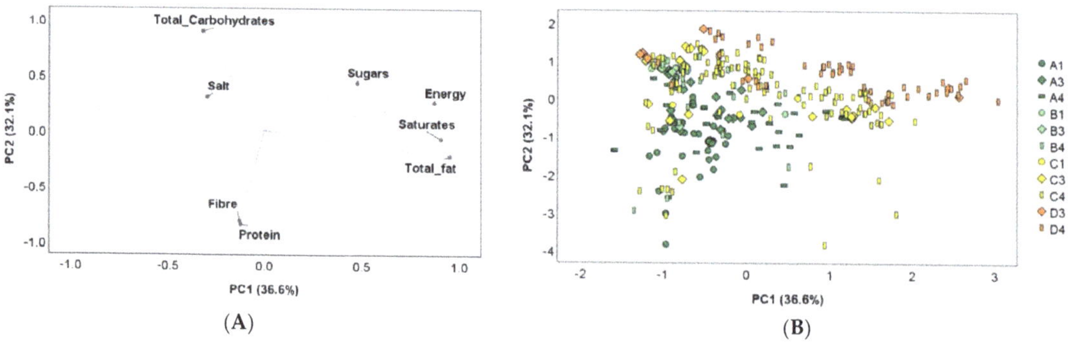

Figure 3. Principal component (PC) analysis describing the intra-group variability of products based on their nutrient composition (energy (kcal/100 g), total fat (g/100 g), saturates (g/100 g), total carbohydrates (g/100 g), sugars (g/100 g), protein (g/100 g), fibre (g/100 g), and salt (g/100 g)). Loading plots (**A**) of PC1 versus PC2; score plots (**B**) of the nutrient composition of each product analyzed organized according to Nutri-Score. Legend: legend is composed of letters (A to D) indicating the letters of the Nutri-Score and numbers (1, 3, and 4) indicating the NOVA groups.

Regarding the NutrInform battery, Table 3 reports the batteries of products classified based on the NOVA groups and grouped for the breakfast cereal typologies. Data confirmed the evidence of a tight variability for daily energy contribution of a 30 g serving of the products, from 5 to 7%, with few distinctions among the NOVA groups. On the contrary, most of the variability referred to the contribution to the daily amounts of total and saturated fats, but just for a few typologies. Interestingly, NOVA 3 "other cereals" accounted for up to 14% of daily saturated fats vs. 3% of the NOVA 4; nevertheless, it is worth to remember that, among "other cereals", only three NOVA 3 items were retrieved compared to fifty-six NOVA 4. On the whole, NOVA 1 products were almost not contributing to the daily salt amount; the items with the most impact were bran cereals and flakes, both NOVA 3 and 4, with up to 8% of the daily salt amount. Finally, despite all the products not being largely different for the daily sugar amounts, NOVA 3 and NOVA 4 puffed cereals contributed to 8% and 11% of sugar daily amounts, respectively.

Table 3. NutrInform battery of breakfast cereals stratified on the basis of the NOVA group.

	NOVA 1	NOVA3	NOVA 4
All	Ciascuna porzione (30g) contiene: ENERGIA 465 kJ 110 kcal — 6 % GRASSI 1,5 g — 2 % GRASSI SATURI 0,3 g — 2 % ZUCCHERI 0 g — 0 % SALE 0 g — 0 % delle Assunzioni di Riferimento di un adulto medio (8.400kJ / 2.000kcal) per 100g: 1549 kJ / 366 kcal	Ciascuna porzione (30g) contiene: ENERGIA 475 kJ 113 kcal — 6 % GRASSI 0,6 g — 1 % GRASSI SATURI 0,2 g — 1 % ZUCCHERI 4,5 g — 5 % SALE 0,1 g — 2 % delle Assunzioni di Riferimento di un adulto medio (8.400kJ / 2.000kcal) per 100g: 1584 kJ / 375 kcal	Ciascuna porzione (30g) contiene: ENERGIA 500 kJ 119 kcal — 6 % GRASSI 1,7 g — 2 % GRASSI SATURI 0,5 g — 3 % ZUCCHERI 6 g — 7 % SALE 0,2 g — 3 % delle Assunzioni di Riferimento di un adulto medio (8.400kJ / 2.000kcal) per 100g: 1668 kJ / 396 kcal
Muesli	Ciascuna porzione (30g) contiene: ENERGIA 45[illegible] kJ 107 kcal — 5 % GRASSI 2 g — 3 % GRASSI SATURI 0,5 g — 3 % ZUCCHERI 5,6 g — 6 % SALE 0 g — 0 % delle Assunzioni di Riferimento di un adulto medio (8.400kJ / 2.000kcal) per 100g: 1499 kJ / 356 kcal	Ciascuna porzione (30g) contiene: ENERGIA 527 kJ 125 kcal — 6 % GRASSI 3,8 g — 5 % GRASSI SATURI 1,3 g — 7 % ZUCCHERI 5,4 g — 6 % SALE 0,1 g — 2 % delle Assunzioni di Riferimento di un adulto medio (8.400kJ / 2.000kcal) per 100g: 1756 kJ / 418 kcal	Ciascuna porzione (30g) contiene: ENERGIA 550 kJ 131 kcal — 7 % GRASSI 4,8 g — 7 % GRASSI SATURI 1,2 g — 6 % ZUCCHERI 6 g — 7 % SALE 0,1 g — 2 % delle Assunzioni di Riferimento di un adulto medio (8.400kJ / 2.000kcal) per 100g: 1834 kJ / 437 kcal
Flakes	Ciascuna porzione (30g) contiene: ENERGIA 465 kJ 110 kcal — 6 % GRASSI 2,1 g — 3 % GRASSI SATURI 0,4 g — 2 % ZUCCHERI 0 g — 0 % SALE 0 g — 0 % delle Assunzioni di Riferimento di un adulto medio (8.400kJ / 2.000kcal) per 100g: 1549 kJ / 366 kcal	Ciascuna porzione (30g) contiene: ENERGIA 473 kJ 112 kcal — 6 % GRASSI 0 g — 0 % GRASSI SATURI <0,1 g — 1 % ZUCCHERI 2,4 g — 3 % SALE 0,3 g — 5 % delle Assunzioni di Riferimento di un adulto medio (8.400kJ / 2.000kcal) per 100g: 1578 kJ / 372 kcal	Ciascuna porzione (30g) contiene: ENERGIA 481 kJ 114 kcal — 6 % GRASSI 0,6 g — 1 % GRASSI SATURI 0,2 g — 1 % ZUCCHERI 4,2 g — 5 % SALE 0,2 g — 3 % delle Assunzioni di Riferimento di un adulto medio (8.400kJ / 2.000kcal) per 100g: 1605 kJ / 379 kcal
Bran Cereals	Ciascuna porzione (30g) contiene: ENERGIA 424 kJ 101 kcal — 5 % GRASSI 1,8 g — 3 % GRASSI SATURI 0,3 g — 2 % ZUCCHERI 0,6 g — 1 % SALE 0 g — 0 % delle Assunzioni di Riferimento di un adulto medio (8.400kJ / 2.000kcal) per 100g: 1412 kJ / 336 kcal	Ciascuna porzione (30g) contiene: ENERGIA 431 kJ 103 kcal — 5 % GRASSI 1,4 g — 2 % GRASSI SATURI 0,3 g — 2 % ZUCCHERI 5,4 g — 6 % SALE 0,5 g — 8 % delle Assunzioni di Riferimento di un adulto medio (8.400kJ / 2.000kcal) per 100g: 1436 kJ / 343 kcal	Ciascuna porzione (30g) contiene: ENERGIA 403 kJ 96 kcal — 5 % GRASSI 1,2 g — 2 % GRASSI SATURI 0,2 g — 1 % ZUCCHERI 5,1 g — 6 % SALE 0,4 g — 7 % delle Assunzioni di Riferimento di un adulto medio (8.400kJ / 2.000kcal) per 100g: 1342 kJ / 321 kcal
Puffed Cereals	Ciascuna porzione (30g) contiene: ENERGIA 480 kJ 113 kcal — 6 % GRASSI 0,8 g — 1 % GRASSI SATURI 0,2 g — 1 % ZUCCHERI 0 g — 0 % SALE 0 g — 0 % delle Assunzioni di Riferimento di un adulto medio (8.400kJ / 2.000kcal) per 100g: 1600 kJ / 378 kcal	Ciascuna porzione (30g) contiene: ENERGIA 486 kJ 114 kcal — 6 % GRASSI 1,3 g — 2 % GRASSI SATURI 0,2 g — 1 % ZUCCHERI 7,5 g — 8 % SALE 0 g — 0 % delle Assunzioni di Riferimento di un adulto medio (8.400kJ / 2.000kcal) per 100g: 1620 kJ / 382 kcal	Ciascuna porzione (30g) contiene: ENERGIA 495 kJ 117 kcal — 6 % GRASSI 1 g — 1 % GRASSI SATURI 0,2 g — 1 % ZUCCHERI 9,9 g — 11 % SALE 0 g — 0 % delle Assunzioni di Riferimento di un adulto medio (8.400kJ / 2.000kcal) per 100g: 1651 kJ / 390 kcal
Other Cereals	No items	Ciascuna porzione (30g) contiene: ENERGIA 598 kJ 143 kcal — 7 % GRASSI 6,6 g — 9 % GRASSI SATURI 2,7 g — 14 % ZUCCHERI 5,7 g — 6 % SALE 0,2 g — 3 % delle Assunzioni di Riferimento di un adulto medio (8.400kJ / 2.000kcal) per 100g: 1992 kJ / 476 kcal	Ciascuna porzione (30g) contiene: ENERGIA 496 kJ 117 kcal — 6 % GRASSI 1,4 g — 2 % GRASSI SATURI 0,5 g — 3 % ZUCCHERI 7,5 g — 8 % SALE 0,2 g — 3 % delle Assunzioni di Riferimento di un adulto medio (8.400kJ / 2.000kcal) per 100g: 1652 kJ / 390 kcal

4. Discussion

The present manuscript analyzed the breakfast cereals currently on the Italian market in terms of both nutritional quality—intended as nutritional values retrieved in the food labelling and as some of the FOPNLs proposed so far (i.e., Nutri-Score and NutrInform battery)—and level of processing, according to the NOVA system developed by Monteiro and colleagues with the intention to classify foods into four groups based on the type of processing [12]. Considering the number and types of items retrieved, some general considerations can be made. First, compared with our previous survey conducted in 2019 [11], we found an increasing number of breakfast cereal products sold on the market; in particular, muesli increased from 54 to 104 and puffed cereals from 29 to 43 products, for a total of 349 retrieved products.

Regarding the nutritional quality, we found several differences between the categorization with the Nutri-Score and that with the NutrInform battery. Specifically, 70% of breakfast cereals were labelled as Nutri-Score C and A (40% and 30%, respectively), followed by D and B (17% and 13%, respectively), while no products scoring E were found. On the Spanish market, Morales et al. found similar percentages of B and C among 53 breakfast cereals sold in 2018, while fewer A-labelled products (19%) and more D-labelled products (30%) were observed compared to the present study [14]. Vermote and colleagues analyzed the distribution of Nutri-Score among breakfast cereals in the Belgian market, in order to compare changes between 2017 and 2018 [15]. In both years, the authors found a prevalence of Nutri-Score C (43.4% and 40.6% in 2017 and 2018, respectively), A (25.0% and 29.7%), and D (22.8% and 17.9%), while only 8.4% and 11.5% scored B and only one item scored E in both years.

When the NutrInform battery was used, the differences among products were lower since in this FOPNL the nutrient content of specific components is expressed considering the serving size of 30 g, as suggested by the Italian food-based dietary guidelines [16]. These discrepancies between Nutri-Score and NutrInform battery in evaluating the nutritional quality of those products further highlight the differences between these two types of FOPNLs in providing information about the nutritional quality of food products.

Regarding the level of processing, we classified the breakfast cereals in three out of the four groups based on the type of the food processing as described by the NOVA system [12]. A large majority of products were classified as NOVA 4, but we also found minimally processed items classified in the NOVA 1 group and others with added culinary ingredients (e.g., salt and sugar), thus falling in the NOVA 3 group. These results are in line with the ones found by Morales et al. [14] who reported that 59%, 30%, and 11% of products were labelled as NOVA 4, 3, and 1, respectively.

In this survey, we also aimed at understanding whether NOVA and Nutri-Score describe the nutritional quality of the food products in a similar way by considering breakfast cereals, a food group whose products belong to many NOVA and Nutri-Score groups. When the products were grouped according to NOVA classification, we found that most of the items belonging to the NOVA 4 group were characterized by C and D Nutri-Score letters, which implies a medium-to-low nutritional quality. However, if from one side we found no NOVA 1 products labelled as Nutri-Score D and more than 80% of them labelled as A, then from the other side we retrieved many B and C products as in NOVA 1 as well as in NOVA 3 and NOVA 4 items. These findings support the previous hypothesis that minimally processed foods show a better nutritional quality than processed and ultra-processed analogues, mainly attributable to lower amounts of added ingredients, i.e., sugar and salt [17]. Nevertheless, the presence of some C Nutri-Score products in the NOVA 1 group underlines the concept that nutritional quality and food processing are not always in agreement to describe food characteristics. Regarding the NutriInform battery and NOVA classification, by considering the single energy and nutrient contents per 30 g serving of breakfast cereals, data showed that, except for specific types (i.e., muesli) and nutrients (i.e., total fats), the change of the battery loads across the different NOVA groups was pretty tight. Consequently, these data point out that there is no absolute consensus that the lowest nutritional quality, described by means of Nutri-Score and NutrInform battery, can be totally ascribed to the technological process of the food. For this reason, we do believe that the consumer should be carefully taught how to read and understand all the information on the food pack for making conscious shopping choices, independently of the FOPLNs.

As to the relationship between NOVA and Nutri-Score, our findings are in line with those of a survey concerning almost 10,000 various products sold on the Spanish market, showing that NOVA 3 and NOVA 4 items are widely characterized by all kinds of Nutri-Score letters and the NOVA 1 group is poorly represented by C to E Nutri-Score items [10]. Similar results have been found also by a Chilean study performing a crossing-ranking analysis between Nutri-Score and NOVA FOPLNs of 736 food products sold on the Chilean market [18]. Data showed that (i) for the NOVA 4 products, 70% out of the total were labelled as Nutri-Score B; (ii) for the NOVA 3 products, 25% and 2% of the products were characterized by Nutri-Score A and E, respectively. Additionally, in this case, the majority of the NOVA 1 products fell in the classifications A and B of the Nutri-Score [18]. The wide variability of the nutritional quality—evaluated by means of the Nutri-Score—of more than 220,000 products classified as NOVA 4 sold in France in 2020 was well described in the paper of Galán et al. [19]. In fact, despite only 21% of the products being labelled as A or B, the remaining ones were almost equally characterized by a C, D, or E Nutri-Score, underlying the wide variability in terms of the nutritional quality of ultra-processed foods. The same authors also showed that, by considering 2,036 products used in the NutriNet-Santé study, 58% to 86% of the products were ultra-processed ones, independent of the Nutri-Score letter [19].

Therefore, we are here to argue whether both the NOVA and Nutri-Score systems may converge in a unique definition of the healthiness of the product. This uncertainty in defining the healthiness of breakfast cereals has been deeply considered by Dickie et al. [20], who analyzed—within the "cereal and cereal products" group—221 breakfast cereals present on the Australian market. The authors calculated a percentage value of agreement between the classification of "healthy" and "unhealthy" by different FOPLNs, among which were Nutri-Score and NOVA. The value for breakfast cereals was 23%, which has been classified as a "high degree of disagreement" between the classification of healthiness for the two FOPNLs [20]. Authors attributed these disagreements between the two FOPNL to the different aspects considered for the definition of the healthiness of the products: while Nutri-Score has a nutrient-based scheme for the calculation of the different values, NOVA does not profile nutrients, but just the processing. This means that, for example, the presence of high amounts of salt or sugars in breakfast cereals—as also confirmed by data from our group [11,21]—are differently taken into account by Nutri-Score and NOVA algorithms, i.e, low Nutri-Score and minimally processed items. On the contrary, the presence of industrial ingredients and additives, which classifies the product as "not healthy" by the NOVA system, is not considered for the Nutri-Score. However, these last characteristics of the NOVA classification—ingredient addition and healthiness of the product—should be carefully contemplated. An Australian study examined different breakfast models that include or do not include ultra-processed breakfast cereals, categorized according to fortification/addition with vitamins, minerals, or fibre [22]. The authors evaluated whether those dietary models met the nutrient requirement by the Australian Dietary Guidelines, which discourages the consumption of ultra-processed foods at the expense of minimally or not processed foods. Data showed that the exclusion of such ultra-processed foods—among which are breakfast cereals—resulted in a significantly lower intake of key nutrients, such as some vitamins and iodine, with potentially harmful health consequences [22].

One of the main future goals to pursue in the nutrition field is to educate the customer in reading and understanding the whole information present on the food pack and, particularly, the FOPLNs boasted on the products and their differences in depicting the nutritional quality. We demonstrate here, for example, that i) many items rated as A or B had similar energy, total fats, and sugars regardless of the NOVA group and ii) within each NOVA group, many products with different Nutri-Score values have a similar nutritional profile, especially for serving, highlighting that the technological process of the food cannot be a descriptor of its nutritional quality.

5. Conclusions

The presence of FOPLNs on food packs for several different food groups is globally increasing, with the intention to help the consumer to recognize the key nutritional and technological aspects of the food and invite him to consider the global healthiness of the product. If from one side this effort may result in a global improvement of the dietary habits of the citizens—as confirmed by many epidemiological surveys relating the better scores of Nutri-Score and NOVA products to the lowering of risks of obesity and chronic diseases—from another side, our data show that the agreement of the different FOPLNs in describing the whole healthiness of the product is not valid for all the food groups. This disagreement has been deeply discussed in this paper, mainly explained by the different characteristics of the food considered for the development of the Nutri-Score and Nutri-Inform battery (mainly energy, nutrients, and some ingredients) and NOVA system (mainly the degree of process of the food). Not least, the concept of "healthiness" should not only be attributed to a single food but also to the quantity and the frequency of consumption as well as the influence of such food on the whole diet of the single individual.

In conclusion, the findings of the present survey suggest that neither the NOVA system nor the Nutri-Score or NutriInform battery are capable to describe the healthiness of the breakfast cereal products in a similar way. The three different FOPLNs give information on different characteristics of the products. We here rebate that the simple presence of symbols or colors on the food pack cannot drive the intention-to-buy of the customers. On the contrary, they should be carefully trained in how to read and understand the nutritional information present on the food pack and in how to interpret the FOPLNs on the item, and they should be left with the choice of which breakfast cereals satisfy their nutritional and health needs.

Supplementary Materials: The following supporting information can be downloaded at: https://www.mdpi.com/article/10.3390/nu15082013/s1, Table S1: energy, nutrients, and salt content of retrieved breakfast cereals, stratified based on the presence of whole grain ingredients and the NOVA group; Table S2: energy, nutrients, and salt content of retrieved breakfast cereals, stratified based on tertiles of sugar, fibre, and salt content and the NOVA group.

Author Contributions: Conceptualization, D.M.; methodology, D.A. and D.M.; formal analysis, D.A., N.P., and D.M.; investigation, B.G. and D.M.; data curation, D.M.; writing—original draft preparation, D.A., M.D., and D.M.; writing—review and editing, N.P. and B.G.; visualization, D.A. and M.D.; supervision, D.M. All authors have read and agreed to the published version of the manuscript.

Funding: This research received no external funding.

Institutional Review Board Statement: Not applicable.

Data Availability Statement: The data presented in this study are available on request from the corresponding author.

Acknowledgments: The authors wish to thank all students who participated to the development of the dataset.

Conflicts of Interest: The authors declare no conflict of interest.

Appendix A

Table A1. SINU Young Working Group.

Margherita Dall'Asta	Department of Animal Science, Food and Nutrition, Università Cattolica del Sacro Cuore, Piacenza, Italy
Gaetana Paolella	Department of Chemistry and Biology "A. Zambelli", University of Salerno, Fisciano, Italy
Marika Dello Russo	Institute of Food Sciences, National Research Council, Avellino, Italy
Stefania Moccia	Institute of Food Sciences, National Research Council, Avellino, Italy
Carmela Spagnuolo	Institute of Food Sciences, National Research Council, Avellino, Italy
Daniele Nucci	Veneto Institute of Oncology IOV-IRCCS, Padova, Italy
Veronica Pignone	Freelance Nutritionist, San Giorgio del Sannio, Italy
Emilia Ruggiero	Department of Epidemiology and Prevention, IRCCS Neuromed, Pozzilli, Italy
Alice Rosi	Department of Food and Drug, University of Parma, Italy
Giorgia Vici	School of Biosciences and Veterinary Medicine, University of Camerino, Italy

References

1. Thorning, T.K.; Bertram, H.C.; Bonjour, J.-P.; de Groot, L.; Dupont, D.; Feeney, E.; Ipsen, R.; Lecerf, J.M.; Mackie, A.; McKinley, M.C.; et al. Whole dairy matrix or single nutrients in assessment of health effects: Current evidence and knowledge gaps. *Am. J. Clin. Nutr.* **2017**, *105*, 1033–1045. [CrossRef]
2. Monteiro, C.A.; Cannon, G.; Levy, R.; Moubarac, J.-C.; Jaime, P.; Paula Martins, A.; Canella, D.; Louzada, M.; Parra Also with Camila Ricardo, D.; Calixto, G.; et al. NOVA. The star shines bright. *World Nutr.* **2016**, *7*, 28–38.
3. Moradi, S.; Entezari, M.H.; Mohammadi, H.; Jayedi, A.; Lazaridi, A.-V.; Kermani, M.A.H.; Miraghajani, M. Ultra-processed food consumption and adult obesity risk: A systematic review and dose-response meta-analysis. *Crit. Rev. Food Sci. Nutr.* **2023**, *63*, 249–260. [CrossRef] [PubMed]
4. Taneri, P.E.; Wehrli, F.; Roa-Díaz, Z.M.; Itodo, O.A.; Salvador, D.; Raeisi-Dehkordi, H.; Bally, L.; Minder, B.; Kiefte-de Jong, J.C.; Laine, J.E.; et al. Association between ultra-processed food intake and all-cause mortality: A systematic review and meta-analysis. *Am. J. Epidemiol.* **2022**, *191*, 1323–1335. [CrossRef] [PubMed]
5. Pagliai, G.; Dinu, M.; Madarena, M.P.; Bonaccio, M.; Iacoviello, L.; Sofi, F. Consumption of ultra-processed foods and health status: A systematic review and meta-analysis. *Br. J. Nutr.* **2021**, *125*, 308–318. [CrossRef] [PubMed]
6. Lane, M.M.; Davis, J.A.; Beattie, S.; Gómez-Donoso, C.; Loughman, A.; O'Neil, A.; Jacka, F.; Berk, M.; Page, R.; Marx, W.; et al. Ultraprocessed food and chronic noncommunicable diseases: A systematic review and meta-analysis of 43 observational studies. *Obes. Rev.* **2021**, *22*, e13146. [CrossRef] [PubMed]
7. Martini, D.; Godos, J.; Bonaccio, M.; Vitaglione, P.; Grosso, G. Ultra-processed foods and nutritional dietary profile: A meta-analysis of nationally representative samples. *Nutrients* **2021**, *13*, 3390. [CrossRef] [PubMed]
8. Nohlen, H.; Bakogianni, I.; Grammatikaki, E.; Ciriolo, E.; Pantazi, M.; Dias, J.; Salesse, F.; Moz Christofoletti, M., Wollgast, J.; Bruns, H.; et al. Front-of-Pack Nutrition Labelling Schemes—Publications Office of the EU. Available online: https://op.europa.eu/en/publication-detail/-/publication/2339cf79-2f25-11ed-975d-01aa75ed71a1/language-en (accessed on 25 February 2023).
9. Cutroneo, S.; Angelino, D.; Tedeschi, T.; Pellegrini, N.; Martini, D. Nutritional quality of meat analogues: Results from the food labelling of italian products (FLIP) project. *Front. Nutr.* **2022**, *9*, 852831. [CrossRef] [PubMed]
10. Romero Ferreiro, C.; Lora Pablos, D.; Gómez de la Cámara, A. Two dimensions of nutritional value: Nutri-Score and NOVA. *Nutrients* **2021**, *13*, 2783. [CrossRef]
11. Angelino, D.; Rosi, A.; Dall'Asta, M.; Pellegrini, N.; Martini, D. Evaluation of the nutritional quality of breakfast cereals sold on the Italian market: The food labelling of italian products (FLIP) study. *Nutrients* **2019**, *11*, 2827. [CrossRef] [PubMed]
12. Monteiro, C.A.; Cannon, G.; Moubarac, J.-C.; Levy, R.B.; Louzada, M.L.C.; Jaime, P.C. The UN Decade of Nutrition, the NOVA food classification and the trouble with ultra-processing. *Public Health Nutr.* **2018**, *21*, 5–17. [CrossRef] [PubMed]
13. Italian Society of Human Nutrition (SINU). *Livelli di Assunzione di Riferimento di Nutrienti ed Energia per la Popolazione Italiana*; SICS: Milan, Italy, 2014.
14. Morales, F.J.; Mesías, M.; Delgado-Andrade, C. Association between heat-induced chemical markers and ultra-processed foods: A case study on breakfast cereals. *Nutrients* **2020**, *12*, 1418. [CrossRef] [PubMed]
15. Vermote, M.; Bonnewyn, S.; Matthys, C.; Vandevijvere, S. Nutritional content, labelling and marketing of breakfast cereals on the belgian market and their reformulation in anticipation of the implementation of the Nutri-Score front-of-pack labelling system. *Nutrients* **2020**, *12*, 884. [CrossRef] [PubMed]
16. CREA Linee Guida per Una Sana Alimentazione Italiana. Revisione 2018. Available online: http://www.salute.gov.it/imgs/C_17_pubblicazioni_2915_allegato.pdf (accessed on 25 February 2023).
17. Visioli, F.; Marangoni, F.; Fogliano, V.; Del Rio, D.; Martinez, J.A.; Kuhnle, G.; Buttriss, J.; Da Costa Ribeiro, H.; Bier, D.; Poli, A. The ultra-processed foods hypothesis: A product processed well beyond the basic ingredients in the package. *Nutr. Res. Rev.* **2022**, 1–11, *in press*. [CrossRef] [PubMed]

18. Valenzuela, A.; Zambrano, L.; Velásquez, R.; Groff, C.; Apablaza, T.; Riffo, C.; Moldenhauer, S.; Brisso, P.; Leonario-Rodriguez, M. Discrepancy between food classification systems: Evaluation of Nutri-Score, NOVA classification and chilean front-of-package food warning labels. *Int. J. Environ. Res. Public Health* **2022**, *19*, 14631. [CrossRef] [PubMed]
19. Galán, P. Nutri-Score and ultra-processing: Two different, complementary, non-contradictory dimensions. *Nutr. Hosp.* **2021**, *38*, 201–206. [CrossRef] [PubMed]
20. Dickie, S.; Woods, J.; Machado, P.; Lawrence, M. Nutrition classification schemes for informing nutrition policy in australia: Nutrient-based, food-based, or dietary-based? *Curr. Dev. Nutr.* **2022**, *6*, 6008005. [CrossRef] [PubMed]
21. Martini, D.; Strazzullo, P.; Serafini, M.; Porrini, M.; Pellegrini, N.; Angelino, D. Sodium content in cereal-based products sold in italy: How far are we from the global benchmarks? *Nutrients* **2022**, *14*, 3088. [CrossRef] [PubMed]
22. Estell, M.L.; Barrett, E.M.; Kissock, K.R.; Grafenauer, S.J.; Jones, J.M.; Beck, E.J. Fortification of grain foods and NOVA: The potential for altered nutrient intakes while avoiding ultra-processed foods. *Eur. J. Nutr.* **2022**, *61*, 935–945. [CrossRef] [PubMed]

nutrients

Article

Ultra-Processed Food Intakes Are Associated with Depression in the General Population: The Korea National Health and Nutrition Examination Survey

Sunghee Lee * and Myungjin Choi

Department of Food and Nutrition, College of Health Science, Kangwon National University, Samcheok 25949, Republic of Korea; soon6920@naver.com
* Correspondence: sunglee@kangwon.ac.kr; Tel.: +82-33-540-3315

Abstract: Depression is the most common mental illnesses worldwide. The consumption of ultra-processed food (UPF) has increased globally due to its affordability and convenience; however, only a few studies have investigated the link between UPF intake and depression in the general population. We investigated the associations between UPF and depression using the Korea National Health and Nutrition Examination Survey. A total of 9463 individuals (4200 males and 5263 females) aged above 19 years old participated in this study. The prevalence of depression was identified using the Patient Health Questionnaire-9. Dietary intake was assessed through a 24-h recall interview. The percentage of energy from UPFs was ascertained based on the NOVA classification. The associations between the quartile ranges of UPF intake and depression were estimated using logistic regression models. Individuals in the highest quartile had a 1.40 times higher likelihood of having depression, with marginal significance (95% confidence intervals (CIs) = 1.00–1.96). In a sex-specific stratification, only females demonstrated a significant association (odds ratio (OR) = 1.51, 95% CI 1.04–2.21), even after adjusting for confounders (p-value for trend = 0.023). Our findings revealed a significant association between higher UPF intake and depression among females but not among males in the Korean general population.

Keywords: ultra-processed food; depression; general population; mental health

Citation: Lee, S.; Choi, M. Ultra-Processed Food Intakes Are Associated with Depression in the General Population: The Korea National Health and Nutrition Examination Survey. *Nutrients* **2023**, *15*, 2169. https://doi.org/10.3390/nu15092169

Academic Editor: Jose Lara

Received: 26 March 2023
Revised: 23 April 2023
Accepted: 28 April 2023
Published: 1 May 2023

1. Introduction

Depression is among the most prevalent mental disorders worldwide [1]. A meta-analysis demonstrated that the burden of depression was 12.9%, particularly higher in females by 14.4% [2]. A general population-based study among adults in the United States (US) showed that the prevalence of depression increased from 8.5% in 2017–2018 to 27.8% in 2020 [3]. Among Korean adults, the prevalence of depression has been increased from 4.3% in 2018 to 5.2% in 2020 [4]. Also, depression increases not only the risk of suicide deaths [1] but also the risk of metabolic syndrome [5].

As a modifiable risk factor, dietary intervention mitigates the risk of depression. The recent advances in food technologies including food packaging, preparation, and extension of shelf-life made possible convenient food packages such as home meal replacements, meal kits, and ready-to-cook meals. Due to its palatability and affordability, the intake of ultra-processed foods (UPFs) intake has increased globally. Particularly, previous studies reported persistent increases in UPFs intake among adults in the US from 53.5% in 2001–2002 to 57.0% in 2017–2018 [6], and among adults in the United Kingdom (UK) from 48.6% in 2009–2010 [7] to 56.8% in 2008–2014 [8]. By contrast, Asian countries reported lower proportions. For example, previous studies have reported the percentage of the UPF intake among Korean adults was 26.8% [9] and 25.1% [10] in 2016 and 2018, respectively. Among Italians, a median UPF intake was approximately 10% [11] due to their characterized Mediterranean diet. Evolving evidence demonstrated that high UPF intake has been associated with mortality [12], obesity [13,14], hypertension [15], diabetes [16], cardiovascular disease [17,18], and dementia [19]. This increased risk of developing chronic diseases was attributed to high sodium intake, excessive sugar intake, high fat intake, and use of food additives [20]. Furthermore, these highly processed foods are less likely to contain fiber, vitamins, and minerals that are commonly found in fresh vegetables and fruits. Therefore, high UPF intake affects the physiology of individuals.

However, existing evidence supporting the association between the psychological aspects of individuals and UPF intake, especially depression, is limited. In addition, only a few studies have examined this link. In France, a previous study among 26,730 individuals aged from 18–86 years old with over 5.4 years of follow-up reported a link between UPF and a higher risk of developing depressive [21]. Particularly, this study demonstrated that the risk of depression was enlarged by 1.21 times as 10% of UPF intake increased (hazard ratio (HR) = 1.21, 95% confidence intervals (CIs) = 1.15–1.27) [21]. Another study in Spain examining 14,907 young adults aged about 36.7 years showed that participants in the highest quartile of UPF intake had a 1.33 times higher likelihood of developing depression than those in the lowest quartile (HR = 1.33, 95% CI 1.07–1.64) [22]. These studies suggested the potential biological links between UPF intake and depression through not only nutritional aspects but also non-nutritious food additives. Hence, more epidemiological evidence needs to be accumulated to elucidate this link.

Thus, we aimed to examine the associations between UPF intake and depression among 9463 participants (4200 males and 5263 females) aged 20 years or older in a general population using the Korea National Health and Nutrition Examination Survey (KNHANES).

2. Materials and Methods

2.1. Study Population

This study used the data derived from the KNHANES by the Korea Centers for Disease Control and Prevention. The KNHANES was regularly performed to monitor the dietary intakes, health status, and health-related factors in a nationally representative sample. All the participants signed an informed consent. We followed the Declaration of Helsinki guidelines and received approval from the Institutional Review Board (2018-01-03-2C-A, 2018-01-03-P-A). This survey has an exemption by the Bioethics Act of 2016 with the purpose for the public well-being.

Of the 23,501 participants included in 2016, 2018, and 2020 KNHANES, 3589 with no data on dietary intake, 327 with implausible daily energy intake (such as <500 or >5000 kcal), and 5559 with no Patient Health Questionnaire-9 (PHQ-9) scores were excluded. Of the remaining 14,026 participants, 3810 who were receiving dietary therapy (n = 3761) and were pregnant (n = 49) at the time of the study were excluded. In addition, 753 with missing data on body mass index (BMI, n = 74), physical activity (PA, n = 21), education (n = 4), smoking (n = 18), hypertension (n = 48), diabetes (n = 460), and UPF (n = 128) were excluded. Hence, 9463 adults were in the study. A flowchart of this process is illustrated in Figure 1.

2.2. Definition of Depression

The PHQ-9 was used to identify cases of depression. It comprised nine questions. The scores of PHQ-9 ranged from 0 to 27, with a score of 10 or higher indicating depression [23]. The validity and reliability of the PHQ-9 were previously confirmed [24,25]. The PHQ-9 was administered during the 2016, 2018, and 2020 KNHANES.

2.3. Dietary Measurement

To order to determine the dietary intake, professionally trained research staff interviewed those participants and helped them to complete a 24-h recall examination. The nutrient and energy intakes were calculated using the dietary intake data from the Rural Development Administration of Korea database [26]. More than 4000 food items were classified according to the NOVA classification to examine the UPF intake [27,28]. As highlighted by Monteiro and colleagues, the NOVA system has classified food items into four groups: (1) minimally processed or unprocessed foods, (2) culinary ingredients, (3) processed foods, and (4) ultra-processed foods [27,28]. After ascertaining the UPF group, the percentages of total energy intake from the consumption of UPF (%UPF) were calculated.

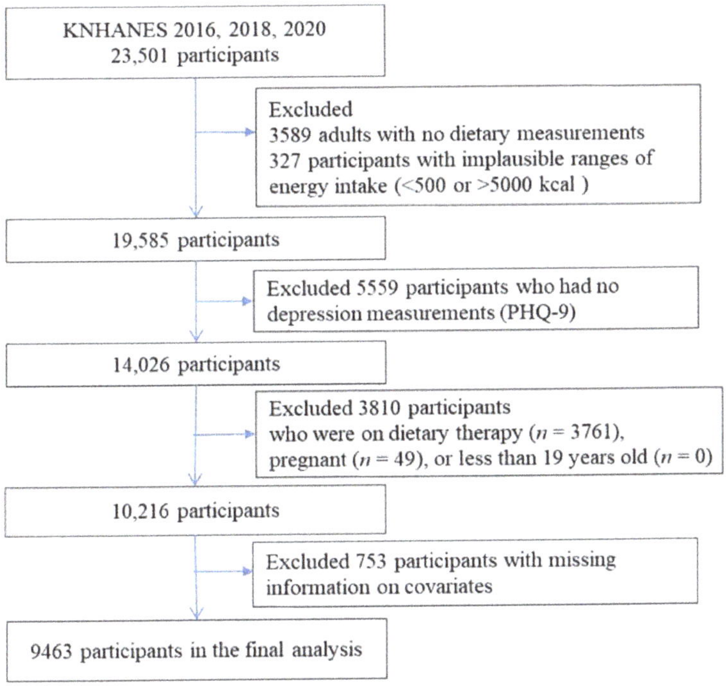

Figure 1. Flow diagram of the study participants.

2.4. Covariates

All participants completed the study questionnaires and underwent physical examinations in two large-sized buses as mobile examination centers. Physical activity (PA) was determined using the Global Physical Activity Questionnaire (GPAQ), with moderate or vigorous activities and total activity [29]. Based on the time spent performing these activities and the intensity of moderate or vigorous activities, the participants were asked about activities at work, for recreation, and when traveling to and from certain places, and the frequency and duration of walking during a usual week [29,30]. The validity and reliability of the Korean version of the GPAQ have been verified [30]. According to the PA guidelines [31], adults should perform more than 150 min of moderate–intensity, 75 min of vigorous–intensity, or the corresponding analogous combined time [30]. Smoking status was classified as nonsmoker, ex-smoker, or current smoker. "Alcohol drinker" was defined as an individual who consumed alcohol more than once per month during the past year; "alcohol non-drinker" was defined as an individual who consumed alcohol less than once per month. Individuals with "hypertension" were identified as those with a systolic blood pressure of $\geq$140 mmHg, with a diastolic blood pressure of $\geq$90 mmHg, or using antihypertensive medications. Participants with "diabetes mellitus" were defined as those with a fasting glucose level of $\geq$126 mg/dL, who were diagnosed by a medical doctor or were using antidiabetic medications.

2.5. Statistical Analysis

To investigate the KNHANES data derived using the complex sampling method, weighted survey analyses such as surveymeans, surveyfreq, and surveyreg, were used to examine the data. To evaluate the different characteristics according to the quartiles of UPF, survey regression models were used with a p-value for trend. Additionally, to assess the associations between UPF intake and depression, multivariable logistic regression models (e.g., surveylogistic) were used after adjusting for confounding variables including age, BMI, education, PA, alcohol drinking, smoking, hypertension, and diabetes. All analyses were performed using SAS version 9.4, with the significant level as p-value < 0.05.

3. Results

Table 1 demonstrates the general characteristics of the 9463 participants and the comparisons of sex-specific characteristics among 4200 males and 5263 females between those with and without depression. Among males, smoking status ($p < 0.001$) and diabetes ($p = 0.003$) were significantly associated with depression; among males with depression, 53.79% were current smokers, and a significantly higher proportion had diabetes. On the other hand, among females, BMI ($p = 0.008$), education ($p = 0.003$), smoking ($p < 0.001$), diabetes ($p < 0.001$), and UPF intake ($p = 0.002$) were significantly associated with depression.

Table 1. General characteristics of the study participants according to depression (n = 9463).

	Total (n = 9463)			Male (n = 4200)			Female (n = 5263)		
	Depression	No Depression	p	Depression	No Depression	p	Depression	No Depression	p
	n = 445 (4.40%)	n = 9018 (95.60%)		n = 133 (3.04%)	n = 4067 (96.96%)		n = 312 (5.90%)	n = 4951 (94.10%)	
Age, years	45.81 ± 1.07	46.97 ± 0.29	0.276	44.06 ± 1.61	46.17 ± 0.36	0.195	46.80 ± 1.35	47.87 ± 0.33	0.433
BMI, kg/m^2	24.01 ± 0.24	23.76 ± 0.05	0.324	24.58 ± 0.47	24.47 ± 0.07	0.825	23.68 ± 0.27	22.96 ± 0.07	0.008
Education, %									
<High school	31.57	21.12		23.72	16.54		36.01	26.30	
High school	36.08	36.95	<0.001	39.82	39.19	0.094	33.96	34.42	0.003
>High school	32.35	41.93		36.46	44.27		30.03	39.28	
Physical activity, %									
Active	37.02	44.04	0.021	39.79	46.88	0.171	35.45	40.83	0.125
Inactive	62.98	55.96		60.21	53.12		64.55	59.17	
Smoking status, %									
Current smokers	32.75	21.31		53.79	35.33		20.82	5.43	
Ex-smokers	19.21	22.48	<0.001	30.35	37.19	0.001	12.90	5.84	<0.001
Non-smokers	48.05	56.21		15.87	27.48		66.28	88.73	
Alcohol drinker, %	55.92	59.11	0.298	74.01	71.32	0.557	45.68	45.29	0.919
Hypertension, %	27.05	26.10	0.696	31.40	29.15	0.623	24.59	22.66	0.463
Diabetes mellitus, %	14.80	8.22	<0.001	18.12	9.61	0.003	12.92	6.64	<0.001
UPF, %	31.11 ± 1.28	27.32 ± 0.29	0.003	33.29 ± 2.41	29.07 ± 0.41	0.080	29.87 ± 1.41	25.34 ± 0.35	0.002
UPF energy, kcal	615.01 ± 35.52	590.64 ± 8.36	0.504	739.06 ± 68.58	713.87 ± 12.63	0.717	544.72 ± 40.21	451.16 ± 7.87	0.023
Total energy, kcal/day	1875.81 ± 48.46	2037.47 ± 12.21	0.001	2209.56 ± 76.30	2326.86 ± 17.07	0.131	1686.71 ± 56.51	1709.92 ± 11.87	0.686

Mean ± SE; BMI, body mass index.

Table 2 lists the general characteristics of 4200 males according to the quartile ranges of UPF intake. Among the 133 males with depression, those in the higher quartile groups were younger ($p < 0.001$). In contrast, among the 4067 males without depression, those in the higher quartile groups were significantly younger ($p < 0.001$), had a higher BMI ($p = 0.004$), had higher education level ($p < 0.001$), were current smokers ($p < 0.001$), were alcohol drinkers ($p < 0.001$), and were less likely to have diabetes ($p = 0.012$) and hypertension ($p < 0.001$).

Table 2. Means and frequencies of the general characteristics of the males according to ultra-processed food.

	Males (n = 4200)									
	Depression (n = 133)					No Depression (n = 4067)				
	Q1	Q2	Q3	Q4	p	Q1	Q2	Q3	Q4	p
	n = 28 (13.32%)	n = 27 (23.94%)	n = 32 (25.30%)	n = 46 (37.44%)		n = 891 (18.55%)	n = 943 (22.81%)	n = 1069 (26.30%)	n = 1164 (32.34%)	
Age, years	52.52 ± 5.53	39.30 ± 2.80	42.66 ± 3.28	45.03 ± 2.18	<0.001	53.94 ± 0.71	48.54 ± 0.64	45.25 ± 0.57	40.81 ± 0.49	<0.001
BMI, kg/m^2	22.65 ± 0.78	25.82 ± 0.83	25.05 ± 0.99	24.15 ± 0.66	0.891	24.14 ± 0.14	24.28 ± 0.13	24.67 ± 0.13	24.65 ± 0.14	0.004
Education										
<High school	35.87	15.91	14.75	30.46		28.42	18.85	12.58	11.32	
High school	39.58	38.30	46.70	36.24	0.581	32.57	35.86	39.86	44.78	<0.001
>High school	24.55	45.78	38.56	33.30		39.01	45.29	47.55	43.90	
Physical activity, %										
Active	19.09	45.61	39.36	43.74	0.370	45.71	47.38	46.62	47.42	0.915
Inactive	80.91	54.39	60.64	56.26		54.29	52.62	53.38	52.58	
Smoking status, %										
Current smokers	45.19	36.08	52.58	68.98		26.63	28.36	37.58	43.40	
Ex-smokers	35.13	43.87	29.20	20.77	0.306	44.23	41.72	35.85	31.04	<0.001
Non-smokers	19.68	20.06	18.22	10.25		29.14	29.92	26.57	25.56	
Alcohol drinker, %	71.39	74.72	67.71	78.73	0.789	64.00	67.70	72.39	77.19	<0.001
Hypertension, %	41.35	31.30	20.22	35.46	0.452	36.88	27.81	27.03	27.38	<0.001
Diabetes mellitus, %	13.56	11.04	20.34	22.77	0.565	12.76	9.80	9.15	8.05	0.012

Mean ± SE; BMI, body mass index.

Table 3 shows the characteristics of 5263 females according to the quartile of UPF intake. Among the 312 females with depression, those in the higher quartile groups were significantly younger ($p < 0.001$), had higher education level ($p < 0.001$), were current drinkers ($p = 0.007$), and were less likely to have hypertension ($p = 0.001$) and diabetes ($p = 0.023$). Among 4951 females without depression, those in the higher quartile groups were significantly younger ($p < 0.001$), had a lower BMI ($p < 0.001$), had higher education level ($p < 0.001$), were more physically active ($p = 0.026$), were current smokers ($p < 0.001$), were current drinkers ($p < 0.001$), and were less likely to have diabetes ($p < 0.001$) and hypertension ($p < 0.001$).

Table 3. Means and frequencies of the general characteristics of the females according to ultra-processed food.

	Females (n = 5263)									
	Depression (n = 312)					No Depression (n = 4951)				
	Q1	Q2	Q3	Q4	p	Q1	Q2	Q3	Q4	p
	n = 73 (21.60%)	n = 72 (21.13%)	n = 76 (23.84%)	n = 91 (33.44%)		n = 1357 (24.39%)	n = 1317 (25.27%)	n = 1193 (25.41%)	n = 1084 (24.93%)	
Age, years	60.39 ± 2.66	48.28 ± 2.33	46.87 ± 2.40	37.03 ± 1.75	<0.001	56.94 ± 0.51	50.11 ± 0.55	44.82 ± 0.51	39.84 ± 0.57	<0.001
BMI, kg/m^2	24.55 ± 0.46	22.87 ± 0.71	23.74 ± 0.57	23.60 ± 0.47	0.445	23.51 ± 0.11	23.01 ± 0.12	22.77 ± 0.13	22.57 ± 0.14	<0.001
Education										
<High school	66.91	33.81	33.43	19.29		46.64	27.63	17.53	13.97	
High school	17.65	41.09	26.79	45.10	<0.001	30.05	34.82	34.59	38.12	<0.001
>High school	15.44	25.10	39.78	35.61		23.31	37.55	47.88	47.90	
Physical activity, %										
Active	41.47	42.15	31.06	30.45	0.392	38.05	40.20	44.91	40.04	0.026
Inactive	58.53	57.85	68.94	69.55		61.95	59.81	55.09	59.96	
Smoking status, %										
Current smokers	15.15	12.11	22.09	29.09		3.81	4.69	4.06	9.18	
Ex-smokers	6.05	15.60	16.20	13.27	0.103	3.43	6.09	5.94	7.82	<0.001
Non-smokers	78.80	72.29	61.71	57.64		92.76	89.22	90.00	83.00	
Alcohol drinker, %	26.60	47.49	43.84	58.17	0.007	30.90	43.81	48.46	57.66	<0.001
Hypertension, %	41.61	29.34	22.64	11.99	0.001	35.59	24.68	17.22	13.50	<0.001
Diabetes mellitus, %	20.68	15.81	14.74	4.78	0.023	12.06	5.68	4.87	4.11	<0.001

Mean ± SE; BMI, body mass index.

Table 4 presents the nutrient intake of 4200 males and 5263 females according to the quartile of UPF intake. In both males and females, the intakes of sugar, fat, saturated fat, and dietary sodium were positively increased as the consumption of UPFs increased ($p < 0.001$, respectively), except for the carbohydrate and protein intakes. By contrast, the intakes of vegetables and fruits were significantly decreased as the UPF intake increased ($p < 0.001$, respectively). The nutrient intake according to individuals with and without depression within 4200 males and 5263 females were examined and were shown in consistent results (Supplemental Tables S1 and S2).

Table 4. Means of nutrient intakes of the study participants according to the quartile range of ultra-processed food intakes (n = 9463).

	Males (n = 4200)					Females (n = 5263)				
	Q1	Q2	Q3	Q4		Q1	Q2	Q3	Q4	
	n = 919 (18.39%)	n = 970 (22.84%)	n = 1101 (26.27%)	n = 1210 (32.50%)	p	n = 1430 (24.22%)	n = 1389 (25.03%)	n = 1269 (25.32%)	n = 1175 (25.43%)	p
UPF energy, kcal	93.27 ± 3.08	333.32 ± 5.91	698.12 ± 8.99	1347.65 ± 19.51	<0.001	64.04 ± 1.74	251.52 ± 3.58	511.46 ± 6.86	978.01 ± 16.87	<0.001
Total energy, kcal/day	2001.36 ± 27.92	2193.04 ± 32.99	2398.40 ± 28.44	2536.30 ± 29.36	<0.001	1547.14 ± 18.17	1685.45 ± 20.61	1762.20 ± 21.22	1831.62 ± 25.94	<0.001
Carbohydrate, g	322.26 ± 4.59	328.48 ± 4.58	338.22 ± 4.14	324.77 ± 4.07	0.639	263.91 ± 3.29	267.1 ± 3.46	262.62 ± 3.14	264.62 ± 3.50	0.868
Sugar, g	49.59 ± 1.33	62.16 ± 1.57	69.19 ± 1.55	70.22 ± 1.46	<0.001	47.35 ± 1.13	57.53 ± 1.36	61.55 ± 1.23	64.45 ± 1.37	<0.001
Protein, g	79.67 ± 1.55	85.76 ± 1.64	89.20 ± 1.44	82.75 ± 1.23	0.213	56.74 ± 0.96	63.15 ± 1.03	64.96 ± 0.99	59.02 ± 1.02	0.056
Fat, g	41.73 ± 1.41	53.24 ± 1.77	59.09 ± 1.29	58.04 ± 1.12	<0.001	29.04 ± 0.75	38.84 ± 0.93	45.44 ± 0.94	46.91 ± 1.12	<0.001
Saturated fat, g	12.29 ± 0.48	16.55 ± 0.59	18.83 ± 0.43	20.03 ± 0.43	<0.001	8.37 ± 0.24	12.05 ± 0.33	14.70 ± 0.32	17.06 ± 0.47	<0.001
Dietary sodium, mg	3712.80 ± 80.11	3863.48 ± 76.64	4214.23 ± 70.02	4271.37 ± 72.22	<0.001	2557.88 ± 49.32	2852.45 ± 55.58	3053.09 ± 61.80	3182.66 ± 61.09	<0.001
Food Groups										
Vegetables, g	394.40 ± 8.49	353.28 ± 7.77	342.59 ± 7.16	275.66 ± 5.99	<0.001	309.21 ± 6.18	282.97 ± 5.34	253.56 ± 5.50	203.41 ± 5.59	<0.001
Fruits, g	286.27 ± 14.47	254.31 ± 12.29	224.84 ± 11.68	164.82 ± 9.74	<0.001	292.48 ± 10.51	268.06 ± 13.27	223.78 ± 10.00	179.66 ± 8.20	<0.001

Mean ± SE; BMI, body mass index.

Table 5 demonstrates the association between UPF intake and depression after considering the confounding factors. In the total population, individuals in the highest quartile were 1.40 times more likely to have depression, with marginally significance (95% CI = 1.00–1.96). Males showed no association, whereas females demonstrated a significant association, indicating that females in the highest quartile of UPF were 1.51 times more likely to have depression (95% CI 1.04, 2.21), even after adjusting for age, BMI, education, PA, alcohol drinking, smoking, diabetes, and hypertension (p for trend = 0.023).

Table 5. Odds ratios associated with the quartile ranges of the ultra-processed foods on depression (n = 9463).

	Odds Ratio (95% Confidence Intervals)				p-Value for Trend
	Quartile Range				
	Q1	Q2	Q3	Q4	
Total					
n (%)	2349 (21.17%)	2359 (23.89%)	2370 (25.81%)	2385 (29.13%)	
UPF% range	[0, 9.17]	[9.17, 21.22]	[21.22, 37.76]	[35.76, 100]	
OR (95% CI)	1.00 (Ref)	1.18 (0.83, 1.66)	1.21 (0.84, 1.75)	1.40 (1.00, 1.96)	0.066
Males					
n (%)	919 (45.47%)	970 (50.06%)	1101 (53.27%)	1210 (58.40%)	
UPF% range	[0, 9.18]	[9.18, 21.22]	[21.22, 37.75]	[35.75, 100]	
OR (95% CI)	1.00 (Ref)	1.47 (0.74, 2.91)	1.23 (0.64, 2.39)	1.34 (0.70, 2.55)	0.624
Females					
n (%)	1430 (54.53%)	1389 (49.94%)	1269 (46.73%)	1175 (41.60%)	
UPF% range	[0, 9.17]	[9.17, 21.21]	[21.21, 37.76]	[35.76, 100]	
OR (95% CI)	1.00 (Ref)	1.04 (0.70, 1.54)	1.24 (0.81, 1.91)	1.51 (1.04, 2.21)	0.023

Adjusted for age, body mass index, education, physical activity, smoking, alcohol drinking, hypertension, and diabetes mellitus.

4. Discussion

In this cross-sectional study, we found that the burden of depression in the Korean population from 2016 to 2020 was 4.40%, indicating that it was more prevalent in females than males (5.90% vs. 3.04%). Moreover, we identified that the average percentage of energy from UPFs in the Korean general population was approximately 27.49%. Furthermore, participants with higher UPF intake had higher sugar, fat, saturated fat, and dietary sodium intake (p for trend < 0.001, respectively) but lower intakes of vegetables and fruits (p for trend < 0.001, respectively). Furthermore, females in the highest quartile of UPF intake had a 1.51 times higher likelihood of having depression, after adjusting for confounders (OR = 1.51, 95% CI 1.04–2.21, p for trend = 0.023).

Our study was the first to examine the associations between UPF intake and depression in a general population from an Asian country. The percentage of energy intake from the consumption of UPFs varies among different cultural backgrounds; the US reported an energy intake of 57.0% in 2017–2018 [6], the UK indicated 56.8% in 2008–2014 [8], and Italy had approximately 10% due to the Mediterranean diet [11].

Few studies that previously assessed the association between UPF intake and depression reported results that were concurrent with our findings [21,22,32,33]. A previous study of 26,730 participants with a follow-up of 5.4 years reported the association of increased risk of depressive symptoms with UPF intake, showing that a 10% increase in UPF intake was associated with 1.21 times higher risk of developing depression (95% CI = 1.15–1.27) [21]. Another study among 14,907 adults demonstrated an increased risk of depression in the highest quartile of UPF compared with that in the lowest quartile of UPF (HR = 1.33, 95% CI 1.07–1.64) [22]. A recent study among Italian participants demonstrated a significant association between UPF intake and depressive symptoms (OR=2.04, 95% CI 1.04–4.01) [33]. Furthermore, a meta-analysis indicated the link between UPF intake and higher risk of depression (RR = 1.28, 95% CI = 1.19–1.38) [32]. However, different from the two studies, their results adopted various methods in defining UPFs, including not only the NOVA classification of food items but also fast foods, Western dietary patterns, and sweetened foods.

Our study participants showed a relatively lower prevalence of depression. To ascertain those with depression among the study participants, we used the PHQ-9, which has been confirmed to be valid and reliable [24,25]. A study using the NHANES data with PHQ-9 among 34,963 participants aged 18 years or older showed that the prevalence of depression was 8.1% (males: 6.5% and females: 9.6%) between 2015 and 2016 [23]. Based on the responses to the PHQ-9, this current study demonstrated that the prevalence of depression was 4.40% (males: 3.04% and females: 5.90%) between 2016 and 2020. Furthermore, females consistently demonstrated a higher prevalence of depression, which may explain in part the differences in the significance of depression associated with high UPF intake.

We also observed sex-specifically differences in the profiles of risk factors for depression and their associations with UPF intake. The general characteristics among males with or without depression were not different except for smoking status and prevalence of diabetes. However, females with and without depression tended to show significant differences in BMI, education level, smoking status, the prevalence of diabetes, and UPF intake. Sex differences in depression were consistent with a previous study [34].

More importantly, in addition to the significantly higher intake of UPFs among those with depression that we observed above, our findings revealed that according to the quartiles of UPF intake, both males and females demonstrated significantly increasing trends in nutrient intakes in total energy, sugar, fat, and saturated fat, dietary sodium, but significantly decreasing trends in food groups of vegetables and fruits. However, no significant trends were shown in carbohydrate and protein intakes.

The mechanisms underlying the psychological aspects linked to UPFs have not yet been elucidated. However, previous studies have attempted to explain this association. First, high UPF intakes are deficient in bioactive micronutrients such as minerals and vitamins due to limited whole food constituents from vegetables or fruits. This deficiency may subsequently be attributable to the increased risk of depression [21]. Second, high UPF intakes aggravate the disturbance of gut microbiota balance, leading to gut dysbiosis, followed by detrimental effects on the gut–brain axis, which reduce the production of neurotransmitters such as serotonin [32,35]. Moreover, high UPF intake causes hypothalamic-pituitary adrenal (HPA) dysregulation, which affects appetite hormones and regulatory neurotransmitter signals [13,22]. Third, gut dysbiosis from high UPF intake stimulates pro-inflammatory cytokines [36], leading to increased concentrations of high-sensitivity C-reactive protein [37]. Finally, food additives for flavoring, coloring, palatable enhancers, and emulsifiers [38], or byproducts and contaminants in the production of UPFs may induce detrimental effects by disrupting endocrine signals or homeostatic regulatory pathways [39,40]. Thus, further studies on these links are warranted.

This study has several strengths and limitations. Firstly, the study sample was relatively large with 9463 participants; therefore, it had enough statistical power to detect the differences even in the sex-specific stratified groups. Secondly, the data were obtained from the general population through a nationally representative survey, which enables the generalizability of the study results to other populations. However, this study also has limitations. It used a cross-sectional design which limited to establish a causal relationship. Moreover, data on dietary intake were obtained through a single 24-h dietary recall interview, which might have day-to-day variations. However, the day-to-day variations in dietary intake could be mitigated with the person-to-person variations due to the large sample size used in the study.

5. Conclusions

In conclusion, our findings revealed a significant association between high UPF intake and depression among 5263 females but not in 4200 males in the general Asian population, even after adjusting for age, BMI, alcohol drinking, smoking, education, hypertension, diabetes, and PA. Based on these findings, further studies are warranted.

Supplementary Materials: The following supporting information can be downloaded at: https://www.mdpi.com/article/10.3390/nu15092169/s1; Table S1: Means of Nutrients of the Males according to Ultra-processed Food Intakes; Table S2: Means of Nutrients of the Females according to Ultra-processed Food Intakes.

Author Contributions: Conceptualization, S.L.; method and statistical analysis, M.C. and S.L.; writing—draft, review and editing, S.L.; visualization, S.L.; supervision, S.L. All authors have read and agreed to the published version of the manuscript.

Funding: This research received no external funding.

Institutional Review Board Statement: The study was conducted in accordance with the Declaration of Helsinki, and approved by the Institutional Review Board of Korea Centers for Disease Control and Prevention (2018-01-03-P-A and 2018-01-03-2C-A). Additionally, the approval was waived for this study due to the implementation of the public well-being as a national surveillance system, so that it was exempted from the Bioethics Act for 2016.

Informed Consent Statement: An informed consent was obtained from all participants in the study.

Data Availability Statement: The data presented in this study are openly available at http://knhanes.kdca.go.kr.

Conflicts of Interest: The authors declare no conflict of interest.

References

1. WHO. Depression. Available online: https://www.who.int/news-room/fact-sheets/detail/depression (accessed on 10 January 2023).
2. Lim, G.Y.; Tam, W.W.; Lu, Y.; Ho, C.S.; Zhang, M.W.; Ho, R.C. Prevalence of Depression in the Community from 30 Countries between 1994 and 2014. *Sci. Rep.* **2018**, *8*, 2861. [CrossRef] [PubMed]
3. Ettman, C.K.; Abdalla, S.M.; Cohen, G.H.; Sampson, L.; Vivier, P.M.; Galea, S. Prevalence of Depression Symptoms in US Adults Before and During the COVID-19 Pandemic. *JAMA Netw. Open* **2020**, 3, e2019686. [CrossRef]

4. Lee, E.J.; Kim, S.J. Prevalence and Related Factors of Depression Before and During the COVID-19 Pandemic: Findings From the Korea National Health and Nutrition Examination Survey. *J. Korean Med. Sci.* **2023**, *38*, e74. [CrossRef] [PubMed]
5. Ferriani, L.O.; Silva, D.A.; Molina, M.; Mill, J.G.; Brunoni, A.R.; da Fonseca, M.J.M.; Moreno, A.B.; Bensenor, I.M.; de Aguiar, O.B.; Barreto, S.M.; et al. Depression is a risk factor for metabolic syndrome: Results from the ELSA-Brasil cohort study. *J. Psychiatr. Res.* **2022**, *158*, 56–62. [CrossRef]
6. Juul, F.; Parekh, N.; Martinez-Steele, E.; Monteiro, C.A.; Chang, V.W. Ultra-processed food consumption among US adults from 2001 to 2018. *Am. J. Clin. Nutr.* **2022**, *115*, 211–221. [CrossRef]
7. Rauber, F.; Chang, K.; Vamos, E.P.; da Costa Louzada, M.L.; Monteiro, C.A.; Millett, C.; Levy, R.B. Ultra-processed food consumption and risk of obesity: A prospective cohort study of UK Biobank. *Eur. J. Nutr.* **2021**, *60*, 2169–2180. [CrossRef] [PubMed]
8. Rauber, F.; da Costa Louzada, M.L.; Steele, E.M.; Millett, C.; Monteiro, C.A.; Levy, R.B. Ultra-Processed Food Consumption and Chronic Non-Communicable Diseases-Related Dietary Nutrient Profile in the UK (2008–2014). *Nutrients* **2018**, *10*, 587. [CrossRef]
9. Sung, H.; Park, J.M.; Oh, S.U.; Ha, K.; Joung, H. Consumption of Ultra-Processed Foods Increases the Likelihood of Having Obesity in Korean Women. *Nutrients* **2021**, *13*, 698. [CrossRef]
10. Shim, J.S.; Shim, S.Y.; Cha, H.J.; Kim, J.; Kim, H.C. Association between Ultra-processed Food Consumption and Dietary Intake and Diet Quality in Korean Adults. *J. Acad. Nutr. Diet.* **2022**, *122*, 583–594. [CrossRef]
11. Bonaccio, M.; Di Castelnuovo, A.; Costanzo, S.; De Curtis, A.; Persichillo, M.; Sofi, F.; Cerletti, C.; Donati, M.B.; de Gaetano, G.; Iacoviello, L. Ultra-processed food consumption is associated with increased risk of all-cause and cardiovascular mortality in the Moli-sani Study. *Am. J. Clin. Nutr.* **2021**, *113*, 446–455. [CrossRef]
12. Schnabel, L.; Kesse-Guyot, E.; Alles, B.; Touvier, M.; Srour, B.; Hercberg, S.; Buscail, C.; Julia, C. Association Between Ultraprocessed Food Consumption and Risk of Mortality Among Middle-aged Adults in France. *JAMA Intern. Med.* **2019**, *179*, 490–498. [CrossRef]
13. Lane, M.M.; Davis, J.A.; Beattie, S.; Gomez-Donoso, C.; Loughman, A.; O'Neil, A.; Jacka, F.; Berk, M.; Page, R.; Marx, W.; et al. Ultraprocessed food and chronic noncommunicable diseases: A systematic review and meta-analysis of 43 observational studies. *Obes. Rev.* **2021**, *22*, e13146. [CrossRef] [PubMed]
14. Beslay, M.; Srour, B.; Mejean, C.; Alles, B.; Fiolet, T.; Debras, C.; Chazelas, E.; Deschasaux, M.; Wendeu-Foyet, M.G.; Hercberg, S.; et al. Ultra-processed food intake in association with BMI change and risk of overweight and obesity: A prospective analysis of the French NutriNet-Sante cohort. *PLoS Med.* **2020**, *17*, e1003256. [CrossRef]
15. Shim, S.Y.; Kim, H.C.; Shim, J.S. Consumption of Ultra-Processed Food and Blood Pressure in Korean Adults. *Korean Circ. J.* **2022**, *52*, 60–70. [CrossRef] [PubMed]
16. Srour, B.; Fezeu, L.K.; Kesse-Guyot, E.; Alles, B.; Debras, C.; Druesne-Pecollo, N.; Chazelas, E.; Deschasaux, M.; Hercberg, S.; Galan, P.; et al. Ultraprocessed Food Consumption and Risk of Type 2 Diabetes Among Participants of the NutriNet-Sante Prospective Cohort. *JAMA Intern. Med.* **2020**, *180*, 283–291. [CrossRef]
17. Zhang, Z.; Jackson, S.L.; Martinez, E.; Gillespie, C.; Yang, Q. Association between ultraprocessed food intake and cardiovascular health in US adults: A cross-sectional analysis of the NHANES 2011–2016. *Am. J. Clin. Nutr.* **2021**, *113*, 428–436. [CrossRef] [PubMed]
18. Chen, X.; Zhang, Z.; Yang, H.; Qiu, P.; Wang, H.; Wang, F.; Zhao, Q.; Fang, J.; Nie, J. Consumption of ultra-processed foods and health outcomes: A systematic review of epidemiological studies. *Nutr. J.* **2020**, *19*, 86. [CrossRef] [PubMed]
19. Li, H.; Li, S.; Yang, H.; Zhang, Y.; Zhang, S.; Ma, Y.; Hou, Y.; Zhang, X.; Niu, K.; Borne, Y.; et al. Association of Ultraprocessed Food Consumption With Risk of Dementia: A Prospective Cohort Study. *Neurology* **2022**, *99*, e1056–e1066. [CrossRef] [PubMed]
20. Lustig, R.H. Ultraprocessed Food: Addictive, Toxic, and Ready for Regulation. *Nutrients* **2020**, *12*, 3401. [CrossRef]
21. Adjibade, M.; Julia, C.; Alles, B.; Touvier, M.; Lemogne, C.; Srour, B.; Hercberg, S.; Galan, P.; Assmann, K.E.; Kesse-Guyot, E. Prospective association between ultra-processed food consumption and incident depressive symptoms in the French NutriNet-Sante cohort. *BMC Med.* **2019**, *17*, 78. [CrossRef]
22. Gomez-Donoso, C.; Sanchez-Villegas, A.; Martinez-Gonzalez, M.A.; Gea, A.; Mendonca, R.D.; Lahortiga-Ramos, F.; Bes-Rastrollo, M. Ultra-processed food consumption and the incidence of depression in a Mediterranean cohort: The SUN Project. *Eur. J. Nutr.* **2020**, *59*, 1093–1103. [CrossRef] [PubMed]
23. Kauffman, K.; Horvat Davey, C.; Dolata, J.; Figueroa, M.; Gunzler, D.; Huml, A.; Pencak, J.; Sajatovic, M.; Sehgal, A.R. Changes in Self-Reported Depressive Symptoms Among Adults in the United States From 2005 to 2016. *J. Am. Psychiatr. Nurses Assoc.* **2021**, *27*, 148–155. [CrossRef]
24. Han, C.; Jo, S.A.; Kwak, J.H.; Pae, C.U.; Steffens, D.; Jo, I.; Park, M.H. Validation of the Patient Health Questionnaire-9 Korean version in the elderly population: The Ansan Geriatric study. *Compr. Psychiatry* **2008**, *49*, 218–223. [CrossRef] [PubMed]
25. Park, S.J.; Choi, H.R.; Choi, J.H.; Kim, K.W.; Hong, J.P. Reliability and Validity of the Korean Version of the Patient Health Questionnaire-9 (PHQ-9). *Anxiety Mood* **2010**, *6*, 119–124.
26. Park, S.-H.; Kim, S.-N.; Lee, S.; Choe, J.-S.; Choi, Y. Development of 9 th Revision Korean Food Composition Table and Its Major Changes. *Korean J. Community Nutr.* **2018**, *23*, 352. [CrossRef]
27. Monteiro, C.A.; Cannon, G.; Levy, R.B.; Moubarac, J.C.; Louzada, M.L.; Rauber, F.; Khandpur, N.; Cediel, G.; Neri, D.; Martinez-Steele, E.; et al. Ultra-processed foods: What they are and how to identify them. *Public Health Nutr.* **2019**, *22*, 936–941. [CrossRef]

28. Monteiro, C.A.; Cannon, G.; Moubarac, J.C.; Levy, R.B.; Louzada, M.L.C.; Jaime, P.C. The UN Decade of Nutrition, the NOVA food classification and the trouble with ultra-processing. *Public Health Nutr.* **2018**, *21*, 5–17. [CrossRef]
29. WHO. Global Physical Activity Questionnaire (GPAQ). Available online: https://www.who.int/teams/noncommunicable-diseases/surveillance/systems-tools/physical-activity-surveillance (accessed on 2 February 2023).
30. Lee, J.; Lee, C.; Min, J.; Kang, D.W.; Kim, J.Y.; Yang, H.I.; Park, J.; Lee, M.K.; Lee, M.Y.; Park, I.; et al. Development of the Korean Global Physical Activity Questionnaire: Reliability and validity study. *Glob. Health Promot.* **2020**, *27*, 44–55. [CrossRef]
31. Piercy, K.L.; Troiano, R.P.; Ballard, R.M.; Carlson, S.A.; Fulton, J.E.; Galuska, D.A.; George, S.M.; Olson, R.D. The Physical Activity Guidelines for Americans. *JAMA* **2018**, *320*, 2020–2028. [CrossRef]
32. Mazloomi, S.N.; Talebi, S.; Mehrabani, S.; Bagheri, R.; Ghavami, A.; Zarpoosh, M.; Mohammadi, H.; Wong, A.; Nordvall, M.; Kermani, M.A.H.; et al. The association of ultra-processed food consumption with adult mental health disorders: A systematic review and dose-response meta-analysis of 260,385 participants. *Nutr. Neurosci.* **2022**, 1–19. [CrossRef]
33. Godos, J.; Bonaccio, M.; Al-Qahtani, W.H.; Marx, W.; Lane, M.M.; Leggio, G.M.; Grosso, G. Ultra-Processed Food Consumption and Depressive Symptoms in a Mediterranean Cohort. *Nutrients* **2023**, *15*, 504. [CrossRef] [PubMed]
34. Kuehner, C. Why is depression more common among women than among men? *Lancet Psychiatry* **2017**, *4*, 146–158. [CrossRef]
35. Ortega, M.A.; Alvarez-Mon, M.A.; Garcia-Montero, C.; Fraile-Martinez, O.; Guijarro, L.G.; Lahera, G.; Monserrat, J.; Valls, P.; Mora, F.; Rodriguez-Jimenez, R.; et al. Gut Microbiota Metabolites in Major Depressive Disorder-Deep Insights into Their Pathophysiological Role and Potential Translational Applications. *Metabolites* **2022**, *12*, 50. [CrossRef] [PubMed]
36. Martinez Leo, E.E.; Segura Campos, M.R. Effect of ultra-processed diet on gut microbiota and thus its role in neurodegenerative diseases. *Nutrition* **2020**, *71*, 110609. [CrossRef] [PubMed]
37. Lane, M.M.; Lotfaliany, M.; Forbes, M.; Loughman, A.; Rocks, T.; O'Neil, A.; Machado, P.; Jacka, F.N.; Hodge, A.; Marx, W. Higher Ultra-Processed Food Consumption Is Associated with Greater High-Sensitivity C-Reactive Protein Concentration in Adults: Cross-Sectional Results from the Melbourne Collaborative Cohort Study. *Nutrients* **2022**, *14*, 3309. [CrossRef]
38. Kliemann, N.; Al Nahas, A.; Vamos, E.P.; Touvier, M.; Kesse-Guyot, E.; Gunter, M.J.; Millett, C.; Huybrechts, I. Ultra-processed foods and cancer risk: From global food systems to individual exposures and mechanisms. *Br. J. Cancer* **2022**, *127*, 14–20. [CrossRef]
39. Montera, V.; Martins, A.P.B.; Borges, C.A.; Canella, D.S. Distribution and patterns of use of food additives in foods and beverages available in Brazilian supermarkets. *Food Funct.* **2021**, *12*, 7699–7708. [CrossRef]
40. Buckley, J.P.; Kim, H.; Wong, E.; Rebholz, C.M. Ultra-processed food consumption and exposure to phthalates and bisphenols in the US National Health and Nutrition Examination Survey, 2013–2014. *Environ. Int.* **2019**, *131*, 105057. [CrossRef]

nutrients

MDPI

Systematic Review

Ultra-Processed Food Consumption and Incidence of Obesity and Cardiometabolic Risk Factors in Adults: A Systematic Review of Prospective Studies

Sara Paola Mambrini [1,2], Francesca Menichetti [2], Simone Ravella [2], Marta Pellizzari [3], Ramona De Amicis [2,3], Andrea Foppiani [2], Alberto Battezzati [2,4], Simona Bertoli [2,3] and Alessandro Leone [2,*]

1 IRCCS Auxologico, Laboratory of Metabolic Research, San Giuseppe Hospital, 28824 Piancavallo, Italy; sara.mambrini@unimi.it

2 International Center for the Assessment of Nutritional Status and the Development of Dietary Intervention Strategies (ICANS-DIS), Department of Food, Environmental and Nutritional Sciences (DeFENS), University of Milan, 20133 Milan, Italy

3 IRCCS Auxologico, Laboratory of Nutrition and Obesity Research, Department of Endocrine and Metabolic Diseases, 20145 Milan, Italy

4 IRCCS Auxologico, Clinical Nutrition Unit, Department of Endocrine and Metabolic Diseases, 20145 Milan, Italy

* Correspondence: alessandro.leone1@unimi.it

Abstract: Ultra-processed foods (UPF) are energy-dense, nutritionally unbalanced products, low in fiber but high in saturated fat, salt, and sugar. Recently, UPF consumption has increased likewise the incidence of obesity and cardiometabolic diseases. To highlight a possible relationship, we conducted a systematic review of prospective studies from PubMed and Web of Science investigating the association between UPF consumption and the incidence of obesity and cardiometabolic risk factors. Seventeen studies were selected. Eight evaluated the incidence of general and abdominal obesity, one the incidence of impaired fasting blood glucose, four the incidence of diabetes, two the incidence of dyslipidemia, and only one the incidence of metabolic syndrome. Studies' quality was assessed according to the Critical Appraisal Checklist for cohort studies proposed by the Joanna Briggs Institute. Substantial agreement emerged among the studies in defining UPF consumption as being associated with the incident risk of general and abdominal obesity. More limited was the evidence on cardiometabolic risk. Nevertheless, most studies reported that UPF consumption as being associated with an increased risk of hypertension, diabetes, and dyslipidemia. In conclusion, evidence supports the existence of a relationship between UPF consumption and the incidence of obesity and cardiometabolic risk. However, further longitudinal studies considering diet quality and changes over time are needed.

Keywords: ultra-processed foods; NOVA; obesity; cardiometabolic risk; adults; cohort study systematic review

Citation: Mambrini, S.P.; Menichetti, F.; Ravella, S.; Pellizzari, M.; De Amicis, R.; Foppiani, A.; Battezzati, A.; Bertoli, S.; Leone, A. Ultra-Processed Food Consumption and Incidence of Obesity and Cardiometabolic Risk Factors in Adults: A Systematic Review of Prospective Studies. *Nutrients* **2023**, *15*, 2583. https://doi.org/10.3390/nu15112583

Academic Editor: Helmut Schroder

Received: 17 April 2023
Revised: 26 May 2023
Accepted: 29 May 2023
Published: 31 May 2023

1. Introduction

Obesity is a growing worldwide health problem. It is characterized by excessive adiposity that can compromise health status. According to the World Health Organization, obesity affects more than 1 billion people worldwide, 650 million of whom are adults [1]. Obesity is closely linked with metabolic syndrome [2], defined by the National Institutes of Health as a cluster of interconnected metabolic abnormalities, including central adiposity, dyslipidemia, high blood pressure, and impaired fasting glucose [3]. Both obesity and metabolic syndrome are associated with increased risk for mortality and many non-communicable diseases (NCDs) [2].

Obesity and metabolic syndrome are complex, multifactorial diseases whose causes are not yet fully elucidated. However, it is well known that dietary habits play a crucial role in influencing cardiometabolic risk [4]. Several epidemiological studies support an inverse association between adherence to healthy dietary patterns, such as the Mediterranean diet, and cardiometabolic risk [5–7]. In contrast, a diet rich in highly processed foods is strongly associated with obesity and related metabolic comorbidities [8–10].

The NOVA food system was proposed in 2010 to classify foods according to the level of processing [11]. According to this system, foods are classified into four different food groups according to the type, extent, and scope of industrial processes to which foods have been subjected. The first group refers to unprocessed or minimally processed foods. This group includes edible parts of plants or animals and natural foods altered by processes aimed to make them edible, suitable for preservation, safe, or more palatable. The second group refers to processed culinary ingredients including lard, butter, oils, salt, and sugar. They are generally used in combination with foods to make meals and dishes more palatable. The third group refers to processed foods. This group includes food products obtained by adding substances from group 2 to group 1 foods in order to increase their shelf life and enhance sensory qualities. They mostly contain two or three ingredients. The last group references ultra-processed foods (UPF). This group includes formulations mainly made of unmodified and modified substances extracted from foods and assembled with few, if any, whole foods. They also contain food additives to increase palatability, sensory characteristics, and shelf-life. They generally contain five or more ingredients. Examples of UPFs are breakfast cereals, packaged savory and sweet snacks, packaged bread, margarine, reconstituted meat products, pre-prepared frozen dishes, instant soups, sweet and carbonated beverages, and distilled alcoholic beverages.

Several studies reported that UPF consumption is rising, accounting now for more than half of the daily calories of US [12], Canadian [13], or British [14] diets. Moreover, it has been shown that high UPF consumption leads to a nutritionally unbalanced diet, rich in energy, saturated fat, sugar, and salt and poor in fiber, vitamins, and minerals [14], potentially affecting the risk for obesity and cardiometabolic risk factors [15]. Therefore, we conducted a systematic review aimed to summarize the available literature on the association between UPF consumption and the incidences of obesity and cardiometabolic risk factors among adults.

2. Materials and Methods

2.1. Search Strategy

Preferred Reporting Items for Systematic Reviews and Meta-Analysis guidelines (PRISMA) were followed to carry out the study [7]. Studies included in the present review were identified by searching in two electronic databases, including PubMed and Web of Science, using the following search string: (ultraprocessed food* OR ultra-processed food* OR ultra processed food* OR NOVA food*) AND (obesity OR overweight OR waist circumference OR blood pressure OR hypertension OR dyslipidemia OR triglycerides OR cholesterol OR impaired fasting glucose OR diabetes OR metabolic syndrome OR cardiovascular disease OR cardiovascular risk). Electronic search was carried out in September 2022. This systematic review was registered in PROSPERO with registration number CRD42023423112.

2.2. Study Selection, Inclusion and Exclusion Criteria

Initially, we proceeded to exclude duplicates. Then, two independent investigators (S.P.M. and S.R.) selected articles based on title and abstract. The selected articles were then evaluated for eligibility. To be included in the present review, articles needed to be original, include healthy participants aged 18 years or older, written in English, have a prospective cohort study design, use NOVA classification to define UPF, and have as outcomes general or central obesity and cardiometabolic risk factors. No country/region/ethnicity nor date restrictions were applied. Cross-sectional and case-control studies were excluded. Studies limiting the evaluation only to a specific food category included in the definition of UPF, such as reconstituted meat products or sugar-sweetened beverages, or that assess household availability or purchase of UPF were excluded. We further excluded meta-analyses, review articles, congress abstracts, letters, and comments. Disagreements in study selection were resolved through consensus or by seeking the opinion of a third investigator (A.L.) if consensus could not be reached.

2.3. Data Extraction

From each article, we extracted the following data: main author, country, year of publication, number of participants, outcomes, dietary assessment method, confounding factors, and main results. Two independent investigators (S.P.M. and F.M.) reviewed selected articles and performed data extraction. A third investigator (A.L.) supervised data extraction and solved inconsistencies and disagreements.

2.4. Quality Assessment

Two independent investigators (S.P.M. and M.P.) conducted the quality assessment. The Critical Appraisal Checklist for cohort studies proposed by Joanna Briggs Institute was used to assess the methodological quality of the selected studies [16]. The checklist included 11 items related to the following critical domains: population characteristics, follow-up, outcomes, exposure, confounders, and statistical analysis. For each item, it was possible to respond with "no", "yes", "unclear", or "not applicable". Based on the responses, an overall critical assessment of the quality of the study was obtained. In cases where the two investigators disagreed in answering individual items, the opinion of a third investigator (A.L.) was sought. Studies that received a positive score in at least half of the items were considered to be of acceptable quality for inclusion in this Review [9].

3. Results

A total of 2662 articles were initially found on Pubmed and Web of Science (Figure 1). We then removed 717 duplicates and discarded an additional 1852 articles based on title and/or abstract, as they were deemed irrelevant to the review. The remaining 93 records were evaluated for eligibility. Of these, 2 articles were not written in English, 42 were review, meta-analysis, editorial, commentary, or congress abstracts, and 32 were original studies but with a study design different from the cohort study (mainly cross-sectional), and therefore were removed. At the end of the evaluation process, 17 studies were included in this systematic review. The quality assessment of the selected studies is shown in Figure 2.

3.1. Study Characteristics

The 17 studies included a total of 822,213 adults of both sexes (Table 1). The sample size ranged from a minimum of 652 to a maximum of 348748. Four studies were conducted in Brazil [17–20], two in France [21,22] one in Mexico [23], one in the Netherlands [24], five in Spain [25–29], two in the UK [30,31], and another in China [32]. One study used data from the EPIC study cohort, which collects data from several European countries such as Denmark, France, Germany, Italy, and Norway [33]. Regarding dietary assessment, nine studies used food frequency questionnaires (FFQ) consisting of a different number of questions [18–20,23,24,26–28,33], six studies used the 24 h recall [17,21,22,30–32], and two studies used dietary history [25,29]. The cohort study published by Cordova et al. uses both Food Frequency Questionnaires (FFQ) and dietary interviews to collect data on UPF consumption. UPF consumption (exposure variable) was assessed as the percentage of energy from UPF (%UPF_{energy}) in six studies [17,18,23,25,29,31], as servings of UPF consumed per day in two studies [26,27], as grams of UPF (UPF_g/day) per day in five studies [19,22,28,32,33] and as the proportion of UPF intake in the total weight of food consumed (%UPF_{intake}) in three studies [21,30]. In one study, UPF consumption was expressed both as a %UPF_{energy} and a %UPF_{intake} [20].

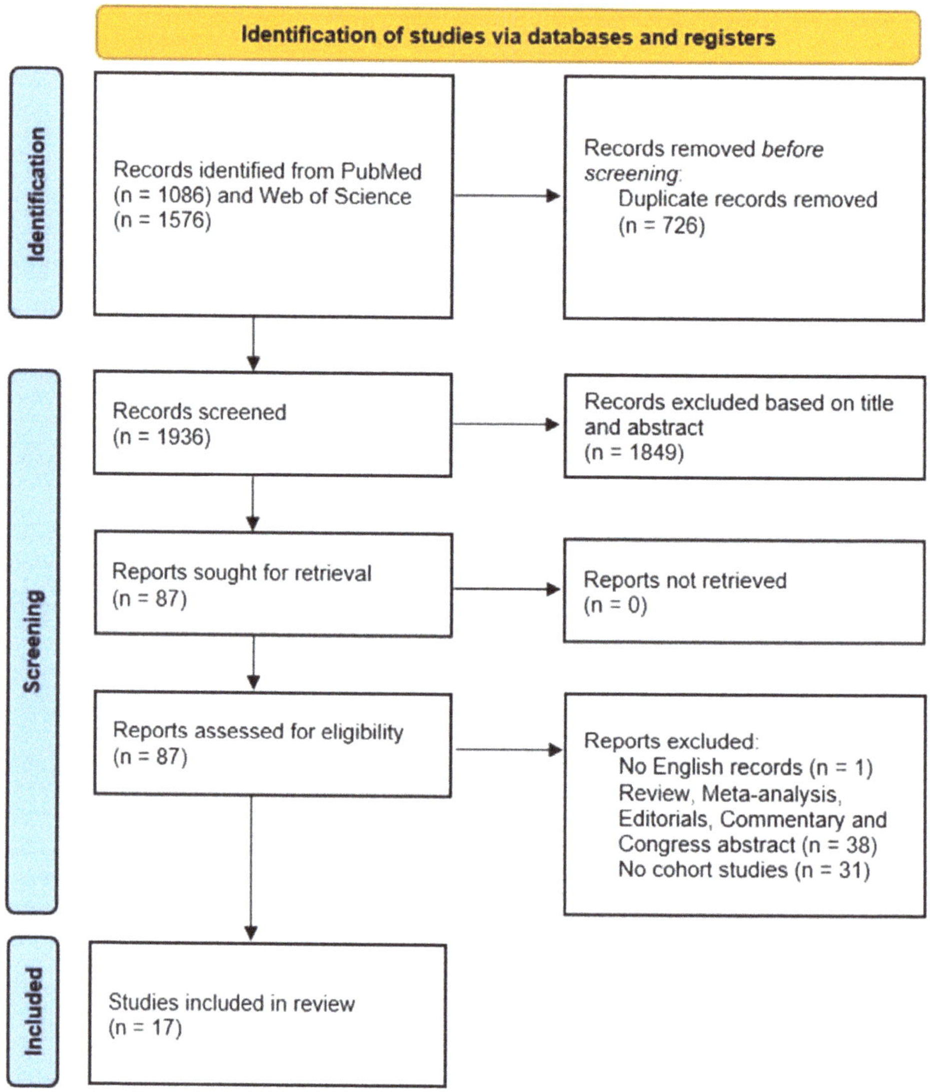

Figure 1. Flow chart of studies' selection process.

	Q1	Q2	Q3	Q4	Q5	Q6	Q7	Q8	Q9	Q10	Q11	Include
Mendonça et al. (2016)	+	+	+	+	+	+	+	+	+	+	+	+
Canhada et al. (2019)	+	+	+	+	+	+	+	+	X	+	+	+
Beslay et al. (2020)	+	+	+	+	+	+	+	+	X	+	+	+
Sandoval-Insausti et al. (2020)	+	+	+	+	+	+	+	+	X	+	+	+
Cordova et al. (2021)	+	+	+	+	+	+	+	+	+	+	+	+
Li et al. (2021)	+	+	+	+	+	+	+	+	-	+	+	+
Rauber et al. (2021)	+	+	+	+	+	+	+	+	X	+	+	+
DaSilva Magalhães et al. (2022)	+	+	X	+	+	+	+	+	+	+	+	+
Mendonça et al. (2017)	+	+	+	+	+	+	+	+	+	+	+	+
Monge et al. (2021)	+	+	+	+	+	+	+	+	+	+	+	+
Scaranni et al. (2021)	+	+	+	+	+	+	+	+	+	+	+	+
Srour et al. (2019)	+	+	+	+	+	+	+	+	+	+	+	+
Duan et al. (2022)	+	+	+	+	+	+	+	+	X	X	+	+
Donat-Vargas et al. (2021)	+	+	+	+	+	+	+	+	X	+	+	+
Scaranni et al. (2022)	+	+	+	+	+	+	+	+	+	+	+	+
Levy et al. (2021)	+	+	+	+	+	+	X	+	-	+	+	+
Llavero-Valero et al. (2021)	+	+	+	+	+	+	+	+	+	+	+	+

+ Yes
- Unclear
X No

Q1: Were the two groups similar and recruited from the same population?
Q2: Were the exposures measured similarly to assign people to both exposed and unexposed groups?
Q3: Was the exposure measured in a valid and reliable way?
Q4: Were confounding factors identified?
Q5: Were strategies to deal with confounding factors stated?
Q6: Were the groups/participants free of the outcome at the start of the study?
Q7: Were the outcomes measured in a valid and reliable way?
Q8: Was the follow up time reported and sufficient to be long enough for outcomes to occur?
Q9: Was follow up complete, and if not, were the reasons to loss to follow up described and explored?
Q10: Were strategies to address incomplete follow up utilized?
Q11: Was appropriate statistical analysis used?

Figure 2. Quality checklist. Studies receiving a positive score in at least half of the items were considered to be of acceptable quality for inclusion [17–33].

Table 1. Summary of the selected studies investigating the association between UPF consumption and obesity and cardiometabolic risk factors in adults.

Author (Year)	Country (Cohort)	Subjects (n) and Baseline Characteristics	Outcome	Follow-Up Time	Dietary Assessment	Covariates Included in the Fully Adjusted Model	Type of Exposure	Results
Mendonça et al. (2016) [26]	Spain (SUN cohort)	8451 participants 35.1% men 64.9% women Age: 37.6 ± 11.0 years	Overweight/obesity	Median follow-up: 8.9 years	Semi-quantitative FFQ (136 items)	Sex, age, baseline BMI, educational status, marital status, physical activity, smoking status, siesta sleep, television watching, following a special diet at baseline, snacking between meals, and consumption of fruit and vegetables.	servings/d	Participants in the fourth quartile of UPF consumption had a higher risk of developing overweight or obesity (HR = 1.26, 95% CI: 1.10, 1.45, P_{trend} = 0.001) than participants in the first quartile.
Canhada et al. (2019) [17]	Brazil (ELSA cohort)	11,827 participants 45% men 55% women Age: 51.3 ± 8.7 years	Overweight/obesity	Mean follow-up: 3.8 years	FFQ (114 items)	Age, sex, school achievement, center, and color/race, as well as smoking and physical activity, waist/weight gain, incidence of overweight/obesity, baseline BMI, and baseline waist circumference.	$\%UPF_{energy}$	Participants in the fourth quartile of UPF consumption (>30.8 %) presented 20% greater risk (RR:1.20; 95% CI: 1.03, 1–40) of incident overweight and obesity than participants in the first quartile (<17.8%). No association between UPF quartiles and risk of incident obesity among overweight participants was observed (RR:1.02; 95% CI: 0.85, 1.21).
Beslay et al. (2020) [21]	France (French NutriNet-Santè cohort)	110,260 participants 22.8% men 78.2% women Age: 43.1 ± 14.6 years	Overweight/obesity	Median follow-up: 4.1 years	24 h dietary record	Age, sex, marital status, BMI, educational level, physical activity, smoking status, alcohol intake, number of 24 h dietary records, energy intake, health, and Western dietary pattern.	$\%UPF_{intake}$	Normal-weight participants with low UPF consumption had a lower risk of developing overweight or obesity during follow-up ($HR_{Q4\ vs.Q1}$ = 1.22, 95% CI: 1.14, 1.31, P_{trend} < 0.001) than those with a higher intake. Moreover, a 10% increment of UPF intake was associated with a higher risk of developing overweight or obesity (HR = 1.10, 95% CI: 1.07, 1.13; P < 0.001). Non-obese subjects with low UPF consumption had a lower risk of developing obesity during follow-up ($HR_{Q4\ vs.Q1}$ = 1.20, 95% CI: 1.08, 1.33, P_{trend} < 0.001) than those with a higher intake. Moreover, a 10% increment of UPFs intake was associated with a higher risk of developing obesity (HR = 1.11, 95% CI: 1.07, 1.15; P < 0.001).
Sandoval-Insausti et al. (2020) [25]	Spain (Seniors-ENRICA-1)	652 participants 55.7% men 44.3% women Age: 67.08 ± 5.8 years	Abdominal obesity	Median follow-up: 6 years	Dietary history (DH-ENRICA) record	Age, sex, educational level, marital status, ex-drinker status, smoking, physical activity in the household, physical activity during leisure time, prevalence of chronic disease, number of medications consumed daily, and adherence to Mediterranean diet.	$\%UPF_{energy}$	Participants in the first tertiles of UPF consumption had a higher risk of developing abdominal obesity (RR: 1.61; 95% CI: 1.01, 2.56, P_{trend}=0.048) than participants in the first tertile.

Table 1. *Cont.*

Author (Year)	Country (Cohort)	Subjects (n) and Baseline Characteristics	Outcome	Follow-Up Time	Dietary Assessment	Covariates Included in the Fully Adjusted Model	Type of Exposure	Results
Cordova et al. (2021) [33]	Denmark, Germany, Italy, France, Greece, the Netherlands, Spain, Norway, Sweden and the UK (EPIC cohort)	348,748 participants 26.6% men 73.4% women Age: 51.7 ± 9.0 years	Overweight/obesity	Median follow-up: 5 years	(a) Quantitative FFQ (Italy, Spain, the Netherlands, Germany, and France) (b) Semi-quantitative FFQ (Denmark, Naples (Italy), Norway, and Umeå (Sweden), (c) A combination of semi-quantitative FFQ and 7- and 14-day records in the UK and Malmo (Sweden).	Age, sex, BMI baseline, education level, smoking history, physical activity, alcohol intake, Mediterranean diet score, and plausibility of dietary energy reporting.	g/day	Normal-weight participants in the fifth quintile of UPF consumption had a 15% higher risk (RR = 1.15, 95% CI: 1.11, 1.19, P_{trend} <0.001) of becoming overweight or obese during follow-up than participants in the first quintile. Similarly, participants with overweight in the highest quintile of UPF consumption had a 16% higher risk (RR = 1.16; 95% CI: 1.09, 1.23, P_{trend} <0.001) of becoming obese during follow-up than participants in the lowest quintile.
Li et al. (2021) [32]	China (CNHS cohort)	12,451 participants 48.7% men 51.3% women Age: 43.7 ± 14.7 years	Overweight/obesity and abdominal obesity	10 years	3-day 24 h dietary recall	Age, sex, income, urbanization, education, smoking, alcohol drinking, and physical activity, energy intake, fat intake, and dietary patterns.	g/day	Participants consuming 1–19 g/day, 20–49 g/day, or ≥ 50 g/day of UPF were at a higher risk of developing overweight and obesity and abdominal obesity than non-consumers. Adjusted ORs for overweight and obesity were 1.45 (95% CI: 1.26, 1.65), 1.34 (95% CI: 1.15–1.57), and 1.45 (95% CI: 1.21–1.74), respectively. Adjusted ORs for abdominal obesity were 1.54 (95% CI: 1.38, 1.72), 1.35 (95% CI: 1.19, 1.54), and 1.50 (95% CI: 1.29, 1.74), respectively.
Rauber et al. (2021) [31]	England, Scotland and Wales (UK Biobank)	22,659 participants 47.9% men 52.1% women Age: 55.9 ± 7.4 years	General and abdominal obesity	Median follow-up: 5 years	24 h dietary recall	Sex, BMI, waist circumference or body fat at baseline, smoking status, level of physical activity, sleep duration, Index of Multiple Deprivation (IMD).	$\%UPF_{energy}$	Non-obese participants in the uppermost quartile of UPF consumption were at a higher risk of developing obesity (HR = 1.79, 95% CI: 1.06, 3.03) than participants in the lowest quartile. Similarly, participants with normal waist circumference at baseline but in the first quartile of UPF consumption were at a higher risk of developing abdominal obesity (HR = 1.30, 95% CI: 1.14, 1.48) than participants in the lowest quartile.

Table 1. *Cont.*

Author (Year)	Country (Cohort)	Subjects (n) and Baseline Characteristics	Outcome	Follow-Up Time	Dietary Assessment	Covariates Included in the Fully Adjusted Model	Type of Exposure	Results
DaSilva Magalhães et al. (2022) [20]	Brazil (Ribeirão Preto cohort)	896 particpants 44.3% men 55.7% women Age: 23–25 years	MetS and its components	14–16 years	Semi-quantitative FFQ (83 items)	Sex, age, education, marital status, skin color, family income, smoking, level of physical activity, and alcohol consumption. In the analyses with the consumption of UPF in %g, total energy intake was additionally included.	$\%UPF_{energy}$ and $\%UPF_{intake}$	UPF consumption was not associated with the risk of metabolic syndrome (%kcal PR: 1.00; 95% CI: 0.99–1.01; %g PR: 1.00; 95% CI: 0.99–1.01). However, women with higher UPF consumption were at a higher risk of developing abdominal obesity (%kcal: RR = 1.01, 95% CI: 1.00, 1.02, $p = 0.030$; %g: RR = 1.01, 95% CI: 1.00, 1.02, $p = 0.003$) and low HDL-cholesterol (%kcal: RR = 1.02, 95% CI: 1.01, 1.04, $p = 0.041$). No significant associations between UPF consumption and other metabolic syndrome components were observed.
Mendonca et al. (2017) [27]	Spain (SUN cohort)	14,790 36.3% men 63.7% women Age: 36.3 ± 10.3 years	Hypertension	Mean follow-up: 9.1 years	Semi-quantitative FFQ (136 items)	Sex, age, baseline BMI, physical activity, hours of television watching, smoking status, following a special diet at baseline, use of analgesics, alcohol consumption, family history of hypertension, hypercholesterolemia, total energy intake, fruit and vegetable consumption, and olive oil intake.	servings/d	Participants in the third tertile of UPF consumption were at a higher risk of developing hypertension (HR = 1.21, 95% CI: 1.06, 1.37, $P_{trend} = 0.004$) than participants in the first tertile.
Monge et al. (2021) [23]	Mexico (Mexican Teachers' Cohort)	64934 participants (only women) Age: 41.7 ± 7.2 years	Hypertension	Median follow-up: 2.2 years	Semi-quantitative FFQ (140 items)	Age, smoking status, physical activity, menopausal status, ethnicity, internet access and insurance for serious conditions, family history of hypertension, total energy intake, and multivitamin supplementation.	$\%UPF_{energy}$	No association between categories of $\%UPF_{energy}$ (≤20%, 21–25%, 26–35%, 36–45% >45% energy/d) and incident hypertension was found. Compared with the first category, IRRs were 0.96 (95% CI: 0.86, 1.07), 0.92 (95% CI: 0.84, 1.02), 0.95 (95% CI: 0.85, 1.06), and 0.98 (95% CI: 0.84, 1.14).
Scaranni et al. (2021) [18]	Brazil (ELSA cohort)	8754 participants 42% men 58% women Median age: 49.0 years	Hypertension	Mean follow-up: 3.9 years	114-item FFQ	Sex, age, self-declared color/race, education, smoking, alcohol consumption, antihypertensive drug use, Na consumption, physical activity, total daily energy intake, and BMI.	$\%UPF_{energy}$	Participants with higher UPF consumption had a marginally significant greater risk of developing hypertension (OR = 1.17; 95% CI: 1.00, 1.37) than participants with lower UPF consumption.

Table 1. *Cont.*

Author (Year)	Country (Cohort)	Subjects (n) and Baseline Characteristics	Outcome	Follow-Up Time	Dietary Assessment	Covariates Included in the Fully Adjusted Model	Type of Exposure	Results
Srour et al. (2019) [22]	France (French NutriNet-Santè cohort)	1047,07 participants 20.8% men 79.2% women Age: 42.7 ± 14.5 years	Type 2 Diabetes	Median follow-up: 6 years	24 h dietary record	Sex, age, BMI, weight change during follow-up, educational level, smoking status, physical activity level, number of 24 h dietary records, alcohol intake, energy intake without alcohol, overall diet quality, family history of diabetes, baseline dyslipidemia and hypertension, and treatments for these conditions.	g/day	An increment of 10% of UPFs in diet was associated with an increased risk of T2D (HR = 1.13, 95% CI: 1.03, 1.23, $p = 0.04$). Similarly, a 100g/day increment in UPF consumption was associated with the risk of T2D (HR = 1.05; 95% CI: 1.02, 1.08, $p = 0.003$).
Duan et al. (2022) [24]	Netherlands (Lifelines cohort)	70,421 participants 41.4% men 58.6% women Age 49.1 ± 8.8 years	Type 2 Diabetes	Median follow-up: 3.4 years	Semi-quantitative FFQ (110 items)	Sex, age, BMI, educational level, energy intake, alcohol intake, Life diet score, smoking status, physical activity, and TV-watching time.	$\%UPF_{intake}$	An increment of 10% in UPF consumption was associated with a 25% higher risk of developing T2D (OR = 1.25; 95% CI: 1.16, 1.34).
Levy et al. (2021) [30]	England, Scotland and Wales (UK Biobank)	21,730 participants 47.1% men 52.9% women Age: 55.8 ± 7.4 years	Type 2 Diabetes	Mean follow-up: 5.4 years	24 h dietary recall	Sex, age, BMI, smoking, physical activity level, ethnicity, family history of T2D, Index of Multiple Deprivation (IMD), and total energy intake.	$\%UPF_{intake}$	Participants in the highest quartile of UPF consumption were at a higher risk for T2D (HR = 1.44; 95% CI: 1.04, 2.02, $P_{trend} < 0.028$) than participants in the lowest quartile. Moreover, a 10%-point increment in UPF consumption was associated with a 12% increased risk of T2D (HR = 1.12, 95% CI: 1.04, 1.20).
Llavero-Valero et al. (2021) [28]	Spain (SUN cohort)	20,060 participants 38.5% men 61.5% women Age: 37.4 ± 12.2 years	Type 2 Diabetes	Median follow-up: 12 years	Semi-quantitative FFQ (136 items)	Age, sex, BMI, educational level, smoking status, 8-item active + sedentary lifestyle score, following a special diet at baseline, snacking, and family history of diabetes.	g/day	Participants in the highest tertile of UPF consumption were at a higher risk of T2D than participants in the lowest tertile (HR = 1.53, 95% CI: 1.06, 2.22, $P_{trend} = 0.024$). After using repeated measurements of UPF consumption, the association remained significant (HR = 1.65, 95% CI: 1.14, 2.38).
Donat-Vargas et al. (2021) [29]	Spain (ENRICA cohort)	1082 participants 48% men 52% women Age: 68 ± 6 years	Dyslipidemia	5–7 years	Dietary history (DH-ENRICA) record	Sex, age, BMI, smoking status, physical activity, educational level, marital status, total energy intake, alcohol consumption, fiber intake, consumption of unprocessed or minimal processed foods, number of medications, and number of chronic diseases.	$\%UPF_{energy}$	Participants in the uppermost tertile of UPF consumption were at a higher risk for incident low HDL cholesterol (OR = 2.23; 95% CI: 1.22, 4.05; $P_{trend} = 0.012$) and hypertriglyceridemia (OR = 2.66, 95% CI: 1.20, 5.90; $P_{trend} = 0.011$) than participants in the lowest tertile. However, the consumption of UPF was not associated with the incident risk of high LDL cholesterol.

Table 1. *Cont.*

Author (Year)	Country (Cohort)	Subjects (n) and Baseline Characteristics	Outcome	Follow-Up Time	Dietary Assessment	Covariates Included in the Fully Adjusted Model	Type of Exposure	Results
Scaranni et al. (2022) [19]	Brazil (ELSA cohort)	5275 participants 42.2% men 57.8% women Age: 50.6 ± 8.8 years	Dyslipidemia	4 years	Semi-quantitative FFQ (114 items)	Sex, age, BMI, schooling, smoking, physical activity, alcohol consumption, total energy intake, diabetes and time since baseline, and Brazilian Healthy Eating Index—Revised (BHEI-R).	g/day	Individuals with medium and high consumption of UPF had higher risks of developing isolated hypertriacylglycerolemia (OR = 1.14, 95% CI: 1.03, 1.26 and OR = 1.30, 95% CI: 1.17, 1.45), isolated hypercholesterolemia (OR = 1.12, 95% CI: 1.00, 1.27 and OR = 1.28, 95% CI: 1.12, 1.47), mixed hyperlipidemia (OR = 1.21, 95% CI: 1.05, 1.39 and OR = 1.38, 95% CI: 1.18, 1.62), and low HDL (OR = 1.12, 95% CI: 1.00, 1.24 and OR = 1.18, 95% CI: 1.05, 1.32), respectively, than participants who consumed less UPF.

3.2. Consumption of Ultra-Processed Food, Excess Body Weight, and Abdominal Obesity

Eight cohort studies investigated the relationship between UPF consumption and the risk of weight excess and abdominal obesity, all finding a positive relationship [17,20,21,25,26,32–34]. Four studies focused on the risk of overweight and obesity [17,21,26,33] and two studies on the risk of abdominal obesity [20,25,31], while two other studies investigated both the risk of overweight and obesity and of abdominal obesity [31,32]. Mendonca et al. [26] analyzed data from the SUN cohort, reporting that normal-weight participants consuming higher amounts of UPF, expressed as consumption of servings per day, had a 26% higher risk of developing overweight or obesity during follow-up (HR = 1.26; 95% CI: 1.10, 1.45, P_{trend} = 0.001), than participants with lower UPF consumption. Similarly, data from the ELSA cohort [17] showed that normal-weight participants in the uppermost quartile of UPF consumption had a 20% higher risk of overweight and obesity during follow-up than participants in the lowest quartile (RR = 1.03, 95% CI: 1.0, 1.40). However, no association between UPF consumption and incident risk of obesity was found among participants who were overweight at baseline (fourth vs. first quartile RR = 1.02, 95% CI: 0.85, 1.21). Results from the French NutriNet-Santè cohort [21], including 110260 adults, reported an 11% increase in the risk of developing overweight or obesity among normal-weight participants (HR = 1.11, 95% CI: 1.08, 1.14; $p < 0.001$) and a 9% increase in the risk of developing obesity among overweight participants (HR = 1.09, 95% CI: 1.05, 1.13; $p < 0.13$), associated with a 10% increase in the % of energy from UPF. Data from the EPIC cohort [33], including a multi-national population of 348748 adults, also reported that normal-weight participants in the fifth quintile of UPF consumption had a 15% higher risk of developing overweight or obesity (RR = 1.15, 95% CI: 1.11, 1.19; $P_{trend} < 0.001$) than participants in the first quintile of UPF consumption. Similarly, overweight participants in the highest quantile of UPF consumption had a 16% higher risk of developing obesity (RR = 1.16, 95% CI: 1.09, 1.23; $P_{trend} < 0.001$) than overweight participants with low consumption of UPF. Data from the China Health and Nutrition Survey (CHNS) [32], including 12451 adults of both sexes, showed a higher risk of overweight and obesity (OR = 1.45, 95% CI: 1.21, 1.74) and abdominal obesity (OR = 1.50, 95% CI: 1.29, 1.74) in participants consuming ≥50 g/day of UPF than non-consumers. Additionally, Rauber et al. [34] found that participants in the fourth quartile of UPF consumption presented a 79% and 30% greater risk of developing obesity (HR = 1.79, 95% CI: 1.06, 3.03) and abdominal obesity (HR = 1.30, 95% CI: 1.14, 1.48), respectively, than participants in the first quartile of UPF consumption. Sandoval et al. [25] reported that, in the Seniors-ENRICA-1 cohort, the incidence of abdominal obesity in elders was significantly higher in participants in the uppermost tertile of UPF consumption than participants in the lowest one (OR = 1.61; 95% CI: 1.01, 2.56; P_{trend} = 0.048). Finally, DaSilva Magalhães et al. [20] assessed UPF consumption in 896 men and women aged 23–25 years and related it to the incidence of metabolic syndrome and its components at ages 37–39. They found that UPF consumption was associated with a higher risk of abdominal obesity in women (RR = 1.01, 95% CI: 1.00, 1.02) but not in men.

3.3. Consumption of Ultra-Processed Food, Impaired Fasting Glucose, and Diabetes Mellitus

The association between UPF consumption and incident risk of impaired fasting glucose was investigated in only one study [20]. On the other hand, four studies focused on the relationship between UPF consumption and the risk of type 2 diabetes (T2D) [24,28,30,35]. New cases of diabetes were identified by self-reported data supported by medical records [20,28,35] or nurse interviews [30] or blood glucose and HbA1c measurements [24]. Silva Magalhães et al. [20] reported that UPF consumption at 23–25 years was not associated with impaired fasting glucose at 37–39 years (%UPF_{energy} RR = 1.00, 95% CI: 0.99, 1.01; %UPF_{intake} RR = 0.99; 95% CI: 0.98, 1.00). Concerning the incident risk of T2D, in the NutriNet-Santé cohort, Srour et al. [35] found a 15% higher risk of T2D associated with an increment of 10% of UPF consumption (grams per day) (HR = 1.15, 95% CI, 1.06–1.25; p = 0.001). Similarly, for each 100g/d increment in the absolute amount of

UPF, the risk of T2D increased by 5% (HR = 1.05; 95% CI: 1.02, 1.08). In the Lifelines cohort, including participants aged 35–70 years, Duan et al. [24] found that a 10% increment in UPF consumption was associated with a 25% higher risk of T2D (OR = 1.25, 95% CI: 1.16, 1.34). Levy et al. [30], in the UK Biobank cohort, found that participants in the fourth quartile of UPF consumption had a 44% higher risk of T2D than participants in the first quartile of UPF consumption (HR = 1.44; 95% CI: 1.04, 2.02). Moreover, they observed a significant 12% increased risk of T2D per 10%-point increments in UPF consumption (HR = 1.12; 95% CI: 1.04, 1.20). Finally, in the SUN cohort, Llavero-Valero et al. [28] found a 53% increased risk of T2D (HR = 1.53; 95% CI: 1.06, 2.22; $P_{trend} < 0.001$) in participants in the third tertile of UPF consumption as compared with participants in the first one.

3.4. Consumption of Ultra-Processed Food and Hypertension

Four studies focused on the relationship between UPF consumption and the incidence of hypertension [18,20,23,27]. Three studies [18,20,27] evaluated this relation both in men and women, whereas only one [23] did so for women. Additionally, two studies evaluated the outcome as self-reported diagnoses of hypertension [23,27] while in the other two [18,20], the outcome was defined by measuring the systolic and diastolic blood pressure during the follow-up. Mendonça et al. [27] observed a 21% higher risk of hypertension among participants in the uppermost tertile of UPF consumption compared with participants in the first tertile (HR = 1.21, 95% CI: 1.06, 1.37, $P_{trend} = 0.004$). Similarly, Scaranni et al. [18] found participants of the ELSA-Brasil cohort with high UPF consumption to have a 17% increased risk of developing hypertension (OR = 1.17; 95% CI: 1.00, 1.37) than participants with low UPF consumption. In contrast, in the Mexican Teachers' Cohort (MTC), including 64934 women, Monge et al. [23] did not find UPF consumption significantly associated with the incident risk of hypertension when comparing extreme categories of UPF consumption (IRR = 0.96; 95% CI: 0.79, 1.16; $P_{trend} = 0.95$). Finally, DaSilva et al. [20] reported that the %UPF at 23–25 years is marginally associated with the risk of hypertension at 37–39 years old (%kcal adjusted RR = 1.01; 95% CI: 1.00, 1.02).

3.5. Consumption of Ultra-Processed Food and Lipid Profile

Among the studies selected, three studies investigated the association between UPF consumption and the incidence of dyslipidemia [19,20,29]. Two of them focused on adults [19,20] and the other one on elders (≥60 years old) [29]. Of the 1821 participants from the Seniors-ENRICA cohort, Donat-Vargas et al. [29] reported that participants in the third tertile of energy intake of UPF had a higher risk for hypertriglyceridemia (OR = 2.66; 95% CI: 1.20, 5.90; $P_{trend} = 0.011$) and low HDL cholesterol (OR = 2.23; 95% CI: 1.22, 4.05; $P_{trend} = 0.012$) than participants in the first tertile. No association between UPF consumption and high LDL cholesterol emerged. Scaranni et al. [18,19], in the ELSA-Brasil cohort, observed that participants with medium and high UPF consumption had a higher risk of developing isolated hypertriglyceridemia (OR = 1.14, 95% CI: 1.03 and 1.26; OR = 1.30, 95% CI: 1.17 and 1.45), isolated hypercholesterolemia (OR = 1.12, 95% CI: 1.00 and 1.27; OR = 1.28, 95% CI: 1.12 and 1.47), low HDL cholesterol (OR = 1.12, 95% CI: 1.00 and 1.24; OR = 1.18, 95% CI: 1.05 and 1.32), and mixed hyperlipidemia (OR = 1.21, 95% CI: 1.05 and 1.39; OR = 1.38, 95% CI: 1.18 and 1.62) than participants consuming lower amounts of UPF. However, the association with low HDL cholesterol was lost when BMI was included in the model. On the contrary, DaSilva et al. [20] reported that UPF consumption at 23–25 years old was not associated with the risk of hypertriglyceridemia at the age of 37–39. On the other hand, UPF was associated with a higher risk of low HDL only in women (RR = 1.02, 95% CI: 1.01, 1.04).

3.6. Consumption of Ultra-Processed Food and Metabolic Syndrome

Only one study evaluated the relationship between UPF consumption and the incident risk of MetS [20]. The authors [20] reported that UPF consumption at 23–25 years was not associated with the risk of MetS at 37–39 years (RR = 1.00; 95% CI: 0.99, 1.01).

4. Discussion

In this systematic review, we summarized all available prospective studies focused on the association between UPF consumption and the incidence of obesity and cardiometabolic risk factors in adults. All studies included reported UPF consumption associated with the risk of developing overweight and obesity [17,21,26,31–33]. Moreover, although more limited in number, the studies included in this review agreed on the association between UPF consumption and abdominal obesity [17,20,25]. Much more limited and heterogeneous were the prospective studies investigating the association between UPF consumption and cardiometabolic risk factors. However, most evidence supports the existence of a relationship with an increased risk of dyslipidemia, hypertension, and diabetes.

Traditionally, UPFs are energy-dense products with low nutritional quality. They contribute to increasing dietary intakes of saturated and trans fatty acids, sugars, refined carbohydrates, and sodium, and to reducing dietary intakes of fiber, micronutrients, and other protective bioactive compounds naturally present in foods [14]. In addition, it has been reported that these products are less satiating and characterized by a greater glycemic response than minimally processed foods [36]. Because of the higher energy density, low satiating effect, and large portion packing [37], consumption of these products may promote excess energy intake [38]. The minimal preparation skills required for UPF consumption can then alter eating patterns, leading to the rapid and unconscious consumption of food while engaged in routine alternative activities [39,40], altering neural and digestive functions that signal hunger and satiety, leading to overconsumption [41,42]. In addition, given their high fat and sugar contents, they can alter the reward neurocircuit mechanism, leading to increased food cravings and further exacerbating overconsumption [43,44]. To this it should be added that, under conditions of energy excess, the increased glucose response induced by UPF consumption may alter the insulin response, favoring the storage of excess nutrients in adipose tissue rather than their oxidation [45]. Excessive energy intake and obesity resulting from UPF consumption are certainly reasons for the development of cardiometabolic risk factors. However, this cannot entirely explain the associations observed between UPF and cardiometabolic risk factors, as many studies controlled their models for BMI and total energy intake. Many UPFs, such as condiments, broths, soup powders, and processed meats, have high levels of salt, contributing to higher sodium intake, a known risk factor for developing hypertension [46]. Added sugar could also alter fructose metabolism in the liver, promoting insulin resistance in the liver and throughout the body. Added fructose has been found to contribute to low-grade inflammation and oxidative stress, potentially causing β-cell damage and reducing insulin secretion [47]. Moreover, excess dietary fructose has been reported to impair the catabolism of very low-density lipoprotein cholesterol (VLDL-C) and increase VLDL-C synthesis, leading to an increase in triglycerides [48]. UPFs are also a source of trans and saturated fatty acids, which may contribute to an increased risk of dyslipidemia. Several RCTs found trans fatty acids having adverse effects on lipid profile [49], such as decreasing HDL cholesterol [50]. In addition, the intake of saturated fatty acids may have a negative impact on lipid metabolism, especially by virtue of the fact that UPFs are simultaneously low in PUFA [51]. Finally, UPFs contain plenty of chemical additives, synthetic antioxidants, and preservatives; many of these have been shown to increase the risk of obesity, deteriorate the lipid [52] and glucose profiles [53], and induce low-grade inflammation and metabolic syndrome [54]. In addition, the packaging of UPF can release known endocrine-disrupting chemicals (e.g., bisphenol A) into the food, increasing the risk of obesity and cardiometabolic risk [55–57]. Finally, it is presumed that those who consume high amounts of UPF have lower consumption of whole grains, fruits, and vegetables, limiting the intake [58] of micronutrients and bioactive compounds that may reduce cardiometabolic risk.

Despite the associations found, some considerations need to be made to evaluate the associations between UPF consumption and the incidence of obesity and cardiometabolic risk factors and to compare the results between studies. Only six studies controlled for dietary patterns or quality [19,21,22,24,25,33]. Considering the overall dietary pattern avoids potential confounding by other aspects of the diet, allows for evaluation of the interaction between synergistic components, and increases the ability to assess stronger effects due to the cumulative effects of many dietary characteristics [59]. An approach focused only on UPF consumption does not take into account the substitution effects of foods and associated foods [60]. Consumption of UPFs in a varied and balanced diet may not have the same effect as when they are consumed in a high-calorie diet, in which consumption of UPFs leads to the reduction of foods of higher nutritional value [9]. In addition, a very limited number of studies have repeatedly measured exposure. Dietary habits may change over time according to the food offered and living or environmental conditions, and consequently, they may influence the risk of obesity and cardiometabolic risk factors. An additional source of bias may be the method used for diet assessment. Most studies used 24 h recall and FFQ, while only two studies used dietary history. Although they are all accepted methods for evaluating dietary consumption, they are susceptible to recall bias and to difficulties in quantifying portions, compared with prospective methods based on recording and weighing foods consumed. Moreover, although a single 24 h recall is generally considered valid for assessing a population's food intake, to have a better estimation of habitual UPF consumption, especially given the wide range of products that are part of it, it may be necessary to consider multiple food days. In addition, although the FFQ is a commonly used method to assess the diet–health association, it suffers from some measurement errors. The dietary assessment is often limited to a specific list of foods that varies according to the questionnaire used and the quantification of intake is not as accurate as with the 24 h recall or food diary [61]. Moreover, it should be remembered that all of these methods are not specifically designed to assess UPF consumption as it is defined by the NOVA classification. This can determine an overestimation or underestimation of UPF consumption. Finally, the use of different units of measurement to assess exposure to UPF (e.g., $\%UPF_{energy}$, servings, g/day, ($\%UPF_{intake}$) may have contributed to increased heterogeneity among studies. Future studies should therefore standardize the units of measurement to facilitate the comparison of results. Since obesity, as well as cardiometabolic risk and other NCDs, is strongly related to caloric intake, it is important to discern the effect of UPFs from that of total energy intake. Using a nutrient density model (%UPFenergy), without further adjustment for total calories, is not sufficient to remove the effect of total energy intake [62]. In addition, this approach does not allow for the consideration of UPFs that do not provide energy (e.g., artificially sweetened beverages). Similarly, the use of daily consumption frequency alone, without portion quantification, does not allow for true quantification of the foods consumed (e.g., many small portions might be equivalent to one large portion). These limitations can be overcome by using the total weight of foods consumed. In addition, using energy-adjusted food weight with the residual method would control for confounding factors by total energy intake and remove extraneous variations due to total energy intake [62].

Among the limitations of the present review is that many of the studies assessed exposure only at baseline. It must be considered that dietary habits assessed at baseline may have changed during follow-up, affecting risk estimates. To obtain a better representation of dietary habits and identify the direction of their relationship with cardiometabolic risk, more longitudinal studies with repeated assessments of food intake are needed. Moreover, several studies had a retention rate during follow-up that was potentially suboptimal (<80%). In addition, although the adjustment for confounders was considered satisfactory overall, several studies did not consider diet quality, which may have influenced the result. Poor geographic representativeness is a further limitation. The majority of the studies were conducted in Brazil, Spain, France, and England, limiting the generalizability of the results for other countries. Although the NOVA classification is internationally recognized, it may not be appropriate in all countries due to different cultural and dietary habits, as well as different industrial food production technology. For example, it was found that 23% of UPFs sold in Italy were of high nutritional quality considering three front-of-pack labeling schemes [63]. Therefore, further studies need to be conducted on other populations in order to develop correct nutrition policies and recommendations.

Nevertheless, this systematic review also has some strengths. We included only prospective cohort studies that, by measuring events in a temporal sequence, allow us to distinguish causes from effects [64]. This also made it possible to limit the variability due to the use of different study designs. In addition, we only included studies that used the NOVA classification, limiting the variability among studies due to different methods of defining UPF.

5. Conclusions

In conclusion, studies currently available in the literature agree that the consumption of ultra-processed foods is associated with the incidence of obesity. Less clear is its relationship with the incidence of outcomes related to cardiometabolic risk. Despite the positive associations found between the consumption of ultra-processed foods and cardiometabolic risk, the studies reported in the literature are still very limited, especially for some outcomes, and some results are conflicting, probably due to the adoption of different methods for assessing dietary habits, adjustment for possible confounders not always optimal, and other methodological limitations. Further longitudinal studies are therefore needed to better compare these associations, possibly considering overall dietary quality and dietary changes over time.

Author Contributions: Conceptualization, A.L.; methodology, S.P.M., F.M., S.R., M.P. and A.L.; writing—original draft preparation, A.L.; writing—review and editing, A.L., F.M., R.D.A., A.F., S.B. and A.B.; supervision, A.L. All authors have read and agreed to the published version of the manuscript.

Funding: This research received no external funding.

Institutional Review Board Statement: Not applicable.

Informed Consent Statement: Not applicable.

Conflicts of Interest: The authors declare no conflict of interest.

References

1. World Health Organization. *World Obesity Day 2022—Accelerating Action to Stop Obesity*; World Health Organization: Geneva, Switzerland, 2022. Available online: https://www.who.int/news/item/04-03-2022-world-obesity-day-2022-accelerating-action-to-stop-obesity (accessed on 25 January 2023).
2. Engin, A. The Definition and Prevalence of Obesity and Metabolic Syndrome. *Adv. Exp. Med. Biol.* **2017**, *960*, 1–17. [CrossRef] [PubMed]
3. Saklayen, M.G. The Global Epidemic of the Metabolic Syndrome. *Curr. Hypertens Rep.* **2018**, *20*, 12. [CrossRef]
4. Estruch, R.; Ros, E. The role of the Mediterranean diet on weight loss and obesity-related diseases. *Rev. Endocr. Metab. Disord.* **2020**, *21*, 315–327. [CrossRef] [PubMed]
5. Babio, N.; Toledo, E.; Estruch, R.; Ros, E.; Martínez-González, M.A.; Castañer, O.; Bulló, M.; Corella, D.; Arós, F.; Gómez-Gracia, E.; et al. Mediterranean diets and metabolic syndrome status in the PREDIMED randomized trial. *CMAJ* **2014**, *186*, E649–E657. [CrossRef]
6. Leone, A.; De Amicis, R.; Battezzati, A.; Bertoli, S. Adherence to the Mediterranean Diet and Risk of Metabolically Unhealthy Obesity in Women: A Cross-Sectional Study. *Front. Nutr.* **2022**, *9*, 858206. [CrossRef] [PubMed]
7. Leone, A.; Bertoli, S.; Bedogni, G.; Vignati, L.; Pellizzari, M.; Battezzati, A. Association between Mediterranean Diet and Fatty Liver in Women with Overweight and Obesity. *Nutrients* **2022**, *14*, 3771. [CrossRef]
8. Medina-Remón, A.; Kirwan, R.; Lamuela-Raventós, R.M.; Estruch, R. Dietary patterns and the risk of obesity, type 2 diabetes mellitus, cardiovascular diseases, asthma, and neurodegenerative diseases. *Crit. Rev. Food Sci. Nutr.* **2018**, *58*, 262–296. [CrossRef]
9. De Amicis, R.; Mambrini, S.P.; Pellizzari, M.; Foppiani, A.; Bertoli, S.; Battezzati, A.; Leone, A. Ultra-processed foods and obesity and adiposity parameters among children and adolescents: A systematic review. *Eur. J. Nutr.* **2022**, *61*, 2297–2311. [CrossRef]
10. Leone, A.; Martínez-González, M.; Craig, W.; Fresán, U.; Gómez-Donoso, C.; Bes-Rastrollo, M. Pre-Gestational Consumption of Ultra-Processed Foods and Risk of Gestational Diabetes in a Mediterranean Cohort. The SUN Project. *Nutrients* **2021**, *13*, 2202. [CrossRef]
11. Monteiro, C.A.; Cannon, G.; Moubarac, J.C.; Levy, R.B.; Louzada, M.L.C.; Jaime, P.C. The UN Decade of Nutrition, the NOVA food classification and the trouble with ultra-processing. *Public Health Nutr.* **2018**, *21*, 5–17. [CrossRef]

12. Martínez Steele, E.; Juul, F.; Neri, D.; Rauber, F.; Monteiro, C.A. Dietary share of ultra-processed foods and metabolic syndrome in the US adult population. *Prev. Med.* **2019**, *125*, 40–48. [CrossRef]
13. Moubarac, J.C.; Batal, M.; Louzada, M.L.; Martinez Steele, E.; Monteiro, C.A. Consumption of ultra-processed foods predicts diet quality in Canada. *Appetite* **2017**, *108*, 512–520. [CrossRef]
14. Rauber, F.; da Costa Louzada, M.L.; Steele, E.M.; Millett, C.; Monteiro, C.A.; Levy, R.B. Ultra-Processed Food Consumption and Chronic Non-Communicable Diseases-Related Dietary Nutrient Profile in the UK (2008–2014). *Nutrients* **2018**, *10*, 587. [CrossRef]
15. Lane, M.M.; Davis, J.A.; Beattie, S.; Gómez-Donoso, C.; Loughman, A.; O'Neil, A.; Jacka, F.; Berk, M.; Page, R.; Marx, W.; et al. Ultraprocessed food and chronic noncommunicable diseases: A systematic review and meta-analysis of 43 observational studies. *Obes Rev.* **2021**, *22*, e13146. [CrossRef]
16. Moola, S.; Munn, Z.; Tufanaru, C.; Aromataris, E.; Sears, K.; Sfetcu, R.; Currie, M.; Qureshi, R.; Mattis, P.; Lisy, K.; et al. *Chapter 7: Systematic Reviews of Etiology and Risk*; Aromataris, E., Munn, Z., Eds.; JBI Manual for Evidence Synthesis: Adelaide, Australia, 2020.
17. Canhada, S.L.; Luft, V.C.; Giatti, L.; Duncan, B.B.; Chor, D.; Fonseca, M.; Matos, S.M.A.; Molina, M.; Barreto, S.M.; Levy, R.B.; et al. Ultra-processed foods, incident overweight and obesity, and longitudinal changes in weight and waist circumference: The Brazilian Longitudinal Study of Adult Health (ELSA-Brasil). *Public Health Nutr.* **2020**, *23*, 1076–1086. [CrossRef]
18. Scaranni, P.; Cardoso, L.O.; Chor, D.; Melo, E.C.P.; Matos, S.M.A.; Giatti, L.; Barreto, S.M.; da Fonseca, M.J.M. Ultra-processed foods, changes in blood pressure and incidence of hy.ypertension: The Brazilian Longitudinal Study of Adult Health (ELSA-Brasil). *Public Health Nutr.* **2021**, *24*, 3352–3360. [CrossRef]
19. Scaranni, P.; de Oliveira Cardoso, L.; Griep, R.H.; Lotufo, P.A.; Barreto, S.M.; da Fonseca, M.J.M. Consumption of ultra-processed foods and incidence of dyslipidemias: The Brazilian Longitudinal Study of Adult Health (ELSA-Brasil). *Br. J. Nutr.* **2022**, *129*, 336–344. [CrossRef]
20. Magalhães, E.; de Oliveira, B.R.; Rudakoff, L.C.S.; de Carvalho, V.A.; Viola, P.; Arruda, S.P.M.; de Carvalho, C.A.; Coelho, C.; Bragança, M.; Bettiol, H.; et al. Sex-Dependent Effects of the Intake of NOVA Classified Ultra-Processed Foods on Syndrome Metabolic Components in Brazilian Adults. *Nutrients* **2022**, *14*, 3126. [CrossRef]
21. Beslay, M.; Srour, B.; Méjean, C.; Allès, B.; Fiolet, T.; Debras, C.; Chazelas, E.; Deschasaux, M.; Wendeu-Foyet, M.G.; Hercberg, S.; et al. Ultra-processed food intake in association with BMI change and risk of overweight and obesity: A prospective analysis of the French NutriNet-Santé cohort. *PLoS Med.* **2020**, *17*, e1003256. [CrossRef]
22. Srour, B.; Fezeu, L.K.; Kesse-Guyot, E.; Alles, B.; Debras, C.; Druesne-Pecollo, N.; Chazelas, E.; Deschasaux, M.; Hercberg, S.; Galan, P.; et al. Ultraprocessed Food Consumption and Risk of Type 2 Diabetes Among Participants of the NutriNet-Sante Prospective Cohort. *JAMA Intern. Med.* **2019**, *180*, 283–291. [CrossRef]
23. Monge, A.; Silva Canella, D.; López-Olmedo, N.; Lajous, M.; Cortés-Valencia, A.; Stern, D. Ultraprocessed beverages and processed meats increase the incidence of hypertension in Mexican women. *Br. J. Nutr.* **2021**, *126*, 600–611. [CrossRef] [PubMed]
24. Duan, M.J.; Vinke, P.C.; Navis, G.; Corpeleijn, E.; Dekker, L.H. Ultra-processed food and incident type 2 diabetes: Studying the underlying consumption patterns to unravel the health effects of this heterogeneous food category in the prospective Lifelines cohort. *BMC Med.* **2022**, *20*, 7. [CrossRef]
25. Sandoval-Insausti, H.; Jiménez-Onsurbe, M.; Donat-Vargas, C.; Rey-García, J.; Banegas, J.R.; Rodríguez-Artalejo, F.; Guallar-Castillón, P. Ultra-Processed Food Consumption Is Associated with Abdominal Obesity: A Prospective Cohort Study in Older Adults. *Nutrients* **2020**, *12*, 2368. [CrossRef] [PubMed]
26. Mendonca, R.D.; Pimenta, A.M.; Gea, A.; de la Fuente-Arrillaga, C.; Martinez-Gonzalez, M.A.; Lopes, A.C.; Bes-Rastrollo, M. Ultraprocessed food consumption and risk of overweight and obesity: The University of Navarra Follow-Up (SUN) cohort study. *Am. J. Clin. Nutr.* **2016**, *104*, 1433–1440. [CrossRef] [PubMed]
27. Mendonca, R.D.; Lopes, A.C.; Pimenta, A.M.; Gea, A.; Martinez-Gonzalez, M.A.; Bes-Rastrollo, M. Ultra-Processed Food Consumption and the Incidence of Hypertension in a Mediterranean Cohort: The Seguimiento Universidad de Navarra Project. *Am. J. Hypertens* **2017**, *30*, 358–366. [CrossRef] [PubMed]
28. Llavero-Valero, M.; Escalada-San Martín, J.; Martínez-González, M.A.; Basterra-Gortari, F.J.; de la Fuente-Arrillaga, C.; Bes-Rastrollo, M. Ultra-processed foods and type-2 diabetes risk in the SUN project: A prospective cohort study. *Clin. Nutr.* **2021**, *40*, 2817–2824. [CrossRef]
29. Donat-Vargas, C.; Sandoval-Insausti, H.; Rey-García, J.; Moreno-Franco, B.; Åkesson, A.; Banegas, J.R.; Rodríguez-Artalejo, F.; Guallar-Castillón, P. High Consumption of Ultra-Processed Food is Associated with Incident Dyslipidemia: A Prospective Study of Older Adults. *J. Nutr.* **2021**, *151*, 2390–2398. [CrossRef]
30. Levy, R.B.; Rauber, F.; Chang, K.; Louzada, M.L.D.C.; Monteiro, C.A.; Millett, C.; Vamos, E.P. Ultra-processed food consumption and type 2 diabetes incidence: A prospective cohort study. *Clin. Nutr.* **2021**, *40*, 3608–3614. [CrossRef]
31. Rauber, F.; Chang, K.; Vamos, E.P.; da Costa Louzada, M.L.; Monteiro, C.A.; Millett, C.; Levy, R.B. Â Ultra-processed food consumption and risk of obesity: A prospective cohort study of UK Biobank. *Eur. J. Nutr.* **2021**, *60*, 2169–2180. [CrossRef]
32. Li, M.; Shi, Z. Ultra-Processed Food Consumption Associated with Overweight/Obesity among Chinese Adults-Results from China Health and Nutrition Survey 1997-2011. *Nutrients* **2021**, *13*, 2796. [CrossRef]
33. Cordova, R.; Kliemann, N.; Huybrechts, I.; Rauber, F.; Vamos, E.P.; Levy, R.B.; Wagner, K.H.; Viallon, V.; Casagrande, C.; Nicolas, G.; et al. Consumption of ultra-processed foods associated with weight gain and obesity in adults: A multi-national cohort study. *Clin. Nutr.* **2021**, *40*, 5079–5088. [CrossRef]

34. Rauber, F.; Steele, E.M.; Louzada, M.L.d.C.; Millett, C.; Monteiro, C.A.; Levy, R.B. Ultra-processed food consumption and indicators of obesity in the United Kingdom population (2008–2016). *PLoS ONE* **2020**, *15*, e0232676. [CrossRef]
35. Srour, B.; Fezeu, L.K.; Kesse-Guyot, E.; Alles, B.; Mejean, C.; Andrianasolo, R.M.; Chazelas, E.; Deschasaux, M.; Hercberg, S.; Galan, P.; et al. Ultra-processed food intake and risk of cardiovascular disease: Prospective cohort study (NutriNet-Sante). *BMJ* **2019**, *365*, l1451. [CrossRef]
36. Fardet, A. Minimally processed foods are more satiating and less hyperglycemic than ultra-processed foods: A preliminary study with 98 ready-to-eat foods. *Food Funct.* **2016**, *7*, 2338–2346. [CrossRef]
37. Sadeghirad, B.; Duhaney, T.; Motaghipisheh, S.; Campbell, N.R.; Johnston, B.C. Influence of unhealthy food and beverage marketing on children's dietary intake and preference: A systematic review and meta-analysis of randomized trials. *Obes. Rev.* **2016**, *17*, 945–959. [CrossRef]
38. Pérez-Escamilla, R.; Obbagy, J.E.; Altman, J.M.; Essery, E.V.; McGrane, M.M.; Wong, Y.P.; Spahn, J.M.; Williams, C.L. Dietary energy density and body weight in adults and children: A systematic review. *J. Acad. Nutr. Diet.* **2012**, *112*, 671–684. [CrossRef]
39. Canella, D.S.; Levy, R.B.; Martins, A.P.; Claro, R.M.; Moubarac, J.C.; Baraldi, L.G.; Cannon, G.; Monteiro, C.A. Ultra-processed food products and obesity in Brazilian households (2008–2009). *PLoS ONE* **2014**, *9*, e92752. [CrossRef]
40. Lam, M.C.L.; Adams, J. Association between home food preparation skills and behaviour, and consumption of ultra-processed foods: Cross-sectional analysis of the UK National Diet and nutrition survey (2008–2009). *Int. J. Behav. Nutr. Phys. Act.* **2017**, *14*, 68. [CrossRef]
41. Robinson, E.; Aveyard, P.; Daley, A.; Jolly, K.; Lewis, A.; Lycett, D.; Higgs, S. Eating attentively: A systematic review and meta-analysis of the effect of food intake memory and awareness on eating. *Am. J. Clin. Nutr.* **2013**, *97*, 728–742. [CrossRef]
42. Robinson, E.; Almiron-Roig, E.; Rutters, F.; de Graaf, C.; Forde, C.G.; Tudur Smith, C.; Nolan, S.J.; Jebb, S.A. A systematic review and meta-analysis examining the effect of eating rate on energy intake and hunger. *Am. J. Clin. Nutr.* **2014**, *100*, 123–151. [CrossRef]
43. Schulte, E.M.; Avena, N.M.; Gearhardt, A.N. Which foods may be addictive? The roles of processing, fat content, and glycemic load. *PLoS ONE* **2015**, *10*, e0117959. [CrossRef]
44. Carter, A.; Hendrikse, J.; Lee, N.; Yücel, M.; Verdejo-Garcia, A.; Andrews, Z.B.; Hall, W. The Neurobiology of "Food Addiction" and Its Implications for Obesity Treatment and Policy. *Annu. Rev. Nutr.* **2016**, *36*, 105–128. [CrossRef] [PubMed]
45. Hall, K.D. A review of the carbohydrate-insulin model of obesity. *Eur. J. Clin. Nutr.* **2017**, *71*, 323–326. [CrossRef] [PubMed]
46. Filippini, T.; Malavolti, M.; Whelton, P.K.; Vinceti, M. Sodium Intake and Risk of Hypertension: A Systematic Review and Dose-Response Meta-analysis of Observational Cohort Studies. *Curr. Hypertens Rep.* **2022**, *24*, 133–144. [CrossRef]
47. Lustig, R.H. Fructose: Metabolic, hedonic, and societal parallels with ethanol. *J. Am. Diet. Assoc.* **2010**, *110*, 1307–1321. [CrossRef]
48. Zakim, D. The effect of fructose on hepatic synthesis of fatty acids. *Acta Med. Scand. Suppl.* **1972**, *542*, 205–214. [CrossRef]
49. Mensink, R.P.; Zock, P.L.; Kester, A.D.; Katan, M.B. Effects of dietary fatty acids and carbohydrates on the ratio of serum total to HDL cholesterol and on serum lipids and apolipoproteins: A meta-analysis of 60 controlled trials. *Am. J. Clin. Nutr.* **2003**, *77*, 1146–1155. [CrossRef]
50. Matthan, N.R.; Welty, F.K.; Barrett, P.H.; Harausz, C.; Dolnikowski, G.G.; Parks, J.S.; Eckel, R.H.; Schaefer, E.J.; Lichtenstein, A.H. Dietary hydrogenated fat increases high-density lipoprotein apoA-I catabolism and decreases low-density lipoprotein apoB-100 catabolism in hypercholesterolemic women. *Arterioscler. Thromb. Vasc. Biol.* **2004**, *24*, 1092–1097. [CrossRef]
51. Wijendran, V.; Hayes, K.C. Dietary n-6 and n-3 fatty acid balance and cardiovascular health. *Annu. Rev. Nutr* **2004**, *24*, 597–615.
52. El-Ezaby, M.M.; Abd-El Hamide, N.-A.H.; El-Maksoud, M.A.E.; Shaheen, E.M.; Embashi, M.M.R. Effect of some food additives on lipid profile, kidney function and liver function of adult male albino rats. *J. Bas Environ. Sci.* **2018**, *5*, 52–59.
53. Bhattacharyya, S.; Feferman, L.; Tobacman, J.K. Carrageenan Inhibits Insulin Signaling through GRB10-mediated Decrease in Tyr(P)-IRS1 and through Inflammation-induced Increase in Ser(P)307-IRS1. *J. Biol. Chem.* **2015**, *290*, 10764–10774. [CrossRef]
54. Chassaing, B.; Koren, O.; Goodrich, J.K.; Poole, A.C.; Srinivasan, S.; Ley, R.E.; Gewirtz, A.T. Dietary emulsifiers impact the mouse gut microbiota promoting colitis and metabolic syndrome. *Nature* **2015**, *519*, 92–96. [CrossRef]
55. Bertoli, S.; Leone, A.; Battezzati, A. Human Bisphenol A Exposure and the "Diabesity Phenotype". *Dose Response* **2015**, *13*, 1559325815599173. [CrossRef]
56. Rolfo, A.; Nuzzo, A.M.; De Amicis, R.; Moretti, L.; Bertoli, S.; Leone, A. Fetal-Maternal Exposure to Endocrine Disruptors: Correlation with Diet Intake and Pregnancy Outcomes. *Nutrients* **2020**, *12*, 1744. [CrossRef]
57. Tonini, C.; Segatto, M.; Bertoli, S.; Leone, A.; Mazzoli, A.; Cigliano, L.; Barberio, L.; Mandalà, M.; Pallottini, V. Prenatal Exposure to BPA: The Effects on Hepatic Lipid Metabolism in Male and Female Rat Fetuses. *Nutrients* **2021**, *13*, 1970. [CrossRef]
58. Bird, S.R.; Hawley, J.A. Update on the effects of physical activity on insulin sensitivity in humans. *BMJ Open Sport Exerc. Med.* **2017**, *2*, e000143. [CrossRef]
59. Schulze, M.B.; Martínez-González, M.A.; Fung, T.T.; Lichtenstein, A.H.; Forouhi, N.G. Food based dietary patterns and chronic disease prevention. *BMJ* **2018**, *361*, k2396. [CrossRef]
60. Fardet, A.; Rock, E. Toward a new philosophy of preventive nutrition: From a reductionist to a holistic paradigm to improve nutritional recommendations. *Adv. Nutr.* **2014**, *5*, 430–446. [CrossRef]
61. Thompson, F.E.; Subar, A.F. Chapter 1 Dietary Assessment Methodology. In *Nutrition in the Prevention and Treatment of Disease*; Academic Press: Cambridge, MA, USA, 2017; pp. 5–48.

62. Willett, W.C.; Howe, G.R.; Kushi, L.H. Adjustment for total energy intake in epidemiologic studies. *Am. J. Clin. Nutr.* **1997**, *65*, 1220S–1228S. [CrossRef]
63. Lorenzoni, G.; Di Benedetto, R.; Silano, M.; Gregori, D. What Is the Nutritional Composition of Ultra-Processed Food Marketed in Italy? *Nutrients* **2021**, *13*, 2364. [CrossRef]
64. Mann, C.J. Observational research methods. Research design II: Cohort, cross sectional, and case-control studies. *Emerg. Med. J.* **2003**, *20*, 54–60. [CrossRef] [PubMed]

MDPI
St. Alban-Anlage 66
4052 Basel
Switzerland
www.mdpi.com

Nutrients Editorial Office
E-mail: nutrients@mdpi.com
www.mdpi.com/journal/nutrients

www.ingramcontent.com/pod-product-compliance
Lightning Source LLC
Chambersburg PA
CBHW061615220926
51850CB00005B/516
* 9 7 8 3 0 3 6 5 8 4 6 4 5 *